Chabert · A History of Algorithms

Springer
*Berlin
Heidelberg
New York
Barcelona
Hong Kong
London
Milan
Paris
Singapore
Tokyo*

Jean-Luc Chabert (Ed.)

A History of Algorithms

From the Pebble to the Microchip

With 100 Figures

Springer

Editor

Jean-Luc Chabert, University of Picardy

Authors

Évelyne Barbin, IUFM, Créteil and IREM, University Paris VII
Jacques Borowczyk, IUFM, Orléans-Tours
Jean-Luc Chabert, University of Picardy
Michel Guillemot, University Paul Sabatier and IREM, Toulouse
Anne Michel-Pajus, Lycée Claude Bernard, Paris and IREM, University Paris VII

For Arab Mathematics

Ahmed Djebbar, University Paris XI, Orsay centre

For Chinese Mathematics

Jean-Claude Martzloff, CNRS and Institut des Hautes Études Chinoises, Paris

Translator of the English Edition

Chris Weeks

Title of the original edition:
Histoire d'algorithmes. Du caillou à la puce.
© Éditions Belin, Paris 1994
Cover figure: Éric Durant (1993)

Library of Congress Cataloging-in-Publication Data
Histoire d'algorithmes. English. A history of algorithms: from the pebble to the microchip/
Jean-Luc Chabert (ed.); [authors, Évelyne Barbin... et al.; English translator, Chris Weeks].
p. cm. Includes bibliographical references and index. ISBN 3-540-63369-3 (softcover: alk.
paper) 1. Algorithms–History. I. Chabert, Jean-Luc. II. Barbin, E. (Évelyne) III. Title.
QA.58.H5813 1998 511'.8–dc21 98-20468 CIP

ISBN 3-540-63369-3 Springer-Verlag Berlin Heidelberg New York

© Springer-Verlag Berlin Heidelberg 1999
Printed in Italy

The use of general descriptive names, registered names, trademarks, etc. in this publication does not imply, even in the absence of a specific statement, that such names are ex-
empt from the relevant protective laws and regulations and therefore free for general use.

Cover design: *design & production* GmbH, Heidelberg
Typesetting by Ulrich Kunkel Textservice, Reichartshausen

SPIN 10508149   41/3143 – 5 4 3 2 1 0 – Printed on acid-free paper

Contents

Introduction

فأما الأموال والجذور التى تعدل العدد فمثل قولك
مال وعشرة أجذاره يعدل تسعة وثلاثين درهما ومعناهأى مال اذا زدت عليه مثل
عشرةأجذاره بلغ ذلك كله تسعة وثلاثين . فبابه⁽؟⁾ أن تنصف الأجذار وهى فى
هذه المسئلة خمسة فتضربها فى مثلها فتكون خمسة وعشرين فتزيدها على التسعة
والثلاثين فتكون أربعة وستين فتأخذ جذرها وهو ثمانية فتنقص منه نصف
الأجذار هو خمسة فيبقى ثلاثة وهو جذر المال الذى تريد والمال تسعة .

As for squares and roots equal to a number, it is as when you say this: a square and ten of
its roots equal thirty-nine dirhams.

Its meaning is that the square, if you add to it the equivalent of ten of its roots [is such
that] it will become thirty-nine.

Its method [of solution] consists in dividing the roots by two, and that is five in this
problem. You multiply it by itself and this will be twenty-five. You add it to thirty-nine.
This will give sixty-four. You then take its square root which is eight and you subtract
from it half [the number] of the roots and that is five. There remains three and that is the
root that you are seeking and the square is nine.

> from Al-Khwārizmī, *al-Jabr wa l-Muqābala* (*c.* 860).
> The text describes an algorithm for solving certain types of
> quadratic equations through using the example $x^2 + 10x = 39$.

Algorithms have been around since the beginning of time and existed well before a
special word had been coined to describe them. Algorithms are simply a set of step
by step instructions, to be carried out quite mechanically, so as to achieve some desired result. Given the discovery of a routine method for deriving a solution to a
problem, it is not surprising that the 'recipe' was passed on for others to use. Algorithms are not confined to mathematics, although that is the focus of this book. The
Babylonians used them for deciding points of law, Latin teachers used them to get
the grammar right, and they have been used in all cultures for predicting the future,
for deciding medical treatment, or for preparing food. Everybody today uses algorithms of one sort or another, often unconsciously, when following a recipe, using a
knitting pattern, or operating household gadgets.

We therefore speak of recipes, rules, techniques, processes, procedures, methods, etc., using the same word to apply to different situations. The Chinese, for example, use the word *shu* (meaning rule, process or strategem) both for mathematics and in martial arts. This has come into our language, via the Japanese, in the term *ju-jitsu* meaning 'procedural rules for suppleness' or 'algorithms for suppleness'. (The 'ju' comes from the Chinese *shu*.)

Our word algorithm derives directly from al-Khwārizmī, the author of the oldest known work of algebra. His full name, Muḥammad ibn Mūsā al-Khwārizmī means, literally, Muḥammad son of Mūsā from Khwarezm, a region of central Asia, south of the Aral Sea. He was a mathematician of the first half of the 9th century and his book *al-Mukhtaṣar fī Hisāb al-Jabr wa l-Muqābala* gave us the word 'algebra' from 'al-Jabr'.

In the 12th century, this work and others were translated from Arabic into Latin and so the denary positional numeration system then began to spread throughout mediaeval Europe. Since the written sources were Arabic, the numerals came to be referred to as 'Arabic', although the signs for the numerals had been adapted from Indian practice. In the Latin texts of the Middle Ages we often come across the conflict between the advantages of using the new positional notation calculation methods compared with the traditional counting table or abacus methods. The new methods were described as algorisms or algorismus or algorithmus, and so the word came to be used to describe particular routine arithmetic procedures.

During the course of time, the meaning of the word became extended, so that d'Alembert, in an article in the *Encylopédie* wrote:

Arab term, used by several authors, and particularly by the Spanish to mean the practice of algebra. It is also sometimes taken to mean arithmetic by digits ... The same word is taken to mean, in general, the method and notation of all types of calculation. In this sense, we say the algorithm of the integral calculus, the algorithm of the exponential calculus, the algorithm of sines, etc.

In the end, the term algorithm has come to mean any process of systematic calculation, that is a process that could be carried out automatically.

Today, principally because of the influence of computing, the idea of finiteness has entered into the meaning of algorithm as an essential element, distinguishing it from vaguer notions such as process, method or technique. Thus, the *Encyclopaedia Britannica* (15th edn.) describes an algorithm as a:
systematic mathematical procedure that produces – in a finite number of steps – the answer to a question or the solution of a problem.

The concern with finiteness arose in a more general context in Hilbert's 10th Problem, posed in 1900 at the Second International Congress of Mathematicians in Paris: given a polynomial equation with arbitrary rational coefficients, "a method is sought by which it can be determined, in a finite number of operations, whether the equation is solvable in rational numbers".

Here we have a finite number of operations, a finite number of input values, but also a finite number of solution procedures, that is that each step should be able to be carried out by a finite process – something which is not possible, for example, in

The spirit of Arithmetic looks down on the quarrel between the new 'algorists' using written numerals and the traditional 'abacists' with their counting table, from an illustration in *Margarita Philosophica*, 1504.

determining the quotient of two incommensurable real numbers. We also refer to an *effective procedure*, that is one that will effectively achieve a result (in a finite time).

The formalisation of these concepts, by logicians from the 1930s, allowed Matijasevič, at the 1970 International Congress of Mathematicians in Nice, to be able to declare a negative answer to Hilbert's problem. The formalisation particularly preoccupied computer scientists, a computer program is simply an algorithm and a computer language is a language for writing algorithms that a computer is able to read and act on. The study of the formalisation of the idea of algorithm and the idea of an ideal computing machine are touched in the final chapter of this book.

In selecting processes of an algorithmic nature to illustrate the history of algorithms, it may appear that we have not always accepted the modern restriction that algorithms must be finite. The way that a calculation procedure is described, may be such that the final step can never be done; but in practice, the process is halted at some previously agreed stage. For example, consider determining limit values for π by Archimedes' method (Section 5.1). The repeated calculations of the lengths of the inscribed and escribed polygons of a circle, doubling the number of sides at each step, will enclose the required number between closer and closer limits. The process can never end at the true value; but we can agree before we start that we shall stop when the difference between the exact value and the approximate value is sufficiently small, and in this way we turn the procedure into a finite one.

There is also another idea present in algorithmic procedures: that of iteration or recurrence. Thus, in the 1950s, the term algorithm was used essentially, and anachronistically, to refer to *Euclid's algorithm*. This was for determining the greatest common divisor of two integers: the calculations involve successive divisions until the remainder becomes zero. Finiteness lies in the automatic arrival at a zero remainder after a finite number of divisions. Iteration appears in the process through repeated division.

Also, although iteration is not an essential element of an algorithm, it has such an importance that we have tried to show its presence when it is not explicitly obvious, sometimes going further than the original thoughts of the author of the algorithm. We see this, for example, in Newton's method for approximating a root (Chapter 6). In the process of formalising the process, recurrence formulas arise. This recurrence, traditionally handled by iteration, is frequently shown today in a recursive way. Iteration, recurrence, recursivity are each distinct notions but intrinsically linked, playing a fundamental role in both the practice of algorithms and in their theoretic treatment.

We have had to make choices. The texts presented here are almost exclusively to do with numerical algorithms. We have chosen not to include algorithms from other fields, such as geometry or logic.

Each chapter is organised around a number of original texts selected to reflect different aspects of the same theme. The criteria used for selecting texts, apart from their originality and historical interest, is that they should be accessible to mathematical students, though the later chapters are more suitable for university undergraduate mathematicians, and to make them easier to use we have sometimes avoided purely literal translations. The texts are set in their historical context and for each text we have explained the mathematics in an attempt to make the original easier to understand. Naturally, any reading of historical texts poses problems of interpretation, and we have sometimes had to decide on a preferred interpretation. The references at the end of each chapter are intended to provide the reader with a starting point for further reading. What we have tried to do, without claiming to be exhaustive, is to provide historical support for contemporary algorithmic procedures, and to do this through the use of original texts. This added cultural underpinning may be used to pose valid questions about problems which are sometimes viewed from a purely abstract perspective.

The themes dealt with are as follows: 1. Elementary Arithmetic Operations, 2. Magic Squares, 3. Methods of False Position, 4. Euclid's Algorithm, 5. Determining the Value of π, 6. Newton's Method, 7. Successive Approximations, 8. Arithmetic Algorithms, 9. Linear Systems, 10. Interpolation, 11. Quadratures, 12. Differential Equations, 13. Function Approximation, 14. Acceleration of Convergence, 15. The Concept of Algorithm.

The first three chapters deal with distinct topics and ways of treating them. Although they deal with relatively elementary problems, they bring together methods, sometimes similar and sometimes quite different, that were used in widely different cultures, and so reflect ways of thinking about mathematics among Egyptian, Babylonian, Indian, Greek, Chinese and Arab civilisations. The next three chapters – Euclid's Algorithm, Newton's Method and Determining π – are concerned with more elaborate methods or problems. These have evolved over time and illustrate increasing sophistication of techniques and raise questions that remain of contemporary concern.

The first eight chapters of *A History of Algorithms* focus on questions whose origins are ancient, and they are essentially problems about number. The later chapters concern algorithms for handling more complex concepts. For example, the chapter on linear systems provides solutions which are arrays of numbers, today called vectors, and the chapter on approximation of functions shows methods for obtaining general functions. A quick glance at the table of contents might suggest that this is a standard work of numerical analysis, with chapters devoted to interpolation, linear systems, quadratures, differential equations, acceleration of convergence, etc. But the perspective of the book is quite different: the emphasis is on the historical introduction of techniques and on the particular value of the historical text as a way of understanding the initial idea and intention of its author. We think this helps to throw new light on certain questions.

This granted, the technical nature of the subject today is such as not to let us adventure into more recent times. We have restricted ourselves to the end of the 19th century, or just in to the 20th century, later developments only being referred to so as to indicate frequently cited texts, which have already come to be regarded as classics. In addition, we have restricted ourselves to algorithms concerned with what might be called 'natural' questions, although this distinction may be sometimes difficult to make. What we have not done is to deal with questions that are raised by the mathematics itself; we do not, for example, look at algorithms for calculating the eigenvectors of a matrix.

This book, then, is addressed to students, to teachers and more generally to anyone interested in algorithms or numerical analysis. It is not intended as a text book, but to provide a cultural context, a sort of 'source book' for the history of mathematics, but one which offers a different perspective by giving pride of place to algorithms.

The group of authors who prepared *A History of Algorithms* would like to express their gratitude to those who have given them help in a number of ways. We should mention, in particular, Colette Bloch, Alain Boyé, Christian Gilain, Maryvonne

Hallez, Jean-Pierre Levet, Jean Masson, Angel Rial, Marie-Françoise Roy, Michel Serfati and Hourya Sinaceur. But we do not know any algorithm for providing us with an exhaustive list.

Finally, the authors wish to give particular thanks to Chris Weeks for the considerable work he has undertaken in preparation this English translation.

1 Algorithms for Arithmetic Operations

> Many people regard arithmetic as a trivial thing that children learn and computers do, but we will see that arithmetic is a fascinating topic with many interesting facets. [...] The way we do arithmetic is intimately related to the way we represent the numbers we deal with.
>
> Donald Knuth
> *The Art of Computer Programming*
> Vol. 2; pp. 178, 179

The basic arithmetic operations of the elementary school, multiplying and dividing, appear to have derived from extremely early economic needs, certainly earlier than the emergence of civilisations using writing. One of the earliest pieces of evidence of an algorithm of this type is to be found on a clay tablet found at Shuruppak, near Baghdad which concerns a problem of sharing. Engraved by a Sumerian at about 2500 BC, this tablet (see Section 1.1) illustrates the first of the ten episodes which we have chosen to illustrate a history which would occupy several volumes if it were to be written up in detail.

Although we have decided to present our story in a chronological order, we should not expect to find a continuity of mathematical progression in these 'snapshots'. We need to remember that, since the transmission of algorithms came about through human contact, necessarily by chance, those that were remembered and handed on were not necessarily the best (from our point of view) of what had then been developed, but rather what was best understood at the time. Furthermore, an algorithmic method became successful when the techniques it employed corresponded to needs of the time.

Algorithms for arithmetic operations are closely dependent on the systems of numeration in which they operate. A numeration system can be additive, like the Roman number system, or it can be positional, as is ours today, it can use one or more bases and, finally, it may or may not use zero. We shall see that these characteristics determine the necessity and practice of these ancient algorithms.

When writing is present in a civilisation, it is no longer necessary to repeat the same operations. Thus we see that the Babylonians and the Egyptians drew up 'tables' for certain basic calculations which they would be able to use for future calculations (see Sections 1.2 and 1.3). At other times and places, people used other ways to help their calculations: pebbles, marks in the dust, knotted strings (quipu), tokens on a counting board in the Middle Ages, not to mention the beads on an abacus frame still used today in many Asian countries (see Sections 1.6 and 1.7).

As for 'long' multiplication using 'tables', this has an astonishingly long history which ranges over place and time. It occurred in Ancient China, in the Arab world and in mediaeval Europe (see Section 1.4). While starting out as a technique associated with writing, it became a technique using tools with Napier's 'bones'. Napier went on to invent logarithms which radically simplified multiplications. Being improved in the 19th century, Napier's bones continued to be manufactured right up to the 20th century, until they were replaced by calculation rules based on logarithms. These last were dethroned before our very eyes with the advent of calculators.

For most purposes, a denary (base ten) numeration system became the most commonly used, but its use for fractions was comparatively late. Although decimals fractions had been in use from the 12th century, the 'decimal point' notation (see Section 1.8) only came into general use with the arrival of tables of logarithms in the 17th century. Yet, despite the advantages of using base ten for numeration, the binary system has proved to have significant advantages. Judged as a contest between denary and binary numbers, it would seem that the advent of computers gives the victory (provisionally?) to the binary system (see Section 1.10), a system proposed by Leibniz at the beginning of the 18th century (see Section 1.9): he noticed how easy it was to carry out arithmetic operations with binary numbers. Finally, we note that an interest in optimising arithmetic algorithms, that is in reducing the number of separate elementary steps needed for a calculation, can already be found in manuscripts from the Middle Ages (see Section 1.5).

1.1 Sumerian Division

The Sumerian civilisation flourished in southern Mesopotamia from about 3500 BC to 2000 BC. The clay tablets found at Shuruppak, in the Euphrates valley to the south of Baghdad, date from 2500 BC and today belong to the Istanbul Museum.

This Sumerian civilisation had developed a numeration system which was additive, having special signs for 1, 10, 60, 600, and 3600. Thus its structure was both denary and sexagesimal, that is both base ten and base sixty.

1	10	60	600	3600	36000
D	o	D	D	O	⊗ or ◎

Thus the number 163 would be written:

D D   60 60

O O O O D D D   10 10 10 10 1 1 1

Sumerian Tablets
Tablettes sumériennes de Shuruppak
Jestin, 1937, Plates XXI and CXLII

Tablet 50

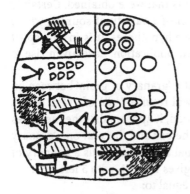

1 granary of grain	36000 36000 36000 36000
silà 7	3600 3600 3600 3600 3600
each man receives	600 600 60 600 600 60
these men are	10 10 10 10 10 1
	remainder 3 silà of grain

Tablet 671

 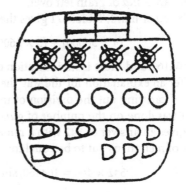

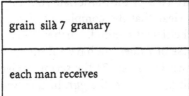

grain silà 7 granary	workers			
	36 000 36 000 36 000 36 000			
	3 600 3 600 3 600 3 600 3 600			
each man receives	600 600 60 60 60			
	600 60 60 60			

The two tablets we show above deal with the same problem: how can we share out a 'granary of grain' between some men? Given the capacity of the granary, and how much each man shall receive, the tablets show how many men can have a share. Thus the two tablets give the result of a division. But although they start out with the same given values, they do not lead to the same result: one is exact, the other is not. This fact is striking, and also very instructive. In fact, although the tablets do not tell us what method was used to find the quotients, the erroneous result suggests the use of an algorithm for division when large numbers occur, which is the case here for the number giving the contents of the granary [9].

In general, the ancient documents we have give no indication of what procedures were used to carry out the operations. Because of this, historians have to try to imagine what algorithms might have led to the results that were obtained. Certainly this process is not without danger, but a comparison of a variety of sources, where they exist, can go some way to support interpretative hypotheses. It is in this spirit that we offer here these two tablets which, since their publication, have given rise to a number of interpretations [12].

The problem is about sharing out the contents of a 'granary of grain' between an unknown number of men in such a way that each will receive 7 *silà* of grain, that is about 7 litres. From other documents in our possession, it appears that the measures used for recording the capacity of a granary were the *gur* and the *silà*: one 'granary of grain' was equal to 2400 *gur*, and one *gur* was equal to 480 *silà*. Thus a 'granary of grain' was equal to 1 152 000 silà. Dividing this by 7 gives 164 571, with 3 left over.

Tablet 50 gives this result exactly since 164 571 is equal to:

$$(4 \times 36000) + (5 \times 3600) + (4 \times 600) + (2 \times 60) + (5 \times 10) + 1$$

and there are 3 *silà* of grain left over.

On the other hand, Tablet 671 gives the quotient as:

$$(4 \times 36000) + (5 \times 3600) + (3 \times 600) + (6 \times 60),$$

that is 164160, and there is no indication of the remainder. This wrong answer allows us to offer an explanation for the way the division was carried out [9]. Suppose that, instead of carrying out the division of the number of *silà*, a very large number, the division was done on the number of *gur*, then this means 2400 had to be divided by 7. This gives an answer of 342 with a remainder of 6. If we neglect the remainder, the number of men turns out to be equal to:

$$342 \times 480 = 164160, \text{ since one gur equal 480 } silà.$$

We do not know what division algorithm was used by the Sumerians, but this example allows us to make some remarks. It is clear that the complexity of the numeration system does not lead to a simple and efficient method. Furthermore, division by 7 is particularly difficult since the inverse of 7 does not have a finite form in sexagesimals (see Section 1.2). It would seem that in Tablet 671 the scribe neglected the remainder and worked with the larger unit of measure, the *gur*. In this way he was able to work with smaller numbers, although neglecting the remainder results in an error. There is no reason why we should conclude that this was a mistake. The 'error' is inevitable, and the scribe may well have been aware of it.

1.2 A Babylonian Algorithm for Calculating Inverses

At the beginning of the second century before Christ, the Babylonians developed a numeration system, traces of which remain with us today in our use of minutes and seconds for time and angle measurement. This system is often referred to as a positional sexagesimal (base 60) number system, although this is not strictly true.

In fact, the Babylonian system keeps traces of the earlier additive system, since the first 59 numbers are written additively, in base 10, using the symbols:

a vertical wedge shape ⟟ for the unit, and

a horizontal wedge shape ⟨ for ten.

Thus, the numbers 32, 24 and 49 would be written, respectively, as:

Up until the Seleucid period, that is up to the third century BC, the Babylonians did not employ a symbol for the absence of certain powers of 60, but simply left a space within the number to show absence. Only the context of the calculations provided a clue to the correct interpretation of the number. Thus, the marks:

which we write today as	2; 13; 20
may stand for	$2 \times 60^2 + 13 \times 60 + 20$,
or could just as easily stand for	$2 \times 60 + 13 + 20/60$,
or even	$2 \times 60^5 + 13 \times 60 + 20/60^3$.

To be completely general, it could stand for $2 \times 60^m + 13 \times 60^n + 20 \times 60^p$ where m, n and p are integers such that $m > n > p$.

In Babylonian texts we find division worked in two stages: first calculating the inverse of the divisor, then multiplying the result by the dividend. Hence, numbers whose inverses have a finite form as sexagesimals are of particular interest. These numbers are those of the form $2^p 3^q 5^r$, where p, q and r are positive or negative integers or zero. Historians refer to these numbers as *regular*.

The Babylonians have left us numerous 'tables of inverses of regular numbers'. The inverse of a regular number being also itself a regular number, the tables consist of pairs of regular numbers each being the inverse of the other. One of these numbers is called *igu* and the other *igibu*, that is its *igu*. The 'standard' tables list numbers from 2 and 30 as far as 1;21 and 44;26;40. We can check that these are certainly inverse pairs, since

$$2 \times 30/60 = 1, \text{ and } (1 \times 60 + 21) \times (44/60 + 26/60^2 + 40/60^3) = 81 \times 160000/60^4 = 1.$$

There are other more important tables: the Tablet A06456 in the Louvre, dating from the Seleucid period, consists of more than 200 pairs of regular numbers and their inverses. The last of these are 2;59;21;40;48;54 and 20;4;16;22;28;44;14;57;40;4; 56;17;46;40. We leave it to the reader to verify that these do in fact form an inverse pair!

The tables were often constructed by duplation (doubling) and mediation (halving): if a and b are inverse pairs, then so are $2a$ and $b/2$. Here we have another algorithm which lets us calculate the inverses of other regular numbers from a 'standard' table of inverses. This algorithm is in evidence in Tablet VAT 6505 in the Berlin Museum. Dating from the Old Babylonian Age, that is between 2000 BC and 1650 BC, it contained originally a dozen calculations of inverses, of which only five remain visible.

Babylonian tablet
Mathematische Keilschrifttexte, Neugebauer, 1935
plate 43

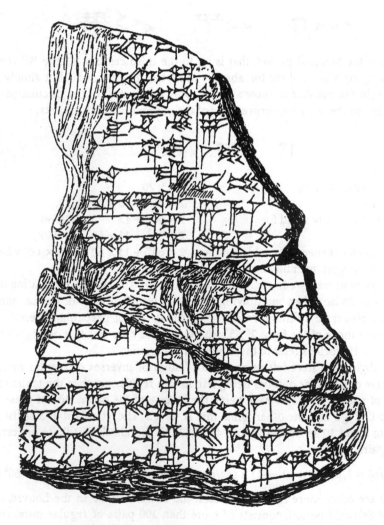

The translation in brackets has been constructed. The numbers of the problems are given in the margin. The term *igu* has been translated as 'number'.

First part of the tablet (beginning section is lost)

2 [The number is 4;10. What is its inverse?
   Proceed as follows.
   Form the inverse of 10, you will find 6.
   Multiply 6 by 4, you will find 24.
   Add on 1, you will find 25.
   Form the inverse of 25], you will find 2;24.
   [Multiply 2;24 by 6], you will find 14;24.
   [The inverse is 14;24]. Such is the way to proceed.

3 [The number is 8;20]. What is its inverse?
   [Proceed as follows.
   Form the inverse of 20], you will find 3.
   Mul[tiply 3 by 8], you will find 24.
   [Add on 1], you will find [25].
   [Form the inverse of 25], you will find 2;24.
   [Multiply 2;24 by 3], you will find [7;12.
   The inverse is 7;12. Such is the way to proceed].

4 [The number is 16;40. What is] its inverse?
   [Pro]ceed as follows.
   Form [the inverse of 6;40], you will find 9.
   Multiply [9] by 10, you will find 1;30.
   [Add] on [1], you will find 2;30.
   Form [the inverse of 2];30, you will find 24.
   [Multiply 24 by 9, you will find 3;36.
   The inverse is 3;36. Such is the way to proceed].

Second part of the tablet (beginning section is lost)

6 [The number is 1;6;40. What is its inverse?
   Proceed as follows.
   Form the inverse of 6;40, you will find 9.
   Multiply] 9 by 1, you will find 9.
   Add on [1, you will find 10].
   Form the inverse of 10, [you will find 6].
   [Multiply] 6 by [9, you will find 54].
   The inverse is 49 (read 54). [Such is the way to proceed].

7 The number is 2;13;20. [What is its inverse?]
   Proceed as follows.
   Form [the in]verse of 3;20, [you will find] 18.
   Multiply 18 by 2;10, [you will find] 3[9].
   Add on 1, [you will find] 40.
   Form the inverse of 40, [you will find] 1;30.
   Multiply 1;30 by 18, you will find 27.
   The inverse is 27. [Such is the way to proceed].

Numbers 8 to 11 are missing. Of number 12, nothing remains except the beginning of the last line: "Such is the way to proceed". The tablet concludes with referring to "12 examples in all".

Despite the condition of the tablet, we can consider that it gives evidence of this same algorithmic step: that is that larger inverse pairs can be calculated from known smaller inverse numbers.

These problems may appear to lack coherence: they are just a sequence of arbitrary calculations on numbers apparently chosen at random. However, a careful scrutiny of the seven problems shows us that the calculations are always carried out in the same order and using the same model. This collection of identical steps allows us to reconstruct the underlying algorithm. To better appreciate its generality and how it works we shall write out the instructions, using letters in the place of numerical values (we use $\bar{y}$ for the inverse of y).

> The number is x. What is its inverse?
> Proceed as follows.
> Form the inverse of y, you will find $\bar{y}$.
> Multiply $\bar{y}$ by z. You will find t.
> Add on 1. You will find u.
> Form the inverse of u. You will find $\bar{u}$.
> Multiply $\bar{u}$ by $\bar{y}$. You will find v.
> The inverse is v. Such is the way to proceed.

The scribe calculates the inverse of a regular number x by writing it as the sum of a regular number y, whose inverse is known, and a second number z. For example, in problem 7, to calculate the inverse of $x = 2;13;20$, he writes this number as the sum of the regular number $y = 3;20$, whose inverse is known, and another number $z = 2;10$. The inverse of y, which is 18, is given in a 'standard' table. Using the result:

$$\frac{1}{y+z} = \frac{1}{\frac{1}{y} \times z + 1} \times \frac{1}{y},$$

we find that the inverse of a number x is the product of the inverses of y and u, where

$$u = t + 1 = \frac{1}{y} \times z + 1.$$

In this example, $u = t + 1 = 18 \times 2;10 + 1 = 39 + 1 = 40$. The number u is regular since it is equal to x/y and x and y are regular numbers. The inverse of u, given in standard tables, is 1;30. Therefore, the inverse of x is equal to $18 \times 1;30 = 27$.

This algorithm can also set up a recursive procedure. In fact, calculating the inverse of a number x can be reduced to calculating the inverses of two smaller numbers y and u, for which we can also apply the algorithm, if necessary.

Thus, in Tablet A06456, the scribe sets out a long list of inverse numbers which allow him to calculate the inverse of $x = 1;5;32;9;36$. At each loop of the algorithm, the scribe chooses a y, whose inverse is known from a standard table of inverses, and applies the algorithm again to u. This is done until a u is found that is sufficiently small to appear in a standard table. Using y_i and u_i for the numbers obtained at the ith loop, the scribe obtains:

$$x = 1;5;32;9;36 = 36 \times 1;49;13;36 \qquad y_1 = 36 \qquad u_1 = 1;49;13;36$$
$$u_1 = 1;49;13;36 = 6 \times 18;12;16 \qquad y_2 = 6 \qquad u_2 = 18;12;16$$
$$u_2 = 18;12;16 = 16 \times 1;8;16 \qquad y_3 = 16 \qquad u_3 = 1;8;16$$
$$u_3 = 1;8;16 = 16 \times 4;16 \qquad y_4 = 16 \qquad u_4 = 4;16$$
$$u_4 = 4;16 = 16 \times 16 \qquad y_5 = 16 \qquad u_5 = 16$$

Thus,

$$\frac{1}{x} = \frac{1}{y_1} \times \frac{1}{u_1} = \frac{1}{y_1} \times \frac{1}{y_2} \times \frac{1}{u_2}$$

....

$$\frac{1}{x} = \frac{1}{y_1} \times \frac{1}{y_2} \times \frac{1}{y_3} \times \frac{1}{y_4} \times \frac{1}{y_5} \times \frac{1}{u_5}$$

$$\frac{1}{x} = 54;55;53;54;22;30$$

Each of the inverses, on the right hand side of the expression for the inverse of x, is known from standard tables, and so the inverse of x can be calculated. This algorithm corresponds, therefore, to a decomposition into regular factors of the number whose inverse is being sought.

1.3 Egyptian Algorithms for Arithmetic

Throughout the four thousand years of their civilisation, the Egyptians used a number system that was additive and denary. Besides integers, the Egyptians also used unit fractions, and the special fraction 2/3, as global entities. (It is interesting to note, that the Chinese also considered 2/3 to be of particular importance, and they referred to it as the 'large half', while 1/3 was called the 'little half'.) The Egyptians, then, would write every 'fraction' as the sum of an integer, together with the fraction 2/3, if necessary, and a number of unit fractions, all of which were distinct. For example, 97/42 would appear in hieroglyphic writing, the form used for inscriptions on monuments, as:

and in hieratic script, as used for writing on papyrus, as:

that is: 1/7   1/2   2/3   1

since: 97/42 = 1 + 2/3 + 1/2 + 1/7.

Hieroglyphic number symbols						
1	10	100	1000	10 000	100 000	1 000 000
vertical stroke	basket handle	coiled rope	lotus flower	bent finger	tadpole	seated god

The additive nature of the number system made the addition and subtraction of integers easy, but the addition of unit fractions caused difficulties, since the unit fractions that made up a 'fraction' had always to be distinct. This meant that twice the unit fraction $1/(2n + 1)$ had to be written as the sum of distinct unit fractions. Furthermore, they had to be able to recognise possible simplifications in a sum of unit fractions. It was for that reason that the Rhind papyrus of 1650 BC - the most important mathematical document in our possession, now held in the British Museum - begins with doubles of unit fractions and simplifying reductions like:

$$2/19 = 1/12 + 1/76 + 1/114 \quad \text{and} \quad 1 + 1/2 + 1/12 + 1/4 + 1/6 = 2$$

These decompositions, doubles of unit fractions and reductions, provide a 'data bank' for carrying out more complicated procedures like multiplications, divisions and 'reducing to the same denominator'. In fact, we find in the Rhind papyrus algorithms that allow these more complicated tasks to be reduced to two 'elementary' ones: doubling or *duplation*, and halving or *mediation*. We present here two extracts from the Rhind papyrus: the first concerns multiplication by doubling, the second uses the 'method of auxiliary reds' to 'reduce to the same denominator' a sum of unit fractions. Each extract shows the hieratic script, as it appears on the papyrus, together with its transliteration into hieroglyphs, after the practice of Egyptologists, in order to facilitate reading the number. We should point out that Egyptian writing goes from right to left. The original is written in black and red ink, and in our version here, we have indicated what was written in red by putting it in bold.

A. Duplation

Rhind Papyrus
Rhind Mathematical Papyrus, translation by A. Chace, 1927–1929
Problem 6, Plate 38

Hieratic text as it appears on the papyrus

Transcription in hieroglyphics

Translation
The making of loaves for man 10.

The doing as it occurs. Make thou the multiplication: $\dfrac{2}{3}\ \dfrac{1}{5}\ \dfrac{1}{30}$ times 10.

$$
\begin{array}{lccc}
1 & \dfrac{2}{3} & \dfrac{1}{5} & \dfrac{1}{30} \\[2ex]
\backslash 2\,1 & \dfrac{2}{3} & \dfrac{1}{10} & \dfrac{1}{30} \\[2ex]
4\,3 & \dfrac{1}{2} & \dfrac{1}{10} & \\[2ex]
\backslash 8\,7 & \dfrac{1}{5} & &
\end{array}
$$

Total loaves 9, it is this.

The scribe has not shown how he obtained the result (2/3 + 1/5 + 1/30), he proposes simply a verification of the result by multiplying by 10. In order to carry out the multiplication, he breaks 10 down into a sum of powers of 2. Since $10 = 2 + 2^3$, he uses successive doubling to obtain the multiplications by 2, 4 and 8, and then sums only the results corresponding to 2 and 8 (marked in the script by oblique lines).

As is his custom, the scribe does not provide us with his intermediary calculations. We are, however, able to reconstruct these. The doubles of unit fractions are given at the beginning of the papyrus. The scribe is therefore able to use the following ones:

$$
\begin{aligned}
2(2/3) &= 1 + 1/3 \\
2(1/5) &= 1/3 + 1/15 \\
2(1/30) &= 1/15 \\
2(1/15) &= 1/10 + 1/30
\end{aligned}
$$

and from this he can find his first duplation:

$$
\begin{aligned}
2(2/3 + 1/5 + 1/30) &= (1 + 1/3) + (1/3 + 1/15) + 1/15 \\
&= 1 + 2/3 + 2/15 \\
&= 1 + 2/3 + 1/10 + 1/30.
\end{aligned}
$$

For the second duplation, we have immediately:

$$
2(1 + 2/3 + 1/10 + 1/30) = 3 + 1/3 + 1/5 + 1/15
$$

But the scribe provides the result: 3 + 1/2 + 1/10, and we can assume he used the result

$$
1/2 = 1/3 + 1/10 + 1/15.
$$

The final duplation is very easy, since we have:

$$
2(3 + 1/2 + 1/10) = 7 + 1/5.
$$

To find the required result, we add the multiplications by 2 and 8 and can see that it certainly comes to 9. Here again, the scribe does not show us how he proceeded, but at the beginning of the papyrus, we find the result:

$$2 = (1 + 1/5 + 1/10) + (2/3 + 1/30)$$

from which we can easily obtain:

$$(7 + 1/5) + (1 + 2/3 + 1/10 + 1/30) = 9.$$

We can see from this that the duplation algorithm is suitable for multiplication, using only the doubles of unit fractions, together with the standard results given at the beginning of the Rhind papyrus. Likewise, mediation, or division by 2, can be used for divisions. The main algorithms that the Egyptians used were directly linked to the operations of duplation and mediation. This is also true for the 'method of auxiliary reds', which we shall present next, and by which the sums of unit fractions can be 'reduced to the same denominator'.

The algorithm of duplation is found in numerous Arab works of arithmetic, and remained strongly in evidence up to the Renaissance. For example, we find it used in a work by Stifel in 1546.

B. The use of red auxiliaries

Rhind Papyrus
Rhind Mathematical Papyrus, translation by A. Chace, 1927-1929
Problem 7, Plate 39

Hieratic text as it appears on the papyrus

Transcription in hieroglyphics

Translation

Example of making complete

1	$\frac{1}{4}$	$\frac{1}{28}$
	7	1
$\frac{1}{2}$	$\frac{1}{8}$	$\frac{1}{56}$
	$3\frac{1}{2}$	$\frac{1}{2}$
$\frac{1}{4}$	$\frac{1}{16}$	$\frac{1}{112}$
	$1\frac{1}{2}\frac{1}{4}$	$\frac{1}{4}$
Total	$\frac{1}{2}$	

We have put in bold the text that was written in red ink. The scribes generally used two sorts of ink: "the black ink was obtained from lampblack; it was used for writing the greater part of the text. The red ink was made with ochre, iron oxide dug from the Egyptian earth. As a contrast with the black, it was used so that series could be distinguished (for example in the earliest documents on computation), to indicate dates, to punctuate a narrative, and above all to mark the heads of chapters, the beginning of paragraphs and the end of a long text. This custom has enriched our own vocabulary with the word 'rubric' (from the Latin *ruber* = red)" [2].

The calculations in problem 7 correspond to calculating the sum:

$$1(1/4 + 1/28) + (1/2)(1/4 + 1/28) + (1/4)(1/4 + 1/28)$$

or $\quad$ $(1/4 + 1/28) + (1/8 + 1/56) + (1/16 + 1/112)$

which gives the result 1/2, as the text indicates. This sum of unit fractions, then, comes into a calculation by mediation.

In order to carry out this calculation, the scribe 'reduces to the same denominator' by taking 28 as the common denominator. Since the smallest unit fraction lower than this is 1/112, this means that the 'numerators' are not necessarily integers. The scribe writes the 'numerators' in red. Thus:

$$1/4 = 7/28 \quad 1/8 = (3 + 1/2)/28 \quad\quad 1/16 = (1 + 1/2 + 1/4)/28$$
$$1/28 = 1/28 \quad 1/56 = (1/2)/28 \quad\quad 1/112 = (1/4)/28$$

The sum of the 'numerators' is equal to:

$$(7 + 1) + (3 + 1/2 + 1/2) + (1 + 1/2 + 1/4 + 1/4).$$

This can be done easily since, the scribe proceeding by mediation, it is made up only of integers or unit fractions with denominators which are powers of 2. We immediately find that the sum is 14, which is half of 28, and the total is therefore 1/2.

The scribe uses 28 as the 'common denominator' since it appears as part of the data of the problem, even though the mediation process will end up with non integer 'numerators'. But in problem 31 of the Rhind papyrus, in order to make it easier to use the method of 'auxiliary reds', the scribe needs to use a 'common denominator' of 42, although the number being handled is 1 + 2/3 + 1/2 + 1/7.

1.4 Tableau Multiplication

There are numerous multiplication algorithms. For the most part they are not completely general but were devised in order to deal with special cases. Thus we find algorithms for multiplying any number whatever by some particular number, for example one containing only 9s, or for calculating squares. We also find algorithms for mental calculations, and others that are restricted to written calculations. These latter were of two types, depending on the surface on which they were inscribed: those that involved 'rubbing out', associated with dust boards, and those that were done 'without rubbing out', restricted to recording on paper. Among these, some involved the need to remember things (for carrying, for example), and some did not.

Among all these many algorithms, we have decided to present here the technique using a tableau, since it provides a good illustration of how some mathematical techniques spread throughout the world. In fact, it is found contemporaneously, and at different times, in China, India, the Arab world and in Europe.

The use of this technique in all civilisations rests on the same basic mathematical principle, but with many variations in the way it is presented and the names given to it: it is known variously as multiplication using a 'tableau', a 'grid', a 'net', or 'the jalousie'. This last was so called because it referred to a type of Venetian blind in the form of a grating, common in Venice, through which nuns or ladies could see out from the inside without being observed from the outside. Numbers can be imagined as appearing in the spaces of the grating. The word 'jalousie' gives us 'Gelosia' by which the method is now commonly known (see Smith [23], vol. II, pp. 114–117).

The calculation technique uses the fact that numbers are written using positional notation. It has the advantage of avoiding the carrying that comes from our usual 'long' multiplication, in that all the results of the subsidiary multiplications are written down and arranged in a geometrical configuration which provides for the different decimal or sexagesimal positions.

To take an example, consider one found in the 16th century Indian astronomer Ganesa's commentary on the 12th century Indian book *Līlāvati* by Bhāskara [5]. The results of the separate products for the multiplication of 135 by 12 appear in the small squares.

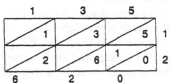

The descriptions of the technique of using a tableau or grid that exist in Chinese, Indian and European works are presumably there because the technique was in current use, but historians have not yet been able to determine the origin of the algorithm with any certainty. What we shall do here, therefore, is simply to note the presence of the algorithm in the different traditions of Oriental and Occidental mathematics, and we shall give some examples drawn from Arabic, Chinese and European contexts.

A. From Arab mathematics

In his book *al-fuṣūl fī-l-ḥisāb al-hindī* (Chapters on Indian calculations), the Arab mathematician al-Uqlīdisī describes a multiplication algorithm, called by him 'multiplication with the aid of cells', which uses a tableau with square cells although it does not, as in other tableau methods [21], use triangular sub-cells for separating the two digits, The same procedure is also described by al-Baghdādī, in the 11th century, in his book *at-Takmila fī-l-ḥisāb* (The Culmination of Calculation), which he calls 'products through geometry' [22].

In the 13th century Maghreb arithmetic book *Talkhīṣ aᶜmāl al-ḥisāb* (A summary of calculation methods) by Ibn al-Bannā we find the earliest known description of the tableau method in Arab mathematics. In his *Talkhīṣ*, Ibn al-Bannā puts arithmetic algorithms into three categories: product by translation, product by semi-translation and product without translation.

The tableau algorithm belongs to the third category. As with products formed by the two other categories, it dispenses with the need to memorise carrying terms, but it has the advantage over the other methods in not requiring rubbing out of intermediary results. We shall come back to the first two categories in the next section.

Ibn al-Bannā

From the *Talkhīṣ* of Ibn al-Bannā,
M. Souissi, University of Tunis publications, 1968, p. 48.
From the French translation of the Arabic by A. Djebbar.

Its description is as follows: you construct a quadrilateral which you subdivide vertically and horizontally in as many strips as there are places in the numbers to be multiplied. Divide the squares [that are formed] by diagonals going from the lower left [corner] to the upper right [corner].

Set out the multiplicand above the quadrilateral, making each of its places correspond to a column. Then set out the multiplier to the left or to the right of the quadrilateral, [such that] it goes down it and also each place corresponds with a line. Then you multiply, one after the other, each of the places of the multiplicand of the square by all the places of the multiplier, and you put the [corresponding partial] product in each position in the square where [its respective column and line] intersect, putting the units above the diagonal and the tens below. Then you start to add, starting from the upper left corner: you add what lies between the diagonals, without rubbing out, putting each number in its position, transferring the tens figure of each [partial] sum to the next diagonal and adding it to what is there. The sum that you get will be the product.

We should remember that the writing, and therefore the reading, of the numbers goes from right to left. To multiply 348 by 27, using the way Ibn al-Bannā describes setting out the work, is shown in the tableau. The result 9396 is written along the bottom and up the left hand side.

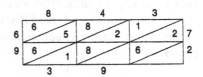

Ibn al-Bannā does not give any examples in his *Talkhīṣ*, but we find them in commentaries on his work. To illustrate his text, we have chosen a later work that uses the technique with the aid of alphabetic numeration. This is found in *Miftāḥ al-ḥisāb* (The key to calculation) by al-Kāshī, a 15th century Persian mathematician, who worked in Asia, in particular at the famous Maragha observatory constructed by Ulug Beg, the grandson of Tamburlaine.

Arab mathematicians were responsible for spreading, and improving, a denary number system inherited from India, written with the use Indian numerals or letters of the alphabet. Alongside this system, Arab mathematicians also used a sexagesimal system, inherited from the Greeks and written using letters of the Arabic alphabet. The tableau from al-Kāshī's work presented here carries out the multiplication of two numbers expressed in sexagesimal form.

The tableau by al-Kāshī is equivalent to the multiplication of two numbers which we would write today as:

$$38 \times 60^{-3} + 40 \times 60^{-2} + 15 \times 60^{-1} + 24 \text{ and } 20 \times 60^{-1} + 51 + 9 \times 60 + 13 \times 60^2$$

The use of triangular cells allows the different sexagesimal places to be seperated and ordered. Thus, $20 \times 38 = 760 = 12 \times 60 + 40$, and so the triangular cells situated to the left of the tableau contain the numbers 40 and 12. Notice also that the multiplication $9 \times 40 = 6 \times 60$ produces a cell containing zero, and this is written out in letters.

The arrangement of the tableau facilitates the addition which completes the multiplication. By adding vertically, we obtain the product:

$$40 \times 60^{-4} + 50 \times 60^{-3} + 27 \times 60^{-2} + 44 \times 60^{-1} + 54 + 22 \times 60 + 19 \times 60^2 + 5 \times 60^3.$$

For the reader who wishes to follow the calculations in the original, step by step, we should point out that the copyist, or the printer, has not followed the usual way of writing these numbers, for example the number 5.

Table by al-Kāshī

From *Miftāḥ al-ḥisāb* (The key to calculation)

A. S. Damirdash and M. H. al-Hafni, 1967, pp. 110–111.

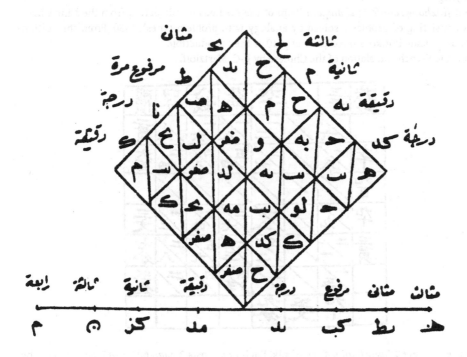

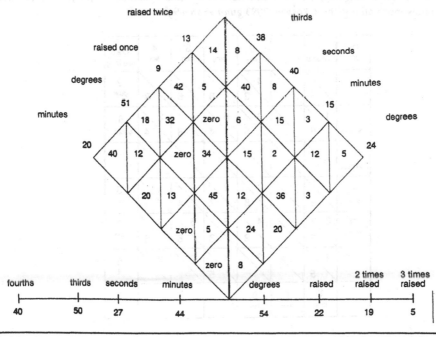

B. From Chinese mathematics

Wu Jing

From *Jiuzhang suanfa bilei daquan* (Sum of the methods of calculation from the Nine Chapters consisting of problems solved by analogy with problem types), 1450. From the edition kept at Seikado bunko, Tokyo. Recto of p. 10 of the Introduction.
From the French translation of the Chinese by J.-C. Martzloff.

Let there be 3069 *liang* 8 *qian* 4 *fen* of silk. Each *liang* costs 2 *guan* 603 *wen* 7 *fen* 5 *li*. For the price: how much all together? Answer: 7993 *guan* 95 *wen* 9 *fen*.

total price	3 thousands			6 tens	9 liang	8 qian	4 fen	price of one liang
		6		1 / 2	1 / 8	1 / 6	/ 8	2 guan
7 thousands	1 / 8			3 / 6	5 / 4	4 / 8	2 / 4	6 hundreds
9 hundreds	/							
9 tens	/	9		1 / 8	2 / 7	2 / 4	1 / 2	3 wen
3 guan	2 / 1			4 / 2	6 / 3	5 / 6	2 / 8	7 fen
	1 / 5			3 /	4 / 5	4	2	5 li
	9 tens	5 wen	9 fen			.		

The first evidence of the use of a tableau for multiplications in China is found in a treatise by Wu Jing, *Jiuzhang suanfa bilei daquan* (Sum of the methods of calculation from the Nine Chapters consisting of problems solved by analogy with problem types). The author of this manuscript, which was completed in 1450, was financial governor in Zhejiang Province.

The technique of using a tableau appears in the preliminary part of the work. The author does not provide any theoretical explanation but simply a sequence of arithmetical problems of increasing complexity, with the corresponding tableau in each case.

The units of measurement used in the problem we illustrate are the following:

length measures for measuring cloth	1 *liang* = 10 *qian*	1 *qian* = 10 *fen*	
monetary measures	1 *guan* ('ligature') = 1000 *wen*	1 *wen* ('sapek') = 10 *fen*	1 *li* = 1/10 *fen*

The tableau shows that the unit called the *fen* was used both for measuring length and for sums of money, so it was a sort of positional marker to record fractions of a main unit, whatever that main unit might have been.

Empty cells are where there are no corresponding numbers. In other words, the idea of zero here only involves the idea of absent units, for which the cell is left unfilled. However, the existence of an 'empty cell' does not appear in the written form. Translated literally, the text states: "Let there be three thousands [and] sixty nine *liang* eight *qian* four *fen* of silk ..." which, to make it easier to read, we have translated as "Let there be 3069 *liang* 8 *qian* ...". The zero, clearly present in the number in the translation, does not appear in the written text, although it is implied by the existence of an empty cell in the tableau used for the multiplication.

It is almost certain that this method of multiplication had been borrowed from the Arab tradition. In fact, in China, calculations are never carried out in written form, and the Chinese name for this technique is *xie suan*, that is 'calculation by writing'.

C. From Europe

As well as in China and the Arab world, the method of multiplication using a tableau is also found in Europe, and has a long history which extends from the Middle Ages right up to the beginning of the twentieth century. The earliest known European example of the use of a tableau appears in a Latin manuscript of about 1300 in England in the reign of Edward II (figure 1).

This method is found later in a number of arithmetic books from the Renaissance period, like the *Treviso Arithmetic* [26], published anonymously in Treviso in 1478 (figure 2) and Luca Pacioli's *Suma* of 1494 (figure 3).

Despite its benefits, this technique does have disadvantages, since it sometimes takes as long to draw up the tableau as it does to carry out the calculations themselves. It is not surprising, then, that the idea of a permanent 'tableau' should have occurred to mathematicians, as we shall see.

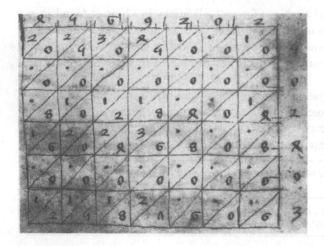

Fig. 1

Multiplication of 4569202 by 502403 in the Latin manuscript *Tractatus de minutis philosophicis et vulgaribus* (A treatise on small measures, philosophical and vulgar), about 1300, Oxford, Bodleian Library, MS Digby 190. fo. 75r.

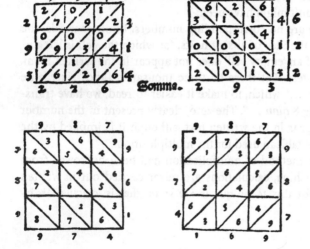

Fig. 2

Gelosia multiplication, in the Treviso Arithmetic (1478). Source: Smith, *History of Mathematics*, vol. II (1925), repub. Dover, New York, 1958, Vol II, p. 115.[23]

Fig. 3

Gelosia multiplication, in the Arithmetic by Luca Pacioli (1494). Source: Smith, *History of Mathematics*, vol. II, p. 116.

Napier's Rods

In 1617, Napier published a small work *Rabdologiae Sev Numerationis per Virgulas* (from the Greek word *rabdos* = rod). These rods were commonly known as Napier's Bones and consisted of a set of four sided rods on which were written out the multiplications from 1 to 9 in a column for each index number from 0 to 9 (figure 4). The products were written with a diagonal line separating the tens and units digits as shown in the figure. To multiply any number all that was required was to arrange the rods in the order of the digits of the multiplicand and to read off the answer, adding the corresponding digits for each place as they were read (figure 5). Naturally, it was sometimes necessary to use the same digit more than once, which was why different index numbers appeared on each of the four faces of the rods.

MILLIONS N°1546	0	1	2	3	4	5	6	7	8	9
1	0/0	0/1	0/2	0/3	0/4	0/5	0/6	0/7	0/8	0/9
2	0/0	0/2	0/4	0/6	0/8	1/0	1/2	1/4	1/6	1/8
3	0/0	0/3	0/6	0/9	1/2	1/5	1/8	2/1	2/4	2/7
4	0/0	0/4	0/8	1/2	1/6	2/0	2/4	2/8	3/2	3/6
5	0/0	0/5	1/0	1/5	2/0	2/5	3/0	3/5	4/0	4/5
6	0/0	0/6	1/2	1/8	2/4	3/0	3/6	4/2	4/8	5/4
7	0/0	0/7	1/4	2/1	2/8	3/5	4/2	4/9	5/6	6/3
8	0/0	0/8	1/6	2/4	3/2	4/0	4/8	5/6	6/4	7/2
9	0/0	0/9	1/8	2/7	3/6	4/5	5/4	6/3	7/2	8/1
	8901	1234	9012	0123	3456	2345	6789	4567	5678	7890

Fig. 4 Napier's Rods for multiplication.

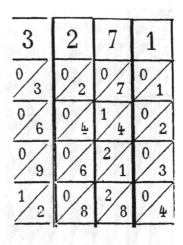

Fig. 5 Multiplication of 3271. Successive products can be read off at sight.

Lucas-Genaille Rods

Ingenious though it was, Napier's idea could be further improved. In 1885 at the French Association for the Advancement of Science, the French mathematician Edouard Lucas, known for his ingenious recreations in mathematics, and also for his many original contributions to the theory of numbers (see Sections 8.4 and 8.5), posed the problem of improving Napier's rods. He remarked that the rods involved a complication since it was always necessary to take account of the carrying digit when reading off a product. Might it be possible, he wondered, to be able to read off an answer directly without any need for additional calculations?

Some years later, a French railway engineer, Henri Genaille, produced some rods of his own invention that allowed the multiplication of any number by a single digit to be read off with ease. This time, the result was to be found by following arrows, carefully marked on the rods (Figure 6). These Lucas-Genaille rods were produced commercially by the publishers Belin in Paris, and were a successful product up until about 1911 [29].

Suppose we want to multiply any number by a single digit, say 3271 multiplied by 4. After arranging the rods in the correct order for the number 3271, preceded by the '0' rod, the result is read off from right to left, following the black arrows, starting from the top of the strip corresponding to the multiplier. The result can be written down immediately, starting with the units, then the tens digit, etc.

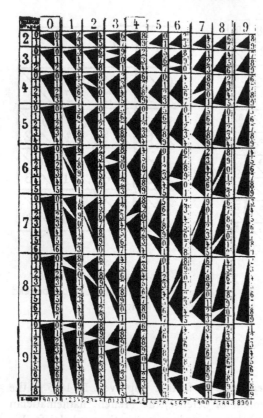

Fig. 6 Lucas-Genaille multiplication rods.

Fig. 7 3271×4 = 13084 using Lucas-Genaille rods.

1.5 Optimising Calculations

The text shown below compares two multiplication methods used in the Middle Ages. Alongside the procedure known as 'product by translation', which was the one generally in use at the time, there existed a method used specifically for calculating squares. This procedure was known as 'product by semi-translation' because, compared with the first, fewer elementary calculations were needed.

The text enumerates the number of elementary multiplications needed in the case where we are to find the square of a number, and emphasises the economy involved in using 'semi-translation': this is effectively an optimisation problem. It is one of those cases of multiplication methods which were specific to the numbers to be multiplied, which we referred to earlier.

We shall explain what is meant by the methods of translation and semi-translation by considering the example of finding the square of 438.

Ibn al-Majdī
From Ḥāwī l-lubāb
Ms. British Museum, N° Add. 7469, ff. 41b–42a.
From the French translation of the Arabic by A. Djebbar.

You need, in [problems] like this to carry out a number of multiplications [equal] to the number of the square [of the number of digits] of the number [you wish to square] [...].

To summarise, we can say: take away from the number of multiplications half of what remains of the square of the number of places of [the number] raised to a square, after having taken [from it] half the number of [the digits of] that number.

[For example], if the number of places of the number is ten, the number of multiplications will be 100. And we take away forty five. If the number is five, the [number of] multiplications will be twenty five; and we take away ten. Understand that.

Translation method

The calculation is carried out in three steps, because the number to be squared has three digits. The number is written on a first line, where it is the multiplier, and again on a second line below where it is the multiplicand, and this is placed so that the units digit, here the 8, is under the digit of the multiplier corresponding to each step of the method.

			1	8	8	3	4		1	9	1	8	4	4			
1	7	5	2			2	4						6	4			
		3	2			9						2	4				
	1	2			1	2					3	2					
1	6		1	7	5	2			1	8	8	3	4				
					4	3	8					4	3	8			
		4	3	8	4	3	8					4	3	8			
		4	3	8													

– Step 1: multiply the digits of the second line by 4 and put the results, one above the other, keeping them in their right places, then add these products to give the number 1752 in the top line.

– Step 2: move the multiplicand one place to the right, write down the part answer 1752 above, keeping the same denary position, and repeat the multiplication, this time multiplying the bottom number by 3.

– Step 3: move the multiplicand again one place to the right, and repeat stage 2.

Semi-Translation method

The method of semi-translation, only used for calculating squares of numbers, is based on the following algebraic expansion. If x is written as a number abc where a,

b and c occupy the hundreds, tens and units places of the number, that is $x = 100a + 10b + c$, then

$$x^2 = 10000a^2 + 1000.2ab + 100(b^2 + 2ac) + 10.2bc + c^2.$$

In practice, the number to be squared is written with its digits separated by dots: $4 \cdot 3 \cdot 8$. Then the terms of the expression above are calculated, and set out one above the other, placing their results conveniently so as to correspond to their denary positions.

The calculation involves five steps:

Step 1: calculate $a^2 = 4 \times 4 = 16$ and place it above the $4 \cdot 3 \cdot 8$ with the 6 above the 4, occupying the ten thousands place.

Step 2: calculate $2a = 2 \times 4 = 8$ and $2ab = 8 \times 3 = 24$. The first answer, which is intermediary in the calculations, is put below the $4 \cdot 3 \cdot 8$, and the second is placed above, with the 4 occupying the thousands place.

Step 3: calculate $b^2 = 3 \times 3 = 9$ and $2ac = 8 \times 8 = 64$

Step 4: calculate $2b = 2 \times 3 = 6$ and $2bc = 6 \times 8 = 48$

Step 5: calculate $c^2 = 8 \times 8 = 64$

The final result is found by adding the terms of each column above $4 \cdot 3 \cdot 8$, giving the answer 191844.

How many elementary multiplications are carried out using each of these methods when squaring a number of n digits? With the translation method, there are n^2 multiplications, but with the semi-translation method there are $1 + 2 + \ldots + n = n(n + 1)/2$ elementary multiplications, not counting the doublings. For $n = 10$ and $n = 5$ we obtain 45 and 15 respectively, the results given by Ibn al-Majdī.

					6	4
				4	8	
			6	4		
			9			
		2	4			
1	6					
	4	.	3	.	8	
		8				
		6				

1.6 Simple Division by Difference on a Counting Board

Different variations of types of abacus, or counting boards, were used in Antiquity. Gerbert of Aurillac, who became Pope Sylvester II in 999, is credited with having introduced a calculation board with arcs and columns, still called the *Pythagorean arc*.

This board was divided into squares for positioning tokens, still known as *apices*, on which were marked the first nine letters of the Greek alphabet or other special

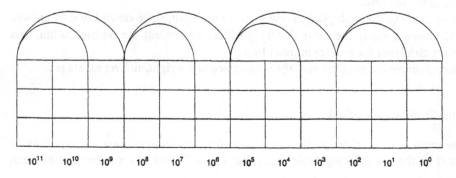

10^{11} $\quad 10^{10}$ $\quad 10^{9}$ $\quad 10^{8}$ $\quad 10^{7}$ $\quad 10^{6}$ $\quad 10^{5}$ $\quad 10^{4}$ $\quad 10^{3}$ $\quad 10^{2}$ $\quad 10^{1}$ $\quad 10^{0}$

marks. The positions for the apices were in accordance with the denary counting system. Starting from the right, the first column was used for units, the second for tens, the third for hundreds, and so on for all the powers of ten required. There was no symbol for zero, an empty place was used to show absence of a digit. The columns were sometimes linked together in threes by arcs. This is why the columns were referred to as arcs, and why the term Pythagorean arc was used to describe the whole abacus.

The example given here is from a translation made by Michel Chasles in 1843 of a manuscript which had been written in about the year 1200. From it we have taken the algorithm of 'simple division by differences' whose origin is not known. The manuscript begins with a description of the abacus and defines two sorts of numbers: *digits* and *articles*. The *digits* were the numbers up to 9 and the *articles* are all the multiples of 10. Thus every whole number is the sum of a *digit* and an *article*.

The manuscript explains that the algorithm of 'simple division by differences' only applies to those cases where the divisors are *digits* and the dividends are *articles*. The example used in the extract was 900 divided by 8.

The principle of the division process is as follows. Let a and b be the first two digits of the dividend, which can then be written as $a.10^n + b.10^{n-1} + k$, where $k < 10^{n-1}$, and let the divisor be d, taken to be less than 10.

We have: $\dfrac{a.10^n + b.10^{n-1} + k}{d} = a.10^{n-1} + \dfrac{(a(10-d)+b).10^{n-1} + k}{d}$, so a partial quotient is $a.10^{n-1}$ and we replace the dividend by $(a(10-d)+b).10^{n-1} + k$. This process can be repeated with the reduced dividend, provided we add up the partial quotients we obtain.

At the start, the abacus is laid out as follows. The first line contains the tens complement of the divisor, the second line contains the divisor, the third line contains the dividend and the fourth line is empty. Then for each iteration of the process just described:

1. the two last lines must be changed so that the reduced dividend appears as the dividend, that is on the third line, and

2. the partial quotient should be added to the line containing the quotient, which is the fourth line.

These two operations on the abacus can be represented symbolically as below:

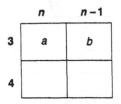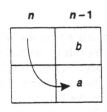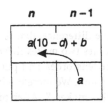

Michel Chasles

From 'Histoire de l'arithmétique, Règles de l'Abacus',
Chasles, *Comptes Rendus de l'Académie des Sciences*, vol. 16 (1843), 228–229.

[In order to make it easier to follow the text, we have divided it up into sections and put accompanying diagrams in the right hand column of the corresponding calculations on the abacus.]

Example of simple division by differences. Divide 900 by 8.

Let us give an example of simple division by differences.
Place VIII for the divisor in the units column, and above it its difference from X, that is two. Put nine for the dividend into the hundreds column

		②
		⑧
⑨		

and now we must say: If a simple divisor, with its difference, is in the units column, the denomination taken all together is taken down one rank. That is the first rule. [...]
Take the denomination, that is the complete dividend as denomination, and put it in the second column, that is put the nine in the tens column, in the lowest section.

		②
		⑧
	⑨	

Multiply the difference by the denomination, and say the following rule: When the multiplier is a number of tens, put the digit in the second column counting from the multiplicand, and the article one rank beyond. So put VIII in the tens column and the unit in the hundreds, and you will then have the product of the multiplication, that is a digit and an article.

		②
		⑧
①	⑧	
	⑨	

Transfer the article to the first denomination, following the rule that says: If there is an article, it is returned to the denomination.

		②
		⑧
	⑧	
	①⑨	

Multiply the difference, that is two by this article, that is a unit: one times two makes II. Now say the rule for the multiplier of the tens order, and put the product, namely two, in the tens column. We therefore have in the tens column, VIII and II which makes X. After picking up the VIII, we put a unit in the hundreds column, then it is carried above the other denominations, following the rule for the article.

		②
		⑧
	①①⑨	

We multiply the difference, that is two, by this unit: one times two makes II. Using the rule for a multiplier of the tens order, this product two has to be placed in the tens column;

		②
		⑧
	②	
	①①⑨	

and as this is a digit, it is put as a denomination of a rank after the denominations already calculated.

		②
		⑧
	①①⑨	②

Now we have to say: Two times II, four; and using the rule for a multiplier of the units column, we put this IIII itself in the units column, and after picking up the difference related to this divisor VIII, it is clear that the divisor, which is greater, can not take anything from four in whole numbers. The division is therefore finished.

		②
		⑧
		④
	①①⑨	②

Now, following the rule of purgation, what must be done is to carry one of the units from the tens column to the hundreds column, and leave one of them and pick up the IX. And what results is two units, one in the hundreds column, and the other in the tens column, and the digit two in the units column.

		②
		⑧
		④
①	①	②

And it can be clearly seen that nine hundred pears being shared between eight soldiers, each will have one hundred and twelve, with four left over.

The algorithm is carried out using a number of 'loops' and requires a number of intermediary calculations, the results of which are to be retained in the 'memory'. The different ranks of the abacus are therefore designed to hold and retain data or particular calculations:

the 1st rank holds the difference $10 - d$
the 2nd rank holds the divisor d
the 3rd rank holds successive dividends
the 4th rank accumulates the successively calculated partial quotients.

At the beginning of the calculation, the 3rd rank holds the dividend. While the number in this rank is greater than 10, each 'loop' of the algorithm consists of two operations:

1. reduce the dividend by $a.10^n$ by placing $a.10^{n-1}$ on the 4th rank.
   This involves moving the token at the extreme left of the 3rd rank, called the 'denomination', one place to the right and putting it on the fourth rank.
2. put on the third rank $a(10-d).10^{n-1}$, which provides a new dividend on this rank.
   We shall simulate this algorithm for the example proposed in the text, namely dividing 900 by 8. Our abacus contains three 'vertical' columns, numbered from the left, and four 'horizontal' ranks, numbered from the top down.

Initialisation: put the difference $(10-d)$, here 2, in the first rank of the right hand column. Then place the divisor d, here 8, on the second rank of the same column. Finally, put the dividend on the third rank, with the units digit in the right hand column. Here, 9 is put in the hundreds column.

		②
		⑧
⑨		

1st loop

1st operation

		②
		⑧
	⑨	

2nd operation

		②
		⑧
①	⑧	
	⑨	

2nd loop

1st operation

		②
		⑧
	⑧	
	①⑨	

2nd operation

		②
		⑧
①		
①		

3rd loop

1st operation

		②
		⑧
①	①	

2nd operation

		②
		⑧
	②	
①	①	

4th loop

1st operation

		②
		⑧
①	①	②

2nd operation

		②
		⑧
		④
①	①	②

The process ends when the number appearing on the third rank contains only a digit. If this digit is less than the divisor d, it is the remainder of the division; the final quotient can be read from the fourth rank. In the case where the digit appearing on the fourth rank is not less than the divisor d, then a final division by d has to be done.

We can check that the operations are correct at each stage, since the sum of the third rank together with the product of the fourth and the second ranks, is always equal to the initial dividend.

1.7 Division on the Chinese Abacus

The Chinese abacus has only been commonly used in China since the second half of the 16th century. In its present day version, it consists of a number of wires set in a rectangular frame. On each of the wires there are two sets of beads, separated by a transverse bar, and the beads can be moved freely. The abacus is based on the principle of denary and positional number notation: each of the two upper beads is worth five units, and each of the five lower beads one unit. Only the beads next to the transversal bar are counted. For example, the number 123456789 would be represented as follows:

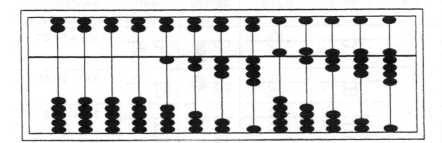

There are a number of operating techniques, depending on the type of abacus and the mathematical method being used. There are, in fact, particular division methods for each of the numbers from 1 to 9, as well as, sometimes, rules for dividing by numbers between 11 and 99. But there are also more general rules, valid for any number. We shall explain here a particular method for dividing by 7. The rules below are adapted from the famous arithmetic *Suanfa tongzong* of 1592.

These division rules are not 'logical' rules, but simple memory techniques for what to do when faced with certain arrangements of beads. The rules start by a reminder of the divisor, here 7, followed by the dividend. The second part of the rule indicates the manipulations, adding or taking away beads, that are required in each case.

Part of the rules are implicit. It must be understood that the dividend has been set out on the abacus, and that the rules must be applied to successive digits, starting with the highest denary order. "Add below" means "put the beads on the wire one denary place lower", and "goes up" means "put the bead one denary place higher than the one being considered".

From *Suanfa tongzong*

From *The principles and practice of the Chinese Abacus*
Lau Chung Him (1958), p. 77.
From the French translation of the Chinese by J.-C. Martzloff.

Rules for dividing by 7

(1) qi-yi xia jia san	seven-one? add three below!
(2) qi-er xia jia liu	seven-two? add six below!
(3) qi-san si sheng er	seven-three? four remainder two!
(4) qi-si wu sheng wu	seven-four? five remainder five!
(5) qi-wu qi sheng yi	seven-five? seven remainder one!
(6) qi-liu ba sheng si	seven-six? eight remainder four!
(7) feng-qi jin yi	'seven' met? 'one' goes up!

To make it easier to understand, we give below a translation of the rules into movements of beads on the abacus. The beads that have been added are shown in black, and the ones removed have been struck through. We only show the beads being used, that is the ones next to the bar.

Rule	Dividend	Divisor	applying the rule	final expression	'translation' into arithmetic
(1)					$10/7 = 1 + 3/7$
(2)					$20/7 = 2 + 6/7$
(3)					$30/7 = 4 + 2/7$
(4)					$40/7 = 5 + 5/7$
(5)					$50/7 = 7 + 1/7$
(6)					$60/7 = 8 + 4/7$
(7)					$7/7 = 1$

To divide a number by 7, put the divisor to the left and the dividend on the right of the abacus. Operate on the successive digits of the dividend, starting with the one of the highest order. The rules are then simply carried out mechanically.

As an example of using the rules consider 1234 divided by 7:

1st stage: set out the divisor and the dividend.

2nd stage: the first digit of the dividend is 1, so use the first rule. Then 'simplify' the five.

3rd stage: the second digit of the new dividend is 5, so apply the fifth rule.

4th stage: the third digit of the new dividend is 4, so apply the fourth rule.

5th stage: the fourth digit of the new dividend is 9, so we apply the seventh rule.

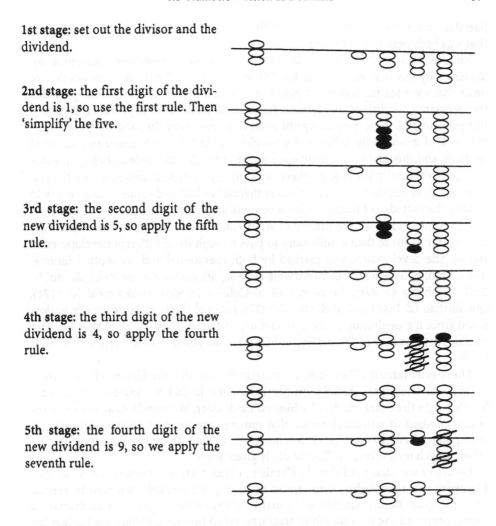

Thus 1234 divided by 7 is 176 with a remainder 2. The operation can be checked by the use of other particular rules for dividing by 7.

1.8 Numbers Written as Decimals

The idea of decimal fractions is very old but, for a long time it existed alongside other, non-decimal, fractional use, like that of the Egyptian's unit fractions and the ever popular use of sexagesimal fractions.

Taking a global view of numeration systems, which were generally neither denary nor positional, we find that decimal fractions made an appearance only at isolated points, not as belonging to a coherent and systematic approach to numbers, but as a happy consequence of some inherent properties of particular numeration or calculation procedures: anything that might make calculations easier would be bound to be successful. For decimal fractions to be thought of as such, and for them to be handled in a systematic way, there had first to be established a numeration system that was both base ten and positional.

Thus in Ancient China, at the time of the unification of weights and measures undertaken by the first emperor of the Qin dynasty (*d.* 210 BC), the introduction of units that were for the most part based on tens, hundreds and thousands, provided the necessary conditions for decimal fractions to make their appearance in China. But prior to this, there were many obstacles to be overcome. In particular, the different units of measure (length, volume, weight, etc.) had a concrete meaning attached to them, and this had to be superseded for units to take on a role of being, in some way, pure decimal positional markers, without any particular concrete significance. Likewise, later on, these decimal markers themselves had to disappear, to give way to a more abstract idea of number, independent of any system of units.

As recent studies on the history of Arab, Chinese and Indian mathematics have shown, it took more than a millenium to pass through these different developmental stages. The development was marked by both operational and conceptual innovations. Among those whose contributions were significant we can mention Liu Hui (*c.* 300), Āryabhata (*b.* 476), Sunzi (*c.* 500), al-Uqlīdisī (*c.* 950), as-Samaw'al (*d.* 1174), Qin Jiushao (*c.* 1247) and al-Kāshī (*d.* 1437). Decimal fractions were clearly understood from the beginning of the 13th century and, from then on, an author like Qin Jiushao, the inventor of the theorem of Chinese remainders, used decimal fractions freely.

These brief remarks show that, contrary to the way that the history of mathematics is sometimes presented, Simon Stevin was not in fact the inventor of decimal fractions. On the other hand, as he himself made clear, his contribution was to introduce a method of writing decimals that removed the onerous burden of having to manipulate them like ordinary fractions. An extract from his famous work *La Disme*, which means literally The Tenth, is given below.

La Disme was first published in Flemish in 1585, then in French in the same year. The work was republished many times in its original Flemish, and also in French and in English. Being experienced in many different areas of practical mathematics, Stevin proposed the elimination of fractions, called broken numbers, and using "an invented arithmetic using the progression of tens, to deal with all the affairs of astronomers, surveyors, cloth merchants, engravers, measurers, mint-masters, and all merchants in general".

Simon Stevin

From *La Disme* in *Œuvres mathématiques de Simon Stevin de Bruges*, ed. Girard, pub. Leyden, 1634, pp. 208–209.
English translation: Vera Sanford in D. E. Smith, *A Source Book in Mathematics*, McGraw-Hill Book Co., 1929, reprinted Dover, New York, 1959, pp. 24–27.

Definition II

Any given number is called the *unit* and has the sign ⓪.

Explanation

In the number three hundred and sixty four, for example, we call the three hundred and sixty four units, and write the number 364⓪. Similarly for other cases.

Definition III

The tenth part of a unit is called a *Prime*, and has the sign ①, and the tenth of a prime is called a *Second*, and has the sign ②. Similarly for each tenth part of the unit of the next higher figure.

Explanation

Thus 3①7②5③9④ is 3 primes, 7 seconds, 5 thirds, 9 fourths, and we might continue this indefinitely. It is evident from the definition that the latter numbers are

$$\frac{3}{10} \ \frac{7}{100} \ \frac{5}{1000} \ \frac{9}{10000} \quad \text{and that this number is} \quad \frac{3759}{10000}.$$

Likewise   8⓪9①3②7③

has the value $8\dfrac{9}{10}\dfrac{3}{100}\dfrac{7}{1000}$ or $8\dfrac{937}{1000}$

And so for other numbers. We must also realise that in these numbers we use no fractions and that the number under each sign except the 'unit' never exceeds 9. For instance, we do not write 7①12② but 8①2② instead, for it has the same value.

Definition IV

The numbers of the 2nd and 3rd definitions are called *Decimal Numbers*.

. [...]

Proposition III – to multiply decimal numbers
Given the number 32⓪5①7② and the multiplier 89⓪4①6②. Required to find their product.

Construction

Place the numbers in order and multiply in the ordinary way of multiplying whole numbers (by the third problem of *l'Arithmétique*). This gives the product 29137122. To find what this is, add the last two signs of the given numbers, the one ② and the other ② also, which together are ④. We say then that the sign of the last figure of the product will be ④. Once this is established, all the signs are known on account of their continuous order. Therefore, 2913⓪7①1②2③2④ is the required product.

```
                    ⓪①②
                3 2 5 7
                8 9 4 6
              1 9 5 4 2
              1 3 0 2 8
            2 9 3 1 3
            2 6 0 5 6
          2 9 1 3 7 1 2 2
                ⓪①②③④
```

Proof

As appears by the third definition of *La Disme*, the given number 32⓪5①7② is

$32\dfrac{5}{10}\dfrac{7}{100}$ or $32\dfrac{57}{100}$, and likewise the multiplier 89⓪4①6② is $89\dfrac{46}{100}$. Multiplying the

aforesaid $32\dfrac{57}{100}$ by this number gives the product $2913\dfrac{7122}{10000}$ (by the twelfth problem of

l'Arithmétique). But the aforesaid product 2913⓪7①1②2③2④ has this value and is, therefore, the true product which we were required to prove. We will now explain why second multiplied by second gives the product fourths, which is the sum of their signs, and why the fourth by fifth gives the product ninth, and why a unit by third gives the product third, and so forth. Let us take for example 2/10 and 3/100 which, by the definitions of *La Disme*, are the values 2① and 3②. Their product is 6/1000, which by the third definition given above is 6③. Hence, multiplying prime by second gives a product in thirds, that is a number whose sign is the sum of the given signs.

Conclusion

Having been given a decimal number to multiply and the multiplier we have found the product, which was to be done.

The operational algorithm is not new. As Stevin himself indicates, it is "the ordinary way of calculating with whole numbers". On the other hand, we see a symbolism here that he had already used for writing polynomials, in *l'Arithmétique*. Thus his "3② + 4 equals 2① + 4" corresponds to our modern form: $3x^2 + 4 = 2x + 4$.

1.9 Binary Arithmetic

Leibniz wrote his memoir on *The explanation of binary arithmetic* in 1703 but he had, for more than forty years, been interested in this system of numeration, which was for him a model of what he calls a *Universal Characteristic*. From 1666, in the treatise *De Arte combinatoria* (The Art of Combinations), Leibniz was seeking a *Universal Characteristic*, that is a sort of algebra that could be used for discovering and proving propositions. This research, for a young man in his twenties, not yet passionately attached to mathematics, took the form of a search for a *Universal Language*.

The idea of a *Universal Language* was current at the time. It would have been suggested to him by the works of many of his contemporaries, particularly those of the Jesuit professor of mathematics and philosophy A. Kircher. Leibniz was critical of earlier attempts to reduce problems of ambiguity in words and diversity of syntactical forms, and of the need for enormous dictionaries. To reduce these disadvantages, he proposed the use of what he called *real characters*, that is characters that represent things and ideas, and not words. What we have here, properly speaking, is the idea of *ideographs*. This has its foundation in logic: concepts can be decomposed into a restricted number of simple concepts. These simple concepts constitute the logical alphabet, they can be represented by hieroglyphs (like those of the Ancient Egyptians or Chinese), and reasoning with them will use arithmetic digits and algebraic signs. Thus, for Leibniz, Algebra and Arithmetic are specimens of his *Universal Characteristic*.

Concepts can be characterised by their decomposition into simple concepts just as numbers can be characterised by their decomposition into prime factors. This point of view led Leibniz to consider numeration systems which were close to those which Pascal had considered before him (see Chapter 8). The extent to which the divisibility characteristics of a number are evident or not depends upon the numeration system being used. The binary system would allow properties of numbers to be deduced from the way they were written. It also has the advantage of only requiring a small number of signs for representing a number.

We also know about the interest that Leibniz had for Chinese hieroglyphics, it being moreover one of the topics dealt with in his 1703 memoir. Leibniz relates that, according to Chinese legend, king Fohy (Fu Xi) introduced 'the figure of eight Cova' consisting of diagrams of the form:

> -- — -- — -- — -- —
>
> -- -- — — -- -- — —
>
> -- -- -- -- — — — —

The meaning of this figure had remained mysterious up till then, but Leibniz provided an explanation by appealing to the binary numeration system. In fact, he remarks that — stands for one, and - - stands for zero. Hence the signs above, read in columns from left to right, represent the integers 0 to 7. Leibniz's interest was increased by the fact that king Fohy had also been the inventor of Chinese characters. He writes, at the end of his memoir, "Now since it is believed in China that Fohy was also the author of the Chinese characters, although they have been much altered through the passage of time, his Essay on Arithmetic shows that he could well have discovered something else of considerable importance concerning numbers and ideas – if the foundation of Chinese writing can be unearthed, to the extent that this is believed in China – that he had a regard for numbers in its establishment" [16].

G. W. Leibniz
From 'L'explication de l'arithmétique binaire', *Mémoires de l'Académie Royale des Sciences*, 5 May 1703,
Matematische Schriften, ed. C. I. Gerhardt, 1863; reprnt. Hildesheim: Olms Verlag, 1962, vol. VII, pp. 223–225.

Explanation of binary arithmetic

Which uses only the characters 0 and 1; with remarks on its usefulness and on how it provides a meaning for the Ancient Chinese figures of Fohy.

[...]

The ordinary calculations in Arithmetic are carried out following a progression in tens. [...] But instead of a progression in tens, I have for several years used the simplest progression of all, which goes up in twos; having found that it serves to perfect the science of Numbers. Thus I do not use for this other characters than 0 and 1, and then going to two, I start again. That is why *two* is written as 10, and two times two, or *four*, is written as 100, and two times four or *eight* is written as 1000, and two times eight or sixteen is written as 10000 and so on. Here is the Table of Numbers in this manner, which can be continued as far as is wished.

At a glance we can see the reason for the *famous property of the double Geometric progression* of whole Numbers, which provides that if we only have one of these numbers to each degree, we can make all the other whole numbers less than double of the highest degree. For here, it is as if we say, for example, that 111 or 7 is the sum of four, two and one. And that 1101 or 13 is the sum of eight, four and one. This property is used by Assayers for weighing all sorts of masses with a few weights, and can be used for currency to provide for several values using a few coins.

1	0	0	4		1	0	0	0	8		
	1	0	2			1	0	0	4		
		1	1					1	1		
1	1	1	7		1	1	0	1	13		

With this expression for Numbers being established, it can easily be used for all sorts of operations.

For *Addition*
for example

110	6
111	7
1101	13

101	5
1011	11
10000	16

1110	14
10001	17
11111	31

For *Subtraction*

1101	13
111	7
110	6

10000	16
1011	11
101	5

11111	31
10001	17
1110	14

For *Multiplication*®

11	3
11	3
11	
11	
1001	9

101	5
11	3
101	
101	
1111	15

101	5
101	5
101	
1010	
11001	25

For Division

15	ꞮꞮ11
	ꞮꞮꞮ1
	ꞯ1

| 101 | 5 |

And all these operations are so easy, that there is never any need to test or divine, as must be done with ordinary division. Here there is no need any more to learn anything by heart, as must be done for ordinary calculations, where we need to know for example that 6 and 7 make 13; and that 5 multiplied by 3 gives 15, following the Table of *one times one is one*; which is called Pythagorean. But here everything can be found out and checked at source, as we can see by the examples above with the signs⁹ and ®.

However, I do not recommend this way of counting to be introduced instead of the usual practice by ten. For besides the fact that we are used to using ten, there is no need for learning here what has already been learnt by heart: thus the practice of tens is shorter, and the numbers less long. And if one were used to counting by twelve or by sixteen, it would be even better. But calculating by two, that is with 0 and 1, as a recompense for its length, is more fundamental to science, even in practice with numbers, and above all for Geometry; the reason for which is, being reduced to the simplest principles, as 0 and 1, what appears throughout is a marvellous order.

Only division requires any commentary. It uses what is sometimes called the 'Spanish' method, or the 'Galley' method, a way of setting out divisions that was used at the time. We shall explain this method using the denary number system, in order to see the ease offered by the binary system.

Suppose we want to divide 8643 by 37. The divisor must be written under the dividend, the quotient being obtained, digit by digit, to the right of the dividend (fig. 1). As for our usual method of division, we first test for dividing 86 by 37, which gives 2, which we place to the right. Then we multiply $2 \times 3 = 6$ and put the difference $8 - 6 = 2$ above the 8. The digits that have been used are struck through (fig. 2). The same is now done for 7: we multiply $2 \times 7 = 14$ and put the difference $26 - 14 = 12$ above the 26 (fig. 3).

At the next step, we replace the 37 under the 8643, this time shifting one place to the right, and we start all over again (fig. 4). At the final stage, the 37 is placed beneath the 43 and we obtain the quotient 233 and the remainder 22 (which are the remaining digits which have not been struck through) (fig. 5).

				1		1̷2̷	
			1	1̷3̷		1̷3̷4̷	
	2		2̷2̷	2̷2̷3		2̷2̷3̷2	
8643	8̷643	2	8̷8̷43	2	8̷8̷4̷3	23	8̷8̷4̷3̷	233
37	3̷7̷		3̷7̷	3̷7̷		3̷7̷3̷7̷	
				3̷7̷		3̷7̷	
Fig. 1	Fig. 2		Fig. 3	Fig. 4		Fig. 5	

In the case of binary division, there is no need for intermediary calculations above the dividend, since either the difference at each stage is zero, and we record it as such, or it is equal to 1 and we can limit ourselves to not striking through the 1 of the dividend. For 15 (1111) divided by 3 (11) there are three steps in the calculation:

Initial situation:	1111		First stage:	1̷1̷11	1
	11			1̷1̷	
				11	

Second stage:	1̷1̷11	10	third stage:	1̷1̷1̷1̷	101
	1̷1̷11			1̷1̷1̷1̷	
	11			1̷1̷	

1.10 Computer Arithmetic

Leibniz's preoccupations with binary arithmetic, which we have referred to already, led him to dream of the possibility of reducing logic to mechanical operations. He made several practical advances in this direction in 1673 and, some thirty years after Pascal's adding machine, he invented a machine for the four operations of arithmetic. He also invented, in 1674, a machine capable of solving equations and even envisaged a type of cylinder that could be used to produce theorems. With the birth of the computer, Leibniz's dream has been, to a certain extent, realised today. The modern computer was preceded by many other calculating machines, like Babbage's 'difference engine' in 1822 (see Chapter 10), and the inventions by Bouchon, Falcon and Jacquard for automating the weaving of fabrics. But the history of computers proper starts in the 1940s in the United States with the development of electronics [18].

During the 1930s, electronic circuitry designed to carry out arithmetic operations were developed in the USA, France and Germany, and these all used the binary number system [19]. The valves invented in 1906 by the physicist Lee De Forest, and electric relays, worked on a binary principle: either the current passed or it did not. The first computers, however, used denary arithmetic. The binary system was adopted following the appearance of an article by von Neumann in 1946 which extolled the virtue of using such a system [13]. Valves, which were bulky and unreliable, were replaced by transistors after their invention in 1947 by W. Shockley. With the invention of the integrated circuit, or micro-chip, by J. St. Clair Kilby in 1958 came the miniaturisation of computers which enormously improved their performance.

A computer recognises and handles information coded as binary words, that is they are composed of the digits 0 and 1, called bits (binary digits). These words are of constant length, usually 8, 16, 24 or 32 bits for a micro-computer, even longer for a large computer. For numerical information, the numbers are represented in binary form.

We can also represent a rational number as a binary word, by including in the word a marker to show the beginning of the fractional part. The position of this marker, designated by a point, can be moved, which is why it is called floating point representation. Thus

$$a_n \ldots a_1 a_0 . f_1 f_2 \ldots f_m \text{ with } a_i = 0 \text{ or } 1, \text{ and } f_j = 0 \text{ or } 1$$

stands for the rational number

$$a_n 2^n + \ldots + a_1 2 + a_0 + f_1 2^{-1} + f_2 2^{-2} + \ldots + f_m 2^{-m}.$$

For example, 1011.1001 represents the number

$$8 + 2 + 1 + 1/2 + 1/16 = 11 + 9/16.$$

Using this system, just as with the denary system, rational numbers do not all have an exact finite representation. For example using 0.00011001100 for 1/10 means that the identity $(1/10) \times 10 = 1$ is no longer satisfied.

In order to improve the ability of the computer to handle numbers, another representation is used, also using a floating point. The floating point separates the binary word into two parts: the mantissa, containing the significant bits, and the exponent, which indicates the position of the binary point. But there are also other ways of representing a number. The binary word can be grouped into threes, with each of the three bits representing a number from 0 to 7, or it can be grouped into fours, with each group representing a number from 0 to 15. In this way, the binary words can be converted to base 8 (octal) numbers, or base 16 (hexadecimal) numbers. The bits can also be regrouped in fours to represent the numbers 0 to 9, and this gives the NBCD code (Natural Binary Code Decimal). In this representation, 0001 0010 stands for 12.

For each of these different ways of representing numbers there are procedures for the machine to carry out addition, subtraction, multiplying and dividing. For example, when two numbers are represented in binary form, their addition can be carried out using binary addition. Binary addition in the computer is carried out using a logical network combining AND and OR gates and this network has now been incorporated into an integrated circuit. Subtraction is carried out using addition by the 2s complement of the number which replaces each 0 bit by a 1, and each 1 bit by a 0, and adding 1 to the result. This operation can be easily done by the machine using a NOT gate. For example, for 8 bit words, the 2s complement of 00101101 is 11010011. Adding, we get

$$00101101 + 11010011 = 00000000,$$

the initial '1' being 'lost'.

This principle can also be used to indicate positive and negative numbers, with the former starting with a 0 and the latter with a 1. The text that follows explains how a machine can carry out the multiplication of two binary numbers.

Hartley & Healy
A First Course in Computer Technology
McGraw-Hill, 1978, p. 48.

The rules for multiplication are given in the table below.

A	B	A × B
0	0	0
1	0	0
0	1	0
1	1	1

Example:	Multiplicand	101010	42
	Multiplier	11011	27
		101010	
		101010	
		000000	
		101010	
		101010	
	Product	10001101110	1134

The five intermediate numbers in the above example are termed *partial products*. Note that the product is longer than the multiplicand and multiplier: a full 32-bit product must be allowed for when multiplying two 16-bit numbers. The 32-bit product must, thus, be allocated two computer words; [...] With binary numbers, since each digit is either 1 or 0, each partial product is either the multiplicand or zero. The individual partial products need not be saved. The first is calculated by copying the multiplicand or zero, dependent on whether the LSB

[Least Significant Bit, that is the bit to the right] of the multiplier is 1 or 0 respectively. The second partial product is similarly obtained and added to the first, but displaced one place to the left. This is repeated n times. This technique of adding a number to the previous result (shifting in this special case) is called *accumulation*. In practice, if the multiplier bit is zero, the addition can be simply skipped, but not the shift of course. This method of binary multiplication is referred to as *shift-and-add*.

The result of multiplying two n bit numbers may be a $2n$ bit number, so a double register is needed to accommodate it. The multiplication can be carried out using five registers: register A stays constant and contains the multiplicand, register B contains the multiplier initially, register BM holds the bit multiplier and the double register P_1-P_2 holds accumulated results. At each stage, the contents of B-M and P_1-P_2 are shifted to the right, so the digit in BM disappears, to be replaced by the right hand digit of the contents of B, and a new 0 is placed at the left of the entry in B. At each addition, P_1 is augmented by 000000 if BM is 0 or by the value in A if BM is 1. For example, for the multiplication given above we have:

A	B	BM	P_1	P_2	operation
101010	011011	0	000000	000000	start
101010	001101	1	000000	000000	shift
101010	001101	1	101010	000000	add
101010	000110	1	010101	000000	shift
101010	000110	1	111111	000000	add
101010	000011	0	011111	100000	shift
101010	000011	0	011111	100000	add
101010	000001	1	001111	110000	shift
101010	000001	1	111001	110000	add
101010	000000	1	011100	111000	shift
101010	000000	1	000110	111000	add
101010	000000	0	100011	011100	shift
101010	000000	0	100011	011100	add
101010	000000	0	010001	101110	shift
					and end

Notice that the fifth addition produces an overflow carrying digit of 1 which appears on the next line after the shift.

Bibliography

[1] Boyer, C., *A History of Mathematics*, John Wiley, New-York, 1968.
[2] Benazeth, D., *Naissance de l'écriture*, in Ed. Réunion des Musées Nationaux, Paris, 1982.
[3] Chace, A. B., *The Rhind Mathematical Papyrus*, 2 vols., Oberlin, Ohio: Mathematics Association of America, 1927–1929, abridged repr. in 1 vol., Classics in Mathematics Education 8, Reston, Virginia: The National Council of Teachers of Mathematics, 1979.
[4] Chasles, M., Histoire de l'Arithmétique, Règles de l'Abacus, *Comptes Rendus de l'Académie des Sciences*, 16, 1843.

[5] Colebrooke, H. T., *Algebra with Arithmetic and Mensuration, from the Sanscrit of Brahmegupta*, John Murray, London, 1817.

[6] Damirdash, A. S. & Al-Hafni, M. H., *Miftāḥ al-ḥisāb* (The Key to Calculation), Cairo, 1967.

[7] Guïtel, G., *Histoire comparée des numérations écrites*, Flammarion, Paris, 1975.

[8] Hartley, M.G. & Healy, M., *A first course in Computer Technology*, McGraw-Hill, 1978.

[9] Høyrup, J., Investigations of an early Sumerial division problem, *c.* 2500 BC, *Historia mathematica* 9, 1982, 19–36.

[10] Ibn Al-Majdī, *Ḥāwī l-lubāb* (Compendium of the Pith [of Calculations]) Ms. British Museum, n° Add. 7469.

[11] Ifrah,G., *Histoire universelle des chiffres*, Seghers, Paris, 1981.

[12] Jestin, R., *Tablettes sumériennes de Shuruppak conservées au Musée de Stamboul*, *Mémoires de l'Institut Français d'Archéologie de Stamboul*, Boccard, Paris, 1937.

[13] Knuth, D., *The Art of Computer Programming*, Addison-Wesley, Reading Massassuchetts,1968.

[14] Lam Lay Yong, *A Critical Study of the Yang Hui Suanfa – A Thirteeth-century Chinese Mathematical Treatise*, Singapore University Press, Singapore, 1977.

[15] Lau Chung Him, *The Principles and Pratice of the Chinese Abacus*, 1958, Lau Chung Him and Co, Hong Kong, 5th ed., 1980.

[16] Leibniz, G.W., Explication de l'arithmétique binaire, *Mémoires de l'Académie Royale des Sciences*, 1703: Mathematische Schriften, ed. by C. I. Gerhardt, 1863; republished Olmes Verlag, Hildesheim, 1962, Vol. VII.

[17] Neugebauer, O., *Mathematische Keilschrifttexte*, 1935, 3 Vols. Springer Verlag, Berlin, Repub. 1973.

[18] Phillips, G.M, & Taylor, P.J., *Computers*, Methuen & Co., London, 1969.

[19] Randall, B., (ed.) *The Origins of Digital Computers*, Springer Verlag, 1973.

[20] Rashed, R., *Entre Arithmétique et Algèbre*, Paris: Les Belles Lettres, 1984.

[21] Saidan, A. S., *al-Fuṣūl fi l-ḥisāb al-hindī* (Chapters on Indian Calculation). Alep, I.H.A.S, 1985.

[22] Saidan, A. S., *at-Takmila fi l-ḥisāb* (The Culmination of Calculation) Kuwait, Institut des Manuscrits Arabes, 1985.

[23] Smith, D. E., *History of Mathematics*, vol. II, McGraw-Hill Book Co., 1929, reprinted Dover, New York, 1959.

[24] Souissi, M., *Talkhīṣ aʿmal al-ḥisāb* (A Summary of Calculation Methods), Tunis, Publications de l'Université de Tunis: Critical edition with French translation, 1968.

[25] Stevin, S., *La Disme*, 1585, in *Les oeuvres mathématiques de Simon Stevin de Bruges. Le tout revu, corrigé et augmenté par Albert Girard*, Elsevir, Leyde, 1634; English translation Vera Stanford in Smith, D. E., *A Source Book in Mathematics*, McGraw-Hill Book Co., 1929, reprinted Dover, New York, 1959, pp. 24–27.

[26] Swetz, F. J., *Capitalism and Arithmetic: The New Math of the 15th Century*, (containing the full text of the *Treviso Arithmetic* translated into English by D. E. Smith), Open Court Publishing Company, La Salle, Illinois, 1987.

[27] Tannery, P., Notice sur les deux lettres arithmétiques de Nicolas Rhabdas, *Notices et extraits des manuscrits de la Bibliothèque Nationale*. 32 (1886), 121–252 and *Mémoires scientifiques*, IV, 61–198, Privat, Toulouse, 1920.

[28] Thureau-Dangin, F. *Textes mathématiques babyloniens*, Brill, Leyden, 1938.

[29] Williams, M. R., Napier's Bones, *Annals of the History of Computing*, Vol.5, n°3 (1983), p.293 et seq.

[30] Yan, Li & Shiran, Du, *Chinese Mathematics: A Concise History*, tr. John N. Crossley & Anthony W. -C. Lun, Oxford: Oxford Science Publications, Clarendon Press, 1987.

[5] Colebrooke, H. T., Algebra with Arithmetic and Mensuration from the Sancrit of Brahmagupta, John Murray, London 1817.

[6] Daressadah, A. S. S. Al.-Ba'in M. H., Miftāh al-hisāb (The Key to Calculation), Cairo, 1967.

[7] Guitel, G., Histoire comparée des numérations écrites, Flammarion, Paris, 1975.

[8] Harrison, M. C. S. Healy, M. A., Performance in Computer Technology, McGraw-Hill, 1972.

[9] Høyrup, J., Investigations of an early Sumerian division problem, c. 2500 BC, Historia mathematica 9, 282, 1982.

[10] Ibn Al-Majdī, Ḥisāb mishāfa (Compendium of the Illm [of Calculation]), Ms. Enrich Museum, n° ADd. 2160.

[11] Itrah G., Histoire universelle des chiffres, Seghers, Paris, 1981.

[12] Jestin, R., Tablettes sumériennes de Shuruppak conservées au Musée de Stamboul, Mémoires de l'Institut Français de l'Archéologie de Stamboul, Ceuthal, Paris, 1937.

[13] Knuth, D., The Art of Computer Programming, Addison-Wesley, Reading, Massachusetts, 1968.

[14] Lam Lay-Yong, A Critical Study of the Yang Hui Suan fa — A Thirteenth-century Chinese Mathematical Treatise, Singapore University Press, Singapore, 1977.

[15] Lao Chang Ming, The Principles and Practice of the Chinese Abacus, 1958, Faui Chang Hom and Co., Hong Kong, Sia, ed., 1980.

[16] Leibniz, G.W., Explication de l'arithmétique binaire, Mémoires de l'Académie Royale des Sciences, 1703, Mathematische Schriften, ed. by C. I. Gerhardt, 1863; republished, Olms-Verlag, Hildesheim, 1962, vol. VII.

[17] Neugebauer, O., Mathematische Keilschrifttexte 1935, 3 vols. Springer Verlag, Berlin, Reprint, 1973.

[18] Phillips, G. M. S. Taylor, P.J., Computers and Numerical Analysis, Academic Press, London, 1973.

[19] Randell, B., (ed.), The Origins of Digital Computers, Springer-Verlag, Verlag, 1973.

[20] Rashed R., Entre arithmétique et algèbre, Les Belles Lettres, 1984.

[21] Saidan, A. S., al-Fusūl fi'l-hisāb al-hindi (Chapters on Indian Calculation), Aleppo, 1985.

[22] Saidan, A. S., al-Hisāb fi'l-hind (The Calculation of Calculation), Kuwait, Institut de sciences Arabes, 1985.

[23] Smith, D. E., History of Mathematics, vol. II, McGraw-Hill Book Co., 1926, reprinted, Dover, New York, 1958.

[24] Souissi, M., Talkhīs a'māl al-ḥisāb (A Summary of Calculation Aleboadol), Tunis, Publications de l'Université de Tunis, Critical edition with French translation, 1968.

[25] Stevin, S., La Thiende 1585, in Les nouveaux mathématiques de Simon Stevin de Bruges, le tout revu corrigé et augmenté par Albert Girard, Elzevir, Leyde, 1634; English translation Vera Sanford in Smith, D. E., A Source Book in Mathematics, McGraw-Hill Book Co., 1929; repr. abc. Dover, New York, 1959, pp. 25–27.

[26] Swetz, F., Capitalism and Arithmetic. The New Math of the 15th Century, the only English text of the Treviso Arithmetic, translation into English by D. E. Smith), Open Court Publishing Company, La Salle, Illinois, 1987.

[27] Tannery, P., Notices sur les Deux lettres arithmétiques de Nicolas Rhabdas, Notices et extraits des manuscrits de la Bibliothèque Nationale 32 (1895), (2) 258 and Mémoires scientifiques, IV (2)-198, Privat, Toulouse, 1920.

[28] Thureau-Dangin, F., Textes mathématiques babyloniens, Brill, Leyden, 1938.

[29] Williams, M. R., Napier's Bones, Annals of the History of Computing, vol. 5, n°4 (1983) p. 295 et seq.

[30] Yan Li S. Shi Ran, Du Chinese Mathematics. A Concise History, tr. John N. Crossley & Anthony W.-C. Lun, Oxford Oxford Science Publications, Clarendon Press, 1987.

2 Magic Squares

The expression *magic square* is commonly used for any arrangement of squares, like the squares on a board for playing chess or draughts, in which the individual square cells contain numbers such that the sum of the numbers along any row, column or diagonal produces the same result. A magic square containing n rows and n columns is called an order n magic square. Mathematically, it is often useful to consider the square as a particular *square matrix* possessing the *magic constant S_n* which is equal to the common sum of its rows, columns and diagonals.

Traditionally, the individual cells of a magic square would contain the first n^2 consecutive integers 1, 2, 3, ..., n^2. Since the sum of the first n^2 consecutive integers is $n^2(n^2 + 1)/2$ and the square contains n rows, the magic constant will be

$$S_n = (1 + 2 + 3 + ... + n^2)/n = n(n^2 + 1)/2.$$

By extension, the term magic is also used to describe other configurations of assorted symbols, numerical or not, which are arranged geometrically, on the plane or in space, in rings, circles, stars, cubes, spheres and so on.

More often used as talismans than as objects of mathematical study, magic squares have been closely connected with daily life and all its worries. They have been used variously as an aid to childbirth, as an antidote to stings or bites by venomous animals, and to help in the cure of illnesses. It is hardly surprising, then, that

Fig. 1 Detail from Albrecht Dürer's *Melancholia,* engraving on wood, 1514. The order 4 magic square, which uses the integers 1 to 16 and contains the date 1514 at the centre bottom, has the magic constant 34.

28	4	3	31	35	10
36	18	21	24	11	1
7	23	12	17	22	30
8	13	26	19	16	29
5	20	15	14	25	32
27	33	34	6	2	9

Fig. 2 Magic square of order 6, written with Arabic numerals, engraved on an iron plaque [11]. It was discovered in 1956 among the ruins of the palace of the Prince of Anxi, one of the sons of the Mongol Emperor Qubilai. The palace lies in the suburbs of Xi'an, the ancient capital of the Chinese empire, now the capital of Shenxi province.

44	54	55	41
49	47	46	52
45	51	50	48
56	42	43	53

Fig. 3 Porcelain plate made in China, probably in the nineteenth century, at Jingdezhen, a town in Jiangxi province famous for its porcelain, and later exported by European traders to South East Asia for use by Islamic communities [6]. The plate is decorated in the centre with a magic square written in Arabic numerals and around the outside with religious inscriptions taken from the Koran.

magic squares are not confined to books: many were engraved on monuments, on medallions and metal plaques, on goblets, and on cups and plates, made of copper or porcelain. Some of these magic squares have an interesting history, providing us with concrete evidence of the movement of objects and ideas between time and place. For example, a magic square from the time of the Mongols, engraved on an iron plaque in Arabic symbols (Fig. 2), supports the theory that the Chinese borrowed this mathematical technique from the Islamic world. In fact the most ancient Chinese magic squares of order greater than 3 go back precisely to the period of the Mongol domination of China. Another example, of a nineteenth century Chinese porcelain plate decorated with a magic square, again with Arabic numerals (Fig. 3), is evidence of the presence in China of an Moslem community with links to Islamic traditions.

Belonging to a rather different order of ideas, the *luo shu* (diagram of the river Luo) (Fig. 4) was commonly found among Chinese numerological drawings of the Song dynasty (960–1279). Also in China, but a little earlier, in the ninth and tenth centuries, we find the *jiu gong* (diagram of the 'nine palaces') (Fig. 5), which is a square in which the cells, or 'palaces', contain the characters for the names of colours. Associated with the calendar, it was used to determine the positions of lucky and unlucky stars, and for determining what should not be done at a certain time. The square had to be interpreted by using the following code: 1 = white, 2 = black, 3 = azure, 4 = green, 5 = yellow, 6 = white, 7 = red, 8 = white, 9 = violet [13].

The use of magic squares can also be found in the Maghreb world of the Middle Ages, with the astrologer al-Būnī (*d.* 1225), who came from Bône, then called Annaba. He deals with beneficial magic squares in his *Shams al-maʿ ārif* [The sun of knowledge]. To this day, in northern India, healers make use of a magic square of order 3 in the treatment of malaria [15]. Many further examples could be given to illustrate the way that magic squares are rooted in the daily, if esoteric, practice.

Magic squares have not, however, always been thought of in this way. Some users of magic squares have simply considered them to be alternative ways of arranging numbers. Mediaeval Arab writers simply called them 'numbers in harmony' [ʿadad al-wafq]. At the same period, the Chinese called them, more prosaically, 'transversal-longitudinal diagrams' [zong-heng tu]. At the beginning of the 18th century, we find Canon Poignard referring to them as 'sublime squares'. The term *magic* itself seems to have been coined about the middle of the 17th century.

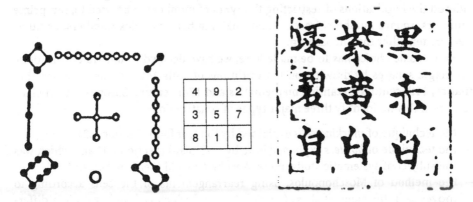

Fig. 4 Diagram of the river Luo, corresponding to the magic square of order 3.

Fig. 5 Diagram of the nine palaces.

Occupying a position that lies between complete irrationality and the most irrefutable rationality, between the real and the imaginary, and between numerology and mathematics, these 'square arrays' also provide a fascination as part of the magic of arithmetic. For this reason they have attracted generations of amateurs of recreational mathematics. Although it is easy to construct these figures, and those who enjoy doing so will find myriads of special cases for each one, those who specialise in listing all possibilities quickly reach a point where they are unable to make further progress. In fact, while there only exist 8 magic squares of order 3, which can all be deduced from each other by means of symmetries and rotation, and although Arnauld, from 1667 onwards, managed to explicitly determine the 16 magic squares of order 4 which, by means of symmetries and rotations, generate all 880 magic squares of order 4, little progress has been made in the subject since then. Martin Gardner [7], the well known American writer on recreational mathematics, wrote in 1961 that the number of magic squares of order 5 is not yet known and must be in the region of 13 million. Some amateurs of recreational mathematics have attempted to restrict the total number of possibilities by insisting on additional properties. This has generated a new procession of such squares, each more 'magic' than the others: cabalistic, panmagical, hypermagical, diabolical, satanic, and so on. We shall not stray into the area of the definitions of these different types here: the reader who wishes to know more may consult [1], [4], [5], [8], [9] and [12].

Although often rejected as worthy of serious consideration, magic squares have nonetheless occasionally attracted the interest of first rate mathematicians, among whom are Ibn al-Haytham, known in Europe as Alhazen (*d.* 1041), Pierre de Fermat in the 17th century, Leonhard Euler in the 18th century and Arthur Cayley in the 19th century. Until quite recently the main questions about magic squares related to their construction, proofs of existence and enumeration of sets of squares possessing some or other particular property. In more recent times, mathematicians have been obliged to examine these strange squares in the context of classical structural algebra, thinking of them in terms of vector spaces or modules and defining their tensor products, the cells of the squares being no longer filled just with consecutive integers, but with any positive or negative numbers. Other investigations have considered the implications of restricting the types of numbers to be used to, say, prime numbers, square numbers, fractions, rational numbers, complex numbers, or number written other than in base ten.

Since some choice has to be made here, we have decided to illustrate work done on magic squares by drawing on the traditions of 13th–14th century Arab mathematics, 14th century Byzantine mathematics and 17th century European mathematics. From this we identify three major types of construction:

- the technique of *marking cells* explained by Ibn Qunfudh (Section 2.2),
- the technique of *borders* explained here by az-Zinjānī (Section 2.1) and which was the subject of systematic study in the West by Arnauld (Section 2.4), and
- the method of Moschopoulos, using rearrangements of the cells according to moves as in the game of chess (Section 2.3), and which is found again in a different presentation in the works of Bachet (see end of Section 2.3).

We shall see also that the choice of different construction algorithms is a condition of the order n of the square, and particularly depends on whether n is divisible by 2. While the techniques of *borders* works just as well for any n, the method of Moschopoulos only works if n is odd, and the method of *marking cells* is only easy to use where n is divisible by 4.

We should also point out that in many cases the 'natural square', that is one composed of consecutive integers written out in order, plays an important part in some of the constructions. Furthermore, in such a square, the sum of the numbers along each diagonal is already equal to the magic constant, as is the sum of the numbers along the central row and column.

2.1 Squares with Borders

Since the 15th century in Europe, a great many mathematicians, such as Michel Stifel, Blaise Pascal and Pierre de Fermat, have considered different methods of construction of magic squares [12]. These all involved a way of filling the squares by using concentric rings of numbers enclosed one within the other and called *frames* or *borders* (Fig. 6).

In fact, the idea of using borders is much older than the 15th century and its traces can be found in the mediaeval Orient in the 10th century ([16], [17]). Also, the 13th century writer az-Zinjānī, whose description of a method of constructing magic squares using borders is given here, is only a fairly late link in the chain of ideas that come down to us

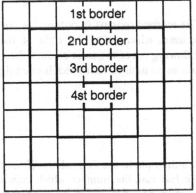

Fig. 6 A square of order 7 composed of borders containing respectively 24, 16, 8 and 1 cells.

from an ancient origin. Late though he may be, he is much earlier than the European mathematicians quoted above, who are usually credited with the discovery of squares using borders.

Just as in all methods using borders, the method of az-Zinjānī consists in first filling all the cells of the border of the square, according to a fixed procedure, then applying the same procedure, iteratively, to each of the borders in turn, leaving alone the borders already filled.

If the initial square is of odd order $n = 2k + 1$, the successive borders have sides composed of $2k + 1, 2k - 1, ..., 5, 3, 1$ cells. The same procedure cannot be used for a square of even order $n = 2k$ since the sides of the successive borders will contain $2k$, $2(k - 1), ..., 6, 4, 2$ cells, and there is no magic square of order 2. All the same, we can use the method of borders if we stop the algorithm when we come to a central square of order 4 and we are able to arrange the contents of the cells in some way so as to produce a magic square. This was the method proposed by Arnauld, as we shall see later. We could also stop the algorithm on coming to a central square of order 6 or even earlier.

Following the practice of other mathematicians of the time, az-Zinjānī explains his procedure without reference to any particular numbers or symbols. Since he wrote in Arabic, the writing is from right to left, and the terms 'first' or 'last' cell, and 'preceding' and 'succeeding' cell correspond exactly to the reverse of how we would usually understand them.

---

az-Zinjānī

Risāla fī aʿdād al-wafq (*Epistle on numbers in harmony*).
Arab edition published by J. Sesiano in Herstellungsverfahren magischer Quadrate aus islamischer Zeit (II'), *Sundhoffs Archiv*, 71 (1987), 80–81.
From the French translation by A. Djebbar.

[To facilitate the reading of the original text, the diagrams have been redrawn and the text divided up into sections.]

The method consists in putting the one in the centre of the first column, which is to the right of the writer, the two in the following cell of the same column, then the three in the next cell, and so on until the cell which is one neighbouring the diagonal.

					1
					2
					3

You then put in the first column [from the left], in the last cell of the last row, the number which comes after the one you have reached. You then put in the cell before it [to the right] the next number, and so on, until you reach the cell next to the one that is next to the middle of that row.

					1
					2
					3
4	5	6			

Then go to the middle cell of the first row and you put there the number you have reached.

		7			
					1
					2
					3
4	5	6			

Then you move to the cell which is next to the middle of the last column. Then move from there to the cell that comes before it and so on until you reach the diagonal.

10			7			
9						
8						
						1
						2
						3
4	5	6				

Then you go to the cell that is next to the middle one of the first row. From there go to the next one and so on until you reach the cell that is next to the diagonal. And now the tour comes to an end.

10			7	11	12	
9						
8						
						1
						2
						3
4	5	6				

Next, you put in each empty cell of this border the number which, together with the one opposite, makes up the square of the number whose magic square you are constructing, plus one. For each diagonal, the opposite one is the cell that you reach when you go along the line of diagonal cells.

10	45	44	7	11	12	46
9						41
8						42
49						1
48						2
47						3
4	5	6	43	39	38	40

When [filling up] this border is accomplished, there remains inside a square which is odd [sided]. You proceed with this according to the steps indicated and you start with the number that comes after the number that you reached in [the course of] the first cycle and which is the number that you had put in the cell next to the first diagonal, that is the second cell of the first line, and you fill up [the cells] according to how we have described it for the cycle already completed.

	19		17	20		
	18					
					13	
					14	
	15	16				

Then you complete the opposite cells in order to obtain the square of the number whose magic square you are constructing – that is to say the initial number, plus one and not the side of the square which remains for the second [cycle] – for, in each cycle, we only take account of the initial number and no other.

	19	34	17	20	35	
	18				32	
	37				13	
	36				14	
	15	16	33	30	31	

When you have also completed this [second] cycle, there will remain in the centre another odd square. You proceed with this one according to the steps indicated until the operation is complete and the result is the perfect square.

10	45	44	7	11	12	46
9	19	34	17	20	35	41
8	18	24	23	28	32	42
49	37	29	25	21	13	1
48	36	22	27	26	14	2
47	15	16	33	30	31	3
4	5	6	43	39	38	40

Let $n = 2k + 1$ be the order of the odd magic square being considered. Number the rows from 1 to n starting at the top, and to be consistent with the Arab method of writing, number the columns from 1 to n starting at the right. With this unconventional notation, the cell (p, q) will now be the one in the pth row and qth column (from the right).

This is how the first border is to be filled up. If $n = 2k + 1$, this first border contains $4.(2k + 1) - 4 = 8k$ cells. The technique described by az-Zinjānī starts off by putting in the consecutive integers $1, 2, 3, \ldots, 4k$ as follows:

(a) Place the integers 1 to k in the first column on the right, starting with the centre cell and going down, that is using the cells $(k + 1, 1), (k + 2, 1), (k + 3, 1), \ldots, (2k, 1)$.

(b) Place the integers from $k + 1$ to $2k$ in the last row, starting with the left most cell, that is successively in the cells $(n, n), (n, n - 1), (n, n - 2), \ldots, (n, n - k + 1)$.

(c) Place the single number $2k + 1$ in the cell $(1, k + 1)$.

(d) Place the numbers from $2k + 2$ to $3k + 1$ in the left column going up, in the cells $(k, n), (k - 1, n), \ldots, (1, n)$.

(e) Place the numbers from $3k + 2$ to $4k$ in the first row, from left to right, in the cells $(1, k), (1, k - 1), \ldots, (1, 2)$.

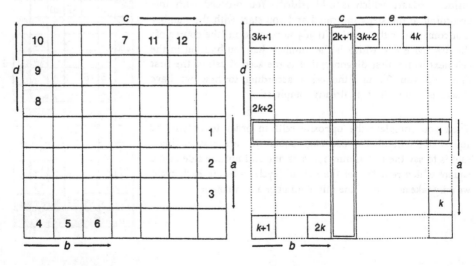

Fig. 7 The beginning of the process of filling the cells: on the left an order 7 square, on the right an order $2k + 1$ square.

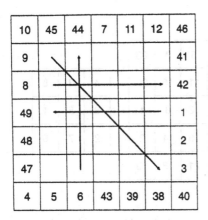

Fig. 8

10	45	44	7	11	12	46
9	19	34	17	20	35	41
8	18	24	23	28	32	42
49	37	29	25	21	13	1
48	36	22	27	26	14	2
47	15	16	33	30	31	3
4	5	6	43	39	38	40

Fig. 9

After completing this cycle, or 'tour' as az-Zinjānī calls it, the number of cells that have been filled is $k + k + 1 + k + (k - 1) = 4k$, being half the total number of cells in the outside border of the whole square.

Then az-Zinjānī begins again in order to fill the $4k$ remaining cells. In looking at figure 7 we can see that the unfilled cells are distributed regularly with respect to the filled cells: all the empty ones are opposite filled ones, either horizontally, vertically or diagonally. Az-Zinjānī asks them to be filled with the 'complement' of the cell opposite (Fig. 8). By the 'complement' of an integer x we are to understand, as he explains in the text, the complement of x with respect to $n^2 + 1$, that is the number equal to $(n^2 + 1) - x$, where n is the order of the square.

Once the first border has been completed, the same procedure is adopted for completing the next border inside, containing $8(k - 1)$ cells. The operations are identical except for two details. First, the initial value taken is $4k + 1$ instead of 1. Next, as az-Zinjānī makes clear, the constant for finding the complement does not change, but remains $n^2 + 1$. We then carry on in the same way for the other successive borders. Figure 9, showing a square of order 7, has 4 'borders', the last of which is a single cell.

As for the initial values to be taken at each stage, we note that if the order of the square is $2k + 1$, the number of borders will be $k + 1$ and, if u_i is the initial value for the ith border, we have:

$$u_1 = 1 \text{ and } u_{i+1} - u_i = 4(k - i + 1).$$

From which: $u_{i+1} = 2i(n - i) + 1$. In particular, the central value will be

$$u_{k+1} = (n^2 + 1)/2.$$

The procedure by borders leads to squares which possess a supplementary property: the successive borders can be removed and there always remains a magic square. However, bearing in mind the way the successive borders were constructed, the number remaining in the central cell is always equal to $(n^2 + 1)/2$, where n stands for the order of the initial square. Now, when r borders are taken away, we are left

10	45	44	7	11	12	46
9	19	34	17	20	35	41
8	18	24	23	28	32	42
49	37	29	25	21	13	1
48	36	22	27	26	14	2
47	15	16	33	30	31	3
4	5	6	43	39	38	40

7	22	5	8	23
6	12	11	16	20
25	17	13	9	1
24	10	15	14	2
3	4	21	18	19

4	3	8
9	5	1
2	7	6

Fig. 10 Starting with the square on the left, the middle square is found by removing a border and subtracting 12 from all the numbers left. The square on the right can be found directly by taking away two borders from the square on the left, and subtracting 20 from each of the numbers left.

with a square of order $n - 2r = m$ and, if we wish the central number to be $(m^2 + 1)/2$ then we must subtract $(n^2 + 1)/2 - (m^2 + 1)/2 = 2r(n - r)$ from all the numbers of this square (Fig. 10).

This supplementary property guarantees, by a sort of descendant recurrence, that the procedure certainly leads to the construction of a magic square. All that is required is to ensure that the sum of the numbers entered into each strip of the exterior border is equal to the required magic sum.

2.2 The Marking Cells Method

The procedure of marking cells was described in the 14th century by Ibn Qunfudh in his book *Lifting the veil from the operations of calculation*. The book is a commentary on Ibn al-Bannā's work *Summary of the operations of calculation*, except for the last chapter which is not a commentary but a chapter on magic squares. Ibn Qunfudh does not, however, claim to be the inventor of the procedure he describes, and it would seem that he has simply collected together material that had been described before, in the Maghreb and elsewhere, in other works of which only the titles have survived.

Ibn Qunfudh identifies four types of squares: (a) the *evenly-evens*, that is those whose sides contain n cells where $n = 2^p$ and $p \geq 2$, (b) the *evenly-even-odds* for which $n = 2^p(2k + 1)$ with $p \geq 2$, (c) the *evenly-odds* for which $n = 2(2k + 1)$ and, finally, (d) the *odds* for which $n = 2k + 1$. He provided an algorithm for the construction of each one of these. The last case, where n is odd, does not admit of the technique of marking cells. In the extracts that follows, we shall only look at the first two cases, those where n is divisible by 4 and where the technique of marking cells is easy to demonstrate. Ibn Qunfudh starts by calculating the value that the sum of the numbers of each side of the square must have.

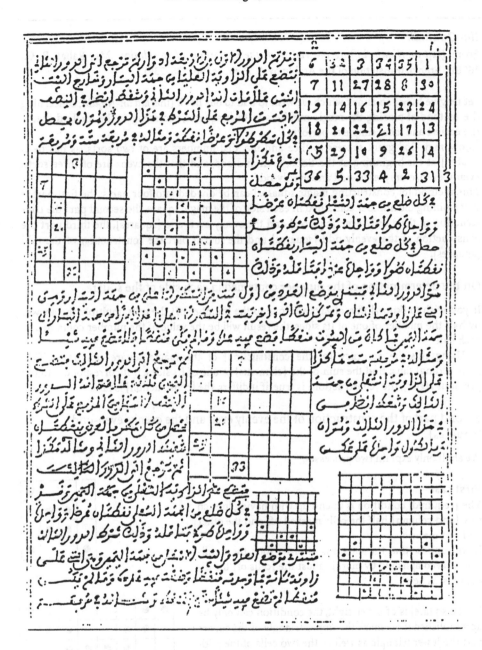

Fig. 11 Extract from Ḥaṭṭ an-Niqāb, *Lifting the veil from the operations of calculation*, Ms., Rabat, B. G., 1678D. We can see, in particular, here the method of marking cells in the case of a square of order 6.

Ibn Qunfudh

From *Ḥaṭṭ an-Niqāb, Lifting the veil from the operations of calculation*, Ms., Rabat, B. G., 1678D. French translation of the Arabic by A. Djebbar (pp. 238–239).

Let us give [now] an annex which we wish to add to our Book [the*Ḥaṭṭ an-Niqāb*], concerning the method of placing the magic numbers in their squares, according to the method that is the easiest and most simple to perform.

Know, may God grant success to thee and me, that numbers are of two sorts, as has been [stated] at the beginning of the Book, even and odd, and that the evens are of three sorts: evenly-even, evenly-even-odd and evenly-odd.

This method [of constructing magic squares] can be divided into four parts: the first part is determining the number contained in each side of the magic square, the second part the construction of the evenly-even [squares] and the evenly-even-odd [squares], the third part the construction of the evenly-odd [squares] and the fourth part the construction of the odd [squares].

On the determination of the number contained in each side of the square

If you wish to [know] this, always add one to the [number of] the cells, multiply the answer by half the number of one of its sides, the result will be the value of the number contained in each side.

Using this the accuracy [of the construction] of the magic square can be verified, [proceeding] thus: find the sum of the [elements] of one of its sides, if it corresponds [to the previous result], the [construction] is true, if it differs, the [construction] is false.

Second part of the construction of the evenly even and the evenly-even-odd [squares]

We shall only explain two processes.

First case

The [required] condition is to end up with having in each row [and column] half the cells marked off and the other half not marked.

If we suppose that the square is of order four, the condition is accomplished by marking just the [cells of] the diagonals, as here:

If [the square] is of order eight, the condition can only be accomplished if we mark the upper triangle of the square and the lower triangle as well as the two cells of the middle of the right diagonal and of the left diagonal, as here:

If the square is of order greater than eight, the condition cannot be accomplished by just adding a [marked] triangle or by marking the right and left [columns]. In order to carry out the procedure, you must arrange that in each row [or column] half the cells are marked and the other half not marked.

For the square of order twelve, for example, the condition can only be accomplished by adding two [marked] triangles in the upper part and two marked triangles in the lower part and by marking four cells of the first column from the right and four cells of the first column from the left as well as two cells of the second column from the right and two cells of the second column from the left.

In general, in order to know what must be put at the two extremes, start from the middle, then complete [the marking of] the two extremes with what remains of half of the cells [to be marked].

Know that it is thanks to observation and reflection that you will acquire the manner of placing [the marks].

Here is the figure for the magic square of order twelve:

If the condition has been set up for the square in which you want to place the numbers, and you have marked half the cells according to the way I have described, start with the first cell and place a one in it, and starting from there, count off the cells using the sequence of integers, and arrange that in each marked cell you come to you write in the [corresponding] number, and that you count each non marked cell without putting anything in it, until you come to the last square of left [column], as in this square:

4			1
	7	6	
	11	10	
16			13

Then starting again, from the last cell which contains a number, count it without putting anything in it, then it's the second [cell's turn]: you do not find anything. You therefore write two in it and you follow on [in the same way] from the left to the right until all the cells have been exhausted, like this:

4	14	15	1
9	7	6	12
5	11	10	8
16	2	3	13

Your filling up of the cells has now been done. If you calculate the horizontal, vertical or diagonal [sum of the elements], you will find thirty four.

If you have understood this example, go on to the evenly-even [squares] and the evenly-even-odds, as before.

Second case

The order of the squares is greater than sixteen.

The rule consists in determining how many order four squares there are contained [in the square], then you mark each of these squares by putting a sign in each of its four corners and you mark off the diagonals of each of the squares. When this marking has been done, start to distribute the numbers starting from the first marked cell that you meet, placing a number [in the marked cells] and counting the non marked cells without putting anything in them, until you reach the last cell of the lowest row. Then starting again with one, starting from this last cell, count those [cells] containing a number without putting anything in them and in those where you do not find a number [put in] the number that corresponds to it.

This case does not differ from the previous one, except for the marking of the cells.

Ibn Qunfudh begins by stating the rule for calculating the sum of numbers for any row, column or diagonal of a square of order n. It is equivalent to

$$S_n = (n^2 + 1)n/2$$

The procedure for completing a magic square described in this extract applies to squares with an order divisible by 4. It can be accomplished in three stages:

First, the cells of the square have to be marked in such a way that half the cells of each row and column are marked.

Then these marked cells are filled in: we start with the top right cell and count 1, 2, 3, ... from cell to cell, to the end of the row: whenever we meet a cell that has been marked, we fill it with its corresponding number; whenever we meet a cell that has not been marked we put nothing in but we do not stop counting. Arriving at the end of the first row, we continue counting along the second row, beginning again from the right, and so on. In this way we end up with a half completed square.

Finally, for filling in the remaining empty cells, we start again counting from 1 to n^2, but this time starting from the bottom left cell and proceeding from left to right: on meeting an empty cell, and therefore a non marked cell, we put in its corresponding number. On coming to the end of the row, we continue the counting along the next row, in the same direction, and so on. At the end of the process, we shall have the required magic square (Fig. 12).

Fig. 12 The three stages: a) marking, b) filling the marked cells, c) filling the non marked cells.

Ibn Qunfudh indicates two ways of marking the cells and he always insists on the condition that for each row and each column half the cells have to be marked and half not marked. The second method, which can be used for squares with more than 16 cells, results, in the case of a square of order 8 for example, in having four marked squares of order 4 (Fig. 13).

Certainly the condition for marking the cells is not sufficient in itself. In fact, in the explanations given, a certain symmetry is always maintained: symmetry with respect to the horizontal and vertical axes. Now, this is sufficient to ensure that the sum of the numbers for each column and row will be the magic number.

We shall verify this for the cth column, from the right following the Arab text.

Consider column c.

If the cell in row p is marked, then the cell of row p', where $p + p' = n + 1$ is also marked. Then

in row p we find the number $x = n(p - 1) + c$, and

in row p' the number $x' = n(p' - 1) + c$, from which:

$$x + x' = n(p + p' - 2) + 2c = n(n - 1) + 2c.$$

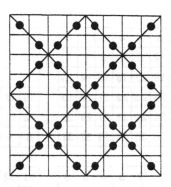

Fig. 13 The marking for a square of order 8 by Ibn Qunfudh's second method.

If however the cell in row p is not marked, then the cell in row p', where $p + p' = n + 1$ is also not marked. In this case,

in row p we find the number $y = n(n - p) + n - c + 1$, and

in row p' we find the number $y' = n(n - p') + n - c + 1$, from which:

$$y + y' = n(2n - p - p') + 2n - 2c + 2 = n(n - 1) + 2n - 2c + 2.$$

Both these totals are independent of the choice of p, and since half the cells are marked and half not marked, the total sum for the column c will be:

$$(n/4)[\{n(n - 1) + 2c\} + \{n(n - 1) + 2n - 2c + 2\}] = n(n^2 + 1)/2.$$

We also note that cells of the diagonals will necessarily be marked and so the sum of each diagonal will be the magic number, since this is the case for the diagonals of the natural square.

As for the construction of magic squares of order $n = 2(2k + 1)$, this also relies on a system of marking cells. However, the difference here is that several different marking methods are required. In the case of constructing a square of order 6, for example, three markings are needed (Fig. 14) and filling the cells has to be carried out in four stages (Fig. 15).

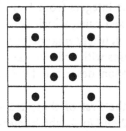 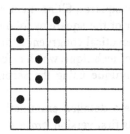 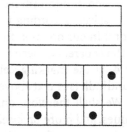

Fig. 14

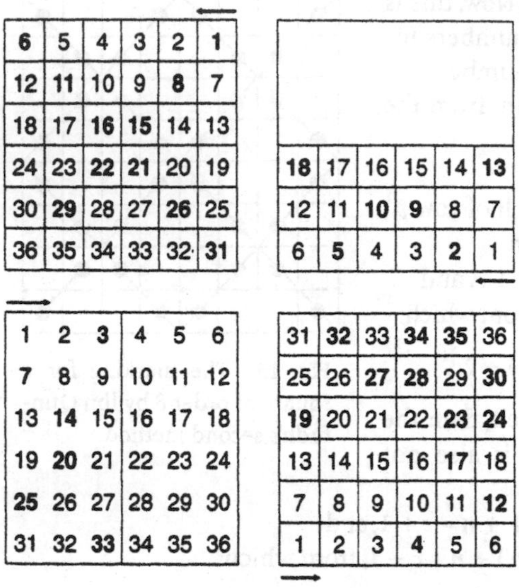

6	5	4	3	2	1
12	11	10	9	8	7
18	17	16	15	14	13
24	23	22	21	20	19
30	29	28	27	26	25
36	35	34	33	32	31

18	17	16	15	14	13
12	11	10	9	8	7
6	5	4	3	2	1

1	2	3	4	5	6
7	8	9	10	11	12
13	14	15	16	17	18
19	20	21	22	23	24
25	26	27	28	29	30
31	32	33	34	35	36

31	32	33	34	35	36
25	26	27	28	29	30
19	20	21	22	23	24
13	14	15	16	17	18
7	8	9	10	11	12
1	2	3	4	5	6

Fig. 15 The first 3 squares are completed using the marking cells technique, and starting from the cell indicated by the arrow, the marked cells are filled in. The final square is filled by starting from the arrow and filling in the cells that remain empty.

2.3 Proceeding by 2 and by 3

The small treatise by the Byzantine mathematician Manuel Moschopoulos, was written in Greek in the 14th century. Originally in manuscript form, it was edited and translated in 1886, with a commentary in French, by Paul Tannery [18].

Moschopoulos, like others before him, passed on techniques that had been given before in Islamic literature. Though perhaps, as Tannery speculates, Moschopoulos was just the last link in a long chain in the transmission of magic square techniques, techniques that might even go right back to the Greeks. This cannot be confirmed however, since we have no earlier evidence of the construction of Greek magic squares. (Nonetheless, the absence of evidence is not the same as evidence of absence.)

Moschopoulos explains basically two types of technique. The first allows the construction of odd ordered magic squares, filling them with the integers from 1 to n^2 successively according to a tour of the square by following fixed rules, which correspond in our notation to mathematical congruences modulo n, where n is the order of the square. The second type of technique, which is for squares with order divisible by 4, is the marking cells technique using the second version described by Ibn Qundfudh (see text above).

We shall therefore only give a detailed examination of the first of these techniques, and more precisely of its first version, *proceeding by 2 and by 3*, which refers to the extent of the displacements used one after the other in carrying out the method. The second version, *proceeding by 3 and by 5*, is entirely equivalent in principle.

Moschopoulos

Original Greek manuscript in Bibliothèque Nationale, Fonds Grec,
edited and translated by Paul Tannery, *Annuaire de l'Association pour l'encouragement des études grecques en France*, 1886, pp. 88–118. Extract from *Mémoires Scientifiques de Paul Tannery*, vol. 4, Gauthier-Villars, Paris, 1920, pp. 26–60: 'Le Traité de Manuel Moschopoulos sur les carrés magiques, texte grec et traduction', (pp. 37–41).

[To facilitate the reading of the original text, the diagrams have been redrawn and the text divided up into sections to accompany the corresponding figures, here of the particular case of order 3.]

[...] we should now consider the arrangement. We start with the first number that can be used; this first number that can be thus arranged to form a square is 3, [...] which we shall talk about in the first place; but the method that will be explained for this number can be applied to all others of the same type (*the odds*). Now we can achieve the disposition which gives equality in all the directions [...]

[Method for the squares of odd]

This is [...] proceeding by 2 and by 3. We first put the unit in the cell in the middle of the three at the bottom.

and we count two cells, the one which has this unit, the other we look for below the first in a direct line, for we always go from top to bottom; since we cannot find one, we go back to the top, still in a direct line, as in going round in a circle (αναχυλουντες), and we come to the second cell; then we put a 2 in the cell to the right of this in a direct line,

and we count two more cells, the one which contains 2, the second below it, and we look for the cell to the right in a straight line to put in the 3; not finding it, we move over to the left in a straight line; for when a row of cells comes to an end, we must always go back to its beginning. We therefore place 3 in the cell that is the last of our steps in the opposite direction, but the first going to the right, that is in the direction we must follow in counting cells as going round in a circle.

Having thus come to 3, which multiplied by itself gives the square, that is which is the side of 9, we do not count two places for putting 4 next to the right; but we count 3 as follows: one, the cell containing 3, two, the one below, three we look below, but not finding any more cells, we go back up to the top in a direct line; we count this as the third cell and place the 4 in it without stepping to the side of the direct line,

then moving off from there as if we were starting anew, we count by 2 and we put the next number to the right following the way indicated;

we continue thus until we come back again to the side of 9, that is 6, the double of 3.

Coming to this number, we begin again to count by 3, and placing the next number in the third cell without stepping aside from the direct line;

then we count by 2 and move to the right, and so on up to the end, always counting by 2, for all the numbers except when we have to pass from one side to another side (*from a multiple of a root to the next multiple of a root*), then we count by three.

We do the same for all the numbers of the same type, by counting according to the rule stated; by two, up to a side (*root*) of the number of cells of the required square; by three, for the next number; and so on right up to the end, taking the cells in a circular way as in the example; in all we observe exactly the same rules, except for the position of the unit; for this will not always be placed in the same cell, but will change its position for each square. For the first square, formed by an odd number, it is put in the middle of the bottom row of cells; for the second square, in the middle of the row immediately above the first, as a general rule, for each passage to a greater number, it has to go up one cell, such that it is always placed in the cell immediately and directly below the cell that is precisely in the centre of all the cells of the required square of that type: all this can be more clearly seen from the figures.

The techniques used by Moschopoulos can very easily be expressed by formulas, particularly if we use congruence notation. We number the columns and rows of the square, from left to right for the columns, and from top to bottom for the rows, with $c(x)$ for the column and $r(x)$ for the row in which we find the number x. Starting with a square of odd order $n = 2k + 1$, the algorithm used by Moschopoulos can be formulated as follows:

(a) Initialisation of the algorithm by placing the unit in the cell immediately below the centre cell. In other words, 1 is placed at the intersection of the $(k + 1)$th column and the $(k + 2)$th row, which gives:

$$r(1) = k + 2$$
$$c(1) = k + 1$$

(b) The application of two conditional rules, as explained below, which are followed to position the number $x + 1$ from the known position of its predecessor x (Fig. 16):
If x is not a multiple of n, use the rule of *proceed by 2*:

$$r(x + 1) \equiv 1 + r(x) \pmod{n}$$
$$c(x + 1) \equiv 1 + c(x) \pmod{n}$$

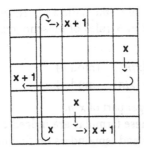 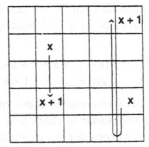

Fig. 16 Proceeding by 2 and by 3.

If x is a multiple of n, use the rule of *proceed by 3*:

$$r(x + 1) \equiv 2 + r(x) \quad (\text{modulo } n)$$
$$c(x + 1) \equiv c(x) \quad (\text{modulo } n)$$

We should point out that with these rules, if the result of taking the modulus gives 0, then we use n and not 0 (the complete residue system that we use is $1, 2, ..., n$ and not the usual $0, 1, 2, ..., n - 1$).

In the case of a square of order 3, for example:

number x	location $(r(x), c(x))$	rule applied
1	$(3, 2)$	initialisation
2	$(4, 3) \equiv (1, 3)$	proceed by 2
3	$(5, 4) \equiv (2, 1)$	proceed by 2
4	$(7, 4) \equiv (1, 1)$	proceed by 3
5	$(8, 5) \equiv (2, 2)$	proceed by 2
6	$(9, 6) \equiv (3, 3)$	proceed by 2
7	$(11, 6) \equiv (2, 3)$	proceed by 3
8	$(12, 7) \equiv (3, 1)$	proceed by 2
9	$(13, 8) \equiv (1, 2)$	proceed by 2

In the case of a square of order 5, we start with the initial conditions $r(1) = 4$, $c(1) = 3$, then apply the *proceed by 2* rule, except when we come to the numbers 5, 10, 15, and 20. These are the multiples of the order of the square, for which we use the rule *proceed by 3* (Fig. 17).

This procedure raises two questions:

11	24	7	20	3
4	12	25	8	16
17	5	13	21	9
10	18	1	14	22
23	6	19	2	15

Fig. 17

- is it certain that we shall end up by filling all the cells of the square, and
- will the square be magic?

Following the approach used by the American mathematician D. N. Lehmer [10], we can replace the 'proceed by 2' and 'proceed by 3' rules of Moschopoulos by the formulas:

for $1 \leq x \leq n^2$, $r(x + 1) \equiv k + 2 + x + [x/n]$ (mod n)
and $c(x + 1) \equiv k + 1 + x - [x/n]$ (mod n)

where $n = 2k + 1$ and where $[x/n]$ is the integer part of x/n.

We leave it to the reader to check that these formulas are indeed equivalent. Another approach is to consider the quotients and remainders on dividing $x - 1$ by n. Let Q, R be the quotient and remainder, with x being an integer from 1 to n^2, that is:

$$x - 1 = Qn + R \text{ where } 0 \leq R \leq n - 1 \text{ (and also } 0 \leq Q \leq n - 1),$$

then the rule can be given as:

$$r(x) \equiv R + Q + k + 2 \pmod{n}$$
$$c(x) \equiv R - Q + k + 1 \pmod{n}.$$

We now turn to the two questions. In order to show that the integers from 1 to n^2 do indeed completely fill the square, it is only necessary to prove that two distinct integers cannot occupy the same place. In fact, if x_1 and x_2 are found in the same cell, then $r(x_1) = r(x_2)$ and $c(x_1) = c(x_2)$ and the new formulation of the rule shows that $R_1 = R_2$ and $Q_1 = Q_2$ and so $x_1 = x_2$.

We now show that the square is magic with respect to its columns. For two numbers x_1 and x_2 to be in the same column, we must necessarily have $Q_1 \neq Q_2$ and $R_1 \neq R_2$, since if $c(x_1) = c(x_2)$, we have: $R_1 - Q_1 \equiv R_2 - Q_2 \pmod{n}$. Now if $Q_1 = Q_2$ we would have $R_1 = R_2$ and so $x_1 = x_2$. Similarly, if $R_1 = R_2$ we would have $Q_1 = Q_2$ and so $x_1 = x_2$. It follows that in the same column, the different numbers x_i will produce all the integers from 0 to $n - 1$ as quotients a_i and all the integers from 0 to $n - 1$ as remainders R_i. Therefore, the sum of the terms in any column will be:

$$\sum_i x_i = \sum_i (Q_i n + R_i + 1) = \left(\sum_i Q_i\right).n + \sum_i (R_i + 1)$$

$$= \left(\sum_{0 \leq Q \leq n-1} Q\right).n + \sum_{0 \leq R \leq n-1} (R + 1) = n.\frac{n(n-1)}{2} + \frac{n(n+1)}{2} = \frac{n(n^2 + 1)}{2}$$

and we obtain the magic constant. For the rows the argument is similar.

Suppose now that we have a number $x = Qn + R + 1$ on the leading diagonal of the square. Then $r(x) = c(x)$ and so $R + Q + k + 2 \equiv R - Q + k + 1 \pmod{n}$. This implies $2Q + 1 \equiv 0 \pmod{n}$, or $2Q + 1 \equiv 2k + 1 \pmod{n}$ and so, finally, $Q = k$. The numbers on the leading diagonal are therefore $nk + 1, nk + 2, ..., nk + n$ and their sum is equal to $(nk + 1 + nk + n)n/2 = (n^2 + 1)n/2$. A similar argument can be used for the trailing diagonal.

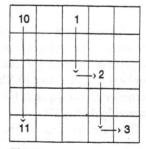

Fig. 18

We mentioned that Moschopoulos gave another method, also for odd magic squares, called 'proceeding by 3 and by 5'. This rule, for which we have not provided the original text, requires 1 to be placed in the centre of the first row and then to move as before, using 3 and 5 for the displacements – the departure and arrival cells being included in counting the moves – as is shown in figure 18.

To put it in formal algebraic language, we have the initial conditions $r(1) = 1$, $c(1) = k + 1$ and then apply the following rules:

If x is not a multiple of n:

$$r(x + 1) \equiv 2 + r(x) \ (\text{modulo } n)$$
$$c(x + 1) \equiv 1 + c(x) \ (\text{modulo } n)$$

If x is a multiple of n:

$$r(x + 1) \equiv 4 + r(x) \ (\text{modulo } n)$$
$$c(x + 1) \equiv c(x) \qquad (\text{modulo } n)$$

Claude-Gaspard Bachet de Méziriac

A variation of the method of Moschopoulos was proposed by Bachet, who added small squares to the sides of the square ABCD to be filled, and then put in the numbers in order along descending diagonals. Then he left the numbers inside ABCD in place, and for the others, he explains: "you put them in the empty places which remain, using only transposition, that is namely that those that are at the top go to the bottom, and those at the bottom you take to the top; those on the left side will pass to the right side, and those on the right side will go to the left side". The reader will recognise this as Moschopoulos's rule of proceeding by 2 and by 3 for the odd squares.

Facsimile Cl.-G. Bachet, *Problèmes plaisants et délectables qui se font par les nombres*, Lyon 1612, 2nd ed., Lyon, 1624 (p. 164).

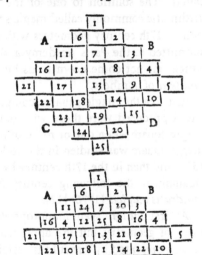

164 *Problemes plaisans & delectables,*
hors des quarrez 5 places ou 7 places plus auant, à cause que 5. est le costé de 25. & 7 est le costé de 49. Pour mieux te faire entendre ceste reigle, i'ay disposé icy en mesme sorte tous les nombres despuis 1. iusques à 25. comme tu vois es deux figures suiuantes.

Ainsi les 25 nombres sont disposez comme il faut dans le quarré A B C D. car la somme de chasque rang est tousiours 65. La mesme reigle sert en tous autres.

Other generalisations are possible. As Lehmer remarks [10], and others before him, the technique used by Moschopoulos is only a special case of a more general rule for constructing odd order magic squares. These are constructions defined by the two congruences:

$$r(x + 1) \equiv p + \alpha x + a[x/n] \ (\text{mod } n)$$
$$c(x + 1) \equiv q + \beta x + b[x/n] \ (\text{mod } n)$$

in which p and q are any integers and α and β and a and b are integers prime to n. The integers p and q indicate the position of 1, and α and β indicate the displacements in general of a cells down and β cells to the right – with the convention that we know –, except when the contents of a cell contain a multiple of n, in which case the displacement is $\alpha + a$ cells down and $\beta + b$ cells to the right. Moschopoulos's 'proceeding by 2 and by 3' therefore corresponds to choosing: $\alpha = 1$, $\beta = 1$, $a = 1$, $b = -1$. For 'proceeding by 3 and by 5' we choose: $\alpha = 2$, $\beta = 1$, $a = 2$, $b = -1$. The necessary condition for the process to completely fill the square is that the number $ab - a\beta$ should be prime to n. The square will also be semi-magic: it is only magic where we can also arrange for the sum of the numbers along both diagonals to be equal to the magic sum.

2.4 Arnauld's Borders Method

As an annexe to his *Nouveaux éléments de géométrie* [2], Arnauld added a section called "The solution to one of the most famous and most difficult problems of arithmetic commonly called magic squares". This provides evidence for the infatuation of 17th century geometers with magic squares. It is known that in 1654 Pascal submitted to the Paris Academy a *Treatise on magically magic numbers*, a treatise which has not come down to us but one which was praised by Fermat, that great *amateur* of numbers.

Just like the magic squares composed by az-Zinjānī, the magic squares that Arnauld produced were those with borders: a magic square with borders remains a magic square when one or more of its borders are removed. This particular type of magic square was studied in the 16th century by Stifel in his *Arithmetica integra* of 1554, and then in the 17th century by Frenicle de Bessy, Fermat, Arnauld, Prestet and Ozanam. In the following century Euler considered their construction in his *De quadratis magicis* (1776) [12].

At first it might seem inappropriate for a treatise on magic squares to appear in a work of geometry, but the approach to the solution of the problem of the magic square presented by Arnauld is certainly in the spirit of the book. This is to present an ordered, clear and natural method for solving the problem that also shows us why it provides a solution. Arnauld therefore provides us with a methodical way for constructing the solution, and it is a way that also leads to a proof of its legitimacy. For the reader should not be satisfied merely with a description of the process.

The method of borders that Arnauld uses is also indicative of the mathematics practice of the time, since he introduces algebraic symbols which indicate the generality of the algorithm. In fact, the method applies to all magic squares, both even or odd, with only small variations needed to take account of the parity of the order. We shall present the method for even squares. The method also has the advantage of producing, for any given order, several different magic squares which cannot be deduced from each other by rotations or symmetries. This advantage derives from the combinatory character of the method. Hence Arnauld's method for the even squares allows him to explain how to obtain the sixteen 'magic dispositions of the square' of order 4, which dispositions are deduced using a combinatory reasoning elegantly presented in a table.

At the end of the treatise, Arnauld states the original aspects of his solution that justify him presenting it to the public. He writes:

I think I am able to conclude from all this that it is not possible to find a method which is easier, shorter and more perfect for creating magic squares, which is one of the most beautiful problems of arithmetic.
What is special here is that:

1. the numbers are only written down twice.
2. we never need to hesitate, but are always sure of what we are doing.
3. the greater squares are no more difficult to construct than the smaller ones.
4. they can be altered as much as we wish.
5. we do nothing which cannot be proved.

Taken separately, the rules are simple and precise. However, we shall make some general remarks to make the reading easier to follow. The procedure is carried out on each border in turn, starting from the outside, except for the last two borders where different rules are to be used. The natural square, that is the square in which the cells are filled by the natural numbers in order, plays an essential role in the method. For each border, Arnauld starts from the natural square: he places letters in certain cells, according to certain rules; this done, he assigns numbers to the letters, which are the values that they take in the natural square. Then, to construct the magic square, he rearranges the letters using given rules, and then replaces the letters with their previously assigned values.

We should point out that, for a square of order $n = 2k$, Arnauld uses the terms 'centre' for the value $(n^2 + 1)/2$, 'small numbers' for the integers 1 to $n^2/2$ and 'large numbers' for the other integers. He uses the small letters α, β, ..., for the small integers and the capital letters A, B, ..., for their complements to $(n^2 + 1)/2$, these capital letters corresponding to his large numbers. Notice also, that in the natural square a number α and its complement A are always found symmetrically with respect to the centre, whereas in the magic square Arnauld places them opposite each other on a row or column, except for the letters at the corners of the square.

Arnauld
Nouveaux éléments de Géométrie, Savreux, Paris, 1667, pp. 327–334.

[To simplify the reading, we use the symbol + where Arnauld uses the symbol . to show addition.]

2. On the natural squares

2. I call natural squares those where the numbers are arranged in arithmetic progression starting from the smallest.

[...]

On even squares

8. There is no centre cell. But we can take the centre to be half the sum of the first and last number.
   And we shall call this whole sum $2c$.
9. Half of the rows, namely those that are the higher ones, will contain the small numbers, and the lower ones the large numbers.
10. The four cells at the centre form the 1st border.
    The cells around these four, the 2nd border.
    Those around the second, the 3rd border.
    And so on.
11. Also, the borders 1, 3, 5, 7, 9, etc., are called the odd borders.
    And the 2, 4, 6, etc. the evens.
12. The small numbers are,
    1. In the top strip of each border.
    2. On the left side from the top strip to where the large numbers start.
    3. And the same on the right.

3. Preparation

13. The greatest mystery in the solution of the problem consists in marking some of the small numbers in each strip by letters.

$$[\ldots]$$

In the even squares

16. Do not mark anything in the first and second borders.
17. In all the other borders in general, mark
The left corner of the top by	e.
The right by	o.
The lowest of the small numbers on the right by	α.
The lowest of the small numbers on the left by	β.
18. Also for the odd borders, starting with the 3rd (which is the one with 6 cells in the top strip) mark 4 of the cells in the top strip,
two by	e'.	o'.
and two by	$\hat{e}$.	$\hat{o}$.

    as has been said in §15. [Pairs of cells of the top strip, equidistant from the cells marked e and o, are marked with the same letters accented.]
    On the left mark the one below e by $\qquad$ ω.
    And on the right the one below α by $\qquad$ γ.

4. Maxims

For the proof of the operation

19. Two numbers, one *small* the other *large*, equally distant from the centre, and which are joined by a line passing through the centre, make a sum equal to two times the centre.
20. When a *small* number is marked by a letter, its *large* number is named (when it is needed to be identified) by the capital of the same letter, whatever letter has been used.
    Thus, $e + E$ makes two times the centre.
    And in the same way $a + A$, or $b + B$, or $o + O$.

Second maxim

21. Four numbers in the same strip, for which the first is as distant from the 2 as the 3 is from the 4, are in arithmetic proportion.
    And consequently the sum of the extremes is equal to the sum of the middle ones.

Examples

22. $e \cdot e' :: o' \cdot o$. Therefore $e + o = e' + o'$.
    From which it follows that everywhere where e', o' are together, or even $è$, $ò$, or their capitals E', O', when it is a matter of finding equalities with other numbers, we can suppose that it is as if they were e, o, E, O, because if the equality is found supposing they were e, o, it would not be disturbed by putting e', o', in their place, which come to as much as e, o.

$$[\dots]$$

In the even squares

24. $e \cdot \omega :: \beta \cdot A$. Therefore $e + A = \omega + \beta$.
    To find A, see §20.

Third maxim

25. When 4 cells make a paralellogram, rectangle or non rectangle, their 4 numbers are in arithmetic proportion. And consequently, the sum of the extremes is equal to the sum of the middle terms.

Examples

$$[\dots]$$

In the evens

29. $e \cdot o :: \beta \cdot \alpha$.   So $e + \alpha = o + b$.
30. $\varpi \cdot \beta :: o \cdot \gamma$.   So $\omega + \gamma = \beta + o$.

5. Method

To magically rearrange the natural square

31. This method requires a very small number of rules; some general, others particular, for the numbers of the natural square to be transposed into the magic square.

First general rule

32. The numbers have to be arranged in borders, those from a border into a similar border, and all the care needed to start with, is to know where to put the small numbers, because where the *small* numbers go determine where the *large* numbers go according to the following rules.

Second general rule

33. When a *small* number is placed in a corner its *large* number must be placed in the diametrically opposite corner.
    So, with α being placed in the left corner of the top strip, A must be put in the right corner of the bottom strip.

Third general rule

34. Except for the corners, the *large* numbers must be placed opposite the *small* numbers of the opposite strip..
    That is why you must watch out never to place two small numbers opposite each other.

Corollary of these rules

The numbers being arranged according to these rules, it follows

35. 1. That the numbers of two opposite strips taken together will come to *c* times the number of numbers in the two strips. For a small and a large are worth two times *c*. Now, there are as many *small* numbers as there are *large* numbers. Therefore it follows,

36. 2. That when it has been proved that the numbers of one strip, following this disposition, are equal to as many times the centre as there are numbers in the strip, that band is equal to its opposite one. It follows,

37. 3. That when there are as many small numbers in one band as there are in the opposite one, and that the sum of any one of them is equal to that of the other, it is a mark of assurance that one band is equal to the other band.

The proof is easy and need not detain us.

Fourth general rule

38. Care is only needed at first for placing the small numbers marked by letters: that having been done, the rest follows without difficulty for this reason.

In the top strip, in whatever squares and whatever borders there may be, besides the cells marked by letters:

Either there are no others,

Or there remain some non marked cells always evenly even in number, that is 4, 8, 12, 16, etc.

And furthermore, they are always 4 to 4 in arithmetic proportion.

So taking the extremes and putting them in one strip, and those of the middle in the opposite strip, they will not disturb the equality which has already been established by the numbers marked by letters.

39. It is the same for the two sides right and left. For the small numbers that remain (if there are any besides the marked ones) are always in number evenly even 4, 8, 12, 16, etc., from 4 to 4 in arithmetic proportion.

Therefore as above.

There is therefore nothing more to do that is troublesome than to arrange the letters. This is done by particular rules.

[...]

7. For the even squares

46. Leave to one side the two first borders, which have their own particular rules.

For the other odd borders

47. The arrangement is like this

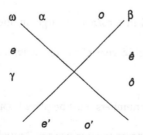

Proof

48. Required 1. to prove that the six numbers of the top strip of which four are small numbers, and two large numbers which come from e' and o' which have been put on the bottom, are equal to six times the centre.

    Which is proved thus.

    $\alpha + A + o + O + e + E = 6c$ by (20)

    Now these six letters are equal to the six, $\omega + E + \alpha + o + O + \beta$.

    For taking away the same which are found in both parts, namely $\alpha + o + O + E$, only $A + e$ remains on the one side and $\omega + \beta$ on the other.

    Now, by (24) $A + e = \varpi + \beta$.

    Hence the six letters $\omega + E + \alpha + o + O + \beta = 6c$.

49. Required 2. to prove that $\omega + e + \gamma = \beta + \acute{e} + \acute{o}$. For if this is so, then the large numbers will also be equal to the large numbers, and the total to the total (37).

    Supposing then that $\acute{e} + \acute{o}$ should be $e + o$ (22), and taking away e and e from both parts, there remains $\omega + \gamma$ in the one part and $\beta + o$ in the other which are equal sums by (30).

    Therefore $\omega + e + \gamma = \beta + \acute{e} + \acute{o}$.

    Therefore the strip equals the strip by (37).

For the even borders

50. The disposition is very easy, and is arranged thus

Proof

51. This is so easy by 22, 29 and 37 that I shall not take the pleasure of explaining it.

    This border can also be made by transposing the corners.

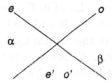

8. Particular rules

For the first and second borders of even squares

[...]

55. And this is how they are found.

    The numbers must always be taken 4 by 4 in that order.

    1. The fours of the insides or interiors.

    2. The four exterior corners.

    3. The two of the middle of the top strip, with the two of the middle of the lower one.

    4. The two of the middle of the strip to the left, with the two of the middle of the strip to the right.

    Now, each of these numbers taken thus 4 by 4 (and which we shall name by the sequence 1.2.3.4.) can be

    Either left in their own position; which will be marked by   *o.*

    Or transposed as a St. Andrew's cross; which will be marked by   *c.*

    Or transposed directly left to right; which will be marked by   *g.*

    Or transposed directly top to bottom; which will be marked by   *h.*

56. Given these remarks, and remembering what the four numbers (1.2.3.4.) and the four letters (*o.c.g.h.*) stand for, the two following tables can be used to find the sixteen dispositions of the magic square of order 4 without difficulty: or, what is the same thing, the two first borders of all even squares.

	I	II	III	IV	V	VI	VII	VIII	IX	X	XI	XII	XIII	XIV	XV	XVI
1	o	o	o	o	c	c	c	c	g	g	g	g	h	h	h	h
2	o	c	g	h	o	c	g	h	o	c	g	h	o	c	g	h
3	c	g	c	g	h	o	h	o	h	o	h	o	c	g	c	g
4	c	h	h	c	g	o	o	g	c	h	h	c	g	o	o	g

[...]

58. Here is an example of the 6th disposition, and another for the 16th. The others are left to be found.

16	2	3	13
5	11	10	8
9	7	6	12
4	14	15	1

13	3	2	16
8	10	11	5
12	6	7	9
1	15	14	4

By way of an example, we shall use Arnauld's method to find a magic square of order 10. According to Arnauld's description this has 5 borders, the outside border being the fifth. The centre c is equal to $(n^2 + 1)/2 = 101/2$, the small numbers are from 1 to 50 and the large numbers are those from 51 to 100.

In the natural square, filled with the numbers from 1 to 100, place the small letters e, o, α, β following the instructions 16 and 17, for the borders 3, 4 and 5. Then place the letters e', o', $\hat{e}$, $\hat{o}$, ω, γ using instruction 18 for the odd borders 3 and 5 (Fig. 20).

1_e	$2_{e'}$	$3_{\hat{e}}$	4	5	6	7	$8_{\hat{o}}$	$9_{o'}$	10_o
11_ω	12_e	13	14	15	16	17	18	19_o	20
21	22	23_e	$24_{\hat{e}}$	$25_{\hat{e}}$	$26_{\hat{o}}$	$27_{\hat{o}}$	28_o	29	30
31	32	33_ω	34	35	36	37	38_γ	39	40_γ
41_β	42_β	43_β	44	45	46	47	48_α	49_α	50_α
51	52	53	54	55	56	57	58	59	60
61	62	63	64	65	66	67	68	69	70
71	72	73	74	75	76	77	78	79	80
81	82	83	84	85	86	87	88	89	90
91	92	93	94	95	96	97	98	99	100

1_e	$2_{e'}$	$3_{\hat{e}}$	4	5	6	7	$8_{\hat{o}}$	$9_{o'}$	10_o
11_ω									20
21									30
31									40_γ
41_β									50_α
51_A									60_B
61									70
71									80
81									90
91_o	$92_{o'}$	$93_{\hat{o}}$	94	95	96	97	98_E	$99_{E'}$	100_E

Fig. 20 Placing the small letters in the natural square.

Fig. 21 The outside border of the natural square.

Note that the position of the accented letters e', o', $\acute{e}$, $\acute{o}$, on the first row is not completely fixed: e' and o' are only at an equal distance from e and o, and the same for $\acute{e}$ and $\acute{o}$.

The maxims 19 to 30 are simply elementary observations about the properties of the arrangements of numbers in the natural square. In particular, the capital letters E, O, A, ..., which represent the complements to $2c = 101$ of their corresponding small letters, occupy cells which have central symmetry with their counterparts e, o, α, ... (Fig. 21).

We now fill in the magic square. We start with the outside border. This is the fifth border, and it is odd; we therefore apply instruction 47 which changes, in an apparently arbitrary way, the disposition of the letters placed in the natural square (Fig. 22). Here again, this reorganisation is not completely fixed. For example, e and γ in the first column, and $\acute{o}$ and $\acute{e}$ in the last column are only constrained in that they must occupy entirely different rows.

Furthermore, from the procedure, we can see that the only thing that is important is to fix the diagonals, the others can move around within the same strip of the border, provided we always have regard for certain conditions of symmetry.

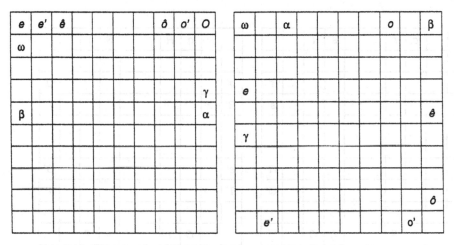

the natural square   the magic square

Fig. 22 A reorganisation of the letters for an odd border.

We now replace the letters in the exterior border by the values that they held in the natural square. Then we put the complement to 101 in each of the cells diametrically opposite the corner ones by instruction 33, and in the directly opposite cell for each row and column by instruction 34 (Fig. 23). There now remain in each side of the border, 4 unfilled cells. In general, there will always be a number of empty cells and these will be a multiple of 4. Arnauld makes the remark that the numbers that have not been placed can be arranged in fours such that, for each group of four, they can be in arithmetic proportion. Suppose that the four small numbers q, r, s, t of the top row of the border of the natural square are thus in arithmetic proportion (that is $q + t = r + s$), then, by instruction 38, we replace the two mean terms r and s of the corresponding row of the magic square – or in the opposite row – with the corresponding capital letters R and S. In the case of a column, an arithmetic proportion can also be formed with two small letters and two capital letters q', r', S', T' (where $q' + T' = r' + S'$), and in a similar way we replace the two means r' and S' by respectively R' and s' in the corresponding column of the magic square – or in the opposite column. Then the complements of these numbers are entered opposite, following instruction 34 (Fig. 24).

11_ω	99	50_α				100_o	92	41_β
1_θ								100
98								3_θ
40_γ								61
93								8_δ
60	$2_{\theta'}$	51				91	$9_{o'}$	90

Fig. 23

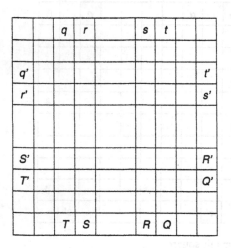

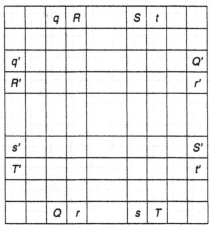

Fig. 24　natural square　　　　　　　magic square

Applying the rule to the sequence 4, 5, 6, 7 for the rows and the sequence 20, 30, 70, 80 for the columns leads to putting 4, 96, 95, 7 in order in the top row and 20, 71, 31, 80 in the right hand column, with their complements opposite. This outside border is now complete (Fig. 25).

11_ω	99	50_α	4	96	95	7	100	92	41_β
81									20
30									71
1_e									100
98									3_δ
40_γ									61
70									31
21									80
93									8_δ
60	2	51	97	5	6	94	91	9	90

Fig. 25 An exterior border for a magic square of order 10.

Now we move to the fourth border which is even. For placing the letters in the magic border, instruction 50 replaces instruction 47 (Fig. 26).

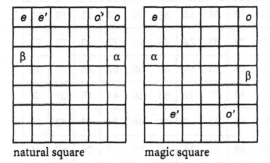

natural square   magic square

Fig. 26 A reorganisation of the letters in the case of an even border.

By using the values of the letters in the corresponding border of the natural square and completing the empty cells with numbers in arithmetic proportion, we obtain the fourth border (Fig. 27).

Now we turn to the third border: all the cells are filled straight away (Fig. 28). There now remains a central square of order 4 for which Arnauld has already given all the possibilities in sections 55 and 56. Choosing one of these, we complete the construction (Fig. 29).

12	88	14	86	85	17	83	19
22							79
69							32
49							52
59							42
39							62
72							29
82	13	87	15	16	84	18	89

33	74	48	28	77	43
23					78
76					25
38					63
75					26
58	27	53	73	24	68

Fig. 27 A fourth border of a magic square of order 10.

Fig. 28 A third border of a magic square of order 10.

11	99	50	4	96	95	7	10	92	41
81	12	88	14	86	85	17	83	19	20
30	22	33	74	48	28	77	43	79	71
1	69	23	64	36	35	67	78	32	100
98	49	76	47	55	56	44	25	52	3
40	59	38	57	45	46	54	63	42	61
70	39	75	34	66	65	37	26	62	31
21	72	58	27	53	73	24	68	29	80
93	82	13	87	15	16	84	18	89	8
60	2	51	97	5	6	94	91	9	90

Fig. 29 A magic square of order 10 with borders.

Bibliography

[1] Andrews, W. S., *Magic Squares and Cubes*, La Salle, Illinois: Open Court Publishing Co., 1908, repr. New York: Dover, 1960.

[2] Arnauld, A., 'Solution d'un des plus célèbres et des plus difficiles problèmes d'arithmétique appelé communément les quarrés magiques', in *Nouveaux Eléments de Géométrie*, Paris: Charles Savreux, 1667.

[3] Bachet de Méziriac, C.-G., *Problèmes plaisants et délectables qui se font par les nombres*, Lyon, 1612; 2nd (augmented) ed., Lyon, 1624; 3rd ed. with additions by Labosne A., Paris: Gauthier-Villars, 1879; reprnt, with forward by J. Itard, Paris: Blanchard, 1959.

[4] Benson, W. H. & Jacoby, O., *New recreations with magic squares*, New York: Dover, 1976.

[5] Cazalas, E., *Carrés magiques au degré n, séries numérales de G. Tarry, avec un aperçu historique et une bibliographie des figures magiques*, Paris: Hermann, 1934.

[6] Ch'en Te-K'un, 'Some Chinese Islamic magic squares porcelain', *Journal of Asian Art*, Nanyang University, Singapore, Inaugural Issue (1972), 146–159.

[7] Gardner, M., *More mathematical puzzles and diversions*, 1961; repub. Harmondsworth: Penguin, 1971.

[8] Hermelink, H., 'Zur Frühgeschichte der magischen Quadrate in Westeuropa' [On the early history of magic squares in Western Europe], *Sudhoffs Archiv*, vol. 4 (1981), 313–338.

[9] Kraitchik, M., *Mathematical Recreations*, New York: Norton, 1942; 2nd edition Dover, 1953.

[10] Lehmer, D. N., 'On the congruences connected with certain magic squares', *Transactions of the American Mathematical Society*, vol. 31; 2 (1929), 529–551.

[11] Martzloff, J.-C., *A History of Chinese Mathematics*, New York: Springer Verlag, 1997, tr. Stephen S. Wilson from *Histoire des mathématiques chinoises*, Paris: Masson, 1987.

[12] Molk, J., (ed.), *Encyclopédie des Sciences mathématiques pures et appliquées*, t. 1, vol. 3, fasc. 1, 'figures magiques', pp. 62–75, Paris: Gauthiers-Villars, 1906; repub. Paris: Gabay, 1992.

[13] Morgan, C., 'Les "neuf palais" dans les manuscrits de Touen-Houang', in *Nouvelles contributions aux études de Touen-Houang* sous la direction de Michel Soymié, Geneva: Droz, 1981, pp. 251–260.

[14] Ollerenshaw, K. & Bondi, H., 'Magic Squares of Order Four', *Philosophical Transactions of the Royal Society of London*, series A, 306 (1982), 443–532.

[15] Rosu, A., 'Les carrés magiques indiens et l'histoire des idées en Asie', *Zeitschrift der Deutschen Morgenländischen Gesellschaft*, vol. 139, 1 (1989), 119–158.

[16] Sesiano, J., 'An Arabic Treatise on the Construction of Bordered Magic Squares', *Historia Scientiarum*, 42 (1991), 13–31.

[17] Sesiano, J., 'Herstellungsverfahren magischer Quadrate aus islamischer Zeit (I & II')' [The methods of constructing magic squares in Islamic times], *Sudhoffs Archiv*, vol. 64, 2 (1980), 187–196 and vol. 71, 1 (1987), 78–89.

[18] Tannery, P., 'Le Traité de Manuel Moschopoulos sur les carrés magiques, texte grec et traduction', *Annuaire de l'Association pour l'encouragement des études grecques en France*, 1886, pp. 88–118. *Mémoires Scientifiques de Paul Tannery*, vol. IV, Paris: Gauthier-Villars, 1920, 27–60.

3 Methods of False Position

A number must be chosen in which the proposed
parts appear whole and this is to avoid fractional
numbers, and not because it could not just as
well be done with another number but with
more difficulty.

Francès Pellos, *Compendion de l'abaco*, 1492

"A lance has a half and a third in the water and 9 palms outside. I ask you how long is it?" This question proposed by Francès Pellos, a fifteenth century nobleman from Nice, would not be thought to present any particular difficulty nowadays for anyone who knows a little algebra. If we let x stand for the length of the lance, we have to solve the equation:

$$x - \frac{1}{2}x - \frac{1}{3}x = 9$$

and deduce that the length of the lance is 54 palms.

Such questions, before algebraic methods became available, were handled by algorithms which, since the end of the fifteenth century, have been known as *methods of false position*. These procedures involved 'trying out' the problem with one or more numerical values for the unknown and then using these values to obtain the correct result.

To answer the above question, Pellos reasons just as we do in 'algebra'. He starts by taking the length of the lance in palms as the unknown. But instead of using a letter for this unknown quantity, he chooses a precise numerical value and invites us to "put 12 for it as we please", that is to carry out the calculations using 12 as a false position. Starting with 12 and using the conditions set out in the problem, he subtracts half of it, and then a third, and ends up with 2, instead of the 9 that was required. He continues: "if 2 comes from 12, from how many would 9 come? You will find 54". This final step refers to the rule of proportionality: $\frac{12}{2} = \frac{x}{9}$ which given in the form of the well known *rule of three* is written: $x = \frac{12 \cdot 9}{2}$.

In all similar problems, it appears that the required quantity is linked to the given values in a linear way. Algebraically, it always reverts to an equation of the type: $ax = b$. So that, if a false position x' leads to a result b', in other words, if $ax' = b'$, then by proportion we have $\frac{x}{x'} = \frac{b}{b'}$ and so:

$$x = \frac{b.x'}{b'}.$$

Except for the notation, this equality corresponds to the final step of Pellos's argument. In this way, a single false position leads to the solution of the proposed problem: the method is called *simple false position*. The algorithm avoids the need to determine explicitly the 'coefficient' a of x in the reduced equation.

The simplified algebraic form, which we are able to use, masks the difficulties that some authors had to face when applying the method of false position. The techniques they used did not take advantage of algebraic simplification; for example, in tackling the question above, Pellos does not use the result:

$$x - \frac{1}{2}x - \frac{1}{3}x = (1 - \frac{1}{2} - \frac{1}{3})x = \frac{1}{6}x.$$

On the other hand, calculations with fractions could present difficulties, and the method of false position aims as far as possible to operate only with whole numbers and a judicious choice of the initial false position is very important. As Pellos himself says: "A number must be chosen in which the proposed parts appear whole and this is to avoid fractional numbers, and not because it could not just as well be done with another number but with more difficulty" ([15], p. 179). In this example, Pellos chose 12 for the false position, which is a multiple of both 2 and 3.

As for the method called *double false position*, its full value is to be found in dealing with problems that, in algebraic form, reduce to the solution of equations of the type:

$$ax + b = cx + d.$$

Using two false positions x' and x'' leads to 'errors' e' and e'':

$$e' = (ax' + b) - (cx' + d) = (a - c)x' + b - d$$
$$e'' = (ax'' + b) - (cx'' + d) = (a - c)x'' + b - d$$

The algebraic solution is found by using the formula:

$$x = \frac{x'e'' - x''e'}{e'' - e'}.$$

This is because we have $x = (b - d)/(c - a).$

and from the simultaneous equations we obtain

$$x'e'' - x''e' = (b - d)(x' - x'')$$

and $$e'' - e' = (a - c)(x'' - x').$$

Here again the algorithm of double false positions avoids the need to calculate the coefficients a, b, c and d explicitly in the reduced equation.

It is difficult to be sure about the origin of methods of false position, and we need to take great care in interpreting methods used by, for example, the Egyptians, since it might very well be that entirely different considerations led to the methods they employed. The application of the method, as we have seen, is to linear problems; however, some have tried to extend the method beyond its strict limits as, for example, in certain problems considered by Chinese mathematicians, and an attempt to use the method for the solution of quadratic equations in the 16th century was proposed by Mellema [18].

The names accorded to methods of false position through the ages reflect both the nature of the problems being worked on and their relation to the mathematical culture of the time. In the 12th century, the Indian mathematician Bhāskara calls the simple method of false position *ista-karma*, that is 'operating with a trial number'. The Chinese called the techniques of double false position *ying bu zu shu*, which means 'rules of too much and not enough'. Fibonacci in the 12th century talks of *regula augmenti et diminucionis*, that is 'the rule of increase and decrease'. Here he is drawing on the work of those Arab mathematicians who used the technique of the *ḥisāb al-khata'ayn*, that is the 'calculation of the two errors'. Ibn al-Bannā, as we shall see, uses what he calls *the method of scales*. It is not until the end of the 15th century that the *rule of the false* or *Regula Falsi* gives place to *the rule of one position* or *two positions* as Chuquet [8] describes it in 1484; the distinction between the two techniques is now quite clearly established. Right up until the beginning of the 20th century a method of false position or *false supposition* is found in works on arithmetic. It represented the summit of success for the young scholar, for it involved fractions, the four rules and, depending on the authors, proportion, the rule of three, reasoning by difference, and so on. Today, the technique of double false position can be seen as linear interpolation (see Chapter 10) and was understood as such, perhaps, by the Chinese.

The explanation of the technique of the method of false position was often presented with the pretext of a pseudo situation involving examples from commerce or inheritance. But beyond the particularity of the problem, the real interest in the different algorithmic procedures that were expounded lies in the detail of the numeration and operational techniques used and with the use of linearity and proportion. In addition, we should not forget the role that underlying images of representation played in the techniques: see, for example, Ibn al-Bannā's scales method. Finally, when using methods of double false position ([18], p. 95), we can see the beginnings of the use of signs attached to numbers so as to take account of the sense of the errors.

As we can see, the history of techniques of false position covers a period of several centuries and many civilisations. It would be impossible to try to do justice to all of this. What we have done is to highlight some aspects, in choosing texts which are more accessible, or appear more easily comprehensible, and we discuss the significance of these examples in the accompanying commentaries.

To conclude, here is a typical 'false position' problem from a 19th century French school text book.

CHAPTER VIII
SOLUTION TO SOME PROBLEMS

377. A. PROBLEM I (**false position**). – *A bag contains 154 fr. in coins of 5 fr. and 2 fr.; there are 41 coins, find the number of 5 fr. coins.*

If all the coins were 2 francs, the bag would only contain 41 × 2 or 82 francs, whereas it contains 154: the difference is 72 francs.

Each time a 2 franc coin is replaced by a 5 franc coin the total number of coins does not change, but the sum increases by 3 francs. So that the sum can be increased by 72 francs, you must replace as many 2 franc coins by 5 franc coins as 3 is contained in 72, that is 24.

Answer: there are 24 coins of 5 francs.

REMARK. – Many problems can be solved by the method we have just used: it is called the method of false position.

Combette, *Cours moyen et Supérieur Arithmétique, système métrique et géométrie usuelle*
Paris: Alcide Picard & Kaan, 10th edition, (n. d.), p. 219.

3.1 Mesopotamia: a Geometric False Position

There is a close connection between geometric similarity, arithmetic proportion and a change of 'variable'. For this reason, the techniques of false position can be used to solve certain 'geometric' problems.

We know that the Babylonians considered 'arithmetic' problems of a type that others, later, were able to solve using the method of false position [20]. One example is the problem 3 of tablet YBC 4669 dealing with provisions ([20], p. 205):

I have eaten two thirds of my provisions: there is left 7.
What was the original (amount) of my provisions?

Another example is problem 3 of the tablet AO 6770 ([20], p. 73):

I took a stone: I did not know its weight. I took away 1/7, the third of a shekel and 15 grains. I put back 1/11 of what I had taken and five sixths of a shekel: my stone was restored to its original state. What was the original (weight) of my stone?

Unfortunately, only an incorrect answer is given to the first of these and what we have of the solution to the second is too imprecise to give an indication of what algorithm was used. Similarly, with tablets dealing with questions about commerce [14] or problems of inheritance [11], the results, while they include the use of fractions, are sometimes too inexact or imprecise to enable us to be certain about the algorithm that was used to find the solutions. Although a considerable number of tablets have been found and transcribed, we still cannot be sure that the Babylonians used a method of false position to solve this type of arithmetic problem. On the other hand, we do have a tablet that that clearly shows a method of false position being used for a problem of geometry and this is the one we present here.

The tablet comes from what is known as the Old Babylonian period, that is about 1800 BC. The problem is to find the dimensions of a rectangle, given its diagonal and a linear relation between the sides. This problem is from the collection of mathematical tablets from Susa provided by Bruins and Rutten [5]. As to the translation given here, the numbers have been given in the form normally used by Babylonian scholars and as described in Chapter 1. So we should remember that while the relative order of the figures in the number 33;45 is sexagesimal, no particular sexagesimal value is attached to the 33 or the 45.

Tablet from Susa
Textes mathématiques de Suse, ed. and translated into French by E. M. Bruins & M. Rutten, Paris: Geuthner, 1961 (p. 102), Problem xix, c.

Let the breadth (of the rectangle) measure a quarter less compared with the length.
40 (is the dimension of the diagonal). What are the length and breadth?
Thou, take 1 as a length, and put 1 for the length (of the rectangle).
15 the quarter, subtract from 1, you will find 45.
Put 1 as the length, put 45 as the breadth, square 1 the length, 1 you find.
Square 45, the breadth: 33;45 you find. From 1 and 33;45 (make) the sum: 1;33;45 you find. What is the square root? 1;15 you find.
Expecting 40, the diagonal that was indicated to you, unravel the inverse of 1;15 the diagonal. 48 (you find). Take 48 to 40 [48×40] the diagonal which you were told, 32 you find.
Take 32 to 1 [32×1] the length that you put: 32 you find, 32 (that is) the length. Take 32 to 45 [32×45] the breadth that you put: 24 you find. 24 (that is) the breadth.

Using ℓ, b and d for the length, breadth and diagonal respectively of the rectangle, we can restate the problem with the equations:

$$b = \ell - \ell/4 \text{ and } d = \sqrt{\ell^2 + b^2} = 40.$$

The tablet says: put 1 for the length. Let ℓ_0 be this position, and b_0 and d_0 the corresponding positions for the breadth and diagonal. The calculations in the tablet are as follows:

1. We put: $\ell_0 = 1$ [or 60]

2. Calculate b_0: $\ell_0/4 = 15$,

                            $b_0 = \ell_0 - \ell_0/4 = 45$

3. Calculate d_0: $\ell_0^2 = 1$ [or 60^2]

                            $b_0^2 = 45^2 = 2025 = 33 \times 60 + 45$ (recorded as 33;45)

                            $d_0^2 = \ell_0^2 + b_0^2 = 5625 = 1 \times 60^2 + 33 \times 60 + 45$ (recorded as 1; 33; 45)

                            $d_0 = \sqrt{\ell_0^2 + b_0^2} = 75 = 1 \times 60 + 15$ (recorded as 1;15)

4. Calculate d/d_0: $1/d_0 = 1/75 = 48/60^2$ (recorded as 48)

                            $d \times (1/d_0) = 40 \times (48/60^2) = 32/60$ (recorded as 32)

5. Calculate ℓ: $\ell = \ell_0 \times (d/d_0) = 60 \times (32/60) = 32$

6. Calculate b: $b = b_0 \times (d/d_0) = 45 \times (32/60) = 24$

To show that this algorithm is valid, and to understand its value, it is easy for us to generalise the method to the solution of the equations:

$$b = k\ell \text{ and } d = \sqrt{\ell^2 + b^2},$$

from which $\ell\sqrt{1 + k^2} = d.$

Hence, $\ell/\ell_0 = b/b_0 = d/d_0$

and so $\ell = \ell_0 \times (d/d_0)$ and $b = b_0 \times (d/d_0)$

which corresponds to the calculations recorded on the tablet.

Note, finally, that the calculations to find the inverse of d_0 require that the number should be regular in the sense that we described in Chapter 1. The choice of 75 for d_0 was not accidental!

3.2 Egypt: Problem 26 of the Rhind Papyrus

The main mathematical texts that we possess from Ancient Egypt date from the 18 century BC. Among these are the Kahun papyrus, the Moscow papyrus and, most importantly, the Rhind papyrus, named after the nineteenth century English Egyptologist who had the good fortune to buy it. The Rhind papyrus [6] was written by the scribe Ahmes in about 1650 BC and is a copy of a two centuries earlier document. The papyrus consists of about a hundred 'problems'; in reality, to quote its introduction, it is about "accurate reckoning" and is the "entrance into the knowledge of all existing things and all obscure secrets". This is not simply a collection of clearly prescribed rules. What we find are examples of arithmetical practice being adapted to particular circumstances. Often the explanations are missing or lack precision. Furthermore, the texts can sometimes be interpreted in different ways.

This is the case with the 'problem' we present here. We may consider that the solution algorithm corresponds to a method of simple false position, but some writers suggest that it involves a change of variable, while others argue that it is nothing more than a division procedure. According to whether we want to focus on the technique (false position), on what it depends (change of variable), or on the result (the division), these different interpretations may differ, without being contradictory.

Many Egyptian 'problems' are related to certain 'quantities' without any further precise detail. Also we are tempted to translate the problems into an algebraic formulation although such usage was quite unknown to the Egyptians. Having done this, we need to take care to be aware of the particular circumstances that make certain operations possible. For example, to solve the equation $ax = b$, we know that all that is required is to divide b by a when a and b are integers, as with problem M 25 of the Moscow papyrus. The question becomes more complicated when a is not an integer. Problems K 3 of the Kahun papyrus, and M 19 of the Moscow papyrus, correspond, 'algebraically', respectively to the equations:

$$\left(\frac{1}{2}-\frac{1}{4}\right)x = 5 \quad \text{and} \quad \left(1+\frac{1}{2}\right)x + 4 = 10 .$$

Following 'reductions', the number a is simple, and easily inverted. From this, the division of b by a is carried out by multiplying b by the inverse of a. On the other hand, problems R 24 to R 27 of the Rhind papyrus are of a different 'arithmetical' type; they correspond to the equation: $x+\frac{1}{n}x = b$, where $n = 7, 2, 4$ and 5. Arithmetically, we cannot in theory 'reduce to the same denominator' the number $1+\frac{1}{n}$ and, even if we wished to write the equation in the form $\left(1+\frac{1}{n}\right)x = b$, the number $1+\frac{1}{n}$ will not, in general, be easily invertible. We must therefore find ways of overcoming these obstacles to produce an algorithm that is simpler than the 'traditional' one of the division of b by $\left(1+\frac{1}{n}\right)$.

We have selected problem R 26 from the Rhind papyrus because it is the one which has been explained the most. Algebraically, it corresponds to solving the equation: $x+\frac{1}{4}x = 15$. In the text, we see the original in hieratic script, and beneath this is the translation into hieroglyphic writing. Here we use bold for the parts of the text that the scribe originally wrote in red ink.

Rhind Papyrus: Problem 26

The Rhind Mathematical Papyrus, by A. B. Chace, 2 vol., Ohio: Oberlin, 1927–1929, plate 49.

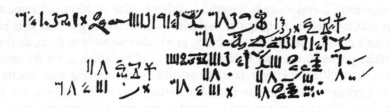

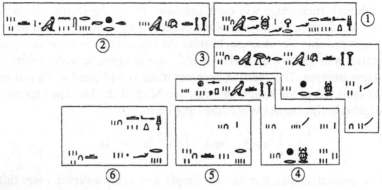

1. **A quantity, $\frac{1}{4}$ of it added to it,** becomes it: 15.

2. Operate on 4; make thou $\frac{1}{4}$ of them, namely 1; The total is 5.

3. Operate on 5 for the finding of 15.

   \1   5
   \2   10
   There become 3.

4. Multiply: 3 times 4.

   1   3
   2   6
   \4   12
   There become 12

5. 1   12
   $\frac{1}{4}$ 3
   Total   15

6. **The quantity is 12.**

   $\frac{1}{4}$ of it is 3; the total is 15.

If we take this to be a simple false position 'problem', we can identify the following steps:

1. The problem statement: algebraically, this corresponds to solving the equation

$$x + \frac{1}{4}x = 15,$$

or, more generally

$$x + \frac{1}{n}x = b.$$

2. Calculating the value of the 'first term' b' starting from the false position $x' = 4$:

$$b' = 4 + 1 = 5.$$

3. Dividing b by b': $b : b' = 15 : 5 = 3.$
4. Multiplying the $(b : b')$ by x': $3 \times 4 = 12.$
5. The 'verification' $12 + 3 = 15.$
6. Stating the result.

This procedure also appears in problems R 24, R 25 and R 27, but the explanations are less detailed and can therefore give rise to other interpretations. The problems contain no explicit opening statement: "Operate on n; make thou $1/n$ of them namely 1; the total is $(n + 1)$", which can be taken to imply a method of false position. Instead we just have the calculations required to find $(n + 1)$. This could simply be a particular division technique for replacing the fractional divisor $(1 + 1/n)$ by the integer divisor $(n + 1)$. If that is the case, then the method is more akin to an arithmetical technique rather than an illustration of a method of false position.

We should end here with a note of caution. The choices of n in these problems is not always random. For example, in problem R 24 the number 7 appears. Now, division by $(1 + 1/7)$ does not appear at first sight to be easy, while division by $(n + 1)$, here 8, is very simple. Also, the principal measure of length, the cubit, was divided into 7 palms, which could be used for calculations in place of cubits. In other words, the choice of numbers in the problems may be influenced by purely numerical considerations, or by practical measurement needs, and the algorithms that are generated only give the appearance of being false position algorithms.

3.3 China: Chapter VII of the *Jiuzhang Suanshu*

The first millennium Chinese method of double false position, *ying bu zu shu* means, literally, the 'rule of too much and not enough'. Nothing could be clearer: the arbitrary choice (or false position, in Chinese *jia she*, *jia* = false, *she* = supposition) of the value of the unknown in a problem will, after carrying out the calculations given in the story of the problem, usually lead to a value that is *too* large (*ying*) or *not* large *enough* (*bu zu*), compared with a given value intended as the numerical 'answer' to the calculations to be carried out on the unknown. If we let x_1, x_2 be, respectively, the two false positions and e_1, e_2 the two errors of excess or deficiency arising from using them, where e_1 is an excess and e_2 a deficiency (or the other way round), then the Chinese obtained the correct value of the unknown by carrying out two multiplications and two additions, as given by the rule:

$$x = \frac{x_1 e_2 + x_2 e_1}{e_1 + e_2}.$$

The oldest known explanation of this rule and its variants is dealt with in Chapter 7 of the *Jiuzhang Suanshu* (Computational Prescriptions in Nine Chapters). The fact that a whole chapter of their arithmetic 'bible' was given over to this rule gives us some indication of the importance the Chinese attached to it.

Most histories of mathematics place the *Jiuzhang Suanshu* at the time of the Han dynasty (*c.* 206 BC–220 AD). If we were also sure that the editions of this work now available to us represent a faithful reproduction of the earliest form of the text, then we could claim that the treatment of the method of double false position given in the *Jiuzhang Suanshu* is certainly richer than that which appears in other ancient and mediaeval arithmetics. Whatever the actual date of the original work, what appears in Chapter 7 of the *Jiuzhang Suanshu* is itself sufficiently rich in content to make it worth studying here.

For historians, there is the richness of detail provided by the constructions of the 20 problems themselves. Leaving aside the improbable scenarios described by the problems, we are nonetheless brought into contact with concrete, if rather unrealistic, examples of trade in cows, wine and lacquer oil, loans at compound interest, and journeys on horseback.

For historians of mathematics, the chapter holds other interests, both in the teaching approach and in the originality of the application of the rule of double false position to novel mathematical situations.

From a pedagogic point of view, we find the presentation of the rule of double false position very carefully laid out in a sequence of problems of increasing difficulty. In the first eight problems, the pupil has to apply the rule but does not himself have to make the necessary false assumptions. He then has to consider the case where both errors are of the same sense, then where they are of opposite sense and finally, where there is no error. The apprentice mathematician is thus led successively to apply the 'rule of too much and not enough' to cases where the supposed initial choices lead to a 'not enough' and a 'too much', to two values of 'too much', to two values of 'not enough', and finally, though improbably, to the case where it is 'exactly enough', the supposed choice having, by luck, turned out to give the correct answer. These different cases each give rise to *ad hoc* rules, the details of which we shall not go into here. But taken as a whole, the chapter also provides the opportunity for considerable revision of the arithmetical skills previously taught, including the conversion of units and the manipulation of fractions, as well as a general revision of the rules of arithmetic. In the fact, the whole of the pupil's arithmetical knowledge, old and new, is mobilised in the task of solving problems of double false position.

From a mathematical point of view, the Chinese text presents us with what we might normally consider to be types of arithmetic problems that are apparently simple. Yet we can see within it mathematics that is by no means trivial. To describe it terms of the mathematics we use today, we could say that the Chinese text proposes problems whose different stages of solution require the use of, not only the method of double false position itself, but also quadratic equations, sums of arithmetic series, simultaneous equations with two unknowns, and even piece-wise defined linear functions. These different aspects of mathematics emerge unexpectedly from problems of the rule of 'too much and not enough'.

Jiuzhang Suanshu: Problem 18

From Qian Baocong (ed.), *Suanjing shishu* (Ten Mathematical Manuals) Peking, 1963, p. 215.
Problem 18 of Ch. 7 of the *Jiuzhang Suanshu* (Computational Prescriptions in Nine Chapters).
From the French translation by J.- C. Martzloff.

9 gold coins weigh as much as 11 silver coins. If, in each pile, one gold coin is replaced by a silver coin, and conversely, the gold pile becomes lighter by 13 *liang*. How much do a gold and silver coin weigh respectively?

Answer: one gold coin weighs 2 *jin* 3 *liang* 18 *zhu*; one silver coin weighs 1 *jin* 13 *liang* 6 *zhu*.

Method: If the gold coin weighs 3 *jin* and the silver coin 2 *jin* and 5/11 of a *jin*: the 'not enough' of the right hand column comes to 49. If the gold weighs 2 *jin* and the silver 1 *jin* and 7/11 of a *jin* the 'too much' of the left hand column comes to 15 [...]

We should first make the remark that the left and right hand columns mentioned in the text refer to the particular way in which the numbers to be worked on are arranged on the counting surface.

It would appear that the author of the text wants his pupils to work on fraction calculations and the exchange of units, as well as the problem of double false position. In order for us to follow the calculations we need to know the equivalences:

$$1 \, jin = 16 \, liang \text{ and } 1 \, liang = 24 \, zhu$$

(the *jin* and the *liang* correspond approximately to the pound and the ounce in weight measures).

We have here a problem involving two unknowns. If we let x be the weight of a gold coin and y the weight of a silver coin, then the problem can be given by the system of equations:

$$9x = 11y \tag{1}$$
$$(10y + x) - (8x + y) = 13 \tag{2}$$

As is explained in the long and complicated commentary by Liu Hui (not translated here), the idea used for the solution is to take one of the unknowns as the principal unknown, with the other playing a subordinate role. The principal unknown here is the weight of a gold coin, and the false positions of the problem are taken to be 2 *jin* and 3 *jin*, being the supposed weights of the gold coin. If we already know that the exact weight of a gold coin lies between 2 *jin* and 3 *jin*, then these false positions are not randomly chosen but represent the best possible interval containing the solution, for integer units of *jin*. It would seem then that the author is inviting his pupil to begin by reflecting on the best order of magnitude for the solution (in *jin*).

Starting from here, the solver now needs to arrange for one of the conditions of the problem to be exactly satisfied while the other is subject to the procedure of double false position. Taking $x_1 = 3$, and using the fact that $9x_1 = 11y_1$ (condition (1)), he deduces that the weight of a silver coin would be $y_1 = 9 \times 3/11 = 2 + 5/11$. Similarly, starting with $x_2 = 2$, he finds that $11y_2 = 2 \times 9 = 18$, from which $y_2 = 18/11$ or $1 + 7/11$. We note that in each case the writer follows the Chinese notation which consists of writing the whole number of units, followed by the fractional part, always less than a unit, relative to the unit of measure being considered as the principal unit (that is the *jin* in this example). The solver now uses the values x_1, x_2, y_1, y_2 to determine the errors e_1 and e_2 that arise from using condition (2) of the problem.

With $x_1 = 3$ and $y_1 = 2 + 5/11$, he calculates

$$(10y_1 + x_1) - (8x_1 + y_1) = 12/11,$$

and since the solution should be 13 *liang* = 13/16 *jin*, we have a 'too much' error e_1, equal to

$$\frac{12}{11} - \frac{13}{16} = \frac{49}{16 \times 11} \ jin.$$

Similarly, using $x_2 = 2$ and $y_2 = 1 + 7/11$, he obtains a 'not enough' error $e_2 = 15/(16 \times 11)$ *jin*.

Using these values and applying the rule gives the exact value for the weight of a gold coin as

$$[3 \times 15/(16 \times 11) + 2 \times 49/(16 \times 11)]/[15/(16 \times 11) + 49/(16 \times 11)].$$

But the denominators simplify, which is why the author gives the values 15 and 49 in his solution, instead of $15/(16 \times 11)$ and $49/(16 \times 11)$ respectively, for the 'not enough' and 'too much'. In other words, the writer replaces the true errors e_1, e_2 of the problem with the values ke_1, ke_2, where $k = 16 \times 11$. Thus the 'normal' rule of double false position is replaced by

$$x = \frac{x_2(ke_1) + x_1(ke_2)}{(ke_1) + (ke_2)}$$

which is equivalent (and we could say that the errors have been defined except for the coefficient of proportionality).

From which we have the simplified calculation recommended in the original text:

the weight of a gold coin = $x = (3 \times 15 + 2 \times 49)/(15 + 49) = 143/64$ *jin*;

and following the conversion of the units

$$x = 143/64 \ jin = 2 \ jin + 15/(16 \times 4) \ jin = 2 \ jin + 3/16 \ jin + 18/(16 \times 24) \ jin$$
$$= 2 \ jin \ 3 \ liang \ 18 \ zhu.$$

Finally, the writer calculates the weight of a silver coin by finding $9/11x$, using condition (1).

To summarise, we could say that the process involves both the 'method of substitution' and the 'method of double false position', the first being used three times (twice at the beginning and once at the end), and the second once. But this way of analysing the Chinese method is not entirely correct, for the substitution made at each stage only uses specific numbers and not variables. Furthermore, with the use of variables, a single calculation would have been sufficient at the first step. We do not therefore have true substitution but rather a recognition of the constraint imposed by condition (1) within a context that remains throughout purely numerical. And this is the difference between algebra and arithmetic.

Jiuzhang Suanshu: Problem 19

From Qian Baocong (ed.), *Suanjing shishu* (Ten Mathematical Manuals), Peking, 1963, p. 216.
Problem 19 of Ch. 7 of the *Jiuzhang Suanshu* (Computational Prescriptions in Nine Chapters).
From the French translation by J.- C. Martzloff.

Suppose there are two horses, one good and the other poor, who both leave Chang'an to travel to Qi. The distance from Chang'an to Qi is 3000 *li*. The good horse goes 193 *li* the first day and increases the distance he travels by 13 *li* on every succeeding day. The poor horse goes 97 *li* on the first day and every day thereafter the distance he travels reduces by half a *li*. The good horse arrives at Qi first and then retraces his journey to meet the poor horse. How many days will have elapsed until they meet, and how far will each horse have travelled?

Answer: they meet after 15 days and 135/191 day; the good horse will have travelled 4534 *li* and 46/191 *li*; the poor horse will have travelled 1465 *li* and 145/191 *li*.

Method: Suppose [that the answer is] 15 days. The deficiency will then be 337 *li* and a half. Suppose 16 days, the excess will then be 140 *li*. Cross multiply the values of the 'too much' and the 'not enough', add [the results obtained] and take [total of these two values] as the dividend. On carrying out the division you will find the number of days [...].

This problem reminds us of those traditional, and almost impenetrable, problems about trains that were so loved by elementary arithmetic text book writers. Those writers may well have been amazed to learn that such problems already existed in ancient Chinese arithmetics.

According to the various Chinese commentaries, from the 3rd to the 13th centuries AD, the solution is obtained by the following procedure in which, in addition to the method of double false position, we need to use the formula for the sum of the first n terms of an arithmetic progression. Being given that the two horses cover distances that vary each day as the successive terms of an arithmetic progression with constant terms respectively 13 and ½ and with first terms respectively 193 and 97, we can deduce that after x days the good and the poor horse will have travelled, respectively:

$$d_g(x) = [193 + (x - 1)13/2]x \; li \qquad (1)$$
$$d_p(x) = [97 - \tfrac{1}{2}(x - 1)/2]x \; li \qquad (2)$$

Hence:

$$d_g(15) = 4260 \; li; \quad d_p(15) = 1402\tfrac{1}{2} \; li$$
$$d_g(16) = 4648 \; li \quad d_p(16) = 1492 \; li$$

After 15 days, the total distance travelled by the two horses comes to $d_g(15) + d_p(15) = 5662\tfrac{1}{2} \; li$ which, compared with the total distance the horses should have travelled when they meet, $2 \times 3000 = 6000 \; li$, is a deficiency of $e_1 = 337\tfrac{1}{2} \; li$; similarly, after 16 days, we have an excess $e_2 = d_g(16) + d_p(16) - 6000 = 140 \; li$. So by applying the double false position formula,

total journey time $= (15 \times 140 + 16 \times 337\tfrac{1}{2})/(140 + 337\tfrac{1}{2}) = 15 + 135/191$ days,

as given in the original text.

In this example, the particular supposed false positions of 15 and 16 play a crucial role since, as the reader can easily verify, if we start from initial values other than 15 and 16 we shall obtain a result other than 15 + 135/191 days. In fact, the algebraic expressions giving the distance travelled by each horse after x days are not linear but quadratic. In such cases, the rule of double false position cannot be used, except for providing an approximate solution. After all, the rule of double false position can be thought of as a method of linear interpolation. However, if we consider the sum of the distances as varying *linearly* over every interval bounded by consecutive integers, then the rule can nonetheless be applied quite rigorously, provided that it is always used with two false positions that most closely enclose the correct solution.

More precisely, $d_g(x)$ and $d_p(x)$ are given by a quadratic expression only when x takes values equal to a whole number of days. Given that the mathematics of the *Jiuzhang Suanshu* is based fundamentally on proportionality, and therefore linearity, the variations of $d_g + d_p$ over time in an interval $[t, t + 1]$ where t is the elapsed time in days, t an integer, must necessarily be taken to be linear. Also, if the graph of $d = d_g + d_p$ is drawn we shall obtain an assemblage of straight line segments connecting the points $(t, d(t))$ to $(t + 1, d(t + 1))$, with $t = 0, 1, 2, 3, ...$ We therefore have a piecewise defined linear function.

Given how old the text is, recourse to such a technique may appear surprising, However, as Neugebauer has shown, such functions had already appeared in much older Babylonian texts on astronomy. Such texts as these were based on tables of previously calculated values. So we can view our Chinese horse problem as an exercise in introducing the pupil to a method of calculating certain non-tabulated intermediary values within tables by the use of linear interpolation between successive tabulated values.

There remains the question of how we could rigorously solve the problem if we did not know in advance that the supposed false positions of 15 days and 16 days enclosed the solution and so would be suitable. To do this we could use trial and error or, less simply, but more systematically, we could take the equations (1) and (2) to be valid for non integer x. The problem is then to solve

$$d_g(x) + d_p(x) = 6000$$

which, after simplification, becomes

$$25/4\, x^2 + 1135/4\, x - 6000 = 0.$$

Since the positive root of this equation is approximately 15.7, we can see that the required integers are indeed 15 and 16. We can now apply the rule of double false position quite rigorously.

3.4 India: Bhāskara and the Rule of Simple False Position

The rule of false position was usually given through the use of numerical example. The first statement of the rule given explicitly appears in the 12th by the Indian mathematician Bhāskara.

From as early as the seventh century Indian mathematicians had calculated with fractions (see [7]) and used a positional decimal notation with zero. They also introduced the first rudiments of algebra: abbreviations, operation symbols and negative numbers. Bhāskara wrote a book entirely devoted to algebra, the *Bijaganita* or 'calculus applied to unknowns', which was a compilation of the results of his predecessors, but also contained original material. Part of that book was republished in the work we consider here, the *Līlāvati*, an arithmetic and algebra text dedicated to a young woman of that name.

The first two chapters of the *Līlāvati* deal with aspects of numeration: the eight operations of arithmetic (addition, subtraction, multiplication, division, squaring, finding the square root, cubing and finding the cube root), fractions and zero. The method of simple false position is found in the third chapter, which the author calls *ista karma*, meaning 'operation with a given number'.

In contrast, the only evidence we have of the use of the rule of double false position in India comes from a 14th century Latin manuscript, by an otherwise unknown author, with the title *The book on the augmentation and diminution, called the calculus of the conjecture, after what the Sages of India had established ....* Some historians, such as Youschkevitch [22], consider the similarity of the expression 'augmentation and diminution' with the Chinese expression 'too much and not enough' argues in favour of the introduction of the method of double false position into Arab literature from the Chinese via a detour through India. However we have no knowledge of an Indian text that deals with this procedure; furthermore, Arab sources known to date remain silent on this subject, added to which Arab scholars do not usually omit to give the credit to the Indians where it is due.

Here then is an extract from the *Līlāvati*, which deals with the method of supposition. We only reproduce the statement of the rule and first example to which it is applied.

Bhāskara
From H. T. Colebrooke, *Algebra with Arithmetic and Mensuration from the Sanscrit of Brahmegupta and Bhāscara*, London: John Murray, 1817, p. 23.

Rule of supposition: one stanza.
Or any number, assumed at pleasure, is treated as specified in the particular question; being multiplied and divided, raised or diminished by fractions: then the given quantity, being multiplied by the assumed number and divided by that [which has been found,] yields the number sought. This called the process of supposition.

Example
What is that number, which multiplied by five, and having the third part of the product subtracted, and the remainder divided by ten, and one-third, a half and a quarter of the original quantity added, gives two less than seventy?

Statement: Mult. 5. Subtractive 1/3 of itself. Div. 10. Additive 1/3, 1/2, 1/4 of the quantity. Given 68.

Putting 3; this, multiplied by 5, is 15; less its third part, is 10; divided by ten, yields 1. Added to the third, half and quarter of the assumed number three, viz. 3/3, 3/2, 3/4, the sum is 17/4. By this divide the given number 68 taken into the assumed one 3; the quotient is 48. The answer is the same with any other assumed number, as one, &c.

The algebraic translation of the example will help us to see the type of problem that the method of simple false position was used for. Here, we are required to find the solution to the equation:

$$\frac{1}{10}\left[5x - \frac{1}{3}(5x)\right] + \frac{1}{3}x + \frac{1}{2}x + \frac{1}{4}x = 70 - 2.$$

It is apparent that the left hand side contains the unknown, subject to linear transformations with rational coefficients: multiplication, division, increase or decrease by fractions. The 'constant terms' only appear on the right hand side and constitute the 'given quantity'.

In contrast to the Ancient Egyptians, for example, who almost invariably used unit fractions, Bhāskara uses general fractions. Nor does he try as much as possible, to operate with integers. Here, he takes 3 as the supposed value and obtains the result 17/4, corresponding to the value of the left hand side of our equation. This becomes the divisor for the division which provides the required value, the dividend being the supposed value, here 3, multiplied by the given quantity, here 68:

$$x = (3 \times 68) : \frac{17}{4}.$$

In the general case of the equation $ax = b$, and starting from a false position x' giving a value b' for the left hand side, the rule stated by Bhāskara corresponds to the relation $x = (bx') : b'$. Of course, in the case where x' is equal to 1, then b' is equal to a and we come back to $x = b : a$.

In practice, in the other examples, Bhāskara chooses 1 for the false position. Further, in one of his examples, he goes as far as reducing all the fractions to the same denominator. The rule then is no longer useful in the general form of a false supposition.

The existence of this rule could however be interpreted as the survivor of older techniques whose traces have vanished. Speaking generally, it is very difficult to know whether India has or has not borrowed aspects of its mathematics from an earlier civilisation, even more so since the work by Bhāskara is relatively late, being from the 12th century.

3.5 Qusṭā Ibn Lūqā: A Geometric Justification

The Arab mathematicians used the method of false position a great deal. There are a number of reasons to explain their interest in the method. First of all, there was the difficulty involved with handling the different types of fractions they used (see [7]);

in fact a glance at the list of contents of the three oldest Maghreb writings known to us will show the remarkable importance attached to fractions, in that almost half of each of the works dealt with this topic. Further, we find that problems of inheritance must have been considered important. For example, in the 10th century, al-Ḥubūbī, in his book *Kitāb al-istiqṣā* (the book of donations), explains "calculation methods, in problems of wills, by problems in algebra, methods of geometry, the method of the two errors, and the method of the dinar and the dirham".

The Arabs concentrated principally on techniques of the method of double false position. Some works were entirely devoted to the topic, such as the 9th century work *Kitāb al Khāṭa' ayn* (the book of the two errors) by Abū Kāmil.

Quṣṭā ibn Lūqā, a Christian Arab living at Balbek at the end of the 9th century, appears to have been the first to have provided a geometric justification for a method of double false position, working within the Euclidean tradition. By considering a number of areas, he justifies the 'formula':

$$x = \frac{x'e'' - x''e'}{e'' - e'}.$$

The proof requires Quṣṭā ibn Lūqā to consider each of the different cases which arise, where the errors are by excess or by deficiency. The extract that follows considers the case where both errors are deficiencies.

Quṣṭā ibn Lūqā

From Suter, H., 'Die Abhandlung Quṣṭā ben Lūcās und zwei andere Anonyme über die Rechnung mit zwei Fehlern und mit der angenommenen Zahl', *Bibliotheca mathematica*, 9 (1908), 111–122.
From the French translation of the German by C. Dulac-Fahrenkrug.

In what has gone before, we have explained how, with the aid of this method, we are able to solve problems of calculation in which there are no roots. But now, we want to prove, by means of geometry, everything that has been illustrated by the application of this rule and to explain it with the aid of figures.

To this end, we first draw a straight line of indeterminate length; this represents the required number. Let *ad* be that line and let *od* be the answer for that supposed length [that is what we obtain when carrying out the operations indicated on *x*] which we draw perpendicular at *d* above *d*. Next we draw *ao*. Now, if we wish to know the required number, which is the line *ad*, we have to consider [the problem] for two different numbers; the numbers can either be too large or too small, or the one too large and the other too small, as we have already shown above.

First of all, let the numbers be smaller than the unknown and let them be represented by the lines *ab* and *ag*. At *b* and *g* we draw two verticals which will meet the line *ao* at the points *h* and *t*. Thus we have the same ratios:

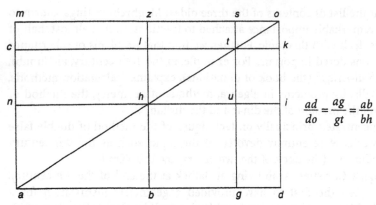

$$\frac{ad}{do} = \frac{ag}{gt} = \frac{ab}{bh}$$

and *bh* is the answer for *ab* and *gt* is the answer for *ag*. We now complete the rectangle *md* and we draw parallels *ni* and *ck* to *ad* through the points *h* and *t*. Extend *bh* and *gt* to *z* and *s*. Now, the first supposed number *ab* is known, as well as its answer *bh*, and its error with respect to the given answer [*od*] is *hz* and this is also known. The second supposed number *ag* is likewise known, as well as its answer *gt* and its error *ts*. We multiply the error of the first number by the second number; we obtain the rectangle *mu*. And if we multiply the error of the second number by the first number we obtain the rectangle *ml*. And if we subtract the small rectangle *ml* from the large rectangle we shall be left with the gnomon *nuszlc*. Since the rectangle *zt* is equal to the rectangle *ti*, the complement is equal to the complement, as Euclid proved in the first book [proposition 43]; then the rectangle *ci* is equal to the gnomon *nuszlc*. This rectangle being known, its breadth *cn* is also known, equal to the difference between the two errors. And, when we divide the rectangle *ci* by its breadth, we obtain *ni*, which is therefore also known, and being equal to *ad*, *ad* which is the required magnitude is also known.

When Qusṭā ibn Lūqā talks of "problems of calculation in which there are no roots", he is referring to what we would call linear problems, whose equations would be linear. For us, a geometrical solution of the equation $ax = b$ can be found from the intersection of the straight line $y = ax$ (Lūqā's line of indeterminate length) and the line $y = b$ (the supposed line). The second line is a given value of the problem, and we determine the first by means of using two false positions.

The intention of ibn Lūqā in this passage is to prove the validity of the 'formula of two false positions'. The proof is clear and we shall restrict ourselves to restating with algebraic notation.

If we obtain b, b', b'' (that is do, gt, bh) when carrying out the operations indicated on x, x', x'' [that is ad, ag, ab], then we shall have "the same ratios":

$$\frac{x}{b} = \frac{x'}{b'} = \frac{x''}{b''} \quad \text{(that is } \frac{ad}{do} = \frac{ag}{gt} = \frac{ab}{bh} \text{).}$$

In the extract we give here, the author chooses to examine the case where the false positions are less than the unknown, so that: $(b - b').x'' =$ area of the rectangle with diagonal *mu*, $(b - b'').x' =$ area of the rectangle with diagonal *ml*, $(b - b').x'' - (b - b'').x' =$ area of the gnomon *nuszlc* (see the figure below) – the gnomon, in the shape of the letter L, being an instrument much used in astronomy.

Now, by proposition 43 of Book I of Euclid's *Elements*, "In any parallelogram the complements of the parallelograms about the diameter are equal to one another". Therefore the areas of the hatched rectangles *zstl* and *tkiu* are equal. Hence $(b - b').x'' - (b - b'').x' =$ area of the rectangle with diagonal *ci*.

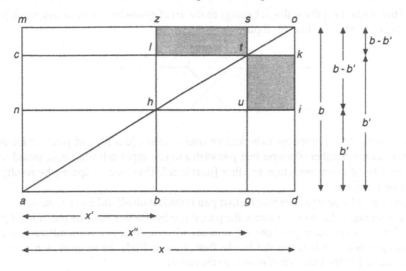

This rectangle has a breadth *cn*, being the difference between the two errors $[(b - b') - (b - b'')]$, and length *ni* or *ad* which is equal to the unknown *x*.

Finally,
$$x = \frac{\text{area of the rectangle with diagonal } ci}{\text{length of the segment } cn} = \frac{x'e'' - x''e'}{e'' - e'}.$$

3.6 Ibn al-Bannā: The Method of the Scales

This is a technique for carrying out the method of double false position. Since we consider two false values for the unknown x' and x'', being either an excess or a deficiency, the model of a balance seems quite appropriate: the pans of the scales rise up or down to indicate the direction of the error, which can therefore be shown above or below the scale pan according to the case being considered. It would seem that we owe this *method of the scales* to the Maghreb mathematician Ibn al-Bannā who lived in Marrakesh in the 13th century.

In actual fact little is known about when and how the techniques taught in the Maghreb arrived there. But we assume that they came to the Maghreb by three distinct routes: the first, and most important was Byzantine and Pre-Islamic, which would have provided for the introduction and dissemination of Byzantine calculation methods, and Ibn al-Bannā himself devoted a whole work to the subject; the second route was from Baghdad to Kairouan which was opened up from the end of the 8th century; the third route, and the last to be established, was through Mohammedan Spain.

Ibn al-Bannā

From the *Talkhīṣ* by Ibn al-Bannā, M. Souissi, University of Tunis publications, 1968, pp 70–71.
From the French translation of the Arabic by A. Djebbar.

As for [the method of] the scales, it belongs to the art of geometry. It can be described [thus]:
you draw a pair of scales like this figure:

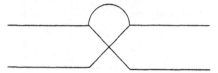

You put the [number] supposed to be known over its dome [centre] and you take for one of
the two pans any number whatever that you wish and you apply to it what is supposed [in the
problem] like addition, reduction or other [operation]. Then you compare [the result] with
what is on the dome:

If it is exactly [what you have found], that pan is the [required] unknown number.

If you are wrong, write the error above the pan, if it is by excess, below the pan if it is by deficiency. Then, for the other pan take any number whatever that you wish, different from the
first, and proceed with it as you did for the first. Then multiply the error of each pan by the
exact [number] of the other. Then examine [the result]:

If the errors are both by excess or both by deficiency, take away the smaller of the two from
the larger and the smaller of the two products from the larger, and divide the result [of the
difference] of the two products by the result [of the difference] of the two errors.

If one of the two [errors] is by excess, and the other by deficiency, divide the sum of the two
products by the sum of the two errors.

If you wish, you [can] take for the second pan the first number or another, take its part that
you compare with what is on the dome, multiply it by the exact [number] of the first [pan],
multiply the error of the first by the exact [number] of the second [pan]; then, if the error of
the first [pan] is by deficiency, add the two products, if it is by excess, take their difference.
The result you divide by the part of the second pan and the [required] result comes out.

The figure given in the 'method of the scales' is a good mnemonic for carrying
out the prescribed operations: the false positions are placed on the two pans, the
'number supposed to be known' in the middle ('on the dome'), the errors above if
they are by excess, and below if they are by deficiency.

errors by excess   errors by deficiency e' by excess,
                                                            e'' by deficiency

$$\frac{x'e''-x''e'}{e''-e'} \qquad \frac{x'e''-x''e'}{e''-e'} \qquad \frac{x'e''+x''e'}{e''+e'}$$

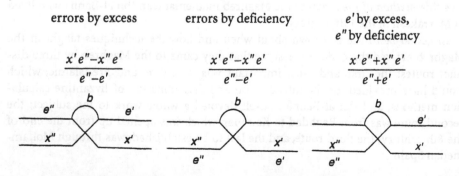

The way it is presented here, can however only be applied to a specific type of problem: in its algebraic formulation, the unknown can only appear on the left hand side of the equation in linear form, and the supposed number on the other side. This condition is clearly apparent in the following part where Ibn al-Bannā proposes a particular form of the method of simple false position. To solve the equation $ax = b$, in the case of a false position x' by deficiency for example, the process comes down to the following formula:

$$x = \frac{x'(ax') + (b - ax')x'}{ax'}.$$

This formula only looks complicated, and like the other formulas of false position it can be used to find x without having to calculate a.

In the Middle Ages, some authors adopted the diagrammatic form but without the dome. In that case only the numbers needed for the calculations need to be recorded. The corresponding method of double false position can then be used where the unknown appears on 'the right hand side'.

3.7 Fibonacci: the *Elchatayn* rule

Leonardo of Pisa, better known by his other name of Fibonacci, is the first known original mathematician of the Western world. His father held the post of representative of the Republic of Pisa at Bougie on the North African coast where Leonardo joined him in 1192 to be initiated into the theory and practice of business and particularly calculating methods. Leonardo therefore received a solid foundation to his mathematics, including learning the Indian calculating methods, that is our present positional arithmetic. After his studies he travelled extensively about the Mediterranean, visiting Egypt, Syria, Greece, Sicily and Provence. On his return to Pisa, in 1202, he composed his *Liber Abaci* or 'Book of the Abacus' [12].

This is a book of fifteen chapters which deals with arithmetic and algebra and which contains a mixture of 'Indian' arithmetic methods, rules for commercial practice, and 'Arab' algebraic methods. Right from the beginning, there is an emphasis on the nine Indian symbols for numbers and the *zephirum*, from which we derive the word zero. In chapters 7 and 8, which deal with fractions, Leonardo uses common fractions, sexagesimal fractions and the Egyptian unit fractions. He even provides a table of the decomposition of fractions into their unit fraction parts; for example, he writes 97 parts of 100 in the form:

$$\frac{1}{50} \frac{1}{5} \frac{1}{4} \frac{1}{2},$$

these particular fractions often turning up in commercial calculations. All the same, Fibonacci knew nothing of decimal fractions, and he decomposed fractions into their whole part and their fractional part: thus, he wrote $\frac{1}{5}7$ for $7 + \frac{1}{5}$, that is the fraction $\frac{36}{5}$.

Following chapters that were more 'commercial' than arithmetical, he deals with problems of false position in chapters 12 and 13. We reproduce here the beginning of chapter 13, where Fibonacci gives the rules for *elchatayn* (the two errors), that is double false position, by means of which "almost all the questions of the abacus are resolved", and a particular example whereby "one can subtly understand the solutions to other questions".

Fibonacci's idea was to approximate to the exact value by considering the differences between the errors and the false positions, to which he applied 'the rule of the fourth proportional'.

Fibonacci

From: *Il liber abaci di Leonardo Pisano* by B. Boncompagni, Tipografia delle scienze matematiche e fisiche, Rome, 1857, vol. I, p. 318.

From the French translation of the Italian by J. Delattre and A. Michel-Pajus

Ch. 13. The rules of elchatayn and the manner of using them to solve almost all the problems of the abacus.

Elchatayn, in Arabic, translates into Latin as the 'rule of two false positions', (*duarum falsarum posicionum regula*) by means of which we can find a solution of almost all questions. Among these is one which we learnt to solve in the third part of chapter twelve: the rules of trees and similar things. For these questions, it is not opportune to use the *elchatayn* in its totality, that is with two positions, since they can be solved with a single position. And yet, we wish to show how these questions, and a great many besides, may be solved with *elchatayn*. Suppose two false positions are chosen at random. Wherever these values are found, either both too small for the truth, or both too great, or the one too great and the other too small, the truth of the solutions can be found from the proportion of the difference of one position to the other, that is from what is found from using the rule of the fourth proportional, when three numbers are known, by which the fourth unknown is discovered, that is the truth of the solution. The first of these numbers is the difference of number from one false position to the other. The second is the approximate to the truth that comes from using this very difference. The third is what still remains for approximating to the truth. How this is done, is what we wish to show first of all in the rule of the cantharus so that, after having shown these three differences for the cantharus, in all its details you will be able to understand other questions for the *elchatayn*.

livres	sous
5	1
3	$\frac{3}{5}$

For example, suppose that a cantharus, that is 100 rotuli, is worth 13 livres and we want to know how much 1 rotulus is worth. Let us suppose, randomly, that 1 rotulus is worth 1 sou. Then 100 rotuli, that is 1 cantharus would, in these circumstances, be worth 100 sous, that is 5 livres. But, seeing that the price of the cantharus is 13 livres, this first position is false and 8 livres distant from the truth, that is the difference that there is between 5 livres and 13 livres. From here we now put for the price of this rotulus 2 sous, that is 1 sou more than for the first position, and in these circumstances, the cantharus becomes worth 200 sous, that is 10 livres. In the same way this value is false and is distant from the truth by 3 livres, that is the difference there is between 10 livres and 13 livres.

Now, in the first position we were 8 livres from the truth, in the second 3 livres. Thus, thanks to the difference between the first and second positions, that is 1 sou, we have become closer to the truth by 5 livres, that is the difference between 8 livres and 3 livres. And there remain 3 livres further to go. This is why we say: for 12 deniers that I have added to the price of a rotulus, I have approached the price of the cantharus by 5 livres. What must I then add to the price of this rotulus to make up the 3 livres remaining to obtain the price of the cantharus from the second position? Thus, multiply the extreme numbers and divide by the middle one, just as we showed with the rules of the trees and similar things, in other words say multiply 12 by 3, and divide by 5 which is the middle. From this we get $\frac{1}{5}7$ deniers. After adding them to the 2 sous which was put in the second position, you will have for the price of a rotulus 2 sous and $\frac{1}{5}7$ deniers.

This is an explanation of this example of the cantharus, that is 100 rotuli, which is worth 13 livres. Not being experienced in using of decimal fractions, Fibonacci does not obtain the price of a rotulus directly by dividing the price of the cantharus by 100. Also, he uses a rule of double false position in his proof, even though a simple false position would do. He uses false positions in sub-units of a livre, that is in sous. We need to know that 1 livre = 20 sous, and 1 sou = 12 deniers, just like the old British pounds, shillings and pence.

If a rotulus is worth $x' = 1$ sou, then a cantharus will be worth 5 livres, giving a "distance from the truth" of $e' = 13 - 5 = 8$ livres.

If a rotulus is worth $x'' = 2$ sous, then a cantharus will be worth 10 livres, giving a "distance from the truth" of $e'' = 13 - 10 = 3$ livres.

The change in the "distances from the truth" is proportional to the change in the false positions, and Fibonacci uses the rule of the fourth proportional: by making a change that gives a difference of 1 sou, or 12 deniers, between the false positions x' and x'', we get closer to the true value by 5 livres, the difference between e' and e''. There is still 3 livres further to go, equal to the second error e''; to obtain the true value x, we must then add to x'' the value of the fourth proportional corresponding to the three numbers 5, 12 and 3. A diagram for this rule of the fourth proportional would look like this:

5 livres	3 livres
$e' - e''$	e''
$x' - x'$	$x - x''$
12 deniers	

The true value then becomes:

$$2 \text{ sous} + \left(\frac{3 \times 12}{5}\right) \text{ deniers} = 2 \text{ sous} + \left(7 + \frac{1}{5}\right) \text{ deniers}.$$

To sum up, these operations correspond to the algebraic formula:

$$x = x'' + e'' \frac{x'' - x'}{e' - e''}.$$

We may also note that in following through the chapter, Fibonacci provides the rule corresponding to the 'standard' formula $x = \dfrac{x''e'-x'e''}{e'-e''}$, with the general method as shown in the following diagram:

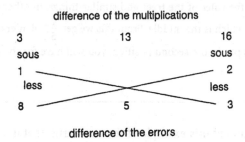

difference of the multiplications

3	13	16
sous		sous
1		2
less		less
8	5	3

difference of the errors

3.8 Pellos:
The Rule of Three and The Method of Simple False Position

The method of simple false position and the rule of three are intrinsically linked through proportions. The rule of three (see [21]) has an interesting history. The Egyptians commonly used a unit for their calculations such as the *pesu* representing, for example, the amount of bread that could be baked from a given quantity of flour. But the use of a unit of this type was not used in the Chinese *Jiuzhang Suanshu* (Computational Prescriptions in Nine Chapters) where the author uses the rule of three for calculations of problems of exchange of cereals. The Indians also invariably used the rule of three, in its simple, inverse or compound form, for work with fractions. Naturally, from this, we find the rule of three in many Arab texts and in most of the arithmetic treatises of the European Middle Ages.

Here we present an extract from the first published Occitan arithmetic (the Occitan, with its own language, is a region that now occupies the south-western region of France): *le Compendion de l'abaco* (Compendium of the abacus). Composed about 1460 by a gentleman of Nice, Francès Pellos, and published in 1492, it was most probably drawn from an earlier Occitan text, the manuscript of Pamiers, of about 1430 [17]. Just as the Indians had done before him, Pellos used the rule of three for fractional numbers. Following this, he presented the method of simple false position and he used it for carrying out the rule of three.

Pellos
From *Compendion de l'abaco* (1492), after the edition by R. Laffont, Montpellier: Editions de la Revue des Langues Romanes, 1967, pp. 104, 181.
From the French translation of the Occitan by M. Guillemot.

Similarly, if three and a half are worth 6 and a half, how much is 4 and a third worth? Multiply 4 and a third by 6 and a half, it comes to 28 and a sixth, which you divide by 3 and a half, and you find 8 and a twenty-first, and this you do in all similar 'examples'.

[...]

Similarly, find a number such that if it is divided by 7 it comes to 3 and a half. Do thus: choose as you wish 14, which you divide by 7, it comes to 2. Then, say this, if 2 comes from 14, what does 3 and a half come from? Use, as you know, the rule of three and you will find 24 and a half. And this is the required number.

The first part, concerning the rule of three, corresponds to the following operations:

$$\frac{\left(4+\frac{1}{3}\right)\left(6+\frac{1}{2}\right)}{3+\frac{1}{2}} = \frac{28+\frac{1}{6}}{3+\frac{1}{2}} = 8+\frac{1}{21}.$$

The second part corresponds to the solution of the equation: $\frac{x}{7}=3+\frac{1}{2}$. Pellos suggests we use the false position $x' = 14$. From which: $b'=\frac{x'}{7}=\frac{14}{7}=2$ instead of $b=3+\frac{1}{2}$, and by the rule of three we obtain:

$$\frac{14.\left(3+\frac{1}{2}\right)}{2} = \frac{49}{2} = 24+\frac{1}{2}.$$

As well as the fact that they are used for linear problems, the rule of three and the method of simple false position have a number of common features. The rule of three is used in the situation: if p times y comes to q, what will r times y come to? The answer is r times y is equal to rq/p. Simple false position is used for the question: if a times x equals b (where a is not explicitly given), what is the value of x? The method is to say: if a times x' equals b' then x is equal to bx'/x. In the last part, in each case we operate on three numbers, in one case on p, q, r and in the other on b, x', b'. Also, in both cases we can work entirely with integers: for the false position, x' is chosen so that bx' and b' will be integers (here, Pellos used 14, being a multiple of both 7 and 2); for the rule of three, it is less systematic, but automatic when r is a multiple of p. Finally, and most importantly, the two methods avoid the need to calculate certain intermediate terms: the value of the coefficient a in the equation $ax = b$, as we have already said, and the value of y in the rule of three.

3.9 Clavius: Solving a System of Equations

The *Epitome Arithmeticae Practicae* of 1583 by Christophore Clavius was not the first work in which a method of false position was used to deal with problems that, algebraically, are equivalent to solving a system of linear equations. Such examples can be found in Chinese mathematics and certainly among the works of Arab mathematicians. With Clavius, however, we find 'actual' systems with several unknowns.

Clavius's work deals with operations on integers, fractions, the simple rule of three and the compound rule of three. The rules of simple and double false position occupy chapters 12 and 13. We give below example 6 from chapter 13, which contains 24 examples. To make it easier to follow the text, we have given an algebraic rendering of Clavius's calculations in the margin.

In the case of a system of three equations with three unknowns, which we shall call x, y and z, and starting from a false position x_1, Clavius starts by solving the 'system formed by the two first equations', using the method of double false position. He therefore finds the false positions y_1, z_1 of the other two unknowns. He begins again from another false position x_2 and finds false positions y_2 and z_2. The 'third equation' is used to determine the errors e_1 and e_2. All that is now needed, as we shall see, is to apply the method of double false position to each of the unknowns with the errors now obtained in order to complete the solution of the problem.

Clavius

Epitome arithmeticae practicae (Practical Arithmetic), Dominici Basae, Rome, 1583, pp. 231–235.
From the French translation of the Latin by M. Speisser.

The corresponding algebraic calculations are given to the right of the text.

Find three numbers such that the first added to 73 makes the double of the other two; the second with 73 makes triple the two others, the third, finally, with 73, makes quadruple the two others.

$$x + 73 = 2(y + z)$$
$$y + 73 = 3(z + x)$$
$$z + 73 = 4(x + y)$$

2.   **5.**
35.   M   M   **32.**
33.   **21.**

1 2.
diuiſor.

Let the first number be 1, or any other odd number, such that added to 73, it makes an even number, which can have a half, since the first together with 73 must make a number double the two others.

$x_1 = 1$

Therefore, since 1 with 73 makes 74 which must, in agreement with the proposed example, be double the two others, it must be that the two others together come to 37.

$37 = y_1 + z_1$

And since the second with 73 has to make a number triple the first (which is 1) and the third, the number 37, following the preceding question, will be divided into two parts of which the first with 73 makes triple the number made up of the second part and 1. Thus, before solving the proposed question it is first necessary to solve another which appears in this operation.

$y_1 + 73 = 3(1 + z_1)$

Thus let the first part be 2, and so the second 35. The first with 73 makes the number 75, the second with 1 makes 36, whose triple is not 75 but 108. We have therefore lacked 33 units with respect to the truth, since our number 75 is less than 108 by as many units. Rather then let the first part be 5, and so the second is 32. But the first with 73 makes 78, and the second with 1 makes 33, whose triple is not the number 78, but 99. So, we have again a lack of 21 units [from the truth].

$y_1' = 2$
$z_1' = 35$
$y_1' + 73 = 75$
$3(1 + z_1') = 108$
$e_1' = 108 - 75 = 33$
$y_1'' = 5$
$z_1'' = 32$
$e_1'' = 21$

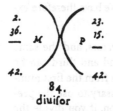

2.
36.
—
  M
23.
15.
—
  P
42.
42.
84.
diuiſor

Using the precepts of the rule of false, we find that the first part is 10¼ and the second is 26¾. This is why, if the first number of the proposed question is 1, the second will be 10¼ and the third 26¾.

$$y_1 = \frac{y_1'' e_1' - y_1' e_1''}{e_1' - e_1''}$$
$$= 10\tfrac{1}{4}$$
$$z_1 = 26\tfrac{3}{4}$$

Now, in fact, the first number with 73 makes the double of the other two and the second with 73 makes triple the two others. So if the third with 73 makes the quadruple of the two others, the question will be satisfied. But the third with 73 makes the number 99¾ which is not the quadruple of the number 11¼, the union of the first with the second, but 45 is the quadruple of 11¼. We have therefore exceeded the truth by 54¾.

$z_1 + 73 = 99\tfrac{3}{4}$
$4(x_1 + y_1) = 45$
$e_1 = 54\tfrac{3}{4}$

Now, let the first number be 3, which with 73 makes 76, which must be a number double the two others. The two remaining numbers therefore make 38. And because the second with 73 has to make triple the first (which is 3) and the third, the number 38 will be divided, following the preceding question, into two parts, of which the first with 73 makes the triple of the number which is made up of the second part and 3.

$x_2 = 3$
$y_2 + z_2 = 38$
$y_2 + 73 = 3(3 + z_2)$

Now let the first part be 2 and so the second is 36. But the first with 73 makes 75 and the second with 3 makes 39 whose triple is not the number 75 but the number 117. We have therefore lacked 42 from the truth.

$y_2' = 2$
$z_2' = 36$
$e_2' = 42$

Rather then let the first part be 23 and so the second is 15. The first with 73 makes 96, and the second with 3 makes 18 whose triple is not 96 but the number 54. We have therefore exceeded the exact number by 42.

$y_2'' = 23$
$z_2'' = 15$
$e_2'' = 42$

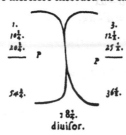

1.
10¼.
10¾.
—
  P
3.
12¼.
25¾.
—
  P
54¾.
36¾.
18¾.
diuiſor.

Carrying out the rule of false, you will find that the first part is 12½ and so the last part is 25½. So if the first number of the proposed question is 3, then the second will be 12½ and the third 25½.

$x_2 = 3$
$y_2 = 12½$
$z_2 = 25½$

Now, in fact, the first number with 73 makes the double of the two others and the second with 73 makes triple the two others. So if the third with 73 makes the quadruple of the two others, the question will be satisfied. But the third with 73 makes the number 98½ which is not the quadruple of the number 15½, which is 62. We have therefore exceeded the true number by 36½.

$e_2 = 36½$

If you cross multiply the first numbers by the errors, and the same with the second and third [numbers] (the second and third can be found more easily this way than by deriving them from the first number that has been found; because it will be necessary to use the preceding question) and having done the subtraction, if you divide the answers by the required divisor 18¼,, that is the difference of the errors, since in each situation we have an excess, you will find that the first number is 7, the second 17 and the third is 23.

$$x = \frac{x_2 e_1 - x_1 e_2}{e_1 - e_2} = 7$$

$$y = \frac{y_2 e_1 - y_1 e_2}{e_1 - e_2} = 17$$

$$z = \frac{z_2 e_1 - z_1 e_2}{e_1 - e_2} = 23$$

From this, the first with 73 makes 80, which is the double of the two others, and the second with 73 makes 90 which is the triple of the two others. Finally, the third with 73 makes 96 which is the quadruple of the two others.

The remark made by Clavius that "the second and third can be found more easily this way than by deriving them from the first number that has been found" has important implications in practice. In fact, we have only chosen two false positions, x_1 and x_2 with the intention of deriving the value of the unknown x from them. But, in the course of the calculations, false positions y_1 and y_2 as well as false positions z_1 and z_2 have also been found. The method of false positions can now be applied to these to obtain the values of the unknowns y and z. The validity of this remark can be shown as follows.

Consider the system of equations:

$$\begin{cases} ax + by + cz = d \\ a'x + b'y + c'z = d' \\ a''x + b''y + c''z = d'' \end{cases} \text{ where both } A = \begin{bmatrix} a & b & c \\ a' & b' & c' \\ a'' & b'' & c'' \end{bmatrix} \text{ and the sub-matrix } \begin{bmatrix} b & c \\ b' & c' \end{bmatrix} \text{ are non-singular.}$$

By taking false positions x_i ($i = 1, 2$), we have to consider the false positions y_i and z_i which are solutions of the system of equations:

$$\begin{cases} by + cz = d - ax_i \\ b'y + c'z = d' - ax_i \end{cases}$$

and these lead to errors

$$e_i = a''x_i + b''y_i + c''z_i - d''.$$

In other words, (x_i, y_i, z_i) is the solution of the system:

$$\begin{cases} ax + by + cz = d \\ a'x + b'y + c'z = d' \\ a''x + b''y + c''z = d'' + e_i \end{cases}$$

Therefore, the variables x, y and z behave symmetrically. If we consider, for example, x we have:

$$x_i = \frac{\begin{vmatrix} d & b & c \\ d' & b' & c' \\ d''+e_i & b'' & c'' \end{vmatrix}}{detA} = \frac{\begin{vmatrix} d & b & c \\ d' & b' & c' \\ d'' & b'' & c'' \end{vmatrix}}{detA} + e_i \frac{\begin{vmatrix} b & c \\ b' & c' \end{vmatrix}}{detA} = x + e_i \frac{\begin{vmatrix} b & c \\ b' & c' \end{vmatrix}}{detA}$$

From which:
$$x = \frac{x_2 e_1 - x_1 e_2}{e_1 - e_2}.$$

If we do the same for y and for z we shall obtain the calculations that Clavius carried out. This algorithm can, of course be extended to the solution of systems of linear equations with more than 3 unknowns (see [19]).

Bibliography

[1] Al-Bannā, *Talkhīṣ fi aᶜmāl al-Ḥisāb* (Summary of calculation operations); text edited and translated by M. Souissi, Publications de l'Université de Tunis, Tunis, 1969.

[2] Barrème, *L'arithmétique*, 2nd ed., Avignon, 1810.

[3] Benoit, P., Chemla, K. & Ritter, J., *Histoire de fractions, fractions d'histoire*, Birkhäuser, Basle, 1992.

[4] Bhāskara in H.T. Colebrooke, *Algebra with arithmetic and mensuration from the sanscrit of Brahmegupta and Bhascara*, John Murray, London, 1817.

[5] Bruins, E. M. & Rutten, M., *Textes mathématiques de Suse*, Paul Geuthner, Paris, 1961.

[6] Chemla, K., Djebbar, A. & Mazars, G., Monde arabe, chinois, indien : quelques points communs dans le traitement des nombres fractionnaires, in [3], pp. 263-276.

[7] Chuquet, N., in Marre, A., 'Le triparty en la science des nombres par Maistre Nicolas Chuquet parisien', *Bullettino di bibliografiae di storia delle scienze matematiche e fisiche* t. 13 (1880), 555-659 and 693-814. An extensive translation of Chuquet's manuscript can also be found in Flegg, G., Hay, C. & Moss, B, *Nicholas Chuquet: Renaissance Mathematician*, Dordrecht: Reidel, 1985.

[8] Clavius, C., *Epitome arithmeticae practicae*, Dominici Basae, Rome, 1583.

[9] Combette, E., *Cours Moyen et Supérieur.Arithmétique, système métrique et géométrie usuelle*, 10th edn., Paris: Alcide Picard et Kaan, n.d.

[10] Demare-Lafont, S., 'Les partages successoraux paléobabyloniens', in [3], pp. 103-104.

[11] Fibonacci, in Boncompagni, B., *Il liber abaci di Leonardo Pisano*, Tipographia delle scienze matematiche e fisiche, Rome, 1857.

[12] *Jiuzhang Suanshu* (Computational Prescriptions in Nine Chapters), in Qian Baocong, *Suanjing shishu* (Ten Mathematical Manuals), Peking, 1964.

[13] Michel, C., 'Les fractions dans les tablettes économiques du début du second millénaire en Assyrie et en Babylonie', in [3], p. 101.

[14] Rhind Papyrus in Chace, A., *The Rhind Mathematical Papyrus*, 2 vols.., Oberlin, Ohio: Mathematical association of America, 1927-1929.

[15] Pellos, F., *Compendion de l'abaco* (Compendium of the abacus), 1492. Text edited par R. Laffont, Editions de la Revue des Langues Romanes, Montpellier, 1967.

[16] Qusṭā Ibn Lūqā, in Suter, H., 'Die Abhandlung Qustā ben Lūqās und zwei andere Anonyme über die Rechnung mit zwei Fehlern und mit der angenommenen Zahl', *Bibliotheca mathematica*, t. 9 (1908), 111–122.

[17] Sesiano, J., 'Une arithmétique médiévale en langue provençale', *Centaurus*, t. 27 (1984), 26–75.

[18] Smeur, A., 'The Rule of false applied to the quadratic equation, in three sixteenth century arithmetics', *Archives internationales d'histoire des sciences*, t. 102 (1978), 66–101.

[19] Spiesser, M., *Equations du premier degré, Méthode de fausse position*, IREM de Toulouse, 1982.

[20] Thureau-Dangin, F., *Textes mathématiques babyloniens*, Leyden: Brill, 1938.

[21] Tropfke, J., *Geschichte der Elementarmathematik*, Berlin: Walter de Gruyter, 1980.

[22] Youschevitch, A., *Les mathématiques arabes (VIIIè-XVè siècles)*, tr. M. Cazenave & K. Jaouiche, Vrin, Paris, 1976.

4 Euclid's Algorithm

By 1950, the word algorithm was most frequently
associated with 'Euclid's algorithm'.

Donald Knuth
The Art of Computer Programming, (v. I, p. 2)

We have already remarked in the introduction to this book, that Euclid's algorithm
often represents for the mathematician the prototype of algorithmic procedure, and
that it has relevance right up to today. It can be of use, not only in the search for the
greatest common divisor, as described by Euclid himself (Section 4.1), but also, by
adapting the procedure, in the solution of indeterminate equations, leading to
Bézout's identity (Section 4.3). It allowed al-Khayyām to compare two ratios, or to
show that they were equal (Section 4.2); this appears even more clearly in the writing
of continued fractions which were systematically studied by Euler (Section 4.4). Finally, what may appear surprising, the algorithm can be used in Sturm's method for
determining the number of real roots of an algebraic equation (Section 4.5).

4.1 Euclid's Algorithm

Before Euclid, we do not find any explicit reference to the method that we know as
'Euclid's algorithm', but hints of it appear in earlier work. For example, Aristarchus
of Samos, in his *Treatise on the sizes and distances of the sun and the moon* [2], replaces the ratio 71 755 875 : 61 735 500 with the ratio 43 : 37, and he uses the ratio
88 : 45 in place of 7921 : 4050. At a later date, Archimedes in *The Measurement of a
Circle* [1] replaced the ratio 6336 : 2017 + 1/4 by the number 3 + 10/71 (= 223 : 71) in
his work on finding the circumference of a circle as a ratio of the diameter (see Section 5.1 below). These three approximations are all derived naturally by using
Euclid's algorithm which, to avoid anachronism and keeping closer to the text of the
original, is sometimes called *anthyphairesis* or alterna   )traction. An interesting
presentation of the method of anthyphairesis can be   . in *The Mathematics of
Plato's Academy* by D. H. Fowler [18].

The two propositions given in the extract below for   : basis of the procedure.
They give a method for determining whether two numbers are prime to one another.
(Proposition 1) and if not, for determining their greatest common divisor (Proposition 2). After a number of definitions concerning whole numbers, these two propositions begin Book VII of *The Elements* [14]. The three Books VII, VIII and IX together form 'The Arithmetic Books of Euclid' [21] and established the basis of the
Theory of Numbers.

In the text, Euclid says a number may 'measure' another number, which means that it will divide into it (exactly). Putting it this way simply corresponds to the Greek idea of numbers corresponding to lengths, the first length being a unit for measuring the second. We can also note here, that in Euclid numbers are not referred to by number symbols, but by letters and line segments. This has the advantage that we can talk about numbers of any size, and also that the arguments can be transposed to magnitudes other than lengths.

Euclid
Elements, Book VII (3rd century BC)
T. L. Heath, *The Thirteen Books of Euclid's Elements,* vol. 2, pp. 296–299.

Proposition 1
Two unequal numbers being set out, and the less being continually subtracted in turn from the greater, if the number which is left never measures the one before it until an unit is left, the original numbers will be prime to one another.

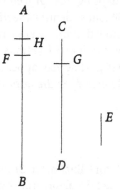

For, the less of two unequal numbers *AB, CD* being continually *subtracted* from the greater, let the number which is left never measure the one before it until an unit is left;
I say that *AB, CD* are prime to one another, that is that an unit alone measures *AB, CD.*

For, if *AB, CD* are not prime to one another, some number will measure them.

Let a number measure them, and let it be *E;* let *CD,* measuring *BF,* leave *FA* less than itself,
let *AF,* measuring *DG,* leave *GC* less than itself,
and let *GC,* measuring *FH,* leave an unit *HA.*

Since, then, *E* measures *CD,* and *CD* measures *BF,* therefore *E* also measures *BF.*

But it also measures the whole *BA;* therefore it will also measure the remainder *AF.*

But *AF* measures *DG;* therefore *E* also measures *DG.*

But it also measures the whole *DC;* therefore it will also measure the remainder *CG.*

But *CG* measures *FH;* therefore *E* also measures *FH.*

But it also measures the whole *FA;* therefore it will also measure the remainder, the unit *AH,* though it is a number: which is impossible.

Therefore no number will measure the numbers *AB, CD;* therefore *AB, CD* are prime to one another.

Q.E.D.

Proposition 2
Given two numbers not prime to one another, to find their greatest common measure.

Let *AB, CD* be the two given numbers not prime to one another.
Thus it is required to find the greatest common measure of *AB, CD.*
If now *CD* measures *AB* – and it also measures itself – *CD* is a common measure of *CD, AB.*
And it is manifest that it is also the greatest; for no greater number than *CD* will measure *CD.*
But, if *CD* does not measure *AB,* then, the less of the numbers *AB, CD,* being continually subtracted from the greater, some number will be left which will measure the one before it.

For an unit will not be left; otherwise *AB*, *CD* will be prime to one another, which is contrary to the hypothesis.

Therefore some number will be left which will measure the one before it.

Now let *CD*, measuring *BE*, leave *EA* less than itself, let *EA*, measuring *DF*, leave *FC* less than itself, and let *CF* measure *AE*.

Since then, *CF* measures *AE*, and *AE* measures *DF*, therefore *CF* will also measure *DF*.

But it also measures itself; therefore it will also measure the whole *CD*.

But *CD* measures *BE*; therefore *CF* also measures *BE*.

But it also measures *EA*; therefore it will also measure the whole *BA*.

But it also measures *CD*; therefore *CF* measures *AB*, *CD*.

Therefore *CF* is a common measure of *AB*, *CD*.

I say next that it is also the greatest.

For, if *CF* is not the greatest common measure of *AB*, *CD*, some number which is greater than *CF* will measure the numbers *AB*, *CD*.

Let such a number measure them, and let it be *G*.

Now, since *G* measures *CD*, while *CD* measures *BE*, *G* also measures *BE*.

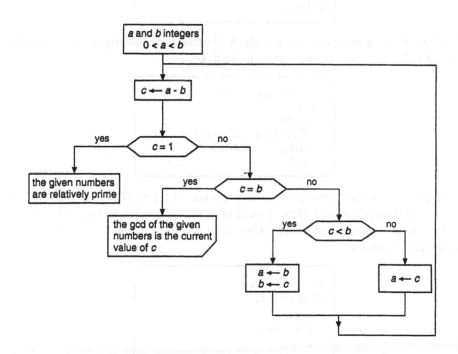

Fig. 4.1

But it also measures the whole *BA*; therefore it will also measure the remainder *AE*.

But *AE* measures *DF*; therefore *G* will also measure *DF*.

But it also measures the whole *DC*; therefore it will also measure the remainder *CF*, that is, the greater will measure the less: which is impossible.

Therefore no number which is greater than *CF* will measure the numbers *AB*, *CD*; therefore *CF* is the greatest common measure of *AB*, CD.

Porism. From this it is manifest that, if a number measure two numbers, it will also measure their greatest common measure.

Q.E.D.

We can set out the procedure involved in Euclid's algorithm by means of a flow chart as in Fig. 4.1.

To express the same process as a computer program we can write:
(*a* and *b* are assumed both > 0)

```
u := a; v := b;
while u ≠ v do
  begin
  if u > v then u := u – v
  else v := v – u
  end
gcd := u
```

To show this as a recursive process which illustrates the underlying reasoning behind the algorithm, we could program it like this:

```
if a = b then gcd := a
  else
    begin
    if a > b then gcd := gcd (a – b, b)
    else gcd := gcd (a, b – a)
    end
```

If we extend our arithmetic beyond the processes of repeated subtraction, then we can use 'Euclidean division'. Here, instead of considering the difference, *c*, between *a* and *b*, we consider the remainder *r* after division of *a* by *b*, and this accelerates the convergence process.

```
u := a; v := b;
while v > 0 do
  begin
  r := u mod v;
  u := v;
  v := r
  end
gcd := u
```

Returning to Aristarchus of Samos and his supposed use of this algorithm, this is the reconstruction proposed by Itard ([21], p. 27) for the first of the two ratios: Let $A = 71\ 755\ 875$ and $B = 61\ 735\ 500$:

$$A - B = C = 10\ 020\ 375$$
$$B - 6C = D = 1\ 613\ 250$$
$$C - 6D = 340\ 875$$

Neglecting this last remainder, we have:

$$C = 6D, B = 6C + D = 37D, A = B + C = 43D, \text{ so } A\ /\ B \approx 43/37.$$

In the 17th century these processes became established with the 'continued fractions' algorithm (see Section 4.4).

If in the previous example, the process is continued to its end, from $D = 1\ 613\ 250$ and $E = 340\ 875$, we get:

$$D - 4E = F = 249\ 750, \ E - F = G = 91\ 125, \ F - 2G = H = 67\ 500,$$
$$G - H = I = 23\ 625, \ H - 2I = J = 20\ 250, \ I - J = K = 3375,$$
$$J - 6K = 0,$$

so the GCD of $A = 71\ 755\ 875$ and $B = 61\ 735\ 500$ is therefore $K = 3375$.

Euclidean ring

Nowadays we use the term *Euclidean ring* [34] for a ring A which possesses a 'Euclidean algorithm' that is, an application g of A in $\mathbb{N}$ such that: given two elements a and b of A, b non-zero, there exist two elements q and r of A satisfying:

$$a = bq + r \text{ and } g(r) < g(b).$$

For example: $A = \mathbb{Z}$ with $g(a) = |a|$ (the least remainder as an absolute value); or $A = \mathbb{R}[X]$ with $g(P(X)) = 1 + \deg (P(X))$ and $g(0) = 0$.

The possibility of an *a priori* upper bound to the number of divisions, in terms of the given numbers ($0 < b < a$), needed to carry out Euclid's algorithm, does not appear to have been studied until the first half of the 19th century. This provides a measure of the 'complexity' of the algorithm. Thus, Reynaud (1811) gives an upper bound of $b/2$, Finck (1841) obtains $2[\log_2 b] + 1$, and Lamé (1844) obtains $5([\log_{10}b] + 1)$ as an upper bound, where the brackets [...] denote the integer part of the argument (Shallit [35]).

We have already seen that, for Euclid, magnitudes were represented using line segments. Here we are dealing with whole numbers and so the procedure will necessarily end with a result for the GCD after a finite number of steps. If, on the other hand, we started with the line segments themselves, as is suggested by Euclid's proof, the procedure might never terminate, and we might end up with an infinite loop in the program. What is at issue here is the possibility that the magnitudes are incommensurable, a possibility envisaged by Euclid in Book X of *The Elements* (Heath [14], vol. 3, p. 17). Proposition 2 of Book X, states:

If, when the less of two unequal magnitudes is continually subtracted in turn from the greater, that which is left never measures the one before it, the magnitudes will be incommensurable.

We shall come back to this with the text by al-Khayyām which follows.

4.2 Comparing Ratios

In Book V of *The Elements*, Euclid gives precise definitions for the equality or inequality of the ratios of magnitudes, whether commensurable or not. So, for example, Definition 5 (Heath [14], vol. 2, p. 114) states:

Magnitudes are said to be in the same ratio, the first to the second and the third to the fourth, when, if any equimultiples whatever be taken of the first and third, and any equimultiples whatever of the second and fourth, the former equimultiples alike exceed, are alike equal to, or alike fall short of, the latter equimultiples respectively taken in corresponding order.

Letting a, b, c, d stand for the four magnitudes, what is meant by the equality of the ratios a/b and c/d? We cannot, initially, compare the products ac and bd since we are considering here magnitudes and not integers. On the other hand, we do know how to compare integer multiples of magnitudes, and it is this that Euclid uses in his Definition 5. The equality of a/b and c/d is established in a way that we may translate as: whatever the integers k and h, ka is greater than (or respectively less than or equal to) hb if and only if kc is greater than (or respectively less than or equal to) hd. In other words, this definition is equivalent to comparing the two ratios under consideration a/b and c/d with all possible rational ratios h/k.

Nowadays, it is easier to understand the subtlety of this approach, following Dedekind's axiomatic treatment of the subject through his use of cuts on the number line, but these theoretical definitions remain difficult to use in practice.

There happens to be another way of describing the ratio of two magnitudes: by the sequence of successive integers obtained by the application of Euclid's algorithm, a sequence that is finite – as we have seen – in the case of integers, or more generally for commensurable magnitudes, but which is infinite for incommensurable magnitudes.

Itard ([21], p. 25) and Caveing ([11], vol. 3, p. 1261) suggest that in *Topics* VII 3 Aristotle ([3]) alludes to alternate subtractions in the context of the equality of ratios. But this is a matter of interpretation.

In the 9th century, al-Māhānī explicitly identified the equality of ratios with the fact that they had the same sequences of corresponding quotients (see [41]). Al-Khayyām went further and derived a definition for the inequality of ratios from these sequences.

Like many Arab mathematicians, al-Khayyām's interest in Euclid's *Elements* focused on the problems of parallel lines in Book I and, as we can see here, the question of ratios in Book V and of irrational magnitudes in Book X. In the extract that follows, al-Khayyām's shows how the equality of two ratios rests on the fact that the two sequences of quotients associated with them are the same and how two unequal ratios can be compared by using the first two quotients which differ and the parity of their numerical position in the sequences. He further claims to show that his definitions are not in conflict with those given by Euclid, and even that they imply them.

In actual fact, although it is easy to see that two ratios are different by the appearance of two quotients that differ, we cannot be sure that sequences of quotients are equivalent, except for finite sequences, which represent rational number, or for periodic sequences, which we now know must be those of algebraic numbers of degree 2 (this is discussed further in Section 4.4).

---

al-Khayyām

The Second Epistle on the Evocation of Proportion, of the Idea of Proportionality and of their true [meaning].
A.I.Sabra, Dār al Maʿārif, Alexandria, 1961, pp. 44–47.
From the French translation of the Arabic by A. Djebbar

[The words appearing in brackets have been added to make the text easier to read.]

[Definition of the equality of two ratios]

Being [given] four magnitudes [such that] the first is equal to the second and the third is equal to the fourth, or [such that] the first is a part of the second and the third that same part of the fourth, or [such that] the first is a number of parts of the second and the third that same number of parts of the fourth, then the ratio of the first to the second must necessarily be the same as the ratio of the third to the fourth; and this ratio is numerical.

If the [magnitudes] are not of these three forms and if, from the second all the multiples of the first [contained in the second] are subtracted until there is left a residue less than the first, and if, in the same way, when from the fourth all the multiples of the third are subtracted until there is left a residue less than the third, and then the number of multiples of the first [contained] in the second is equal to the number of multiples of the third [contained] in the fourth. And if afterwards [from the first] all the multiples of the residue of the second with respect to the first are subtracted, so that there is left a residue less than the residue of the second and if, in the same [manner], are subtracted [from the third] all the multiples of the residue of the fourth with respect to the third, until there remains a residue less than the residue of the fourth, and if then the number of multiples of the residue of the second is equal to the number of multiples of the residue of the fourth; and if, [after this,] when from the residue of the second are subtracted all the multiples of the residue of the first, and if from the residue of the fourth are subtracted all the multiples of the residue of the third, their number is the same; and if, when in the same [manner], all the multiples of the residues are subtracted, successively each from the other, in the way we have shown, the number of the residues of the first and the second is equal to the number of corresponding residues of the third and the fourth [and this] indefinitely, then the ratio of the first to the second will necessarily be as the ratio of the third to the fourth. And this it is that is the true ratio for the geometric type [of magnitudes].

[Definition of the inequality of two ratios]

As for smaller and greater true ratios, it is like when you say: if being [given] four magnitudes, the first is equal to the second and the third is less than the fourth; or the first is greater than the second and the third is not greater than the fourth; or the first is a part of the second and the third is another part of the fourth, smaller than the other part, or several parts [of the fourth] whose total is smaller than that part; or again the first is several parts of the second and the third another part of the fourth less than these parts. Then, [in all these cases,] the ratio of the first to the second is greater than the ratio of the third to the fourth.

[...]

As for the geometric case, if on subtracting from the second all the multiples of the first there remains a residue [less than the first] and from the fourth all the multiples of the third and there remains a residue, and the number of multiples of the first is less than the number of multiples of the third; or if the former number is equal to the latter number, but, when on subtracting [from the first] all the multiples of the residue of the second with respect to the first until there remains a residue and when on subtracting [from the third] all the multiples of the residue of the fourth with respect to the third until there remains a residue, and the number of multiples of the residue of the second is greater than the number of multiples of the residue of the fourth; or again, if the latter number is equal to the former but, when on subtracting all the multiples of the residue of the first with respect to the residue of the second and on subtracting all the multiples of the residue of the third with respect to the residue of the fourth, the number of multiples of the residue of the first is less [than the number of multiples of the residue of the third], or if there is no residue of the second, but there does remain a residue of the fourth or [a residue] of the residue of the fourth; then, [in all these cases,] and in a true sense, the ratio of the first to the second is necessarily greater than the ratio of the third to the fourth.

In general, for these case [of geometric magnitudes], if there is no residue of the second or of one of its [successive] residues, or if its residues are less [than the corresponding residues of the fourth], or if there remains a residue of the first and of its residues, and there remains no residue of the third and of its residues; or again if the residues of the first are greater than the residues of the third, then, [in all cases,] we must necessarily have the ratio of the first to the second is greater than the ratio of the third to the fourth.

This idea has developments, which are more extensive than these, which it is possible for you to come to know through the use of this rule. Understand that.

We shall start with a symbolic transcription of al-Khayyām's text.

Definition of the equality of two ratios

For rational ratios: let a, b, c, d be four magnitudes such that $a = b$ and $c = d$, or $a = b/n$ and $c = d/n$, or again $a = (k/n)b$ and $c = (k/n)d$, where n and k stand for integers, then, in all cases, the ratios a/b and c/d are equal. For whatever magnitudes, if they are not one of these three forms, but we have:

$$b - n_1 a = p_1 \quad \text{where} \quad p_1 < a \quad \text{and} \quad d - n_1 c = q_1 \quad \text{where} \quad q_1 < c,$$
$$a - n_2 p_1 = p_2 \quad \text{where} \quad p_2 < p_1 \quad \text{and} \quad c - n_2 q_1 = q_2 \quad \text{where} \quad q_2 < q_1,$$
$$p_1 - n_3 p_2 = p_3 \quad \text{where} \quad p_3 < p_2 \quad \text{and} \quad q_1 - n_3 q_2 = q_3 \quad \text{where} \quad q_3 < q_2,$$

and so on indefinitely, that is, if at each stage k, the factor n_k is the same in each of the inequalities, then the ratios a/b and c/d are equal.

Definition of the inequality of two ratios

For rational ratios: if $a = b$ and $c < d$, or $a > b$ and $c = d$, or $a = b/n$ and $c = (k/m)d$ with $(k/m) < 1/n$, or again $a = (k/n)b$ and $c = (k'/m)d$ with $(k'/m) < (k/n)$, then $a/b > c/d$. For any magnitudes whatever, we write: $b - n_1a = p_1$ with $p_1 < a$ and $d - m_1c = q_1$ with $q_1 < c$.

If $n_1 < m_1$, then $a/b > c/d$.

If $n_1 = m_1$, then we proceed by subtracting multiples:

$a - n_2p_1 = p_2$ with $p_2 < p_1$ and $c - m_2q_1 = q_2$ with $q_2 < q_1$.

If $n_2 > m_2$, then $a/b > c/d$.

If $n_2 = m_2$, then we proceed by subtracting multiples:

$p_1 - n_3p_2 = p_3$ with $p_3 < p_2$ and $q_1 - m_3q_2 = q_3$ with $q_3 < q_2$.

If $n_3 < m_3$ or if $n_3 = m_3$ but $p_3 = 0$ and $q_3 \neq 0$, then $a/b > c/d$.

More generally, if one of the successive residues of b, namely a term of the form p_{1+2i} is zero without the corresponding q_{1+2i} of d also being zero, or if the "number of multiples" n_{1+2i} of a residue of a contained in the residue of b happens to be less than the "number of multiples" m_{1+2i} of a corresponding residue of c contained in the residue of d, or if one of the successive residues of c, namely a term of the form q_{2i}, is zero without the corresponding residue p_{2i} of a being zero, or again if the "number of multiples" m_{2i} of a residue of b contained in the residue of a happens to be less than the "number of multiples" n_{2i} of a corresponding residue of d contained in the residue of c, then $a/b > c/d$.

Let us check that this true in comparing the ratios a/b and c/d beginning with the sequence of successive quotients $n_1, n_2, \ldots$ for a/b and $m_1, m_2, \ldots$ for c/d, in the case where the two sequences differ at the kth step, that is we have $n_1 = m_1, \ldots, n_{k-1} = m_{k-1}$ and $n_k \neq m_k$. Let us suppose, for example, that $n_k < m_k$.

Consider the first k steps in the Euclid algorithm:

$$p_{t-1} = n_{t+1}p_t + p_{t+1} \text{ with } 0 < p_{t+1} < p_t$$
$$q_{t-1} = m_{t+1}q_t + q_{t+1} \text{ with } 0 < q_{t+1} < q_t$$

where: $0 \leq t \leq k - 1, a = p_0, b = p_{-1}, c = q_0, d = q_{-1}$.

Hence: $p_{k-2}/p_{k-1} = n_k + p_k/p_{k-1} < n_k + 1 \leq m_k < m_k + q_k/q_{k-1} = q_{k-2}/q_{k-1}$.

And so: $p_{k-1}/p_{k-2} > q_{k-1}/q_{k-2}$.

Similarly: $p_{k-3}/p_{k-2} = n_{k-1} + p_{k-1}/p_{k-2} > n_{k-1} + q_{k-1}/q_{k-2} = q_{k-3}/q_{k-2}$.

And so: $p_{k-2}/p_{k-3} < q_{k-2}/q_{k-3}$.

Similarly, $p_{k-3}/p_{k-4} > q_{k-3}/q_{k-4}$.

And so on ... so that, whenever k is odd, $p_0/p_{-1} > q_0/q_{-1}$, that is $a/b > c/d$, and when k is even, $a/b < c/d$. In the case where $n_k = m_k$, but $p_k = 0$ and $q_k \neq 0$, this proof, slightly modified, leads to the same conclusion.

By way of example, consider some ratios, rational to start with. Suppose $a/b = 223/71$ and $c/d = 355/113$. The successive quotients are 0, 3, 7, 10, for the first, and 0, 3, 7, 16, for the second. In this case, $k = 4$ is even and $n_4 < m_4$, hence 223/71 (≈ 3.1408) is less than 355/113 (≈ 3.14153).

It is also possible to compare irrational ratios as, for example $a/b = \sqrt{3}/3$ and $c/d = \sqrt{2}/2$. Now we have:

$$p_1 = b - n_1 a = 3 - \sqrt{3} \qquad\qquad (n_1 = 1);$$
$$q_1 = d - m_1 c = 2 - \sqrt{2} \qquad\qquad (m_1 = 1);$$
$$p_2 = a - n_2 p_1 = \sqrt{3} - (3 - \sqrt{3}) \qquad (n_2 = 1);$$
$$q_2 = c - m_2 q_1 = \sqrt{2} - 2(2 - \sqrt{2}) \qquad (m_2 = 2).$$

Here, $k = 2$ and $n_2 < m_2$, therefore $\sqrt{3}/3 < \sqrt{2}/2$. All of this will appear even more clear when ratios are written in the form of continued fractions (see Section 4.4).

4.3 Bézout's Identity

How can we find integer solutions of the first degree equation:

$$ax - by = c$$

where x and y are the unknowns and a, b and c are three given integers? Does the problem always have a solution? For the particular case where a and b are relatively prime, it is known that it is always possible to find solutions x_0 and y_0 such that:

$$ax_0 - by_0 = 1$$

and this is known as *Bézout's identity*.

The method had already been explained in the 1624 second edition of Bachet de Méziriac's *Problèmes plaisants et délectables* [5]. The theory is found in proposition XVIII: "Two prime numbers being given, to find the least multiple of each of these, so that one surpasses the other by a unit." The author gives a proof that depends upon the parity of the number of divisions to be carried out in Euclid's algorithm in order to produce a remainder 1. The proof, which uses letters, involves a rather laborious type of recurrence which is not easy to follow.

Bachet considered 'practical' applications, like his sixth Problem: "To find the Number some one else is thinking of", and his tenth Problem, a "little subtlety" was: "There were 41 persons at a banquet, men, women and children, who in all spent 40 sous, but each man paid 4 sous, each woman 3 sous, and each child 4 deniers. I ask how many were there of men, women and children?" We may notice that this belongs to the mediaeval tradition of some Arab, Indian and Chinese problems, called '100 chicken' problems (see [32] and [41]).

What follows is a text by Bézout, an extract from his *Cours d'Algèbre* of 1766 [6], chosen because it is more straightforward than Bachet's presentation. The method is explained by using the example:

$$17x - 11y = 542.$$

The extract also illustrates the way in which Bézout proceeds in all of his *Cours*. Rather than a dogmatic approach – like Euclid – Bézout introduces his explanations through the use of worked examples, which are more or less concrete, and through them draws out the general principles that govern them. Bézout's manuals thus enjoy an important role in mathematics education.

Etienne Bézout

Cours de Mathématiques à l'usage des Gardes du Pavillon et de la Marine, in 6 vols., 1764–1769.

Third part containing 'Algebra and the Application of that Science to Arithmetic and Geometry', Paris: Musier, 1766, pp. 118–121.

Indeterminate Problems

First Question. *In how many ways may a sum of 542 livres be paid, giving coins of 17 livres value, and receiving change in 11 livre coins?*

Let x represent the number of 17 livre coins and y the number of 11 livre coins; by giving x coins of 17 livre value, one will pay x times 17 livres or $17x$: on receiving y coins of 11 livre value one will receive $11y$; consequently, one has paid $17x - 11y$; and since one wished to pay 542 livres, we have $17x - 11y = 542$. Taking the value of y, that is the unknown having the least coefficient, we have: $y = \dfrac{17x - 542}{11}$.

As we have only this equation, it can be seen that, on setting an arbitrary value for x, whatever number is wished, we shall have for y a value that certainly satisfies the equation; but since the question requires that x & y should be whole numbers, this is how we should proceed to obtain them directly.

The value of $y = \dfrac{17x - 542}{11}$ can be reduced, by carrying out division as far as is possible, to $y = x - 49 + \dfrac{6x - 3}{11}$; it is therefore necessary for $\dfrac{6x - 3}{11}$ to be a whole number: let u be this whole number; we shall have $\dfrac{6x - 3}{11} = u$, and consequently $6x - 3 = 11u$ & $x = \dfrac{11u + 3}{6}$, or, on carrying out division, $x = u + \dfrac{5u + 3}{6}$; therefore, $\dfrac{5u + 3}{6}$ must make a whole number: let t be that whole number; we shall have $\dfrac{5u + 3}{6} = t$, and consequently $5u + 3 = 6t$ & $u = \dfrac{6t - 3}{5} = t + \dfrac{t - 3}{5}$; it is therefore necessary for $\dfrac{t - 3}{5}$ to be a whole number: let s be that whole number, we have $\dfrac{t - 3}{5} = s$, and consequently $t = 5s + 3$: the operation finishes here, since it is evident that in taking s to any whole number at will, the value of t required to satisfy the equation will always be a whole number, since there is no denominator.

We can now go back to the values of x and y: since we found $u = \dfrac{6t - 3}{5}$; substituting for t its value $5s + 3$, we shall have $u = \dfrac{30s + 18 - 3}{5} = 6s + 3$: and since we found $x = \dfrac{11u + 3}{6}$, in substituting for u, we shall have $x = \dfrac{66s + 33 + 3}{6} = 11s + 6$: finally, since we found $y = \dfrac{17x - 542}{11}$, by substituting the value for x, we shall have $y = \dfrac{187s + 102 - 542}{11} = 17s - 40$; hence, the corresponding values for x and y become $x = 11s + 6$, & $y = 17s - 40$. For the first of these, one is free to choose for s any number one likes; but the second will not allow a value of s smaller than 3; in order that y should be positive, $17s$ must be greater than 40, or s must be greater than $\dfrac{40}{17}$, that is greater than 2.

This equation can then be satisfied in an infinite number of ways, which can all be found by putting in the values for x and y, in the place of s, all the positive whole numbers imaginable from 3 up to infinity; so by using successively $s = 3$, $s = 4$, $s = 5$, $s = 6$, $s = 7$, &c. we shall have the corresponding values of x and y as follows:

$$x = 39 \ldots \qquad\qquad y = 11$$
$$= 50 \qquad\qquad\qquad\quad = 28$$
$$= 61 \qquad\qquad\qquad\quad = 45$$
$$= 72 \qquad\qquad\qquad\quad = 62$$
$$= 83, \&c. \qquad\qquad\quad = 79$$

Each one of which is such, that on giving the number of 17 livre coins, given by x, and receiving as change the number of 11 livre coins, given by y, one will pay 542 livres.

To find integer solutions for an equation of the type:

$$ax - by = c,$$

what we do today is to introduce an intermediary stage. This first stage involves finding a particular solution to the equation:

$$ax - by = d,$$

where d is the greatest common divisor of a and b. The second stage of deducing all solutions of the initial equation follows easily – and this is solvable if, and only if, c is a multiple of d. In a certain way, Bézout's method of 'descent' followed by 'ascent' can be seen in this first stage. Let us look at Bézout's example.

First stage. Find a particular solution (x_0, y_0) of the equation:

(1) $$17x - 11y = 1.$$

In carrying out the successive Euclidean divisions, starting with 17 and 11 we have:

$$17 = 11 \times 1 + 6, \qquad 11 = 6 \times 1 + 5, \qquad 6 = 5 \times 1 + 1,$$

ending with a remainder of 1, the greatest common divisor of 11 and 17, since the numbers 11 and 17 are self prime. In going back up through the procedure, starting from this last remainder 1, we have:

$$1 = 6 - 5 = 6 - (11 - 6) = 2 \times 6 - 11 = 2(17 - 11) - 11 = 2 \times 17 - 3 \times 11,$$

which gives us a particular solution $(x_0, y_0) = (2, 3)$ for equation (1).

Second stage. Find all the solutions of the equation:

(2) $$17x - 11y = 542.$$

Noting that $(x_1, y_1) = (542 \times 2, 542 \times 3)$ is a particular solution of (2), it can be seen that (x, y) is a solution of (2) if, and only if, (x, y) is a solution of the equation:

(3) $$17(x - x_1) = 11(y - y_1).$$

Therefore, with 17 and 11 not having common factors, 17 must divide into $(y - y_1)$, so y is of the form $y_1 + 17k$ where k is an integer. Going back to (3), we can also see that x is of the form $x_1 + 11k$. Conversely, whatever the value of integer k the pair $(x_1 + 11k, y_1 + 17k)$ is a solution of (2). Therefore, the solutions of (2) are of the form: $x = (2 \times 542) + 11k, y = (3 \times 542) + 17k$, where k is an integer.

In order for the solution (x, y) to be made up of positive integers, it is necessary and sufficient that k should satisfy at the same time: $(2 \times 542) + 11k > 0$ and $= (3 \times 542) + 17k > 0$, that is: $k > -95.6$; hence $k = -95, x = 39, y = 1$; $k = -94, x = 50, y = 28$; ... or:

$$x = 39 + 11h, y = 1 + 17h \text{ where } h \text{ is a positive integer.}$$

As we remarked earlier, the calculations carried out by Bézout also reflect the Euclidean algorithm. In practice, Bézout makes the successive substitutions:

$$u = (\boxed{6}x - 3)/11 \; ; \; t = (\boxed{5}u + 3)/6 \; ; \; s = (\boxed{1}t - 3)/5,$$

and we can see that the numbers in bold are the divisors and the numbers in bold inside boxes are the remainders when carrying out the division algorithm.

However, the proof of the algorithm following from Bézout's technique is relatively complex since, written out his way, we can identify at least three sequences in the solution of $ax - by = c$. The initial conditions in the process used by Bézout allow us to understand where these three sequences come from. In practice, from the equation to be solved, he derives $y = (ax + c)/b$ which, just like x, must take integer values; from this comes the idea of considering the remainder a_2 of the division of $a = a_0$ by $b = a_1$ and the remainder c_2 of the division of $c = c_1$ by $b = a_1$, so we have:

$$y = (ax + c)/b = (a_0x + c_1)/a_1 \equiv (ax + c_2)/a_1 \pmod{1};$$

then comes the idea of introducing a quantity $u = (a_2x + c_2)/a_1$ which, just as y must take integer values. This can then be written:

$$a_2x - a_1u = -c_2,$$

and an iteration can be set up by using $x = (a_1u - c_2)/a_2$. We are therefore led naturally to consider:

- the strictly decreasing sequence of the integer remainders $(a_n)_{0 \le n \le k+1}$ of the Euclidean algorithm applied to the pair (a, b): if $a_0 = a$ and $a_1 = b$, then a_{n+1} will be the remainder when a_{n-1} is divided by a_n; hence, $a_{n+1} \equiv a_n \pmod{a_n}, a_{k+1} = 0$ and $a_k = \text{GCD}(a, b)$.
- the decreasing sequence of the integers $(c_n)_{1 \le n \le k+1}$ where $c_1 = c$ and c_{n+1} represents the remainder when c_n is divided by a_n; hence, $c_{n+1} \equiv c_n \pmod{a_n}$.
- a sequence $(x_n)_{0 \le n \le k+1}$ where $x_0 = y$ and $x_1 = x$ and whose values are related by the recurrence relation:

$$a_nx_{n+1} - a_{n+1}x_n = (-1)^n c_{n+1} \text{ for } n \ge 0.$$

By the way it is constructed, if the value of x_n is an integer, then the difference $x_{n+1} - x_{n-1}$ must also be an integer. In fact,

$$x_{n+1} = (a_{n+1}x_n + (-1)^n c_{n+1})/a_n \equiv (a_{n-1}x_n + (-1)^n c_n)/a_n = x_{n-1} \pmod{1}.$$

From this it can easily be verified by recurrence that, if the values of x and y are integers, then the values of $x_2, \ldots, x_{k+1}$ are also themselves integers.

Now, $a_{k+1} = 0$, so $x_{k+1} = (-1)^k c_{k+1}/a_k$ can only be an integer if a_k divides c_{k+1}, and so $c_{k+1} = 0$, and the GCD of a and b divides c.

Conversely, suppose that the GCD of a and b divides c, then a_k divides c_k, $c_{k+1} = 0$ and $x_{k+1} = 0$. It follows that, to each integer value of x_k there correspond integer values, by going back up through the recurrence, $x_{k-1}, \ldots, x_2, x_1 = x$ and $x_0 = y$.

Bézout's identity

Writing 1, the GCD of 11 and 17, as a combination of 11 and 17 with integer coefficients in the form:

$$1 = 2 \times 17 - 3 \times 11$$

is called *Bézout's identity*.

Such an identity exists not only in the ring of integers $\mathbb{Z}$, but also in the ring $K[X]$ of polynomials with coefficients belonging to a field K, like the ring $\mathbb{R}[X]$ (see Section 4.5 below), and in fact, in all Euclidean rings (see Section 4.1 above). So, if A and B are two relatively prime polynomials, there exist two polynomials U and V such that:

$$AU + BV = 1.$$

The polynomials U and V are found by the same method, that is by using Euclid's algorithm to find the GCD of A and B.

More generally, it is now usual to call a ring for which every finite ideal is a principal ideal, a *Bézout ring*. In such a ring, two elements a and b have a GCD, namely an element d, such that Ad is the ideal generated by a and b, and the Bézout identity certainly holds, since:

$$Aa + Ab = Ad.$$

But there may not necessarily be a Euclidean algorithm to help us determine such a GCD, d.

4.4 Continued Fractions

The continued fraction expansions of rational or irrational numbers provide rational approximations to these numbers, which turn out, in certain respects, to be optimal. Once again, the process shows Euclid's algorithm at work.

It is not known if the idea of continued fractions was the basis for the calculations made by early mathematicians: examples are the simplifying of ratios of large numbers by Aristarchus of Samos or Archimedes, an approximation to the Golden Number in *Liber Abaci* (1202) by Fibonacci, and approximations to square roots by Bombelli in *Algebra* (1572) and by Cataldi in *Trattato* (1613). Although, in each case we find nothing more than numerical results

The germ of the idea is, however, present in Book X of the *Elements*, as we have seen, in Euclid's treatment of incommensurable magnitudes in Proposition 2. Its theoretical development continued with al-Māhānī in the 9th century and al-Khayyām in the 12th century, as we have seen, when they demonstrated the equality of two ratios by considering the sequence of successive quotients obtained by applying Euclid's algorithm to each of the ratios, which is the same – as we shall shortly see – as showing that their expansions in continued fraction form are identical.

The earliest example we have of a number being written explicitly in continued fraction form is the number $4/\pi$ (see Fig. 4.2). This formula, due to Lord Brouncker,

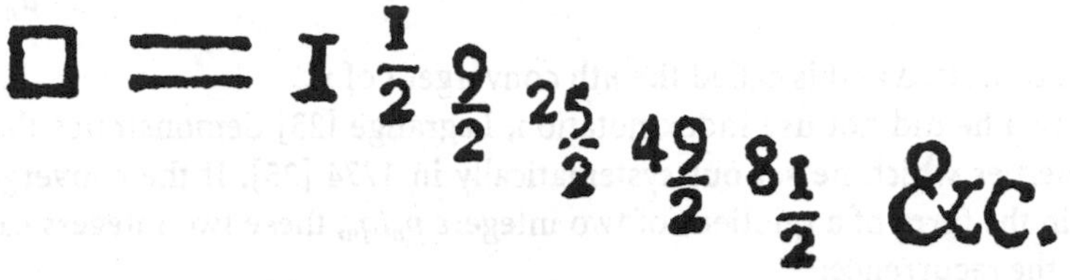

$$\square = 1\frac{1}{2}\ \frac{9}{2}\ \frac{25}{2}\ \frac{49}{2}\ \frac{81}{2}\ \&c.$$

Fig. 4.2

is quoted by Wallis in his *Arithmetica infinitorum* (1656) [39]. Wallis touches here on a theory of continued fractions, which he returns to in his *Algebra* (1685) [40].

We may note, in passing, that Leibniz showed an interest in the idea. Also Huygens was led to use continued fractions in his calculations for the construction of a planetarium (1682). He had to find the number of teeth for intermeshing cogs, and the ratio of these had to be an approximation for the ratios of much larger numbers obtained from astronomical observations. He set out a first approach to the convergence of continued fractions in *De automato planetario* [20].

The theory of continued fractions, as it is known today, was properly established in the 18th century. It was laid down by Euler and pretty well completed by Lagrange. The passages given here are from Euler's first text on the subject, his *De Fractionibus continuis Dissertatio* written in 1737 [15]. Euler returned to the subject in his *Introductio in analysin infinitorum* [16].

The continued fraction expansion of a real number x, is obtained in the following way (we take x to be positive for simplicity). Using $[x]$ for the integer part of x, that is the unique integer a such that $a \le x < a + 1$, the sequences of numbers z_n and a_n are defined by the recurrence relations:

$$z_0 = x,\ a_n = [z_n] \text{ and } z_{n+1} = 1/(z_n - a_n) \text{ provided } z_n \ne a_n.$$

If one of the z_n is an integer, the process ends and this happens if, and only if, x is rational. Hence the number e must be irrational since its continued fraction expansion is infinite (see paragraph 21 in the text by Euler that follows).

For finite continued fractions, we can write:

$$x = a_0 + \cfrac{1}{a_1 + \cfrac{1}{a_2 + \cdots \cfrac{1}{a_{n-1} + \cfrac{1}{z_n}}}}$$

Where $z_n \neq a_n$ we write

$$x = a_0 + \cfrac{1}{a_1 + \cfrac{1}{a_2 + \cdots \cfrac{1}{a_n + \cdots}}}$$

and the fraction written as

$$x_n = a_0 + \cfrac{1}{a_1 + \cfrac{1}{a_2 + \cdots \cfrac{1}{a_n}}}$$

has been truncated and is called the nth convergent of x_n.

Although he did not use index notation, Lagrange [23] demonstrates the following properties which he set out systematically in 1774 [25]. If the convergent x_n is written in the form of a quotient of two integers p_n/q_n, these two integers can be defined by the recurrence:

$$p_n = a_n p_{n-1} + p_{n-2} \text{ with } p_0 = a_0 \text{ and } p_1 = a_0 a_1 + 1$$
$$q_n = a_n q_{n-1} + q_{n-2} \text{ with } q_0 = 1 \text{ and } q_1 = a_1$$

Furthermore, this fraction is irreducible, since:

$$p_{n+1} q_n - q_{n+1} p_n = (-1)^n.$$

Also, the sequence of even convergents is increasing, and the sequence of odd convergents is decreasing, and they both tend towards x:

$$x_{2n} < x_{2n+2} < x < x_{2n+1} < x_{2n-1} \text{ with } |x - x_n| < 1/q_n q_{n+1}.$$

Finally, the fractional representation of x by $x_n = p_n/q_n$ is optimal in this sense, that any fraction that is closer to x than is x_n must have a denominator strictly greater than q_n.

This is why the expansion in continued fraction form produces fractions that are simpler and at the same time approximates to the number being evaluated. This what Huygens did to find simpler approximate ratios for his cog wheels, and this is how we can obtain the approximations found by Aristarchus of Samos and Archimedes. To take the example of Aristarchus, referred to in Section 4.1, we have:

$$\frac{71\,755\,875}{61\,735\,500} = 1 + \frac{10\,020\,375}{61\,735\,500} = 1 + \cfrac{1}{6 + \cfrac{1\,613\,250}{10\,020\,375}} = 1 + \cfrac{1}{6 + \cfrac{1}{6 + \cfrac{1}{\cdots}}} \approx 1 + \cfrac{1}{6 + \cfrac{1}{6}} = \frac{43}{37}$$

In fact, all these theoretical results – without them being stated explicitly – are to be found in paragraph 10 of Euler's text. The connection with Euclid's algorithm is shown in paragraph 12 and the numbers π and e appear in paragraphs 15 and 21.

Leonhard Euler

De Fractionibus continuis Dissertatio (written 1737),
Commentarii academiae scientiarum Petropolitanae, 9 (1744), 98–137.
Opera Omnia, I, xiv, 187–215.

Essay on Continued Fractions

[...]

10. If, then, the following continued fraction is proposed, all of whose
numerators are equal to 1,

$$a + \cfrac{1}{b + \cfrac{1}{c + \cfrac{1}{d + \cfrac{1}{e + \cfrac{1}{f + \text{etc.}}}}}}$$

the fractions of the following sequence

$$\overset{a}{\frac{1}{0}}, \quad \overset{b}{\frac{a}{1}}, \quad \overset{c}{\frac{ab+1}{b}}, \quad \overset{d}{\frac{abc+c+a}{bc+1}}, \quad \overset{e}{\frac{abcd+cd+ad+ab+1}{bcd+d+b}}, \quad \text{etc.,}$$

approach its value, this sequence being uniquely determined by the progression a, b, c, d,
etc.... It can easily be verified, that both the numerator and denominator of each fraction,
multiplied by the index and augmented respectively by the numerator and denominator of
the preceding fraction, will give the numerator and denominator of the following fraction.
Therefore, the value of this continued fraction will be equal to the sum of the following series

$$a + \frac{1}{1.b} - \frac{1}{b(bc+1)} + \frac{1}{(bc+1)(bcd+d+b)} - \frac{1}{(bcd+d+b)(bcde+\cdots)} + \text{etc.}$$

or, equal to the sum of what it can be transformed into

$$a + \frac{c}{bc+1} + \frac{e}{(bc+1)(bcde+de+be+bc+1)} + \text{etc.}$$

namely, a series whose denominators are formed from the denominators of the series above,
in a way that is easy to determine.

[...]

12. Therefore, if it is proposed to transform the fraction A/B into a continued fraction all of
whose numerators are equal to 1, I divide A by B obtaining the quotient a and the remainder
C; the preceding divisor B is divided by the remainder C giving a quotient b and remainder
D, by which C is divided, and so on until one comes to a nil remainder and an infinitely large
quotient. Further, this operation can be represented in the following way:

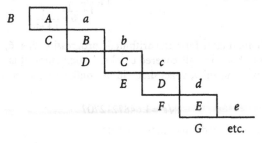

Therefore, by this operation are found the quotients a, b, c, d, e, etc., which, when known, gives:

$$\frac{A}{B} = a + \frac{1}{b + \cfrac{1}{c + \cfrac{1}{d + \cfrac{1}{e + etc.}}}}$$

[...]

15. Let us apply our method for the solution of this problem to the fraction

$$\frac{355}{113},$$

which, according to Metius, expresses approximately the ratio of the circumference to the diameter; consequently, we are looking for fractions, made up of smaller numbers, which differ from this fraction by as little as possible. I therefore divide 355 by 113 and I find

$$\frac{355}{113} = 3 + \cfrac{1}{7 + \cfrac{1}{16}}$$

from which I form the following fractions

$$\frac{3}{1}, \frac{7}{3}, \frac{16}{22}, \frac{355}{113};$$

of which the fractions 3/1 and 22/7 approach the fraction 355/113 closer than any other fractions formed with smaller numbers; further 22/7 is greater, and 3/1 is less, than the proposed fraction, as we have noted above in the general case.

[...]

21. Before calculating other continued fractions, whose denominators are in arithmetic progression, let us consider certain transcendental quantities, whose conversion into continued fractions give denominators that are in arithmetic progression, from which we may disentangle a simpler method for calculating continued fractions of this type. Thus, on inspecting logarithms and other transcendental quantities, I have found that the number whose natural logarithm is 1 and its powers produce continued fractions of this type. Thus I put this number $= e$, namely

$$e = 2.71828182845904,$$

which is transformed into the continued fraction

$$e = 2 + \cfrac{1}{1 + \cfrac{1}{2 + \cfrac{1}{1 + \cfrac{1}{1 + \cfrac{1}{4 + \cfrac{1}{1 + \cfrac{1}{1 + \cfrac{1}{6 + \cfrac{1}{1 + etc.}}}}}}}}}$$

where every third denominator form the arithmetic progression 2, 4, 6, 8, etc, the others being equal to 1. Even if this rule is discovered by observation alone, it is nonetheless probable that it will be valid up to infinity, a fact that will be confirmed for certain later. In the same way, if

$$\sqrt{e} = 1.6487212707$$

is transformed into a continued fraction, this will be

$$\sqrt{e} = 1 + \cfrac{1}{1 + \cfrac{1}{1 + \cfrac{1}{1 + \cfrac{1}{5 + \cfrac{1}{1 + \cfrac{1}{1 + \cfrac{1}{9 + \cfrac{1}{1 + \cfrac{1}{1 + \cfrac{1}{13 + \text{etc.}}}}}}}}}}$$

where the law of progression is similar to the preceding one. We are able to observe similar facts in other continued fractions which are transformations of powers of this same e [...].

Except for the fact that it was written in Latin, Euler's formulation is modern enough for a transcription to be unnecessary. In this same essay, Euler notes that continued fraction expansions which are periodic correspond to numbers which are roots of second degree equations. Lagrange [24] showed the converse to be true.

Lagrange [24] also showed that the speed of convergence of the sequence of residues can be increased by using for a_n, instead of the integer part of z_n, the integer that is nearest to z_n, that is such that $|z_n - a_n| < 1/2$. This corresponds to using the Euclidean algorithm, where the smallest absolute value of the remainder is chosen (or taking $g(a) = |a|$, to use the notation of the Section 4.1). Moreover, this choice of remainder turns out to be optimal, according to Lazard [28].

Lagrange used continued fractions to invent methods of successive approximations for solving numerical equations [23] (see also Section 7.8 below) and for solving differential equations [26]. They also led him to the general solution of Pell-Fermat equations:

$$Ax^2 + 1 = y^2$$

where A is a non-square integer and the unknowns x and y are integers [25] (see also Sections 8.10 and 8.11).

In a way that was analogous to the expression for $4/\pi$ given by Lord Brouncker, Euler considered also so called generalised continued fractions, of the form:

$$a_0 + \cfrac{b_0}{a_1 + \cfrac{b_1}{a_2 + \cfrac{b_2}{\cdots}}}$$

where the b_n are not necessarily equal to 1. However, we are no longer guaranteed the remarkably attractive properties that we have seen above.

It is also possible to introduce variables into continued fractions. For example, Lambert [27] proved the irrationality of π in 1761 and Legendre [29] that of π^2 in 1794, with the aid of the continued fraction expansion of $\tan x$ (see Chap. 5, Epilogue). Note however, the transcendence of e and π were not proved until towards the end of the 19th century (Hermite [19] and Lindemann[30]). Before this, Liouville [31] was able to show the existence of transcendental numbers, again thanks to the use of the continued fraction algorithm. For more information about continued fractions, Brezinski [8] may be consulted.

4.5 The Number of Roots of an Equation

One of the first indications of a rule that gives an upper bound for the number of real roots of an algebraic equation was stated by Descartes in *La Géométrie* of 1637 [13], the famous rule of signs:

We can determine also the number of true and false roots that any equation can have, as follows: An equation can have as many true roots as it contains changes of sign, from + to - or from - to + ; and as many false roots as the number of times two + signs or two - signs are found in succession.

in other words, the number of real positive roots may not exceed the number of changes of sign in the sequence of coefficients in the equation, and the number of real negative roots may not exceed the number of times the signs do not change. Descartes does not prove this rule, but only verifies it using an example where all the coefficients are non zero.

Elsewhere, in his *Algebra* of 1690, Rolle [33] remarks that the real roots of a polynomial must necessarily be separated by those of the polynomial that we now refer to as its derivative, and the roots of that polynomial are separated by the roots of the second derivative, and so on; this is the *Cascade* method.

Lagrange [23] proposed a procedure which in theory would both count the number of roots for any algebraic equation $V = 0$ and separate them. The first stage is to replace the polynomial V with a polynomial V_0 having the same roots as V but with no repeated roots: this only requires dividing V by the GCD of V and its derivative V'. The second stage involves determining an interval that contains all the real roots of the equation: Newton had already given a method for this (Section 6.1 below). For the third stage, a lower bound must be chosen for the distances between the roots: to do this, consider an equation whose roots are the squares of the differences of the roots of the proposed equation, the lower bound can be found from the least positive root of this equation, then take its square root. Finally, the last stage consists in substituting for the unknown x in the polynomial V_0 a sequence of values in arithmetic progression within the interval found at stage 2, and whose difference is less than the lower bound found at stage 3: each change of sign in the sequence of values indicates the existence of a root, and only one, in the interval corresponding to the change of sign. The number of real roots is therefore, in particular, equal to the number of changes of sign in the sequence of values taken by V_0. This procedure is, however, difficult to carry out, since calculating the coefficients of the polynomial whose roots are the squares of the differences of the roots of the proposed equation, has to be done using relations between the coefficients and the roots. The calculation is laborious, and indeed impractical for higher degree equations.

By different routes, Budan [9] in 1811 and Fourier [17] in 1820 obtained an easier rule. According to Budan: "If an equation in x has n roots between 0 and some positive number p, the transformed equation in $(x - p)$ must have at least n variations less than the proposed equation." And Fourier's approach was: consider the sequences of derivatives V, V', V'', ... , $V^{(m-1)}$, $V^{(m)}$; substitute for the unknown x two values A and B and count the number of changes of sign in each of the two sequences $V(A)$, $V'(A)$, ... and $V(B)$, $V'(B)$, ...; the difference between these two numbers is an upper bound for the number of real roots, counted with their multiplicity, which are contained in the interval $[A, B]$. But these two results, which can be considered as equivalent, only provide an upper bound for the number of real roots.

In 1815, Cauchy [10] gave, for the first time, a complete solution to the problem of determining the number of real roots of an algebraic equation $V = 0$ over a given interval, but the calculations are too complicated to have practical use. By contrast, Sturm's method, clearly inspired by Fourier's approach, which he explains in his Memoir of 1835, is relatively simple. Here again, Euclid's algorithm is used, not this time with integers, but with polynomials, namely the polynomial V and its derivative V', and a simple count of sign changes is made. This memoir guarantees the fame of the physicist and mathematician Charles Sturm.

C. -F. Sturm
Mémoire sur la résolution des équations numériques,
Mémoires présentés par divers savants à l'Académie Royale des Sciences,
Sciences mathématiques et physiques, vol. VI (1835), 271–318.

1.

Let

$$Nx^m + Px^{m-1} + Qx^{m-2} + \ldots + Tx + U = 0$$

be a numerical equation of any degree, for which it is proposed to determine all the real roots.

We start by carrying out on this equation the calculations needed to find whether it has any equal roots, by working in the manner we shall indicate. Denoting by V the complete function $Nx^m + Px^{m-1} + $ etc., and by V_1 its derived function (which is formed by multiplying each term of V by the exponent of x and diminishing this exponent by a unit), we have to find the greatest common divisor of the two polynomials V and V_1. First of all, divide V by V_1, and when we come to a remainder of inferior degree to that of the divisor V_1, change the signs of all the terms of this remainder (+ signs to − and − signs to +). Denote by V_2 what this remainder had become after the changes of signs. In the same way, divide V_1 by V_2, and after again changing the signs of the remainder, we shall have a new polynomial V_3 of inferior degree to that of V_2. Dividing V_2 by V_3 will lead in the same way to a function V_4 which will be the remainder of this division after changing the signs. Continue this series of divisions, always taking care to change the signs of the terms in each remainder. This changing of signs, which would be unnecessary if our purpose was only to find the greatest common divisor of the polynomials V and V_1, is necessary to the theory which we are explaining. Since the degree of each successive remainder diminishes, we arrive finally either at a numerical remainder, independent of x and different from zero, or at a remainder function of x which divides exactly into its predecessor. We shall examine these two cases separately.

2.

Suppose, to begin with, that we arrive after a certain number of divisions at a numerical remainder which we shall denote by V_r.

In this case, we are asssured that the equation $V = 0$ has no equal roots, since the polynomials V and V_1 have no function of x in common. Using $Q_1, Q_2, \ldots, Q_{r-1}$ to denote the quotients of the successive divisions which leave the remainders $- V_2, - V_3, ..., - V_n$ we have the sequence of equalities:

$$V = V_1 Q_1 - V_2$$
$$V_1 = V_2 Q_2 - V_3$$
$$V_2 = V_3 Q_3 - V_4$$
$$. \quad . \quad . \quad .$$
$$. \quad . \quad . \quad .$$
$$V_{r-2} = V_{r-1} Q_{r-1} - V_r$$

Having done this, consideration of this system of functions $V, V_1, V_2, ..., V_r$ will provide a sure and easy way of knowing *how many real roots there are for the equation* $V = 0$ *between two numbers A and B whatever their size or sign, B being greater than A*. Here is the rule which does this:

Substitute the number A for x in all the functions $V, V_1, V_2, ..., V_{r-1}, V_n$, *then write out in order, on the same line, the signs of the resultants, and count the number of changes of sign found in that sequence. In the same way, write out the sequence of signs which the same functions take when the other number B is substituted, and count the number of changes of sign in this second sequence. As many as is the number of changes of sign [of the second sequence] fewer, than the number of changes of sign in the first, is the number of real roots of the equation* $V = 0$ *between the two numbers A and B. If the second sequence has the same number of changes as the first, then the equation* $V = 0$ *has no real roots between A and B. Furthermore, B being greater than A, the second sequence cannot have more changes [of signs] than the first.*

$$[...]$$

2nd EXAMPLE

To find the conditions such that the equation

$$x^3 + px + q = 0$$

should have all its roots real.

We have

$$V = x^3 + px + q$$
$$V_1 = 3x^2 + p$$

We find V_2 and V_3 by means of successive divisions. To avoid fractions, multiply the dividend of the first division by 3, and the dividend of the second division by $4p^3$, which is a positive quantity.

We find

$$V_2 = - 2px - 3q,$$
$$V_3 = - 4p^3 - 27q^2.$$

The conditions for real roots [...] for the equation are the two following:

$$- 2p > 0, \qquad - 4p^3 - 27q^2 > 0$$

which reduce to:

$$p < 0, \qquad 4p^3 + 27q^2 < 0$$

The first of these conditions is contained in the second, which is an already well-known condition [for real roots].

In the same way, we could find the necessary conditions for the equation

$$x^4 + px^2 + qx + r = 0$$

to have all its roots real.

This is the basis for the proof in the case where V does not have multiple roots. Using the notation V_1, V_2, ..., V_r of Sturm in his Memoir, the number of changes of signs in the sequence of values of $V(x)$, $V_1(x)$, $V_2(x)$, ... , $V_r(x)$ can only vary when x passes through a value that makes one of the polynomials zero, that is a value of x which is a root of one of the polynomials.

Let us suppose that $x = c$ is a root of V. Then V_1, V_2, ..., V_r can not be zero for $x = c$ since c is a simple root of of V. Therefore V, V_1, V_2, ..., V_r do not have a root other than c in the interval $[c - u, c + u]$ if u is sufficiently small. The formulas:

$$V(c + u) = V(c) + uV'(c) + u^2 V''(c)/2 + ... = u[V'(c) + uV''(c)/2 + ...]$$

and $$V(c - u) = - u[V'(c) - uV''(c)/2 + ...]$$

show that, for sufficiently small u, the sign of $V(c + u)$ is the same as that of $V'(c)$, and so of $V'(c + u)$, while the sign of $V(c - u)$ is opposite that of $V'(c)$, and so of $V'(c - u)$. Consequently, on passing from the sequence $V(c - u)$, $V_1(c - u)$, ... to the sequence $V(c + u)$, $V_1(c + u)$, ... , the number of changes of sign must reduce by 1.

Suppose now that $x = b$ is a root of V_n for a value n lying between 1 and $r - 1$. Then neither V_{n-1} nor V_{n+1} can vanish for $x = b$ unless the constant V_r also vanishes.

From $V_{n-1} = V_n Q_n - V_{n+1}$, we deduce $V_{n-1}(b) = - V_{n+1}(b)$.

Therefore, for u sufficiently small, $V_{n-1}(b - u)$ and $V_{n-1}(b + u)$ will have the same sign, $V_{n+1}(b - u)$ and $V_{n+1}(b + u)$ will also be of the same sign, but this will be of opposite sense, and $V_n(b - u)$ and $V_n(b + u)$ will have different signs. We have therefore, either the sequence + + – twice, or the sequence – + + twice, and finally, no change in the number of changes of sign.

This number, then, counts exactly the number of roots of V. In fact, when V has multiple roots, the procedure still works, but does not count the multiple roots.

Sturm's theorem was perfected by Sylvester [38] in the following way: if W is a polynomial prime to a polynomial V and if we take V_1 to be the polynomial V'W, then the difference between the number of variations of sign in the sequence of values of the polynomials V, V_1, V_2, ... , V_r calculated for A and B is equal to the difference between the number of roots of V lying between A and B for which W is positive, minus the number of roots of V lying between A and B for which W is negative.

Extension of Sturm's Theorem

More generally, while Euclid's algorithm can be used in all polynomial rings with coefficients belonging to a field K, Sturm's theorem remains valid for polynomial equations whose roots belong to any field K provided that the intermediate value theorem can be applied.

For this purpose, K needs, first of all, to possess an order compatible with the operations, that is the idea of a *real field* or ordered field – characterised by the fact that −1 cannot be the sum of squares –, and then there need to be existence conditions for roots in K – that is, that K(i) should be an algebraically closed field –, the real field K is then called a *real closed field* or a maximally ordered field.

Therefore, Sturm's theorem and Sylvester's generalisation of it, remains true if the field ℝ of reals is replaced by a real closed field (Artin and Schreier [4]). On this subject, see [7] or [36].

Although the origin of these algorithms is essentially theoretical, interest in them has become renewed with the development of computer programs for handling the calculus, like MACSYMA, REDUCE, MAPLE and DERIVE. In this context, we may also note the procedure of the automatic proof of theorems in plane geometry proposed in 1977 by the Chinese mathematician Wu Wen-Tsiin, which is based upon repeated use of Euclid's algorithm in polynomial rings with many indeterminates (see Chou Shang-Ching [12]).

Bibliography

[1] Archimedes, Measurement of a Circle, in *The Works of Archimedes Edited in Modern Notation* (1897) with supplement *The Method of* Archimedes (1912), tr. T. L. Heath, Cambridge: Cambridge University Press, 1912, repr., New York: Dover, n.d.

[2] Aristarchus, Treatise on the sizes and distances of the sun and the moon, in *Aristarchus of Samos, The Ancient Copernicus*, ed. & tr. T. L Heath, Oxford: Clarendon Press 1913, repr. New York: Dover, 1981.

[3] Aristotle, Topics, tr. W. A. Pickard-Cambridge, in vol. 1, *The Works of Aristotle translated into English*, Ross W. D. ed., 12 vols., Oxford: Clarendon Press, 1908-52.

[4] Artin E. and Schreier, O., Algebraische Konstruktion reellen Körper, *Abhandlungen aus dem Mathematischen Seminar der hamburgischen Universität*, 1927.

[5] Bachet, Cl.-G., Sieur de Meziriac, *Problèmes plaisans et délectables qui se font par les nombres* (1612, 1624), Lyon: Pierre Rigaud, simplified edn, Paris: Blanchard, 1959.

[6] Bézout E., *Cours de Mathématiques à l'usage des Gardes du Pavillon et de la Marine*, vol. III *Algèbre*, Paris: Musier, 1766.

[7] Bochnak, J., Coste, M. and Roy, M.-F., *Géométrie algébrique réelle*, New York: Springer, 1987.

[8] Brezinski, Cl., *History of Continued Fractions and Padé Approximants*, New York: Springer, 1991.

[9] Budan, Mémoire contenant la démonstration de quelques Théorèmes nouveaux, relatifs aux successions de signes, considérées dans les termes des suites et dans les coefficients des équations, in *Nouvelle méthode pour la résolution des équations numériques d'un degré quelconque*, Paris: Dondey-Dupré, 1822.

[10] Cauchy, A.-L., Mémoire sur la Détermination du nombre de Racines réelles dans les Equations algébriques, *Journal de l'Ecole Polytechnique*, vol. 10 (1815), 457-548, repr., in *Oeuvres*, 2nd series, Paris: Gauthier-Villars, vol. I, 1905, pp. 170-257.

[11] Caveing, M. C, *La constitution du type mathématique de l'idéalité dans la pensée grecque*, Lille: Atelier National de Reproduction des Thèses, 1982.

[12] Chou S.-C., *Mechanical Geometry Theorem Proving*, Dordrecht: Reidel Publishing Company, 1988.

[13] Descartes, R., *La Géométrie*, Leyde: Jan Maire, 1637, English tr. D. E. Smith & M. L. Latham, with a facsimile of the first edition, La Salle, Illinois: Open Court Publishing Company (1925), repr. New York: Dover, 1954.

[14] Euclid: Heath, T. L. *The Thirteen books of Euclid's Elements*, 2nd edn, 3 vols., Cambridge: Cambridge University Press 1926, repr. New York: Dover, 1956.

[15] Euler, L., De Fractionibus continuis Dissertatio, *Commentarii academiae scientiarum Petropolitanae*, 9 (1737), 98-137, repr. in *Opera Omnia*, I xiv, pp. 187-215, English translation in *Math. Systems Theory*, 18 (1985), pp. 295-328.

[16] Euler, L., *Introductio in Analysin infinitorum*, Lausanne, 1748, repr. in *Opera Omnia*, I viii & ix, English tr. J. D. Blanton, *Introduction to Analysis of the Infinite*, New York: Springer, 1988, 1990.

[17] Fourier, J., Sur l'Usage du Théorème de Descartes dans la recherche des limites des racines, *Bulletin des Sciences par la Société philomatique de Paris*, (Oct. 1820), 156–165 & (Dec. 1820), 181–187, repr. in *Oeuvres*, Paris: Gauthier-Villars, vol. II, 1890, pp. 289–309.

[18] Fowler, D. H., *The Mathematics of Plato's Academy: A new Reconstruction*, Oxford: Clarendon Press (1987), repr. with corrections, 1990.

[19] Hermite, Ch., Sur la fonction exponentielle, *Comptes Rendus de l'Académie des Sciences de Paris*, 77 (1873), 18–24, 74–79, 226–233 and 285–293.

[20] Huygens, C., *De automato planetario*, repr. in *Oeuvres Complètes* (1713), vol. xii, pp. 579–652.

[21] Itard, J., *Les Livres Arithmétiques d'Euclide*, Paris: Hermann, 1961.

[22] Knuth, D.E., *The Art of computer programming*, 1st ed., Reading Massachusetts: Addison-Wesley, 1968.

[23] Lagrange, J.-L., Sur la résolution des équations numériques, *Mémoires de l'Académie de Berlin*, 23 (1769) 311–352, repr. in *Oeuvres*, vol. ii, Paris: Gauthier-Villars, 1868, pp. 539–578.

[24] Lagrange, J.-L., Addition au mémoire sur la résolution des équations numériques, *Mémoires de l'Académie de Berlin*, 24 (1770) 111–180, repr. in *Oeuvres*, vol. ii, Paris: Gauthier-Villars, 1868, pp. 581–652

[25] Lagrange, J.-L., Additions à l'analyse indéterminée, in Euler, L., *Elémens d'Algèbre*, French tr. from the German (1774), vol. ii, 369–658, repr. in *Oeuvres*, vol. vii, Paris: Gauthier-Villars, 1877, pp. 5–180,

[26] Lagrange, J.-L., Sur l'usage des fractions continues dans le calcul intégral, *Mémoires de l'Académie de Berlin*, 23 (1776), repr. in *Oeuvres*, vol. iv, Paris: Gauthier-Villars, 1869, pp. 301–332.

[27] Lambert, J.H., Mémoire sur quelques propriétés remarquables des quantités transcendantes circulaires et logarithmiques, *Mémoires de l'Académie de Berlin*, 17 (1761), 265–322, repr. in *Mathematische Werke*, vol. ii, pp. 112–159.

[28] Lazard, D., Le meilleur algorithme d'Euclide pour K[X] et $\mathbb{Z}$, *Comptes Rendus de l'Académie des Sciences de Paris*, 284, Série A (1977), 1–4.

[29] Legendre, A.-M., *Eléments de Géométrie*, Paris: Didot, 1794.

[30] Lindemann, C., Über die Zahl π, *Mathematische Annalen*, 20 (1882), 213–225.

[31] Liouville, J., Sur des classes très étendues de quantités dont la valeur n'est ni algébrique, ni même réductible à des traditionnelles algébriques, *Comptes Rendus de l'Académie des Sciences de Paris*, 18 (1844), 883–885.

[32] Martzloff, J.-C., *Histoire des mathématiques chinoises*, Paris: Masson (1987), English tr. S. S. Wilson, *A History of Chinese Mathematics*, New York: Springer, 1997.

[33] Rolle, M., *Traité d'Algèbre*, Paris: Michallet, 1690.

[34] Samuel, P., About Euclidean Rings, *Journal of Algebra*, 19 (1971), 282–301.

[35] Shallit, J., Origins of the Analysis of the Euclidean Algorithm, *Historica Mathematica*, 21 (1994), 401–419.

[36] Sinaceur, H., *Corps et modèles, Essai sur l'histoire de l'algèbre réelle*, Paris: Vrin, 1991.

[37] Sturm, Ch., Mémoire sur la résolution des équations numériques, *Mémoires présentés à l'Académie Royale des Sciences, Sciences mathématiques et physiques*, vi (1835), 271–318.

[38] Sylvester, J., On a theory of syzygetic relations of two rational integral functions, comprising an application to the theory of Sturm's function, *Philosophical Transactions of the Royal Society of London*, 142 (1853), 407–548, repr. in *Works*, vol. 1, Cambridge, 1904, pp. 429–586.

[39] Wallis, J., *Arithmetica infinitorum*, Oxford, 1656, repr. in *Opera Mathematica*, vol. I Oxford 1695, pp. 355–478.

[40] Wallis, J., *Algebra*, Oxford, 1685, repr. in *Opera Mathematica*, vol. II, 1693, Oxford, pp. 1–482.

[41] Youschkevitch, A., *Les mathématiques arabes (VIIIe–XVe siècles)*, Paris: Vrin, 1976.

5 From Measuring the Circle to Calculating π

וַיַּעַשׂ אֶת־הַיָּם מוּצָק

עֶשֶׂר בָּאַמָּה מִשְּׂפָתוֹ עַד ־שְׂפָתוֹ עָגֹל | סָבִיב

וְחָמֵשׁ בָּאַמָּה קוֹמָתוֹ

וְקָוה שְׁלֹשִׁים בָּאַמָּה יָסֹב אֹתוֹ סָבִיב׃

And he made a molten sea, ten cubits from the
one brim to the other: it was round all about,
and his height was five cubits: and a line of
thirty cubits did compass it round about.
The Bible, I Kings 7, 23 (Authorised Version)

Well before Archimedes showed how to work out the circumference of a circle, empirical values had been used for the ratio of the circumference of a circle to its diameter, which we call π. The quotation from the Bible suggests that it might be taken
to be 3, and the Rhind Papyrus indicates that the Egyptians considered it could be
estimated as equal to 4.$(1 - 1/9)^2 = 3.16...$ We should, however, be careful to avoid
any error at this stage: there is no way that π can ever be considered as a number in a
formula. Right from the start, it was never written, nor ever thought of, in terms of
algebraic expressions. Instead methods of calculations, or 'algorithms' were devised,
which were set out using the language of the times.

For example, the Rhind Papyrus, which was copied at about 1650 BC by the scribe
Ahmes from a much older text, explains, through examples, a rule for calculating the
area of a circle given its diameter. Using this rule, in for example Problem 48, gives
the area of a circle of diameter 9 as that of a square of side 8. The rule can be written
algebraically as the formula:

$$A = (d - d/9)^2$$

and from this we can deduce a value for π s:

$$\pi = 4A/d^2 = 4 (8/9)^2 = 3.16...$$

As regards the quotation from the *Bible* given above, we are the ones who divide
30 by 10 to obtain the value 3. The text is a 'work of literature', and it is often danger-
ous to draw conclusions about the scientific knowledge of a time from works of literature, the more so since, in this case, it is likely that it was written in the sixth
century BC, and we know that other, and better, approximations were known at that
time. The idea that π was a number did not arise until the seventeenth century, and

the use of the Greek letter π for this number by Jones [12] did not gain acceptance until after it had been used by Euler [10].

The different approaches to thinking about π can be put into three categories which developed over time:

Rhind Papyrus, Problem 48 [5]

– Up until the seventeenth century, a geometric approach was used, which was concerned with finding the ratios of lengths or areas: there is no attempt here to evaluate a number as such.

– The advent of the infinitesimal calculus caused a revolution in ways of evaluating π. These included infinite sums and products, using trigonometric functions and even infinite continued fractions.

– Finally there followed more theoretical studies into the nature of the number π. Lambert [15] proved it was irrational and Lindemann [20] succeeded in establishing that it was transcendental, and these studies coincided with achieving values for π to a greater and greater number of decimals.

In this review we are mostly concerned with the first two aspects. We shall illustrate the geometric approach with a text by Archimedes in which the circumference of the circle is shown to lie between regular inscribed and circumscribed polygons (Section 5.1), a Chinese text showing that area of a circle is contained between the areas of these same polygons (Section 5.2) and a text by Descartes which deals with the method of isoperimeters (Section 5.3). The influence of the infinitesimal calculus is shown here by texts by Leibniz (Section 5.4) and then by Euler (Section 5.5), where π has now become a number which can be evaluated as the sum of a series. The first of these attracts us by its beauty, because only integers feature in its formulation, and the second is remarkable for the rapidity of its convergence.

Geometric Approaches

5.1 The Circumference of the Circle

As regards the ratio of the diameter to the circumference, common usage takes it to be as 7 is to 22, but this is something that can not be proved by any demonstration.

Nicolas Chuquet, *La Géométrie*, ms. 1484 ([6], p. 415.)

The book on *The Measurement of a Circle* by Archimedes consists of three propositions only, and is doubtless part of a larger work. The first proposition establishes the result we write today as:

$$A = \frac{rC}{2}$$

where A is the area of a circle with radius r and circumference C. This formula is the same as saying that the ratios:

$$A/r^2 \text{ and } C/2r$$

are equal, ratios which were already known to be independent of the choice of circle. The third proposition, given below, establishes the interval:

$$3\,d + \frac{10}{71}\,d \; < \; C \; < \; 3\,d + \frac{1}{7}\,d$$

where C and d are, respectively, the circumference and diameter of a circle. To obtain this result, Archimedes calculates the perimeters of regular polygons, inscribed and enclosing the circle

Principally concerned with Geometry and Mechanics, Archimedes was capable of producing both ingenious practical inventions and refined theoretical proofs. To a certain extent both these aspects are to be found here: the theory in Proposition 1 in which his proof uses the method of exhaustion, a rigorous method that avoids the idea of the infinite, and his practical skill in Proposition 3 containing the most careful calculations.

It should be pointed out that π never appears explicitly in the text, except in the idea of the ratio of two magnitudes, either lengths or areas. The statement of Proposition 3 by Archimedes himself is that: the circumference of a circle lies between two fractional multiples of its diameter. Heath himself, however, in his translation, says: "the ratio of the circumference of any circle to its diameter is less than $3\frac{1}{7}$ but greater than $3\frac{10}{71}$" ([2], p. 93) which in our notation is:

$$3 + \frac{10}{71} \; < \; \pi \; < \; 3 + \frac{1}{7}$$

We shall see that, not only did Archimedes give a 'good' approximation for π, but he also established upper and lower limits for the value, which gives a measure for the accuracy of the approximation. Furthermore, he works out the values by an iterative method, and this can be continued to obtain an answer to as great a degree of accuracy as is wished.

Archimedes
The measurement of the Circle, in *Les oeuvres complètes d'Archimède*, vol. 1, French tr. by Paul Ver Eecke, 1921; repub. Paris: Blanchard, 1961, pp. 130–134.

Proposition III
The perimeter of any circle is thrice the diameter increased by less than one-seventh part, but by more than seventy-one parts of the diameter.

Let there be a circle, a diameter AΓ, the centre E, a tangent ΓΛZ, and the angle under ZE, EΓ ne third of a right-angle. Then, the ratio of EZ to ΓZ is that of 306 to 153, while the ratio of EΓ to ΓZ is that of 265 to 153. Let us divide the angle ZEΓ into two equal parts by the line EH; then, ZE is to EΓ as ZH is to HΓ. Hence, the sum of ZE, EΓ is to ZΓ as EΓ is to ΓH; so that the ratio of ΓE to ΓH is greater than that of 571 to 153. Then, the ratio of the square of EH to the square of HΓ is equal to that of 349450 to 23409; therefore, the ratio of the roots is equal to that of $591\frac{1}{8}$ to 153.

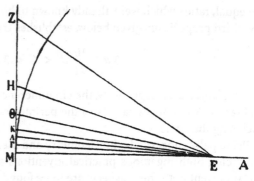

Let the angle HEΓ be again divided into two equal parts by the line EΘ. Then, similarly, the ratio of EΓ to ΓΘ will be greater than that of $1162\frac{1}{8}$ to 153; hence, the ratio of ΘE to ΘΓ will be greater than that of $1172\frac{1}{8}$ to 153.

Let the angle ΘEΓ be again divided into two equal parts by the line EK. Then, the ratio of EΓ to ΓK will be greater than that of $2334\frac{1}{4}$ to 153;

hence, the ratio of EK to ΓK is greater than that of $2339\frac{1}{4}$ to 153.

Let the angle KEΓ be again divided into two equal parts by the line ΛE. Then, the ratio of EΓ to ΛΓ will be greater than that of $4673\frac{1}{2}$ to 153; hence, since the angle ZEΓ, which is a third of a right-angle, has been divided four times into two equal parts, the angle ΛEΓ will be the forty-eighth part of the right-angle. Let us then make at E an angle ΓEM equal to this last angle. Then, the angle ΛEM is the twenty-fourth part of the right-angle and, therefore, the line ΛM is the side of a polygon of 96 sides circumscribing the circle. Hence, since it has been proved that the ratio of EΓ to ΓΛ is greater than that of $4673\frac{1}{2}$ to 153, while AΓ is twice EΓ and ΛM is twice ΓΛ, it follows that the ratio of AΓ to the perimeter of a polygon of 96 sides is also greater than that of $4673\frac{1}{2}$ to 14688. Furthermore, the second number is three times the first with an excess of $667\frac{1}{2}$ which is less than the seventh part of $4673\frac{1}{2}$; so that the polygon circumscribing the circle is less than three plus a seventh of the diameter. Therefore, the circumference of the circle is *a fortiori* smaller than three plus a seventh of the diameter.

Let there be a circle of diamtere AΓ and an angle BAΓ which is the third of a right-angle. Then, the ratio of AB to BΓ is less than that of 1351 to 780, while the ratio of AΓ to ΓB is that of 1560 to 780. Let us divide the angle BAΓ into two equal parts by the line AH. Then, since angle BAH is equal to the angle HΓB as well as to the angle HAΓ, the angle HΓB will also be equal to the angle HAΓ. Furthermore, the right-angle AHΓ is common; and so, the third angle HZΓ will be equal to the third angle AΓH. It

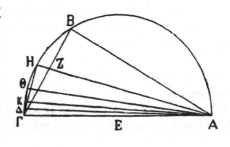

follows that the triangles AHΓ, ΓHZ are equiangular. Hence, AH is to HΓ as ΓH is to HZ and as AΓ is to ΓZ. Now, AΓ is to ΓZ as the sum of the lines ΓA, AB is to BΓ; therefore, the sum of

ΓA, AB is also to BΓ as AH is to AΓ. It follows that the ratio of AH to HΓ is less than that of 2911 to 780, and that the ratio of AΓ to ΓH is less than that of $3013\frac{3}{4}$ to 780. Let us divide the angle ΓAH into two equal parts by the line AΘ; the ratio of AΘ to ΘΓ will, on the same grounds, be smaller than that of $5924\frac{3}{4}$ to 780, or that of 1823 to 240, since these numbers are respectively the $\frac{4}{13}$ of the former numbers. Therefore, the ratio of AΓ to ΓΘ is less than that of $1838\frac{9}{11}$ to 240.

Let us divide again the angle ΘAΓ again into two equal parts by the line KA; then, the ratio of AK to KΓ is less than that of 1007 to 66, since these numbers are respectively $\frac{11}{40}$ of other numbers. Therefore, the ratio of AΓ to KΓ is less than that of $1009\frac{1}{6}$ to 66.

Let us divide the angle KAΓ again by the line ΛA; then, the ratio of AΛ to ΛΓ will be less than that of $2016\frac{1}{6}$ to 66, while the ratio of AΓ to ΓΛ will be less than that of $2017\frac{1}{4}$ to 66.

It follows, by working back, that the ratio of the perimeter of the polygon to the diameter is greater than that of 6336 to $2017\frac{1}{4}$. Now, the first of these numbers is 3 and $\frac{10}{71}$ greater than $2017\frac{1}{4}$; therefore, the perimeter of the polygon of 96 sides inscribed in a circle is also greater than thrice plus $\frac{10}{71}$ diameters; so that the circle is also, *a fortiori*, greater than thrice plus $\frac{10}{71}$ diameters. From this, the circumference of the circle is equal to three times the diameter plus a part, less than a seventh, but greater than $\frac{10}{71}$ of the diameter.

The circumference of a circle is bounded above by the perimeters of all polygons circumscribing the circle. Starting with a regular hexagon, then doubling the number of sides successively, Archimedes obtains more and more precise upper bounds for the circumference, finishing with a 96 sided circumscribing polygon and a limit of 3 + 1/7.

In order to find the ratio of the radius of the circle to the perimeter of a regular polygon, we only need to know the ratio of the radius to a half side of the polygon. In the figure, CZ represents a half side of a regular hexagon circumscribing the circle, and so: EC/CZ = $\sqrt{3}$ > 265/153 (this approximate value appears without explanation, see Section 8.11 below) and EZ/CZ = 2 = 306/153.

If EH is the bisector of angle CEZ, then CH will be the half side of a regular dodecahedron circumscribing the circle. The ratio to be found, EC/CH is equal to EZ/ZH, because the bisector of the angle at E in the tri-

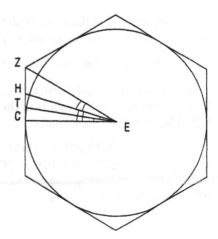

angle ECZ cuts the side CZ in the ratio ZH to HC, and this must be equal to the ratio of the sides EZ and EC (Euclid *Elements*, Book VI, Proposition 3). Therefore:

$$EC/CH = EZ/ZH = (EC + EZ)/(CH + ZH)$$
$$= (EC + EZ)/CZ > (306 + 265)/153 = 571/153.$$

Next, if ET is the bisector of angle CEH, then CT is the half side of a regular 24-sided circumscribing polygon. We now have to find EC/CT starting from EC/CH > a/b, the reasoning being exactly the same, but to do this we need first to calculate the ratio EH/CH. Now, $(EH/CH)^2 = (EC^2 + CH^2)/CH^2 > (a^2 + b^2)/b^2$. Let c be a number such that $a^2 + b^2 > c^2$; then EC/CH > c/b. Hence, EC/CT = EH/HT = (EC + EH)/CH > $(a + c)/b$. Here, we have: $a = 571$, $b = 153$ and $c = 591\frac{1}{8}$.

To state the general case, the ratio x_n of the radius to the half side of a regular circumscribing polygon of 3.2^n sides is greater than the nth term of the sequence a_n/b defined by:

$$a_{n+1}/b = (a_n + c_n)/b,$$

where $a_1 = 265$, $b = 153$ and c_n is a number such that:

$$c_n{}^2 < a_n{}^2 + b^2.$$

Since π s bounded above by $3.2^n.x_n$, considering successive regular circumscribing polygons produces the following upper limits for π:

hexagon	: 6 CZ/EC	< 6 × 153/265	< 3.47
dodecagon	: 12 CH/EC	< 12 × 153/571	< 3.22
24 sides	: 24 CT/EC	< 24 × 153/$1161\frac{1}{8}$	< 3.16
48 sides	: 48 CK/EC	< 48 × 153/$2334\frac{1}{4}$	< 3.147
96 sides	: 96 CL/EC	< 96 × 153/$4673\frac{1}{2}$	< $3\frac{1}{7}$

The second half of the proof is along similar lines. We consider inscribed regular polygons which give a closer lower value for π as they increase in size, and it is the polygon of 96 sides that gives a lower value of $3\frac{10}{71}$.

Now repeat the calculations, replacing the inequalities with equalities. If again we let x_n be the ratio of the radius to the half side of the circumscribing polygon with 3.2^n sides then from the formulas:

$$(EH/CH)^2 = (EC/CH)^2 + (CH/CH)^2 = (EC/CH)^2 + 1$$

and $$EC/CT = EH/HT = EC/CH + EH/CH,$$

we deduce: $x_{n+1} = x_n + \sqrt{1 + x_n{}^2}$.

To take another approach, if we let $2p_n$ and $2q_n$ be the perimeters of regular 3.2^n sided inscribed and circumscribed polygons, respectively, of a circle of unit radius, we have:

$$p_n < p_{n+1} < \pi < q_{n+1} < q_n.$$

Putting $u_n = \pi/3.2^n$ and $c_n = \cos u_n$, we have:

$$p_n/p_{n+1} = \sin u_n/2 \sin u_{n+1} = \cos u_{n+1} = c_{n+1}.$$

Therefore: $\qquad p_{n+1} = p_n/c_{n+1}.$
Similarly: $\qquad q_n = 3.2^n \tan u_n$
and $\qquad p_n/q_n = \cos u_n = c_n.$

From this, for $n > 0$, we have the recurrence relations:

$$c_{n+1}^2 = (1+c_n)/2$$
$$p_{n+1} = p_n / c_{n+1}$$
$$q_n = p_n / c_n$$

with the initial conditions:

$$c_0 = \frac{1}{2} \quad \text{and} \quad p_0 = 3\frac{\sqrt{3}}{2}.$$

As regards convergence, we note that the sequence $\dfrac{\pi - p_{n+1}}{\pi - p_n}$ converges to 1/4.

Concerning the number of iterations needed to achieve an accuracy to a given number of decimals when we take the mean value $(p_n + q_n)/2$, we note that:

$$q_n - p_n = p_n\left[\frac{1}{c_n} - 1\right] < \pi\frac{1-c_n}{c_n} < \pi\frac{u_n^2}{2-u_n^2} = \frac{\pi}{\dfrac{3^2.2^{2n+1}}{\pi^2} - 1}$$

and so, $q_n - p_n < 10^{-k}$ provided that $2n \log 2 > k + 0.24$, that is, approximately, if

$$n > (5k/3) + (2/5).$$

From this we obtain: 3 decimal places if $n = 6$; 4 decimal places with $n = 7$; 9 decimal places with $n = 16$; 16 decimals with $n = 28$.

Other values for π have been calculated since Archimedes. In the second century, Ptolemy produced the sexagesimal 'value' of $3 + 8/60 + 30/3600 = 3.1417...$, a value close to the mean of the two values given by Archimedes. Through the use of a process similar to the one described here, using polygons with more sides, a number of mathematicians obtained approximations, whose accuracy we consider today by the number of exact decimals that they give. For example, c AD 500 the Indian mathematician Āryabhatta obtained the value $62\,832/2000 = 3.1416$ with 3.2^7 sides; in 1424, al-Kāshī of Samarkand obtained a value equivalent to an accuracy of 16 places of decimals using 3.2^{28} sides and, in 1609, in Germany, Ludolph van Ceulen obtained an accuracy of 36 decimal places using 2^{62} sides, to give only the better performances.

As well as the perimeters, the areas of inscribed and circumscribed polygons can be used to approximate to a circle and this was done, for example, by François Viète in 1593 [28] (see Section 5.4), and by James Gregory in 1667 [11]. What follows is a Chinese example using areas that comes from the third century AD.

5.2 The Area of the Circle in the *Jiuzhang Suanshu*

The most influential and best known of all ancient Chinese mathematical works is the *Jiuzhang suanshu* (Computational Prescriptions in Nine Chapters) composed during the Han dynasty (220 BC to AD 206). The work includes 246 problems and provides remarkable logical justifications for its algorithms for calculating the areas of circles, areas of segments, volumes of spheres and pyramids, etc.

For the area of a 'round field', the *Jiuzhang suanshu* commonly uses procedures that are equivalent to taking $\pi = 3$.

Unconvinced by this, the mathematician Liu Hui, (towards the end of the third century), describes in his still famous commentary on the *Nine Chapters*, how to obtain a better value for the ratio of the circumference to the diameter using a sequence of regular polygons of 6, 12, 24, 48, 96 and finally 192 sides, inscribed in the circle. But he differed from Archimedes in considering, not perimeters, but areas.

Besides Liu Hui, the main commentaries on the *Jiuzhang suanshu* are those by Zu Chongzhi (fifth century) and Li Chunfeng (seventh century). The Chinese are indebted to the first of these writers – better known as an astronomer and inventor of ingenious machines – for the approximation to π of 355/113, and to the second the more useful simplification of 22/7. We do not know where Li Chunfeng obtained this second value, but for the value 355/113 it is generally thought that Zu Chongzhi found it by using Liu Hui's method but considering even more polygons.

Problems 31 and 32 of Chapter 1 of the *Jiuzhang suanshu* (not translated here) deal with calculating the area of a circle, given its diameter and circumference, and uses the formula:

$$A = (c/2)(d/2) \qquad\qquad (1)$$

where c and d stand, respectively, for the circumference and diameter. In his commentary on these problems and their solutions, Liu Hui first of all justifies the use of formula (1). Then, recalling that the earlier Chinese took the ratio of the circumference to the diameter to be '3', he explains that this value is in error, and then he determines a more accurate value for the ratio which corresponds to our $\pi = 3.14$.

For us, the structure of the original text by Liu Hui naturally appears somewhat rambling. It will help in reading it to bear in mind the following:

(a) the meaning of the term 'module' (in Chinese *lü*): in the particular context of this text, the 'module' of the diameter and the 'module' of the circumference refer to, respectively, a particular value of the diameter of a circle and the corresponding value of its circumference, both given, explicitly or not, in the same units of length, it being understood that these two lengths vary proportionately with one another. So what we call 'π' was for Liu Hui fundamentally a pair of lengths, one associated with the length of the diameter, the other with the length of the circumference.

(b) units of length: the units in the Chinese text are all based on multiples of ten. Hence, in this sequence, each unit is worth a tenth of its predecessor: *chi* ['foot'], *cun* ['inch'], *fen* ['tenth of an inch'], *li, hao, miao, hu.*

(c) calculations of numbers and their representation: it is generally thought that at the time of Liu Hui calculations were carried out on a 'counting board' consisting of spaces reserved for the different units, but the text does not specify this and so other ways of calculating need not be excluded.

In our translation we have given all numbers in their 'modern' form. Not to do so would have made the text almost illegible. (The numbers of the original text were written in a type of decimal numeration that lacked zero; in practice a zero would probably have appeared as an empty space on the counting board when there were none of a particular unit.)

(d) geometric terms: instead of using letters of the alphabet to identify the different parts of a right angled triangle, as was the custom with the Greeks, Liu Hui used specific terms for the sides of such a triangle: the hypotenuse is *xian* (chord) and the other two sides are *gou* (base) and *gu* (height). When several triangles are being considered in the same figure, Liu Hui differentiates between them by using qualifying adjectives.

The commentary of Liu Hui

From Bai Shangshu, *Jiuzhang suanshu zhushi*, critical edition of the *Jiuzhang suanshu* and of its original commentaries, Peking: Kexue chubanshe, 1983, pp. 35–45.
From the French tr. of the Chinese by J.- C. Martzloff.

By taking the semi-circumference as length and the radius as breadth, the length-breadth product gives the area [of the circle]. Given a circle of diameter 2 *chi*, the side of the inscribed hexagon is exactly equal to the radius of the circle: this corresponds to the 'modules' of the circumference and the diameter having the values 3 and 1 respectively. From the figure, if a side of the hexagon is multiplied by the radius [and the result] multiplied by 3, the area of the [inscribed] dodecagon is found. If the dodecagon is again divided up and the side of the dodecagon is multiplied by the radius [and the result] multiplied by 6 the area of the [inscribed] 24-sided polygon is found. The more finely [the polygon] is divided up, the more the error diminishes. In dividing up the polygon more and more to the point where it is impossible to sub-divide it further, what is arrived at is coincident with the 'body' of the circle and there is no further error. To the exterior of a side of a [given] polygon, there is always a 'remnant diameter'. In multiplying the side of a polygon by its 'remnant diameter', we obtain a shape that 'extends beyond the arc' [of the circle]. This is why, when the polygon results from an extremely fine division, its 'body' coincides with that of the circle and the 'remnant diameter' finally disappears. However, if there is no more 'remnant diameter', the area [of a regular polygon together with small rectangles drawn on the outside of its sides] will no longer extends beyond the arc. Multiplying the radius by the side [that results from division] of the arc, the number of sides of the polygon doubles at each division. This is why the product of the semi-circumference and the radius gives the area of the circle. But since we are now only considering the circumference and the diameter, we can say that the values that we now approach are 'perfectly correct'. It is no longer [therefore] a matter of a circumference and a diameter which have 'modules' respectively of length 3 and 1. [In fact] when we say that the circumference has a value 3, we are only thinking of the hexagon. [Given this] if we use [the 'modules' 3 and 1] to determine the error that is made in finding [the area of] the circle we see that [this error] is similar to that made [when we replace] the arc by the chord. All the same, this method [i.e. using the 'modules' 3 and 1] has been handed down from generation to generation and nobody has thought of finding a more accurate way. Following in the steps of their ancestors, students have learnt the erroneous ancient method. Lacking a firm basis, it is difficult to come to grips [with this question]. In general, things which are not round are square. If one intends to 'understand intimately' the 'modules' of round and of square things one will arrive at knowing them, even though they are not easily accessible. By considering 'modules' in this way, one will come to see that their applications are numerous. I would like to point out, respectfully, that [in what follows] I have used figures to construct 'accurate modules'. [Furthermore] I was afraid of presenting [my] method 'in the abstract' because the numbers [arising in practice might appear] obscure and difficult to understand. Also I freely give here [such numbers] and I make precise observations on them.

Method for dividing the [inscribed] hexagon to obtain a dodecagon

Given the diameter of a circle, 2 *chi*, take the half, that is 1 *chi*: this is the side of a hexagon inscribed in the circle. Take the radius, 1 *chi* as a hypotenuse and the half-side, 5 *cun* for the base. [With these given] find the 'height': subtract the base-squared, 25 *cun* [squared] from the hypotenuse-squared, and there remains 75 *cun* [squared]. Extract the square root [of 75] going as far as [the units called] *miao* and *hu*, then move back the 'divisor' so as to determine the smaller numbers [corresponding to units smaller than the smallest known units of the length, namely the *miao* and the *hu*]. Take the small (*unnamed*) numbers [of the unit] to be numerators having ten as denominator. Simplifying, find 2/5 of *hu*. This gives a 'height' of 8 *cun* 6 *fen* 6 *li* 2 *miao* 5 *hu* and 2/5 *hu*. Subtracting [this value] from the radius, there remains 1 *cun* 3 *fen* 3 *li* 9 *hao* 7 *miao* 4 *hu* and 3/5 *hu*. Call this result the 'little base'. Also, call the half-side of the polygon the 'little height'; [using this 'little base' and 'little height'] find the 'little hypotenuse'. Neglecting the remainder, the square [of the 'little hypotenuse'] has the value 267949193445 (*hu*) [squared]. Extracting the square root [of this number] we obtain the side of the [inscribed] dodecagon.

Method for dividing the [inscribed] dodecagon to obtain a '24-gon'

Again take the radius as hypotenuse and the half-side as base to calculate the [corresponding] 'height'. Write at the top the square of the little hypotenuse and take its quarter. Neglecting the remainder [of the division], we obtain 66987298361 *hu* [squared]. This is the square of the base. Subtracting this from the square of the hypotenuse and extracting the square root of the result, we obtain a 'height' of 9 *cun* 6 *fen* 5 *li* 9 *hao* 2 *miao* 5 *hu* and 4/5 *hu*. Subtracting this from the radius, there remains 3 *fen* 4 *li* 7 *miao* 4 *hu* and 1/5 *hu*. Call [this result] the 'little base'; also call the half-side of the polygon the 'little height'. [Using this 'little base' and this 'little height'] find the 'little hypotenuse'. The square [of this 'little hypotenuse'] has the value 68148349466 *hu* [squared], neglecting the remainder fraction. Extracting the square root, obtain the side of the 24-gon.

Method for dividing the [inscribed] 24-gon to obtain the 48-gon

Again take the radius as hypotenuse and the half-side as base to calculate the [corresponding] 'height'. Write down the square of the little hypotenuse and divide by 4. Neglecting the remainder [of the division], we obtain 17037087366 *hu* [squared]. This is the square of the base. Taking [this square of the base] from the square of the hypotenuse and extracting the square root of the result, obtain a 'height' of 9 *cun* 9 *fen* 1 *li* 4 *hao* 4 *miao* 4 *hu* and 4/5 *hu*. Subtracting [this result] from the radius, there remains 8 *li* 5 *hao* 5 *miao* 5 *hu* and 1/5 *hu*. Call [this result] the 'little base'; also call the half-side of the polygon the 'little height'. [Using this 'little base' and this 'little height'] find the 'little hypotenuse'. Neglecting remainders [of this calculation] its square has the value 17110278813 *hu* [squared]. Extracting the square root, we obtain 1 *cun* 3 *fen* 8 *hao* 6 *hu* for the 'little hypotenuse', neglecting remainders. This is the side of the 48-gon. Multiplying [this value] by the radius 1 *chi*, and then by 24, we obtain an area of 3139344000000 *hu* [squared]. Dividing [this result] by one hundred times one hundred million [10^{10}] we obtain an area of 313 *cun* [squared] and 584/625 *cun* [squared], that is area of the 96-gon.

Method for dividing the [inscribed] 48-gon to obtain the 96-gon

Again take the radius as hypotenuse and the half-side as base in order to calculate the 'height'. Write down the square of the hypotenuse and divide by 4. We obtain 4277569703 *hu* [squared], neglecting the remainder. This is the square of the base. Taking this result from the square of the hypotenuse and extracting the square root, we obtain 9 *cun* 9 *fen* 7 *li* 8 *hao* 5 *miao* 8 *hu* and 9/10 *hu*. Subtracting [this result] from the radius, there remains 2 *li* 1 *hao* 4 *miao* 1 *hu* and 1/10 *hu*, which is what we call the 'little base'. Also call the half-side of the polygon the 'little height'. [Given these values] find the 'little hypotenuse'. Neglecting remainders, its square has the value 4282154012 *hu* [squared]. Extracting the square root, we obtain 6 *fen* 5 *li* 4 *hao* 3 *miao* 8 *hu* for the 'little hypotenuse', neglecting remainders. This is the side of the 96-gon. Multiplying [this value] by the radius 1 *chi*, and then by 48, we obtain an area of 3141024000000 [6 zeros] *hu* [squared]. Dividing [this result] by one hundred times one hundred million *yi*, we obtain an area of 314 *cun* [squared] and 64/625 *cun* [squared]. This is the area of the 192-gon. Subtracting from this value the area of the 96-gon leaves 105/625, which is called the 'difference in area' (*cha-mi*). Doubling [this difference], which gives 210 [as numerator], we have the exterior part that 'goes outside' the circle of the 96-gon. This is the total of the areas corresponding to 'the arrow that multiplies the hypotenuse'. Adding this area to that of the 96-gon, we obtain 314 *cun* [squared] and 169/625 *cun* [squared], and this borders the circle exteriorly. That is why the area of the 192-gon has a total value of 314 *cun* [squared]. This can be considered to be 'the module determining the area of the circle', provided we neglect fractions. Dividing the area of the circle by 1 *chi* and doubling [the result], we obtain 6 *chi* 2 *cun* 8 *fen*, which is the number of the circumference.

Starting from a regular hexagon inscribed in a circle of radius 1 *chi*, that is 1 foot, Liu Hui claims initially that the product of the side of the hexagon, the radius of the circle and the number 3 (= 1/2 × 6) is equal to the area of the inscribed dodecagon. He then shows that the product of the side of a dodecagon, the radius and the number 6 (= 1/2 × 12) gives the area of a regular 24-sided polygon. Having given these two examples, Liu Hui concludes by claiming that by continued application of this process, the polygon will end up coinciding with the circle. In other words, there exists a simple relation connecting the area S_{2n} of a regular 2n-sided polygon, the radius (*d*/2) of the circle and the side c_n of an *n*-sided regular polygon:

$$S_{2n} = c_n (n/2) (d/2) \tag{2}$$

From this we get the formula (1) given above since, intuitively, the limit of nc_n is the length of the circumference *c*.

Accepting the general principle, Liu Hui goes on to work out the actual values for S_{96} and S_{192}, the areas of 96 and 192 sided regular inscribed polygons. To calculate the areas, he starts by finding the lengths of the sides of the two polygons. Starting with c_6 = radius of the circle, he finds, successively, c_{12}, c_{24}, c_{48} and finally c_{96}. On arriving at c_{48} he immediately calculates S_{96} by using the formula (1) given in the first part of his commentary:

$$S_{96} = c_{48} \times r \times 24$$

Again, using the same formula, he obtains

$$S_{192} = c_{96} \times r \times 48.$$

Although the text which we now have available to us contains no figures, it is not difficult to see that Liu Hui's calculations correspond to the following figure:

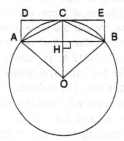

In this figure, OA = OB = r; AB = c_n (the side of a regular inscribed polygon of n sides), AC = CB = c_{2n}, AH = $c_n/2$ and OH = a_n (the apothem of a regular n-sided polygon).

Applying the theorem of Pythagoras to the triangles AOH and ACH, we obtain:

$$a_n = \sqrt{r^2 - \left(\frac{c_n}{2}\right)^2} \qquad c_{2n} = \sqrt{\left(\frac{c_n}{2}\right)^2 + (r - a_n)^2}$$

As can be seen on reading the translation, the algorithm of Liu Hui depends upon these two formulas but not on the systematic calculation of all the square roots. Liu Hui finds a_n and then c_{2n}^2 for n = 6, 12, 24 but it is not until the end of the process that he calculates c_{2n} (n now being equal to 48).

Having found S_{96} =314 584/625 *cun* squared and S_{192} =314 64/625 *cun* squared, Liu Hui also calculates:

$$S_{96} + 2(S_{192} - S_{96}) = 314\ 169/625.$$

As the reader can verify from the figure, the term $2(S_{192} - S_{96})$ corresponds geometrically to the area of 96 small rectangles, all equivalent to ABED which, added to S_{96} completely cover the circle and slightly extending beyond it. Liu Hui's calculations are equivalent to providing bounds for the area S of the circle:

$$314\ 64/625 = S_{192} < S < S_{96} + 2(S_{192} - S_{96}) = 314\ 169/625.$$

Whence the result: $S = 314$ *cun* [squared].

It can be seen that Liu Hui's calculations involved addition, subtraction, multiplication, division and the extraction of square roots. The method of extracting square roots was very similar to that taught in French schools up to 30 years ago. Liu Hui divided the number to be 'square-rooted' into pairs of digits and then found the greatest square number contained in each part. Then he found the digits of the root one after the other. Intermediate calculations required doubling the 'remainder'. Whenever the calculations failed to come out exactly, Liu Hui expressed the remainder as a fraction.

In contrast to Archimedes, who worked systematically, Liu Hui frequently cut out intermediate working. Nevertheless, the final stage of the working produced rigorously correct bounds. However, since the computations are performed with a fixed number of digits and involve numerous roundings, the bounds would not necessarily have been correct with a different number of digits. While Lui Hiu's algorithm is quite correct mathematically, it nonetheless raises difficulties from the point of view of specific calculations and these were overlooked by Chinese mathematicians. Proof is not presented here in the same way as it appears in Greek texts. Often Liu Hui appeals to visual evidence, as when he claims that the side of a regular hexagon is equal to the radius of the circle. He also tends to generalise upon the basis of a few well chosen examples. For example, having explained formula (1) he states the result for the dodecagon, then for the 24-gon and then goes on to generalise without any further justification; the same can be said for his appeal to the limiting process.

5.3 The Method of Isoperimeters

The problem of 'squaring the circle' or the 'quadrature of the circle' (see [4], [8] or [16]), that is the construction 'by ruler and compasses' of a square having the same area as a given circle, arose as early as the fifth century BC. An equivalent problem, the rectification of the circle by ruler and compass, is to construct a line equal in length to the perimeter of the given circle. Whichever problem is chosen, its solution effectively obtains π as the ratio of two rectilinear magnitudes. In the fourth century BC, Dinostratus using the *quadratrix* of Hippias of Elis, and in the third century BC, Archimedes with his *spiral*, demonstrated ways of rectifying a circle. However, one of the two lengths involved in the constructions could not be produced by ruler and compass constructions: they were generated by an infinite limiting process.

In the short extract below, Descartes proceeds in the reverse direction: that is, constructing a circle starting from a given square. More precisely, he shows how to construct, by ruler and compass, the diameter d of a circle having the same perimeter p as the given square, in other words the ratio $p/d = \pi$. In actual fact, in the same way as the methods used by the Greeks, the diameter d, and therefore the value obtained for π, is only found by a limiting process, since the method involves an infinite number of iterations. Nonetheless, the method does allow π to be obtained as accurately as is desired.

← Illustration from the work of Père Grégoire de Saint-Vincent, *Problema austriacum: quadratura circuli*, Anvers, 1647. The shaft of light passing through the square and producing an exact circular pool of light on the floor symbolises the quadrature of the circle, while the sphere resting on a cube evokes the problem of the cubature of a sphere.

The method relies on the principle, that of all plane figures of a given perimeter, the circle has the greatest area. This circle can be found as the limit of a sequence of regular polygons having the same perimeter, the number of sides being doubled at each stage. Producing the sequence of diameters of the circles inscribed in these polygons will lead to the limit diameter *d* of the required circle, whose circumference is equal to the perimeters of the polygons.

The construction for this sequence of diameters is shown in the text reproduced below, which was written by Descartes in about 1640. The text was published posthumously in Amsterdam, in 1701, and appears in a collection of his other works.

The question of *isoperimeters*, that is showing that the circle is indeed the figure with the maximum area for a given perimeter, has a history that goes back a long way. However, although the property appears natural enough, the rigorous proof of it had to wait until the end of the nineteenth century.

René Descartes
Circuli quadratio (On the quadrature of the circle), Excerpta ex MS R. Des Cartes,
Published: Amsterdam, 1701, in *les Oeuvres de Descartes*, Charles Adam & Paul Tannery, t. X, 1908; repub. Paris: Vrin, 1996, pp. 304–305.

As for the *quadrature of the circle* I have found nothing more suitable, than if to the given square *bf* there should be added a rectangle *cg* comprised of the lines *ac & cb*, such that it is equal to a quarter part of the square *bf*; in the same way, the rectangle *dh*, made of the

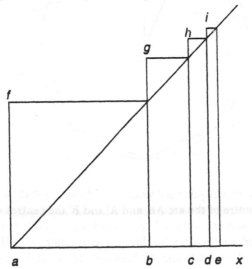

lines *da, dc,* equal to a quarter part of the preceding one; and the same for the rectangle *ei,* and also for an infinity [of them] as far as *x*: all these together shall be equal to a third part of the square *bf.* And this line *ax* will be the diameter of a circle whose circumference is equal to the perimeter of that square *bf*: furthermore *ac* is the diameter of the circle inscribed in an octagon isoperimetric [of equal perimeter] to the square *bf*; *ad* is the diameter of the in-scribed circle of a 16-sided figure, [and] *ae* is the diameter of the inscribed [circle] of a 32-sided figure, isoperimetric to the square *bf*; and so to infinity.

Descartes' text restricts itself to a description of the construction process, to-gether with an indication of why it works. Starting with a square *bf* (referred to by its diagonal), the problem is to find the diameter *ax* of a circle having the same pe-rimeter as the square. The end point, *x*, is the limit of a sequence of points *b, c, d, e,* ... obtained by the succesive construction of rectangles *cg, dh, ei,* ... whose areas are in a geometric progression with common ratio 1/4. Now for the construction. There is no indication of how to construct these rectangles, that is how to find the length *L* of a rectangle, knowing the difference *ℓ* between the length and breadth, and the area *A*/4. This is equivalent to a geometrical solution of the equation $L (L - \ell) = A/4$ (the equation's having just one positive root). The method is given in Euclid's *Ele-ments,* Book VI, Proposition 29. We shall look at this after establishing the recur-rence relations.

As for a justification of the method, Descartes explains that the lengths *ab, ac, ad, ae,* ... are successively the diameters of circles inscribed in regular convex polygons, of the same perimeter, and having 4, 8, 16, 32, ... sides. This can be explained as fol-lows:

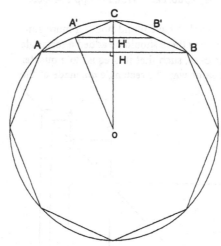

Let AB be the side of a regular polygon of 2^{n+1} sides. Let O be the centre of the circumcircle, C the centre of the arc AB, and A' and B' the centres of AC and BC re-spectively. Then A'B' = AB/2 is the side of a regular polygon having twice the num-ber of sides of its predecessor, yet with the same perimeter *p*. Let H and H' be where OC meets AB and A'B', then OH and OH' are the radii r_n and r_{n+1} of the inscribed circles, respectively, of the first and second polygon. In the right-angled triangle OA'C, we have:

$$A'H'^2 = H'C.H'O, \quad \text{that is: } (r_{n+1} - r_n)r_{n+1} = p^2/4^{n+3}.$$

The successive terms $(r_{n+1} - r_n)r_{n+1}$ are therefore in geometric progression with common ration $1/4$, just as the areas of the rectangles we are considering.

Now, if we put $\quad p/4 = ab = 2r_1, ac = 2r_2, ad = 2r_3, \ldots,$

then we have $\quad\quad ac(ac - ab) = ab^2/4 = (p/4)^2/4,$

which gives $\quad\quad\quad r_2(r_2 - r_1) = p^2/4^4.$

In the same way, $\quad\quad ad(ad - ac) = ac(ac - ab)/4,$

and so $\quad\quad\quad\quad r_3(r_3 - r_2) = p^2/4^5; \text{etc.} \ldots$

Therefore, $ax = 2 \lim r_n$ is the limit diameter of the desired circle.

The preceding calculations allow us to set up the recurrence relations for the isoperimetric method. Using r_n and s_n for the radii, respectively, of the inscribed and circumscribed circles of a regular polygon with perimeter p and 2^{n+1} sides, we have:

$$p/2s_n < \pi < p/2r_n$$
$$r_n < r_{n+1} < s_{n+1} < s_n$$

and as n tends to infinity, $s_n - r_n$ tends to 0. We have just established the recurrence relation:

$$(r_{n+1} - r_n)r_{n+1} = p^2/4^{n+3}$$

and so, putting $r_1 = p/8$, we have:

$$r_{n+1} = \frac{1}{2}\left(r_n + \sqrt{r_n^2 + \frac{r_1^2}{4^n}}\right)$$

This formula shows how Descartes' sequence of rectangles can be constructed. Given $ab = 2r_1$ and $a\ell = 2r_n$ we need to find a point m such that $am = 2r_{n+1}$.

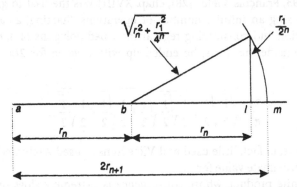

But we can obtain other recurrence relations. For example, $OH' = (OH + OC)/2$, that is:

$$r_{n+1} = (r_n + s_n)/2$$

and $OH'.OC = OH'.(OH' + H'C) = OH'^2 + OH'.H'C = OH'^2 + A'H'^2 = OA'^2,$

that is $s_{n+1} = \sqrt{s_n \cdot r_{n+1}}$

with initial values $r_1 = p/8$ and $s_1 = p\sqrt{2}/8$.

Or again, putting $c_n = \cos[\pi/(2n+1)]$, we can have:

$$c_{n+1} = \sqrt{\frac{1+c_n}{2}}$$

$$r_n = s_n c_n$$

$$s_{n+1} = s_n c_{n+1}$$

with intitial values: $c_1 = \sqrt{2}/2$ and $s_1 = p\sqrt{2}/8$.

The similarity of these formulas with those obtained by Archimedes' method is evident (see Section 5.1 above).

It is now known that the quadrature of the circle is not possible with ruler and compass methods and so the point x that Descartes sought cannot be found in a finite number of steps. In fact Wantzel [30] showed in 1837 that the number π cannot be the solution of a polynomial of degree 2^n, while Lindemann [20] proved in 1882 that π is not an algebraic number.

Analytic Approaches

5.4 Arithmetic Quadrature

As calculations using limiting processes become perfected, analytic formulas for calculating π were developed, initially through the use of infinite products or continued fractions. In 1593, François Viète ([28], chap. XVIII) was the first to give an explicit formula for π using an infinite number of operations. Starting, as always, from a geometric position and considering regular inscribed polygons of 4, 8, 16, ... sides, but this time considering areas, he ended up with a value for $2/\pi$ as the infinite product:

$$\frac{2}{\pi} = \sqrt{\frac{1}{2}} \sqrt{\frac{1}{2} + \frac{1}{2}\sqrt{\frac{1}{2}}} \sqrt{\frac{1}{2} + \frac{1}{2}\sqrt{\frac{1}{2} + \frac{1}{2}\sqrt{\frac{1}{2}}}} \cdots$$

This formula is, in fact, little used and Viète himself used Archimedes' method to calculate an approximate value for π.

Another infinite product, whose convergence is extremely slow, but which contains integers, was given by Wallis [29] in 1656. Using $\square$ for the value $4/\pi$, his formula is:

$$\Box \begin{cases} \text{min or quam} \dfrac{3\times3\times5\times5\times7\times7\times9\times9\times11\times11\times13\times13}{2\times4\times4\times6\times6\times8\times8\times10\times10\times12\times12\times14}\times\sqrt{1\tfrac{1}{13}} \\[2mm] \text{major quam} \dfrac{3\times3\times5\times5\times7\times7\times9\times9\times11\times11\times13\times13}{2\times4\times4\times6\times6\times8\times8\times10\times10\times12\times12\times14}\times\sqrt{1\tfrac{1}{14}} \end{cases}$$

or, more simply, $\dfrac{4}{\pi}=\dfrac{3.3.5.5.7.7....}{2.4.4.6.6.8....}$

Wallis obtained his formula by means of a quadrature of $\sqrt{1-x^2}$ based on Newton's binomial formula, the quadrature of algebraic curves of the form $(1-x^2)^k$ and a principle of induction on the form of algebraic expressions, which amounts to a type of interpolation. Today, this formula occurs when evaluating reduction formulas, sometimes called Wallis integrals, of the type:

$$I_n = \int_0^{\frac{\pi}{2}} \sin^n x\, dx$$

for which we have: $I_n = \{(n-1)/n\}I_{n-1}$ and $I_0 = \pi/2$.

Here is another expression for evaluating π which was derived by Lord Brouncker. The formula is expressed in terms of continued fractions (See Section 4.4) and it also gives π to any desired degree of accuracy, provided the calculations are carried out far enough:

$$\frac{4}{\pi} = 1 + \cfrac{1^2}{2 + \cfrac{3^2}{2 + \cfrac{5^2}{2 + \cfrac{7^2}{2 + \cfrac{9^2}{2+...}}}}}$$

Brouncker's formula was known to Wallis who commented on it. Leibniz also produced a formula which gives π this time in terms of the sum of a series, namely:

$$\frac{\pi}{4} = 1 - \frac{1}{3} + \frac{1}{5} - \frac{1}{7} + \frac{1}{9} - \frac{1}{11} + ...$$

Although it is not known how Brouncker derived his continued fraction formula for $4/\pi$, it can be seen that it is related to the Leibniz series, since the successive convergents of the continued fraction, 1, 3/2, 15/13, 105/76, ... are none other than the inverses of the partial sums of the series.

Leibniz refers to his formula as *arithmetic quadrature* since it "expresses the size of the proposed figure as an infinite sequence of numbers, either rational or commensurable to a given magnitude" [19]. Leibniz derived this formula in 1673, but it did not appear until 9 years later in *Acta Eruditorum* [18]. In this text, he relies on showing the harmony of certain sequences of numbers, without explaining the validity of his reasoning. An idea of his method, the method of *metamorphoses*, can be found in a reading of earlier letters that he had written, for example, his letter to

Oldenburg on 17 August 1673 [27]. For simplicity, we give here extracts from a letter that Leibniz wrote to La Roque, the editor of *Acta Eruditorum* [19], in which he explains the general principles of his method of metamorphoses. This is: the area under a curve is divided up into infinitesimal triangles, each of which has an area equal to half that of an infinitesimal rectangle, and the rectangles together make up the area under another curve whose quadrature can be easily effected. Applying the method to particular cases, it confirms the quadratures that were already known, and in the case of the circle, the method provides an expression for π.

G. W. Leibniz
From a letter to La Roque, *Mathematische Schriften*,
vol. V, 88–92, Hildesheim: Olms, 1962.

Monsieur
    The arithmetic quadrature of the circle and of its segments or sectors, which I have found and communicated to several excellent geometers, some years ago, appeared quite extraordinary to them, and they have exhorted me to make it public. [...]
    The arithmetic quadrature of the circle or of its parts can be found in this theorem: the radius of the circle being a unit, and the tangent BC to the half BD of a given arc BDE being called b, the magnitude of the arc will be: $\dfrac{b}{1} - \dfrac{b^3}{3} + \dfrac{b^5}{5} - \dfrac{b^7}{7} + \dfrac{b^9}{9} - \dfrac{b^{11}}{11}$ etc. Now, with the arcs being found, it is easy to find the areas, and the corollary of this theorem is that the diameter and its square being 1, the circle is $\dfrac{1}{1} - \dfrac{1}{3} + \dfrac{1}{5} - \dfrac{1}{7} + \dfrac{1}{9} - \dfrac{1}{11}$ etc. [...]

    Since there is nothing so important as to see the origins of inventions, which are of greater value, to my mind, than the inventions themselves, on account of their fecundity and because they contain within them the source of an infinity of other [inventions] which can be drawn from them by a certain combination (as I am in the habit of calling it), or application to other subjects, when one knows how it should be done; I believed that I was obliged to let the public know of it. [...]
    Now, the way that the combinations presented themselves to me can be used to find an infinite number of figures commensurable with a given figure. To this purpose, I used this lemma: three parallel lines BC, GE, HF (fig. 1), passing through three angles of a triangle BEF and one of the sides EF being produced as far as to meet one of the parallel lines at C, the rectangle under the interval BC between the point of intersection C and the angle B, through which this parallel line passes, and under GH, the distance of the two other parallel lines GE, HF, that is to say the rectangle PGH (supposing that BGH is normal to BC, and CP equal and parallel to BG) will be twice the triangle BEF. In the same way, if HQ is equal to BM, the rectangle QHN will be equal to twice the triangle BFL. And if these bases EF, FL, etc. are infinitely small, and continued so that the triangles BEF, BFL, etc. fill the whole space EB((E))LFE to the curve EFL((E)), and similarly if GH, HN, etc. are infinitely small so that the rectangles BGH, QHN, etc., fill all the space PG((G))((P))QP to the curve PQ((P)), all that space will be twice the other space. And since FEC, LFM, ((E))((C)) are tangents to the first curve, the theorem can be stated in a general way thus: if from a curve E((E)) is drawn to a side AB of a right-angle ABC the ordinates EG, ((E))((G)), [and] to the other side BC the tangents EC, ((E))((C)), then the sum of the intercepts BC, ((B))((C)) between the point of the angle B and the point of intersection of the tangents C or ((C)) applied normally to the axis AB or GP, ((G))((P)), that

is the figure PG((G))((P))QP will be twice the space EB((E))E taken between a part of the first curve and the lines which join the extremities of that part to the point B.

This theorem is one of the most considerable and one of the most universal in geometry. And I have made many deductions from it which merit being touched on in passing. [...] The third corollary is the arithmetic quadrature of the circle. For the curve E(E)((E)) being an arc of a circle, the curve of the intercepts, namely BP(E)((P)), can be given by this equation $\frac{2az^2}{a^2+z^2} = x$, referred to the right-angle RBC by, calling BG or CP, x and BC or GP, z, that is to say RB will be to BG in double ratio of AC to BC, as it is easy to prove. [...]

Fig. 5.1

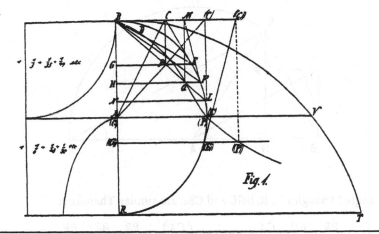

Fig. 1.

Let us review Leibniz' method of metamorphoses. Firstly, we need to show that the area of the triangle BEF is half that of the rectangle GPQH.

Fig. 5.2

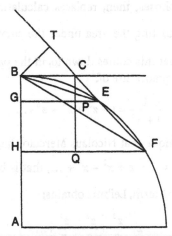

Let BT be the perpendicular to EF through B and let π be the size of the equal angles CBT and PCE (see Fig. 5.2). Then, the area of triangle BEF = EF.BT/2 = EF.(BC.cos α)/2 = (EF.cos α).BC/2 = GH.BC/2 = ½ area of rectangle GPQH.

It follows that the area under the integral curve, bounded by the segments joining B to the extremities of that curve, is equal to the sum of the areas of the triangles BEF, where F is infinitely close to E. Consequently, it is equal to half the sum of the infinitesimal rectangles GPQH, where Q is infinitely close to P, that is half the sum of the area under the curve described by P, bounded by segments parallel to BC and passing through the extremities of the curve. In practice, the point P is found by constructing the tangent at E to replace the line EF.

Let us look at the case where the curve is the quadrant of a circle. Following Leibniz, we use a for the radius, with $z = PG$ and $x = BG$ for the abscissa and ordinate of P.

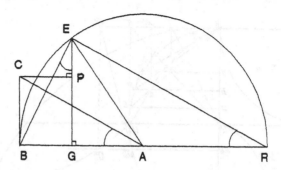

Fig. 5.3

The right-angled triangles BER, BGE and CBA are similar. Therefore:

$$\frac{BR}{BE} = \frac{BE}{BG} = \frac{CA}{CB}; \text{ and so: } \left(\frac{CA}{CB}\right)^2 = \frac{BR}{BE} \times \frac{BE}{BG} = \frac{BR}{BG}$$

from which: $\dfrac{2a}{x} = \dfrac{a^2 + z^2}{z^2}$ and hence $x = \dfrac{2az^2}{a^2 + z^2}$.

The method of metamorphoses, then, replaces calculating of the area under the curve $\sqrt{a^2 - x^2}$ with calculating the area under the curve $\dfrac{2az^2}{a^2 + z^2}$ where x and z take values from 0 to a. What this comes down to, in the particular case where $a = 1$, is having to determine the quadrature of:

$$\frac{1}{1 + z^2} = 1 - z^2 + z^4 - z^6 + \dots$$

By using the same technique that Nicolaus Mercator had used for the quadrature of the hyperbola $y = \dfrac{1}{1 + x} = 1 - x + x^2 - x^4 + \dots$, that is by expanding it as a series and then integrating term by term, Leibniz obtains:

$$z - \frac{z^3}{3} + \frac{z^5}{5} - \frac{z^7}{7} + \dots$$

In actual fact, before Leibniz, Gregory had found the complete series expansion of $\tan^{-1} x$ (or arctan x). For example, in a letter of 15 February 1671 to Collins [27], he states that if the *arcus* is a and the tangent t, then

$$a = t - \frac{t^3}{3} + \frac{t^5}{5} - \frac{t^7}{7} + \frac{t^9}{9} - ...$$

Gregory found this series when he was working on the series expansions of given functions, but there is no record of him using it to find a value for π.

We should also mention that there is an earlier record of this arithmetic quadrature in a Sanskrit manuscript of about AD 1500 by Nilakantha [24].

Finally, we should note that this series is not very effective as a calculating aid for evaluating π. In fact its convergence is extremely slow, since a thousand terms are needed just to achieve an accuracy to three places of decimals. In the next Section we shall see how Euler approached the problem.

5.5 Using Series

Considering again the series expansion of the function arctan t, but this time without using the simplicity that comes from having the arc $= \pi/4$, Euler was interested in seeing if the convergence of the Leibniz series could be improved. To this end, he used the result: $\tan \pi/6 = 1/\sqrt{3}$ and, even better, he considered two series together by replacing the result:

$\pi/4 = $ arctan 1 by: $\pi/4 = $ arctan $(1/2)$ + arctan $(1/3)$.

We should point out that Euler's text reproduced below is not original as regards the technique of procedure, but it is given here because of the clarity of its explanations. In fact, we have already said that Gregory had found the expansion:

$$a = t - \frac{t^3}{3} + \frac{t^5}{5} - \frac{t^7}{7} + \frac{t^9}{9} - ...$$

for the arc a whose tangent is t.

Newton [23] gave a series expansion that started in much the same way, in his *Method of Fluxions*, where he was considering the quadrature of part of a circle given by the equation $y = \sqrt{x - x^2}$. The formula, which has an even better convergence, is:

$$\pi/24 = \sqrt{3}/32 + 2/3.2^3 - 1/5.2^5 - 1/28.2^7 - 1/72.2^9 - ...$$

These analytical formulas considerably increased the number of decimals that could be calculated for π. Abraham Sharp managed to double the performance of van Ceulen in obtaining 72 places, but it was John Machin who made a significant breakthrough with the formula:

$\pi/4 = $ arctan $(1/5)$ – arctan $(1/239)$.

This formula, composed entirely of integers, enabled him to obtain 100 decimal places in 1706. In 1719, M. de Lagny [13] calculated 127 places of decimals starting with the Euler formula:

$$\pi/4 = \arctan(1/2) + \arctan(1/3).$$

Here is the value found by de Lagny:

3,14159265358979323846264338327950 2
8841971693993751058209749445923078 16
4062862089986280348253421170679821480
865132723066470938446 +.

* the 114th digit is not 7 but should be 8!

Leonhard Euler
From *Introductio in analysin infinitorum*, Book I
Translated by John D. Blanton as *Introduction to Analysis of the Infinite*,
Chapter VIII, On Transcendental Quantities Which Arise from the Circle, §§ 140–142.

[...] we obtain

$$z = \frac{\tan z}{1} - \frac{(\tan z)^3}{3} + \frac{(\tan z)^5}{5} - \frac{(\tan z)^7}{7} + \dots.$$ If we let $t = \tan z$ so that z is the arc whose tangent is t, which we will indicate by arctan t, then $z = \arctan t$. When we know the tangent of t, the corresponding arc z is given by $z = \frac{t}{1} - \frac{t^3}{3} + \frac{t^5}{5} - \frac{t^7}{7} + \frac{t^9}{9} - \dots$. If the tangent of t is equal to the unit radius, then the arc z is equal to 45 degrees or $z = \frac{\pi}{4}$ and $\frac{\pi}{4} = 1 - \frac{1}{3} + \frac{1}{5} - \frac{1}{7} + \dots$.
This series, which was first discovered by Leibnitz, can be used to find the value of the circumference of the circle.

141. In order to see the ease with which the length of an arc can be found by means of this series, we should substitute a sufficiently small fraction for the tangent t. For example let us use this series to find the length of the arc z whose tangent is $\frac{1}{10}$. In this case the arc

$$z = \frac{1}{10} - \frac{1}{3000} + \frac{1}{500000} - \dots,$$ and the approximate value of this series is easily expressed by a decimal fraction. However, from such an arc, we cannot conclude anything about the whole circumference of the circle, since the ratio of the arc whose tangent is $\frac{1}{10}$ to the whole circumference is not given. For this reason, in order to find the circumference, we look for an arc such that not only is it some fractional part of the circumference, but also small and easily expressed. For this purpose it is customary to choose the arc to be 30 degrees, whose tangent is equal to $\frac{1}{\sqrt{3}}$, since smaller arcs have tangents which are extremely irrational. Wherefore, since an arc of 30 degrees has length $\frac{\pi}{6}$, we have

$$\frac{\pi}{6} = \frac{1}{\sqrt{3}} - \frac{1}{3 \cdot 3\sqrt{3}} + \frac{1}{5 \cdot 3^2 \sqrt{3}} - \dots \quad \text{and} \quad \pi = \frac{2\sqrt{3}}{1} - \frac{2\sqrt{3}}{3 \cdot 3} + \frac{2\sqrt{3}}{5 \cdot 3^2} - \frac{2\sqrt{3}}{7 \cdot 3^3} + \dots.$$ By means of this series the value of π itself, which was previously exhibited, was determined with incredible labour.

142. The labour involved in this calculation is all the more since, first of all, each term is irrational, but also, since each succeeding term is only about one third of the preceding. In order to avoid these inconveniences, let us take the arc to be 45 degrees, that is of length $\frac{\pi}{4}$.

Although this arc can be expressed by a series which hardly converges, $1 - \frac{1}{3} + \frac{1}{5} - \frac{1}{7} + \dots$,

still we keep this arc and express it by means of two arcs of lengths a and b so that $a + b = \frac{\pi}{4}$,

that is 45 degrees. Since $\tan(a+b) = 1 = \frac{\tan a + \tan b}{1 - \tan a \tan b}$, we have $1 - \tan a \tan b = \tan a + \tan b$

and $\tan b = \frac{1 - \tan a}{1 + \tan a}$. If we let $\tan a = \frac{1}{2}$ then $\tan b = \frac{1}{3}$ and both the arcs a and b can be expressed by rational series which converge much more rapidly than the series above. The sum of these two series gives the value of the arc $\frac{\pi}{4}$. It follows that

$$\pi = 4\left(\frac{1}{1.2} - \frac{1}{3.2^3} + \frac{1}{5.2^5} - \frac{1}{7.2^7} + \frac{1}{9.2^9} - \dots\right) + 4\left(\frac{1}{1.3} - \frac{1}{3.3^3} + \frac{1}{5.3^5} - \frac{1}{7.3^7} + \frac{1}{9.3^9} - \dots\right).$$ In this way

we calculate the length of the semi-circle, π, with much more ease than with the series mentioned before.

A transcription into modern terms is almost unnecessary. Euler is among the first mathematicians whose language and symbolism is close to our own.

Starting with: $$\arctan t = t - \frac{t^3}{3} + \frac{t^5}{5} - \frac{t^7}{7} + \frac{t^9}{9} - \dots$$

and noting that: $$\tan \pi/6 = 1/\sqrt{3}$$

he obtains: $$\pi = 2\sqrt{3}\left(1 - \frac{1}{3.3} + \frac{1}{5.3^2} - \dots + \frac{(-1)^n}{(2n+1).3^n} + \dots\right)$$

Further, using the result that we would write as:

$$\arctan 1 = \arctan(1/2) + \arctan(1/3)$$

Euler obtained:

$$\frac{\pi}{4} = \sum_{n \geq 0} \frac{(-1)^n}{(2n+1)2^{2n+1}} + \sum_{n \geq 0} \frac{(-1)^n}{(2n+1)3^{2n+1}}$$

which has even better convergence. Euler uses here the symbol π which had been used by William Jones [12]. Following this it became widely adopted.

Other formulas can be derived to improve convergence and Euler himself used ones that had more rapid convergence. We recall the famous formula by Machin already mentioned:

$$\pi/4 = \arctan(1/5) - \arctan(1/239),$$

which William Shanks [25] used to calculate π to more than 600 decimal places – although he made a mistake at the 527th place (the error was only discovered in

1945). Furthermore, up until recent times, the formulas that have been used for calculating π have all been variants of Machin's formula.

Euler also found the following elegant formulas:

$$1+\frac{1}{2^2}+\frac{1}{3^2}+\frac{1}{4^2}+\ldots=\frac{\pi^2}{6}$$

$$1+\frac{1}{2^4}+\frac{1}{3^4}+\frac{1}{4^4}+\ldots=\frac{\pi^4}{90}$$

$$1+\frac{1}{3^2}+\frac{1}{5^2}+\frac{1}{7^2}+\ldots=\frac{\pi^2}{8}$$

$$1-\frac{1}{3^2}+\frac{1}{5^2}-\frac{1}{7^2}+\ldots=\frac{\pi^3}{32}$$

etc.

The convergence of these series is even more slow than those used before. They are established nowadays through the use of series expansions of periodic Fourier functions. For example, the functions:

$$f(x) = |x| \text{ and } g(x) = x^2 \text{ for } |x| \le \pi,$$

with period 2π, give $\dfrac{\pi^2}{8}$ and $\dfrac{\pi^2}{6}$ respectively as the sums of infinite series.

5.6 Epilogue

Together with this search for values of π to ever more decimal places, there developed a theoretical study of the nature of the number itself. In 1761, Lambert [15] proved that π was irrational. To do this, he established the general expansion of tan x as a continued fraction in the form:

$$\tan x = \cfrac{x}{1-\cfrac{x^2}{3-\cfrac{x^2}{5-\cfrac{x^2}{7-\ldots}}}}$$

and deduced that tan x is irrational whenever x is rational. It follows that $\pi/4$ is irrational since tan $\pi/4 = 1$, a rational number. In 1794 Legendre [17] used this same expansion to prove that π^2 is irrational.

In 1882 Lindemann [20] proved that π was transcendent, this time using the same method that Hermite had used to prove the transcendence of e. He established, in particular, that if $a, b, c, \ldots, d$ are distinct algebraic numbers and $A, B, C, \ldots, D$ are algebraic numbers, not all zero, then the expression:

$$Ae^a + Be^b + Ce^c + \ldots + De^d$$

cannot be equal to zero. Hence, the Euler formula:

$$e^{i\pi} + 1 = 0$$

shows the transcendence of π for if not, since the coefficients of the equation are algebraic and so the exponent $i\pi$ is not; therefore π is transcendental.

Our knowledge about π continues to grow, yet its mysteries still remain. For example, we mentioned earlier the rather attractive expansion for $4/\pi$ as a continued fraction due to Brouncker; but if we want a simple continued fraction, that is one in which all the numerators are equal to 1 (see Section 4.4 above), then no law for generating one is known. An approximate decimal value for π would lead to the simple continued fraction

$$3 + \cfrac{1}{7 + \cfrac{1}{15 + \cfrac{1}{1 + \cfrac{1}{292 + \cfrac{1}{1 + \dots}}}}}$$

which has the successive convergents:

$$3, \ 22/7, \ 333/106, \ 355/113, \ \dots$$

The second value was given by Archimedes and the fourth by Adrien Metius. Neither is known whether the decimal expansion of π has a random distribution.

On the other hand, more and more decimals in the expansion of π have been found. The methods rely on both more and more refined theories and on more and more powerful computers. The results so far are:

10^4 decimal places in 1958 (F. Genuys)
10^5 decimal places in 1961 (D. Shanks and J. Wrench)
10^6 decimal places in 1974 (J. Guilloud and M. Bouyer)
10^7 decimal places in 1983 (Y. Kanada and associates)
10^8 decimal places in 1987 (Y. Kanada and associates)
6×10^9 decimal places in 1995 (Y. Kanada and associates)

For this last result, it was probably the Borwein quadratically convergent algorithm that was used. On this point, note that the recurrence relations for the Archimedes method described in Section 5.1, which gives the semi-perimeters p_n and q_n of inscribed and circumscribed polygons of $3 . 2^n$ sides, can also be written:

$$p_{n+1} = \sqrt{p_n q_{n+1}} \quad \text{and} \quad \frac{1}{q_{n+1}} = \frac{1}{2}\left(\frac{1}{p_n} + \frac{1}{q_n}\right).$$

In this form, we recognise the appearance of the arithmetic and geometric means and the theory behind this approach was developed by Gauss and has inspired a number of recent fast calculation algorithms, notably the Borwein algorithm (see [3], Chapter II, section 5):

$$x_{n+1} = \frac{1}{2}\left(\sqrt{x_n} + \frac{1}{\sqrt{x_n}}\right), \quad y_{n+1} = \frac{y_n\sqrt{x_n} + \frac{1}{\sqrt{x_n}}}{y_n + 1}, \quad \pi_n = \pi_{n-1}\frac{x_n + 1}{y_n + 1},$$

with $x_0 = \sqrt{2}, \pi_0 = 2 + \sqrt{2}, y_1 = 2^{\frac{1}{4}}$.

The term π_n approaches π incredibly fast:

$$\pi_n - \pi < 10^{-2^{n+1}}.$$

To end, we would like to mention the astonishing result obtained by D. H. Bayley, P. B. Borwein and S. Plouffe in October 1995: "The 10 billionth hexadecimal digit of π is 9" [32]. Any particular digit in the hexadecimal expansion of π can be calculated, without having to calculate the preceding ones, thanks to the formula:

$$\pi = \sum_{n=0}^{\infty}\left(\frac{4}{8n+1} - \frac{2}{8n+4} - \frac{1}{8n+5} - \frac{1}{8n+6}\right)\frac{1}{16^n}.$$

Bibliography

[1] Archimedes, De la mesure du cercle, in *Les Oeuvres complètes d'Archimède*, vol. I., tr. P. Ver Eecke, 1921; repub. Paris: Blanchard, 1961.

[2] Archimedes, Measurement of a Circle, in *The Works of Archmides Edited in Modern Notation* (1897) with supplement *The Method of* Archimedes (1912), tr. T. L. Heath, Cambridge: Cambridge University Press, 1912, repr., New York: Dover, n.d.

[3] Borwein, J. M. and Borwein, P. B., *Pi and the AGM: A Study in Analytic Number Theory and Computational Complexity*, New York: Wiley, 1987.

[4] Carrega, J.-Cl., *Théorie des corps, la règle et le compas*, Paris: Hermann, 1981.

[5] Chace, A. B., *The Rhind Mathematical Papyrus*, 2 vols., Oberlin, Ohio: Mathematics Association of America, 1927–1929. Abridged repr. in 1 vol. Classics in Mathematics Education 8, Reston, Virginia: The National Council of Teachers of Mathematics, 1979.

[6] Chuquet, N., *La Géométrie*, annotated edition by H. l'Huillier, Paris: Vrin, 1979.

[7] Chuquet, N., *Nicolas Chuquet, Renaissance Mathematician*, A study with extensive translation of Chuquet's mathematical manuscript of 1484, ed. Graham Flegg, Cynthia Hay & Barbara Moss, Dordrecht: D. Reidel, 1985.

[8] Dedron, D. & Itard, J., Squaring the Circle, in *Mathematics and Mathematicians*, vol. 2., tr. from the French by J. V. Field, Milton Keynes: Open University Press, 1978.

[9] Descartes, R., *Circuli quadratio*, manuscript published in Amsterdam in 1701; in *Oeuvres*, ed. Ch. Adam & P. Tannery, vol. X, repub. Paris: Vrin, 1996.

[10] Euler, L., *Introductio in Analysin infinitorum*, Lausanne, 1748, repr. as *Opera*, I viii & ix, English tr. J. D. Blanton, *Introduction to Analysis of the Infinite*, New York: Springer, 1988, 1990.

[11] Gregory, J., *Vera circuli et hyperbolae quadratura*, Padua, 1667.

[12] Jones, W., *Synopsis palmariorum matheseos*, London, 1706.

[13] Lagny, T. G. de, *Mémoire sur la quadrature du cercle*, Histoire de l'Académie Royale, Paris, 1719.

[14] Lam, Lay-Yong & Ang, Tian-Se, Circle Measurements in Ancient China, *Historia Mathematica*, 13 (1986), 325–340.

[15] Lambert, J. H., Sur quelques propriétés remarquables des quantités transcendantes circulaires et logarithmiques, *Mémoires de l'Académie de Berlin*, 17 (1761), 265–322; *Mathematische Werke*, vol. II, 112–159.

[16] Lebesgue, H., *Leçons sur les constructions géométriques* professées au Collège de France en 1940/1941, Paris: Gauthier-Villars, 1950; repub. Paris: Gabay, 1988.

[17] Legendre, A. M., *Eléments de géométrie*, Paris: Didot, 1794.

[18] Leibniz, G. W., De vera proportione circuli ad quadratum circumscriptum in numeris rationalibus, *Acta eruditorum*, Leipzig (Feb. 1682); *Mathematische Schriften*, vol. V, 118–122, Hildesheim: Olms, 1962. French tr. Marc Parmentier, Expression en nombres rationnels de la proportion exacte entre un cercle et son carré circonscrit, in *Leibniz, Naissance du calcul différentiel*, Paris: Vrin, 1989.

[19] Leibniz, G. W., Letter to La Roque, *Mathematische Schriften*, vol. V, 88–92, Hildesheim: Olms, 1962.

[20] Lindemann, C. L., Über die Zahl π, *Mathematische Annalen*, 20 (1882), 213–225.

[21] Martzloff, J.-C., *A History of Chinese Mathematics*, New York: Springer Verlag, 1997, tr. Stephen S. Wilson from *Histoire des mathématiques chinoises*, Masson, Paris, 1987.

[22] Montucla, J. -E., *Histoire des recherches sur la quadrature du cercle*, 1754; repb. Paris: Bachelier, 1831.

[23] Newton, I., *Methodus Fluxionum et Serierum infinitarum*, written between 1664 and 1671; English tr. in D. T. Whiteside (ed.) *The Mathematical Papers of Isaac Newton*, vol. III, Cambridge: Cambridge University Press, 1969.

[24] Rajagopal, C. T. and Vedamurthiaiyar, T.V., On the Hindu proof of Gregory's series, *Scripta Mathematica*, 17 (1951), 65–74.

[25] Shanks, W., *Contributions to Mathematics Comprising Chiefly of the Rectification of the Circle to 607 Places of decimals*, London: G. Bell, 1853.

[26] Struik, D.J. (ed.) *A Source Book in Mathematics, 1200-1800*, Harvard: Harvard University Press, 1969; Princeton: Princeton University Press, 1986.

[27] Turnbull, H. W., *The Correspondence of Isaac Newton*, Cambridge: Cambridge University Press, 1959.

[28] Viète, Fr., *Variorum de rebus mathematicis responsorum*, Liber VIII, Tours, 1593; *Opera Mathematica*, Leyde, 1646.

[29] Wallis, J., *Arithmetica Infinitorum*, Oxford, 1656; *Opera Mathematica*, vol. I, Oxford, 1695, 355–478.

[30] Wantzel, L., Recherches sur les moyens de reconnaître si un problème de géométrie peut se résoudre avec la règle et le compas, *Journal de Mathématiques Pures et Appliquées*, 2 (1837), 366–372.

[31] Yan, Li & Shiran, Du, *Chinese Mathematics: A Concise History*, tr. John N. Crossley & Anthony W. -C. Lun, Oxford: Oxford Science Publications, Clarendon Press, 1987.

[32] http//www.mathsoft.com/asolveplouffe/announce.txt

6 Newton's Methods

Newton's method is remarkable both for the simplicity of its principle – based on linear approximation, and for its efficiency – often a very rapid convergence. It is known, in practice, by two names, depending on the circumstances in which it is used. When finding successive approximations to the numerical solution of an equation:

$$P(y) = 0,$$

it is called the *tangent method* or simply Newton's method. When finding y expressed as a series in x, given an equation:

$$F(x, y) = 0,$$

it is called Newton's *polygon method*. These two versions, for finding numerical and algebraic solutions, are both described by Newton in his *Method of Fluxions and Infinite Series*.

The origin of the method can be described as follows. During the course of the period that marked the birth of the infinitesimal calculus, one recurring problem was to find the 'area under curves'. When the curve under consideration is an algebraic curve, whose points satisfy some equation $F(x, y) = 0$, the method was to try to reduce the problem to a known case where y is expressed in the form of a function f of the abscissa x. It is then sufficient to know a primitive of that function f in order to calculate its integral. In fact, at that time, it was relatively easy – particularly so for Newton – to calculate the primitive of functions when they were expressed as a series in x. Newton correctly obtained such expansions of y in terms of x, by a method that is known today as Newton's polygon method. This provides the solution for y from the equation

$$F(x, y) = 0,$$

which solution is in the form $y = f(x)$ where $f(x)$ as a series in x.

But since this "reduction of affected equations in types" is rather abstract, Newton began by explaining his method with "reduction of affected equations in numbers", that is in solving equations of the type

$$P(y) = 0,$$

where P is a polynomial with given coefficients whose roots are to be found. These roots were to be found as the sum of a succession of numbers which, at each stage, improved the previous value. In fact, each method helps us to understand the other: the more concrete numerical version helps us to see what is going on in the algebraic approach, while the algebraic method helps us to understand the principle underlying the numerical method. In both cases the solutions are found by successive approximations.

In what follows, we shall look at extracts in which Newton explains the two methods. These will be followed by other, later, texts which improve on Newton's method or which complement his approach.

The Tangent Method
(The numerical solution of equations)

Newton explained his method in 1669 through the use of numerical examples (Section 6.1). He did not use the geometrical illustration of a curve approximated by its tangent, nor did his work show the use of the recurrence relation that is currently used today. This latter, which avoids the need for meticulous substitutions, was developed by Raphson in 1690 (Section 6.2), whose name is found in the usual description of the method as the *Newton-Raphson method*. The important question of what the initial conditions need to be, so as to guarantee convergence of the successive approximations, was raised explicitly, for the first time, by Mourraille (1768) (Section 6.3) and later by Lagrange. Here, a geometrical approach was used to explore the question. The question of convergence was looked again by Fourier in 1818, who considered the question of the rapidity of convergence, and this brings us back again to numbers. Cauchy, in his turn, studied the question in 1821, and then more particularly in 1829 (Section 6.4). His interest was in what the method implied for complex roots; this again led later to research into the modern idea of fractals, which draws on the approaches of both number and geometry, and is, in fact, a topological treatment of the problem. To illustrate 'tanget methods' then, we look at four texts: by Newton, Raphson, Mourraille and Cauchy, and we shall conclude with a recent text that makes the link with fractals (Section 6.5).

6.1 Straight Line Approximations

Newton's method is a method by successive approximations, and there were those before Newton who used a similar approach. Among these we can cite Heron of Alexandria (*c.* AD 100), for example, whose method for extracting square roots, though starting from entirely different premises, can be viewed as a particular case of Newton's method. We should certainly mention al-Kāshī (15th century) who found sin 1° from sin 3° by solving the equation:

$$3x - 4x^3 = c \ (= \sin 3°)..$$

Viète, in 1600, also used an approach that was in some respects analogous to Newton's. An examination of these different methods will be found in the chapter following this one.

To return to Newton, the first part of his *Method of Fluxions* [17] is an augmented version of *De Analysi per Aequationes numero terminorum infinitas* [16] written in 1669. It consists of a detailed description of the techniques of calculations using series, following which Newton explains his theory of fluxions, namely that the notion of the derivative corresponds to calculating the velocity or the 'fluxion' (geometrically the slope of the tangent), and the notion of integration corresponds to calculating the 'fluent', given the 'fluxions' (geometrically equivalent to calculating the area under a curve). Newton then went on to give many applications of his theory, dealing with the construction of tangents, finding maxima and minima, calculating areas under curves, etc....

Here, we shall concern ourselves with the techniques explained in the first part of *The Method of Fluxions*, and more particularly with the solution of equations. As we have said, Newton starts with explaining his method where he is trying to find numerical solutions to equations of the type $P(y) = 0$, where P is a polynomial with given coefficients. Newton starts by using a value a which is close to a root y. He first of all substitutes the value a in the equation, to determine the difference p, between a and the root y. He then solves this equation by using a linear approximation, that is in neglecting terms whose powers are 2 or more. He then iterates the process using the new approximate value $a + p$, and so on. In this way, Newton considers a series of approximations $a, a + p, a + p + q, a + p + q + r, \ldots$ in a sort of 'arithmetic of the infinite'. It should be understood, that for the present there is no mention of geometry. It was only after considering numerical and algebraic methods, that Newton started to deal with calculating fluxions, and then related this to the tangent to a curve.

Isaac Newton

De Methodus Fluxionum et Serierum infinitorum, written 1664–1671. Methods of Series and fluxions, in D. T. Whiteside, *The Mathematical papers of Isaac Newton*, Cambridge: Cambridge University Press, vol. III (1670–1673), 1969, pp. 43–47.

The reduction of affected equations

When, however, affected equations are proposed, the manner in which their roots might be reduced to this sort of series should be more closely explained, the more so since their doctrine, as hitherto expounded by mathematicians in numerical cases, is delivered in a roundabout way (and indeed with the introduction of superfluous operations) and in consequence ought not to be brought in to illustrate the procedure in species. In the first place, then, I will discuss the numerical resolution of affected equations briefly but comprehensively, and subsequently explain the algebraical equivalent in similar fashion.

Let the equation $y^3 - 2y - 5 = 0$ be proposed for solution and let the number 2 be found, one way or another, which differs from the required root by less than its tenth part. I then set $2 + p = y$, and in place of y in the equation I substitute $2 + p$. From this there arises the new equation $p^3 + 6p^2 + 10p - 1 = 0$, whose root p is to be sought for addition to the quotient. Specifically, (when $p^3 + 6p^2$ is neglected because of its smallness) we have $10p - 1 = 0$, or $p = 0.1$ narrowly approximates the truth. Accordingly, I write 0.1 in the quotient and, supposing $0.1 + q = p$, I substitute this fictitious value for it as before. There results $q^3 + 6.3q^2 + 11.23q + 0.061 = 0$. And since $11.23q + 0.061 = 0$ closely approaches the truth, in other words very nearly $q = -0.0054$ (by dividing 0.061 by 11.23, that is, until there are obtained as many figures as places which, excluding the bounding ones, lie between the first figures of this quotient and of the principal one – here, for instance, there are two between 2 and 0.005), I write – 0.0054 in the lower part of the quotient seeing that it is negative and then, supposing – 0.0054 + r equal to q, I substitute this value as previously. And in this way I extend the operation at pleasure after the manner of the diagram appended.

$$+\,2.10000000$$
$$-\,0.00544852$$
$$2.09455148\ [=y]$$

$2 + p = y.$	y^3	$+8$	$+12p$	$+6p^2$	$+p^3$
	$-2y$	-4	$-2p$		
	-5	-5			
	Total	-1	$+10p$	$+6p^2$	$+p^3$
$0.1 + q = p.$	$+p^3$	$+0.001$	$+0.03q$	$+0.3q^2$	$+q^3$
	$+6p^2$	$+0.06$	$+1.2$	$+6$	
	$+10p$	$+1$	$+10$		
	-1	-1			
	Total	0.061	$+11.23q$	$+6.3q^2$	$+q^3$
$-0.0054 + r = q.$	$+q^3$	$-0\cdot 0000001\cancel{57464}$	$+0.0000\cancel{8748}r$	$-\cancel{0}\cdot\cancel{0162}r^2$	$+\cancel{1}r^3$
	$+6.3q^2$	$+0\cdot 000183\cancel{708}$	$-0.068\cancel{04}$	$+\cancel{6\cdot 3}$	
	$+11.23q$	-0.060642	$+11.23$		
	$+0.061$	$+0.061$			
	Total	$+0.0005416$	$+11.162r$		

$-\,0.00004852 + s = r.$

Near the end, however, (especially in equations of several dimensions) the work will be much shortened by this method. When you have decided how far you wish the root to be extracted, count off as many places from the first figure of the coefficient of the last term but one in the equations resulting on the right side of the diagram as there remain places to be filled up in the quotient, and neglect the decimals which follow after. But in the final term neglect the decimals after as many more places as there are decimal places filled up in the quotient, and in the last term but two neglect all after as many fewer. And so on, progressing arithmetically by that interval of places or, what is the same, cancelling everywhere as many figures as there are in the last but one term provided their lowest places be in arithmetical progression in accord with the series of terms, but alternatively these are to be understood to be filled up with zeros when the circumstances prove otherwise. Thus in the example now propounded, should I desire to complete the quotient to the eighth place of decimals only, while substituting 0.0054 + r for q (at which stage four decimal places in the quotient are entered and the same number remain to be filled in) I could have omitted figures in the five lower places: these I have on that account scored with a small oblique stroke – indeed, even though the first term r^3 had the coefficient 99999, I could have omitted it entirely. Consequently, when those figures are expunged, for the following operation the total comes to 0.0005416 + 11.162r, and this, upon performing division as far as the prescribed term, yields $-0\cdot00004852$ for r, so completing the quotient to the desired period. Finally I subtract the negative portion of the quotient from the positive one, and there arises 2.09455148 for the finished quotient

We should bear in mind that, in this part of his treatise, Newton does not refer to the tangent nor the derivative, and that the following geometrical interpretation of the solution process does not figure in Newton's work in any way.
Consider the function:

$$z = f(y) = y^3 - 2y - 5.$$

Its graph is shown opposite. It cuts the horizontal axis at value between 2 and 3, and 'close to' 2.
Let us see what happens for values of y 'close to' 2. As a first approximation, let a portion of the curve coincide with its tangent at $(2, -1)$. Following the development of the calculus, we know that the tangent has the equation: $z + 1 = (y - 2)$. $f'(2)$, that is:

$$z = -1 + 10(y - 2).$$

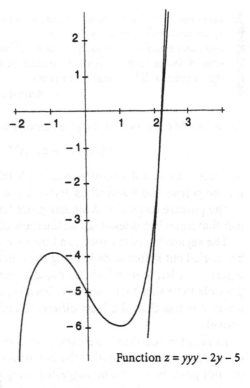

Function $z = yyy - 2y - 5$

Substituting $y = 2 + p$, the curve cuts the horizontal axis for a value of p satisfying:

$$p^3 + 6p^2 + 10p - 1 = 0$$

(see Newton's text) and the tangent cuts the axis when $10p - 1 = 0$.

Replacing the curve by its tangent is equivalent to neglecting the terms $p^3 + 6p^2$, just as Newton does. We therefore find $p = 0.1$ as a first correction.

The approximation is now improved if, in the place of $y = 2$, we now use the new approximate value $y = 2.1$, and so on ....

In order to use Newton's method, the first question to resolve is to find the approximate initial value close to the required root, and to do this, we need first to determine the interval bounds for each root. Newton identified a number of ways of doing this, particularly in his *Arithmetica universalis*, written between 1670 and 1680:

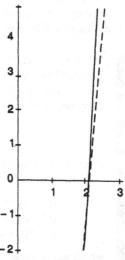

If some number, when substituted, produces in the final place a number having the same sign as the highest term in the equation, but not so in all intermediate places, multiply the equation's terms by the number of their dimensions and divide by its root. Then substitute the previous number in this fresh equation, and if again there finally ensues a number of the same sign as the highest term's, multiply this one's terms by the number of their dimensions and divide through by twice the root, and in it substitute the self-same number; if once more there ensues a number of the same sign as the highest term's, again multiply its terms by the number of their dimensions and divide by three times the root, and substitute the same number afresh. And if on repeating this procedure there should always result a number identical in sign with the highest term of the equation, that number will be a limit to the roots.

Arithmetica universalis, Preliminary drafts, [15, p. 41]

As we would express it today, given a polynomial:

$$P(x) = x^n + a_{n-1}x^{n-1} + \ldots + a_1x + a_0,$$

Newton notes that if a positive number b, substituted for x, produces positive values for the polynomial P and all its derivatives at $x = b$ then the number b is greater than all the positive roots of P. A similar procedure, having changed x into $- x$ will establish that a number is less than all the roots of P.

The separation of the roots, and the choice of initial approximate values, can thus be carried out systematically using the function and its derivatives. First try the integers in order, then refine the process using a decimal search. However, it is not possible to be sure that roots have been separated unless it is known that the search interval is less than the least difference between the roots (on this see Section 4.5 above).

In fact, the conditions necessary for convergence were not properly stated. Newton only required, simply, that the initial value should be within 0.1 of the root; if this did not prove successful he suggested using a second degree approximation, in other words, attempting to find p from $6p^2 + 10p - 1 = 0$. This approach which was non-linear, but only up to the second degree, was taken up again by Halley in 1694 [9].

Finally, it is interesting to consider a method of substituting in polynomial expressions used by Newton in *De Analysi per Aequationes*:

What hard work there is here will be found in substituting one group of quantities for another. That you might accomplish by various methods, but I think the following way exceedingly expedite, especially when the numerical coefficients consist of several figures. Suppose $p + 3$ has to be substituted for y in this equation

$$y^4 - 4y^3 + 5y^2 - 12y + 17 = 0:$$

since that may be resolved into this form $\overline{\overline{y - 4} \times y : + 5} \times y : -12 \times y : +17$, the new equation may be generated thus:

$$(p - 1)(p + 3) = p^2 + 2p - 3; \; (p^2 + 2p + 2)(p + 3) = p^3 + 5p^2 + 8p + 6;$$
$$(p^3 + 5p^2 + 8p - 6)(p + 3) = p^4 + 8p^3 + 23p^2 + 18p - 18; \; \text{and}$$
$$p^4 + 8p^3 + 23p^2 + 18p - 1 = 0,$$

which was required.

De Analysi per Aequationes [16, p. 223]

Newton uses bars over expressions where we would now use brackets, and what he has done is to rearrange the polynomial into what is now sometimes called *Horner's schema*, after W. G. Horner [10]. The method was known earlier to the Chinese and Arabs in the eighth century, in the context of extracting nth roots. The method is extremely quick, making it easy, using a calculator, to obtain the value of a polynomial for a given value of the variable. All that is needed, is to carry out the appropriate operations in the order indicated (see Section 7.9 below).

6.2 Recurrence Formulas

At each stage in his method, Newton produces a new supplementary equation by substituting $a + p$ for the unknown, where a is the current approximation and p is the correction needed. The English mathematician Joseph Raphson offered a simplification of this process in his treatise *Analysis Aequationum universalis* published in London in 1690. Again it is the same equation:

$$a_{n+1} = a_n - f(a_n)/f'(a_n)$$

that produces the successive approximations. Raphson thus opens up the possibility for using algorithmic procedures.

Of course, it was some time before the suffix notation a_n and the use of f' for the derivative came into use, but the basic principle was Raphson's. In the text that follows, Raphson deals with the equation that is connected with trisecting an angle:

$$3y - y^3 = c$$

which appears here as: $ba - aaa = c.$

He solves it by noting that each approximation g, of the sought value a, is improved when g is replaced by $g + x$ where:

$$x = (c + g^3 - bg)/(b - 3g^2).$$

At this time Newton's work on this procedure had not been published. However his ideas were known, since his work was described by John Wallis in his *Algebra* of 1685 [22].

Raphson was an ardent defender of Newton in the controversy with Leibniz over which of the two had been the original inventor of the calculus, and he strongly praised Newton in his posthumously published *The History of Fluxions* (1715).

J. Raphson
Analysis Aequationum universalis, London, 2nd ed., 1697, pp. 8 and 20.

PROP. II
Let it be proposed that $ba - aaa = c$

Let there be chosen any quantity whatever (g) less than (a). I say that (g) obtained closer and closer (by our method), being always greater than its predecessor, but certainly less than (a), will therefore converge to the true value.

By this hypothesis, $g + z = a$. So,

$$bg - ggg + b - 3gg \times z - 3gzz - zzz = ba - aaa = c$$

Therefore $\overline{b - 3gg} \times z - 3gzz - zzz = c + ggg - bg$

Therefore $+z - \dfrac{3gzz + zzz}{b - 3gg} = \dfrac{c + ggg - bg}{b - 3gg} = +x$.

From the convergence Theorems, we have $+z = +x + \dfrac{3gzz + zzz}{b - 3gg}$,

and adding (g) to both parts, produces

$$g + z = a = g + x + \frac{3gzz + zzz}{b - 3gg}.$$

But this new $(g) = g + x$ is greater than the preceding, by the quantity (x), and less however

than (a), by the quantity $\dfrac{3gzz + zzz}{b - 3gg}$, part of its total.   Q.E.D.[...]

PROBLEM XII
Trisection of the angle

Given the Radius of a Circle = r and the Chord of an arc = c, what is the Chord of the Third part of the arc?

The equation $3rra - aaa = crr$ will give the Chord of an arc of 20 degrees, the Third part of sixty.

In the case $c = r = 10.000$ and the equation being $300a - aaa = 1000$, (namely) $ba - aaa = c$, then according to the preceding theorem:

$$g = 3$$
$$c + ggg - bg = + 127$$
$$b - 3gg = + 273) + 127.0 \quad (+.4 = x$$

$$+ \overset{3.}{.4}$$

$$g = 3.4$$
$$c + ggg - bg = + 19.304$$
$$b - 3gg = 265.32) + 19.3040 \quad (+.072 = x$$

$$\overset{3.4}{+.072}$$

$$g = 3.472$$

$$c + ggg - bg = +.254210048$$
$$b. - 3gg = 263.835648) +.2542100480 \quad (+.0009636 = x$$

$$\overset{3.472}{+.0009636}$$

$$g = 3.4729636$$
$$c + ggg - bg = -.0000123100020899.$$
$$b - 3gg = 263.8155715) -.0000123100020899 \quad (-.0000004666 1393 = x$$

$$3.4729636.$$
$$-.0000004666 1393$$

$$a = 3.4729635533338607$$

In problem XII we require to find the solution of the equation:

$$300a - a^3 = 1000,$$

which is a particular case of the equation:

$$ba - a^3 = c,$$

where b and c are coefficients.

The trigonometric identity $\sin 3t = 3 \sin t - 4 \sin^3 t$ is used to produce the equation:

$$c/2r = 3a/2r - 4(a/2r)^3,$$

which reduces to: $$r^2c = 3r^2a - a^3.$$

Raphson uses the result of proposition II by which, if g is an approximate value for a, then the approximation is improved by replacing g by $g + x$ where:

$$x = (c + g^3 - bg)/(b - 3g^2)$$

and therefore the derived sequence of values for g:

$$3 \, ; 3.4 \, ; 3.472 \, ; 3.4729636 \, ; 3.472963553338607$$

will converge to the true value a.

The reason for this is given in earlier explanations by Raphson: when the unknown term of the equation a is replaced by $g + z$ then a new equation is obtained which z must satisfy:

$$b(g + z) - (g + z)^3 = c.$$

Expanding this we have:

$$bg - g^3 + (b - 3g^2)z - 3gz^2 - z^3 = c$$

and so: $$z = (c + g^3 - bg)/(b - 3g^2) + (3gz^2 + z^3)/(b - 3g^2).$$

Neglecting the term $(3gz^2 + z^3)/(b - 3g^2)$ since z is small,

we have: $x = (c + g^3 - bg)/(b - 3g^2)$

from which: $g < g + x < g + z = a.$

In practice, $b - 3g^2$ is not necessarily positive, and it can happen that x is not smaller than z, even where the initial value g is chosen to be less than the root. Thus, in the example of the trisection of the angle, the preceding claims are valid if it is the smallest of the two positive roots that is being sought.

In his work, Raphson sets out a table for the different equations of degree 2, 3 and 4 according to the signs of their coefficients and, for each one, he gives a formula for x and $g + x$. He follows this with another table which gives, what corresponds to, the derived polynomial up to degree 10.

Nowadays we would write the equation in the form:

$$f(a) = 0,$$

and the expression for the correction x would be given as:

$$x = -f(g) / f'(g).$$

In practice, $f(a) = 0$ is written $f(g + x) = 0$. Now,

$$f(g + x) = f(g) + bx + cx^2 + dx^3 + ... + mx^k$$

where $b = f'(g)$. Having regard for the smallness of x, we have the approximate result:

$$f(g) + xf'(g) = 0.$$

And so the new approximation $g + x$ would be written:

$$g + x = g - f(g) / f'(g),$$

and the successive approximations a_n are given by the recurrence formula:

$$a_{n+1} = a_n - f(a_n) / f'(a_n).$$

For a long time, the two methods, Newton's and Raphson's, were set out in distinctly different ways. Newton put the accent on choosing initial values sufficiently close in order to ensure linear approximations, whereas Raphson used intervals, which were obtained by choosing an initial value which was greater or less than the root, according to the type of equation being considered. Lagrange [12], in note V in his *Traité de la résolution des équations numériques*, showed that "these two methods are fundamentally the same, but presented differently".

This process, which had been used by Newton and Raphson only for algebraic equations with rational coefficients, was used by Thomas Simpson [21] for equations with irrational and transcendental coefficients (*Essays on Mathematics*, 1740). However, the initial conditions necessary to guarantee convergence in these cases was not properly considered.

6.3 Initial Conditions

The question of convergence for Newton's method appears to have been discussed properly for the first time in 1768 by Mourraille in his *Traité de la résolution des équations numériques* [14]. Jean-Raymond Mourraille, a mathematician and astronomer, was mayor of Marseille from 1791 to 1793. It is curious to note that his work has passed almost unknown, since it contains many novel ideas. In particular, contrary to Newton and Raphson, it emphasises the geometrical aspect - it is here that the name of the 'tangent method' makes sense - and it is the geometrical representation that is used to explain the behaviour of the iterative sequence produced by the Newton algorithm.

Suppose $f(x) = 0$ is an equation with roots $a, b, c, ...$ to be found, and let ABM be the corresponding graph of the equation.

Let P be a point on the abscissa such that the directed length OP represents an approximate value of the root $b = OB$, and let M be the point on the graph above P. Let Mp be the tangent at M to the graph and let Pp be the corresponding subtangent.

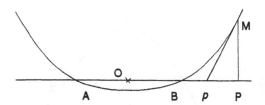

Newton's method consists in replacing the approximate value OP by the new value $Op = OP + Pp$. Now, as Mourraille points out, if the portion of the graph is "convex towards the abscissa", then the point p will be between B and P, and the new value Op will lie between $b = OB$ and OP. Therefore, both the approximation will be improved, and the process can be repeated and will converge. Whenever f' and f'' are of constant sign, this condition, of 'convexity towards the abscissa', can be written symbolically as: $f(OP) \cdot f''(OP) > 0$.

The extract that follows is interesting, since it uses imagery to show what happens when precautions are not taken to ensure that the initial point P is chosen so as to satisfy this condition. Several possibilities of what might happen are given including oscillations, going off to infinity and tending to another root; finally he shows that convergence may be more or less rapid.

J. -R. -P. Mourraille
Traité de la Résolution des Equations en général, London & Marseille, 1768.
Chapter five, section III: On the approximation of incommensurable roots (pp. 340–345.)

Disadvantages of this method

... Supposing that the root OB is to be found and that the point P, lying between A and B, is chosen for the first term OP, I shall examine the pitfalls that may result from using a point that is insufficiently determined, I shall determine the limits within which it should lie in order to avoid these pitfalls, and I shall show that, in order to avoid them, the position must be that which I prescribe.

As the point P moves steadily from B towards A, the point p of the first subtangent Pp moves from B towards N with a very much accelerated movement, it soon passes beyond the point Q, which corresponds to the inflexion R near B, then it rapidly moves as far as the point Π′ which corresponds to the following maximum Π′μ′, then even more rapidly to the point Q′ which corresponds to the inflexion R′, then still more rapidly to the point Π″ where I suppose it reaches a maximum Π″μ″ and so, moving along the whole line of abscissas from B towards N, it will end up finally at an infinite distance from the point B when P reaches the point Π of the maximum Πμ. After this, with P moving from Π towards A, the point p reappears on the same side as I, first of all at an immense distance, and arrives finally at A, as P also arrives there: but when OP is taken greater than OB, and so the second term Pp of the sequence has to be negative with respect to OP, I think it will be seen immediately that the position that makes Pp positive with respect to OP is too close to A, and consequently OP will be reduced.

I suppose then that OP is taken less than OΠ; the second term Pp will become negative with respect to OP: but I suppose also, which can certainly happen, that the part of the curve R′μ″m, instead of having a minimum Π″μ″, has instead a maximum very close to B, that is that it cuts the abscissa at points C and D, very close to B in the portion OB, and that the

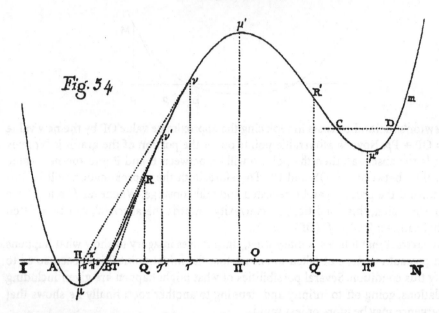

Fig. 54

point p corresponds to a point of the curve along R'C, above the abscissa, which part is convex towards the abscissa. Then the next term pp', or the second subtangent, and all the succeeding terms p'p'', p''p''', etc. will be negative with respect to the first term OP, and the sequence OP – Pp – pp' – p'p'', etc. instead of giving the root OB, will produce the root OC, very close to OB, and perhaps for that reason, without the error being realised. The sequence OP – Pp – pp' – p'p'', etc. will also produce the same root OC, if p corresponds to one of the points of the concave portion μ'R' such that it leads to a succeeding point p' which corresponds to the convex portion R'C; and the sequence OP – Pp + pp' – p'p'' + p''p''' – p'''p^{iv} etc. will again produce the same OC, if p corresponds to a point of the lower concave part of the curve Cμ'', such that it leads to a succeeding point p' which corresponds to the convex part R'C. Yet other sequences may produce the root OC, in several other ways, which do not need to be spelled out. Yet other sequences may, in different ways, produce the next root OD; other sequences can be found to produce other roots, and there is no root which cannot have a sequence, and even many different sequences; sequences can be found which stay in the region of minimum Π''μ'', and continuously hover about it, alternately receding and approaching it. Finally, an indeterminate position for P in the concave portion μB of the curve, can lead to points p, p', p'', p''', etc. along the whole line of abscissas IN, without them ever having a point of convergence.

It is true that, knowing the intermediate limits of successive roots, we can soon come to understand these divergences; but it would be better to avoid them completely, than to have to correct afterwards the position that would give them.

These then are the disadvantages, and here are the limits to these disadvantages.

Limits to the disadvantages

There exists in BΠ a certain point π, where placing the point P of the first term OP, the point p of the first subtangent Pp will be at τ in QΠ', which point τ, corresponding to the point ν of the concave portion Rμ', yields the point p' of the second subtangent pp' at the same position π from where we started.

There exists in Bπ a certain point π', where placing P, the point p will be at τ' in Qτ, which τ', corresponding to the point ν' of the concave portion Rν, yields the point p' at the root B. Finally, there exists in Bπ' a certain point π'', where placing the point P, the point p will be at Q, which point Q, corresponding to the point of inflexion R, produces the point p' at T where the tangent at R meets [the horizontal].

It can therefore be seen, that if the point P moves from π towards B, the point p moves continuously from τ towards B, while the point p' moves from π to B, from B to T, and returns from T to B; and if on the contrary, P moves from π to Π, the point p will move continuously from τ towards N, and the point p' will at first pass through the whole of πI, whereas p only passes along $\tau\Pi'$, etc.

The point π is therefore the limit between the convergence and divergence of the sequence. For the point π, the sequence is Oπ - $\pi\tau$ + $\tau\pi$ - $\pi\tau$ + $\tau\pi$ - $\pi\tau$ + $\tau\pi$, etc. = Oπ, always of the same magnitude; the series is necessarily convergent if P is taken between π and B; and divergent if P is taken between π and Π.

The point π' is the only point of πB which brings the sequence to an end after three terms; and for P at certain positions of $\pi\pi'$, those for which the sequence terminates after a finite number of terms, this sequence will only terminate after 3, 5, 7, 9, etc. terms. For the point π' the sequence is Oπ' - $\pi'\tau'$ + τ' B = OB; and if P lies between π and π' so that the point p', or p''', or p^v, etc. arrives at π' then the sequence OP - Pp + pp' - p'p'' + p''p''' - p'''p^{iv} + p^{iv}p^v - p^{v}p^{vi} + p^{vi}p^{vii}, etc. = OB will only terminate in p''p''' or p^{iv}p^v or p^{vi}p^{vii}, etc. But that never happens with a sequence of numbers, that is in practice, if OB is incommensurable; and if it is commensurable, then much better methods exist for finding it.

If P is taken between π and π', then the sequence may converge slowly if P is very close to π; but it will become more convergent as P approaches π'. The terms of the sequence are alternatively positive and negative up to the point where the term p''' or p^v or p^{vii}, etc. falls beyond B, after which all the terms are positive with respect to the first term.

If P is taken between π' and π'', then the second term of the sequence will be negative with respect to the first, as in the case where P lies between Π and B, but all the following terms will be positive, and the sequence will be the least convergent as P becomes closer to π''.

Finally, if P is taken between π'' and B, then the second term of the sequence will be negative, and all the other terms will be positive, as in the previous case; but the sequence will be ever more convergent as P becomes closer to B.

It follows that, as P moves from π to B, the convergence of the sequence is zero at the point π, it is a maximum or infinite at π', a minimum at π'', and infinite or a maximum at B; this convergence being positive, that is real, throughout the whole of the movement.

But, wherever P is taken between π and B, there will always be a negative term p, or p'', or p^{iv}, etc. where the alternating signs + - will cease, and the following terms will be positive; and consequently when P is taken in the concave part μB, the sequence is convergent, and it always comes back to the case where P is taken on the convex portion BR. Now, it is certainly clear that, in general, taking P in Bπ, the sequence will be less convergent than in taking P in BQ at a distance equal to B. Consider further, that in taking P in BQ at any distance whatever from B, the convergence of the sequence will increase continually, something that never happens when taking P in Bπ. It is therefore always better to take P in BQ, corresponding to the convex portion BR, the more so since it is not difficult to determine that part, while it is much more difficult to determine the points π, π', π'' of which we have spoken.

Here briefly is what, according to Mourraille, may happen when the point P is chosen to lie between B and Π. As P moves from B towards Π, p moves along the half line BN with an 'accelerated movement' and, as P moves from Π towards A, p moves along the half line IA with a 'retarded movement'.

He notes, firstly, that there are positions for P for which the method yields roots other than the one being sought, or which may not lead to any limiting value.

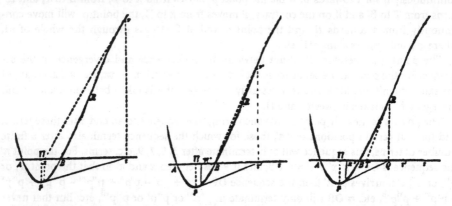

He is then interested in three particular positions for P between Π and B, namely π, π' and π", which are such that at the second iteration we arrive, respectively, at π, B and T. We can summarise what happens as P moves from Π to B in the following way:

$$P: \Pi \rightarrow \pi \rightarrow \pi' \rightarrow \pi'' \rightarrow B,$$
$$p: N \rightarrow t \rightarrow \tau' \rightarrow Q \rightarrow B,$$
$$p': I \rightarrow \pi \rightarrow B \rightarrow T \rightarrow B.$$

Therefore with P lying between π and B we are assured of convergence, but it may be more or less rapid: the convergence 'gets faster' as P approaches π' or B, it 'slows down' as P approaches π or π".

The moral that Mourraille draws from these considerations is the following: for P lying between Π and B - that is for a portion of the curve BM that is concave towards the abscissa - convergence, if it exists, will always lead to the case where P could have been chosen between B and Q - that is where the portion of the curve BM is convex towards the abscissa.

We now find ourselves in the position where the point P is such that the corresponding part of the curve BM is 'convex towards the abscissa'. When does this happen? Suppose that the function f is twice differentiable, and consider the general case where the signs of f' and f'' are constant in an interval containing the sought after root $b = OB$ (like the interval ΠQ in Mourraille's figure). Four cases may occur:

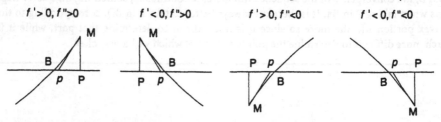

It can be seen that the portion of the curve BM has its convexity turned towards the abscissa whenever $f(OP)$ and $f''(OP)$ are of the same sign.

When this is not so in general, or when f' vanishes at b, the root b is multiple, and the convexity of the curve is still turned towards the abscissa.:

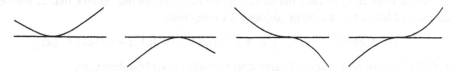

or again, where f'' vanishes at b, but not f', the curve has a point of inflexion at B, the tangent not being horizontal, and the possibilities are as follows:

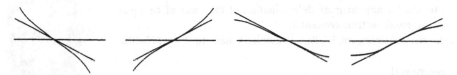

In the last two cases, the curve yet again has its convexity towards the abscissa; however, as Mourraille notes, b being also a root of f'', we are led to the solution of $f''(x) = 0$ and, if needs be, to $f^{(4)}(x) = 0, f^{(6)}(x) = 0, \ldots$ up to the point where we reach a favourable case.

It is this geometrical aspect of the problem that allowed Mourraille to proceed with the question of determining an initial approximate value so as to be sure that the method would lead to convergence. But to be able to say more about the rapidity of that convergence, indicated here only in a quantitative way, we have to return to an analytico-algebraic approach.

6.4 Measure of Convergence

In his *Method of Fluxions* [17], Newton gives an indication of the number of decimal places obtained at each step, but his statement is empirical and not analytical. Mourraille [14], as we have seen, deals with convergence for the case of an algebraic equation. However, his geometric approach, expressed in terms of the convexity of the curve, only allows him to establish the convergence of a sequence of approximations; it does not allow him to specify the rapidity of convergence.

Fourier appears to have been the first to approach this question in a note entitled *Question d'analyse algébrique* (1818) [7], but an error was found in his formula. He returned to the question in *Analyse des équations déterminées* [8], which was published posthumously in 1831. Cauchy studied the subject from 1821 onwards [2], but did not give a satisfactory formulation until 1829 (see text below).

The following passage by Cauchy is taken from a note to *Leçons sur le Calcul différentiel* [3], concerning the approximate determination of the roots of an equation $f(x) = 0$. Cauchy gives the conditions, attached to the values of the derivatives f' and f'', which allow him to determine two new limits within which the exact value lies, and so to determine how the sequence of successive approximate values given by Newton's method converges.

The concern for clarity and rigour, which is found in all of Cauchy's work, tidies up the question of the convergence of Newton's method for the present.

Before beginning to read the extract given here, we should remark that at the beginning of his note, Cauchy introduced the equations:

(1) $f(x) = 0$ (5) $f(a) + if'(a) = 0$ (8) $i = -f(a)/f'(a)$

and that his use of 'numerical value' corresponds to our 'absolute value'.

A. L. Cauchy

Leçons sur le Calcul différentiel, 1829.
Note: On the approximate determination of the roots of an equation, algebraic or transcendental.
Oeuvres, 2nd series, vol. IV, Paris: Gauthier-Villars, 1899, pp. 573–609.

Theorem II

Let us imagine that, the quantity i being determined by equation (5) or equation (8), we let B be a number equal to or greater than the greatest numerical value that the function $f''(x)$ can achieve between the limits $x = a$ and $x = a + 2i$. If the numerical value of the quantity $f'(a)$ is greater than that of the product

(10)   2Bi,
equation (1) will admit of a single real root, enclosed between the limits $a, a + 2i$.

    Proof. - In fact, if one substitutes

(11) $x = a + i + z$,
one will have, by virtue of equation (5) and the formulas (47), (48) of the eighth Lesson,

(12)
$$\begin{cases} f(x) = f(a) + (i+z)f'(a) + \dfrac{(i+z)^2}{2}f''[a+\theta(i+z)] \\[2mm] \quad = zf'(a) + \dfrac{(i+z)^2}{2}f''[a+\theta(i+z)], \end{cases}$$

(13) $f'(x) = f'(a) + (i+z)f''[a+\Theta(i+z)]$

where θ, Θ denote numbers less than one. Now, if the condition stated in theorem II is fulfilled, the function $f'(x)$ will clearly keep the same sign as $f'(a)$ for all the values of z lying within the limits $z = -i$, $z = +i$, and consequently for all the values of x lying within the limits $a, a + 2i$, whereas the extreme values of the function $f(x)$, namely

(14) $-if'(a)$ and $i[f'(a) + 2if''(a+2\theta i)]$,

will possess contrary signs. Hence, by virtue of theorem I, the equation (1) will have a single root within the limits $a, a + 2i$.

Theorem III

Let us imagine that, the quantity i being determined by the formula (8), we put

(15) $b = a + i$

and

(16) $j = -\dfrac{f(b)}{f'(b)}$.

Furthermore, let A be the least numerical value that the function f'(x) can attain within the limits x = a, x = a + 2i, and let B be the greatest numerical value that the function f"(x) can attain between the same limits. If the numerical value of the ratio

(17) $\dfrac{2Bi}{A}$

is less than one, the numerical value of j will not exceed

(18) $\dfrac{B}{2A} i^2$,

and equation (1) will admit of one real root lying within, not only the limits a, a + 2i, but also within the limits b, b + 2j.

Proof. - If, as we suppose, the numerical value of the ratio (17) remains less than one, the numerical value of 2Bi will not become greater than that of $f'(a)$. Therefore, by virtue of theorem II, equation (1) will have a real root, but only one, within the limits $a, a + 2i$. Furthermore, letting θ stand for a number lying between 0 and 1, and having regard for the formula (5), we find

(19) $f(b) = f(a+i) = f(a) + i f'(a) + \dfrac{i^2}{1.2} f''(a+\theta i) = \dfrac{i^2}{2} f''(a+\theta i)$;

then, from equation (19), combined with formulas (15) and (16) we have

(20) $j = -\dfrac{f(a+i)}{f'(a+i)} = -\dfrac{\dfrac{i^2}{2} f''(a+\theta i)}{f'(a+i)}$;

and as, by hypothesis, the numerical values of the quantities

$f''(a + \theta i), \quad f'(a + i)$

will be, the first less than B, the second greater than A, it is clear that the numerical value of j will not become greater than the product

$$\dfrac{B}{2A} i^2 .$$

Therefore the quantity j will lie between the limits

$$-\dfrac{B}{2A} i^2, \ +\dfrac{B}{2A} i^2 ,$$

and the quantity 2j between the following

$$-\dfrac{Bi}{A} i, \ +\dfrac{Bi}{A} i ,$$

and consequently between the limits $-\dfrac{i}{2}, +\dfrac{i}{2}$. Therefore

$$b + 2j = a + i + 2j$$

will be enclosed, as well as b, between the limits a, $a + 2i$; and if we let x take values from $x = b$ to $x = b + 2j$, the numerical values of the functions $f'(x)$, $f''(x)$ will remain, the first greater than A, the second less than B. Finally, since, j being less than i, and A greater than $f'(b)$ (having removed the signs), the numerical value of the product 2Bj will not become greater than that of the product 2Bi which is less than A, nor, even more so, than that of $f'(b)$, we can prove by similar arguments to those that were established in theorem II, that the real root already mentioned is enclosed within the limits b, $b + 2j$.

Corollary I. - It can be seen from the preceding, how, being given an approximate value a of a real root of equation (1), we can, by using formula (5) or (8), obtain new approximate values and restrict more and more the limits within which the root is found to lie. It is the use of this same formula that comprises Newton's method for finding approximate roots of numerical equations. It is good to note that the differences between the sought for root and the two first approximate values a, b will be respectively, the one less than the numerical value of 2i, the other less than the numerical value of 2j, and even more, than the product

$$\frac{B}{A}i^2 = \frac{Bi}{2A}(2i).$$

Therefore, if we use ρ for the numerical value of i, and we let the quantity $\dfrac{Bi}{2A}$ be represented by

(21) $\varepsilon = \dfrac{B\rho}{2A},$

then the differences in question will never become greater than the numbers

$$2\rho, \ 2\rho\varepsilon, \ 2\rho\varepsilon^3, \ 2\rho\varepsilon^7, \ldots$$
$$[\ldots]$$

Cauchy's text can be summarised as follows: let f be a twice differentiable function, and a a number. Let $i = -f(a) / f'(a)$, $\rho = |i|$, $A = \inf_{a \le x \le a+2i}|f'(x)|$ and $B = \sup_{a \le x \le a+2i}|f''(x)|$. If $2B\rho / A < 1$, then the equation $f(x) = 0$ has one and only one root between a and $a + 2i$.

More generally, in letting:

$$i_n = -f(a_n) / f'(a_n), \quad a_{n+1} = a_n + i_n \quad \text{and} \quad a_0 = a,$$

the equation $f(x) = 0$ has one and only one root between a_n and $a_n + 2i_n$ and the corresponding errors in the successive approximations a_n are less than $2\rho\varepsilon^{2^n-1}$, where $\varepsilon = B\rho/2A$, $(<1/4)$.

We note further, with Cauchy [2] and Fourier [7], that if f'' does not change its sign for values of x between a and $a + 2i$, then it is the same for f' which cannot vanish, and so the value of the expression $i_n = -f(a_n) / f'(a_n)$ always has the same sign, and so the sequence of approximations will therefore be either always increasing, or always decreasing.

In the case of Newton's equation:

$$x^3 - 2x - 5 = 0$$

we have: $f'(x) = 3x^2 - 2, f''(x) = 6x, a = 2$ and $i = -f(a)/f'(a) = 0.1$

Over the interval $[2, 2.2]$, we have:

$$A = 10; \quad B = 13.2; \quad 2B\rho/A = 0.264 < 1; \quad \varepsilon = B\rho/2A = 0.066$$

From which: $|x - a| < 0.2; \quad |x - a_1| < 0.0132; \quad |x - a_2| < 0.00006$

If the measure of convergence is taken to be the asymptotic behaviour of the sequence of ratios:

$$|a_{n+1} - r| / |a_n - r|$$

where r is the sought for root, it can be seen that, when the conditions given by Cauchy in theorem III are satisfied, the ratio is bounded above by:

$$|a_n - r|.B/2A.$$

In fact, from: $a_{n+1} = a_n - f(a_n)/f'(a_n),$
we deduce that: $(a_{n+1} - r)/(a_n - r) = 1 - f(a_n)/(a_n - r)f'(a_n),$
and, since: $f(r) = 0 = f(a_n) + (r - a_n) f'(a_n) + (r - a_n)^2 f''(x_n)/2$

where x_n lies between r and a_n, we have:

$$|a_{n+1} - r| / |a_n - r| \leq |a_n - r| |f''(x_n)| / 2 |f'(a_n)| \leq |a_n - r| B/2A.$$

The convergence is said to be quadratic since:

$$|a_{n+1} - r| \leq k |a_n - r|^2, \quad (k \text{ constant})$$

which can be expressed, in Newton's words, as: "for thus, at any stage you will gain twice as many figures in the quotient." [16, Whiteside, p. 220]

Of course, Cauchy's theorem III cannot be applied in cases such as where r is a multiple root, since the derivative f' vanishes there. However, when Cauchy's conditions are satisfied, the Raphson formula will still produce a convergent process, but the convergence will only be linear (the ratio $|a_{n+1} - r| / |a_n - r|$ tending to a limit k where $0 < k < 1$), the coefficient of convergence k depending on the multiplicity of the root r. In the case of algebraic equations, $P(x) = 0$, this case can always be excluded, by replacing the polynomial P with the polynomial obtained by dividing P by the GCD of P and P'.

Finally, we remind ourselves of what must first be done before applying Newton's method: namely, the separation of the roots, that is determining intervals which contain just one root (which may be multiple). Newton [17] (see Section 4.1 above), Mourraille [14], Lagrange [12], Cauchy [2] and Fourier [8] all offer many methods for achieving this in the case of algebraic equations.

At the end of his note, Cauchy makes the changes needed to adapt his theorems to determining imaginary roots. Newton's method applied to complex numbers had already been used by Waring in 1770, in his *Meditationes algebricae* [23], but here we are moving into difficult territory...

6.5 Complex Roots

To find the roots of the equation:

$$P(x) = 0,$$

is to find the fixed points of the function:

$$F_P(x) = x - P(x)/P'(x).$$

Newton's method and Raphson's formula:

$$x_{n+1} = x_n - P(x_n)/P'(x_n) = F_P(x_n)$$

is equivalent to considering the sequence of iterations:

$$x_n = F_P^n(x_0)$$

where x_0 denotes an initial value, preferably close to a root of P.

When P is a polynomial, F is a rational fraction and the search for roots, real or complex, leads to a study of the sequence of iterations of x_0 generated by F.

This area of study was started in the 1920s by Fatou [6] and Julia [11] (see for example [4]). Much more interest has been shown in it during the last two decades, and recently this field of research has become immensely popular, with the opportunities for graphical display afforded by modern computers and the new interest in fractals. The text that follows is a short extract from a lecture given for those not specialised in the topic, a passage in which the link with Newton's method is established.

To make the reading easier, we start with a small glossary:

– fixed super attractive point of F: a value a such that $F(a) = a$ and $F'(a) = 0$.
– attractive cycle of F: a sequence $(x_1, ..., x_k)$ such that $F(x_1) = x_2$, $F(x_2) = x_3$, ..., $F(x_k) = x_1$ and $\Pi_{1 \le i \le k}|F'(x_i)| < 1$.
– critical point of F: a value c such that $F'(c) = 0$.

When $F(z) = z - P(z)/P'(z)$, we have $F'(z) = P(z)P''(z)/P'/z)2$ and so the critical points of F are, on the one hand, the roots a of P, and on the other hand, the points ω such that $P''(\omega) = 0$.

– F hyperbolic: every critical point of F is attracted by an attractive cycle.
– Mandelbrot set: the set M of complex numbers c such that the sequence $(f_c^n(0))_{n \ge 0}$ does not tend to infinity where $f_c(z) = z^2 - c$.

Public Lecture
"The study of quadratic complex polynomials and new applications",
lecture given by Adrien Douady, 26 January 1985,
on the occasion of the Annual Day of the Mathematical Society of France.

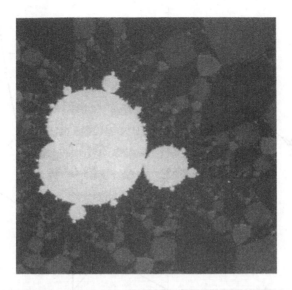

Photo: J. Hubbard and H. Smith, 1982, Cornell University

Let P be a polynomial of degree 3. Starting from a point z_0 we define a sequence (z_n) by z_{n+1} $= F_p(z_n)$ where $F_p(z_n) = z - P(z)/P'(z)$. If z_0 is in the neighbourhood of a root a of P, the sequence z_n tends very quickly to a (which is a fixed super attractive point of F_p). Good polynomials are those such that for almost every choice of z_0 the sequence z_n tends to a root. Among bad polynomials there are those for which F_p has an attractive cycle other than roots of P.

From a theorem of Fatou's, such a cycle necessarily attracts the point ω where $P''(\omega) = 0$, which is a critical point of F_p. If ω is attracted to a root, F_p is hyperbolic and P is good.

Curry, Garnett and Sullivan and also Hubbard and Homer Smith have studied the family $\left(P_\lambda = (z-1)(z+\frac{1}{2}-\lambda)(z+\frac{1}{2}+\lambda) \right); \lambda \in \mathbb{C}$. Every polynomial of degree 3 is equivalent, for this problem, to 6 polynomials belonging to this family.

For each λ, we take $z_0 = \omega = 0$. We colour λ blue [dark grey] if $z_n \to 1$, red [black] if $z_n \to$ $-1/2 + \lambda$, green [mid grey] if $z_n \to -1/2 - \lambda$ and yellow [light grey] if none of these. The figure shows a small region of the plane of λ. It contains a copy in yellow [light grey] of the Mandlebrot set.

Here is an example of an attractive cycle obtained by using Newton's method in the case of:

$$P(z) = z^3 - 2z + 2.$$

We have: $\quad F_p(z) = (2z^3 - 2)/(3z^2 - 2)$;

so: $\qquad\quad F_p(0) = 1$; $F_p(1) = 0$;

$\qquad\qquad F_p'(0) = 0 = P''(0).$

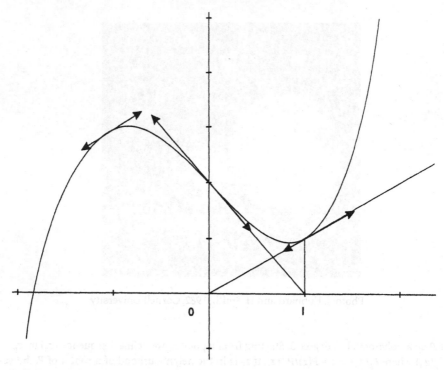

The Mandlebrot set [13] arises naturally in the following way. For every complex parameter c we put:

$$f_c(z) = z^2 - c$$

and let K_c stand for the set of complex numbers z whose sequences of iterations $(f_c^n(z))$ do not tend to infinity. Julia [11] and Fatou [6] showed that the set K_c is connected if, and only if, 0 is a member (if not then K_c is isomorphic to a Cantor set). The set M is the set of parameters c such that K_c is connected, and is moreover itself connected. The set M is the black area shown below, the fine details of which are not visible. It is symmetric about the real axis, and the points of the 'cardioid' on the real axis have real values - 0.75 and 0.25 respectively.

In fact, this definition of M is in agreement with that given in the glossary above: it, also, is the set M of complex numbers c such that the sequence $f_c^n(0)$ does not tend to infinity, where $f_c(z) = z^2 - c$. What is curious, in the extract of the lecture given above, is that the set M can also be obtained as a set of points of the plane for which Newton's method does not work, that is for a set of (complex) initial values which do not converge to a root.

Newton's Polygon
(Algebraic solution of equations)

Immediately following his description of his numerical method for solving equations, Newton used the same principle to show how to obtain algebraic solutions of equations. He explains how the method of successive linear approximations can be adapted by using a ruler and 'small parallelograms' (the text follows below), the first version of what is called Newton's polygon. The method was re-applied in a more general case later, by Puiseux in 1850, both in considering multiple branches and in considering functions of a complex variable.

6.6 The Ruler and Small Parallelograms

... to fit the doctrine recently established for decimal numbers in similar fashion to variables ...
                                                                          Isaac Newton

Newton sets out to solve the algebraic equation $P(x, y) = 0$, where

$$P(x, y) = \sum_{i,j} a_{ij} x^i y^j$$

is a polynomial in x and y. What is required is to express y as a function of x. What Newton does is to find y in the form of a series in x:

$$y = \sum_{k \geq 0} b_k x^k$$

He obtains successive approximations, using polynomials in x of increasing degree. Thus, a series, an infinite sum of single terms, is approached by polynomials, being finite sums of single terms, just as in the numerical method, explained earlier, a real number is approached by decimal numbers, which are in fact truncated forms of the sought for real number.

Now, in order for the series $y = \sum_{k \geq 0} b_k x^k$ to make sense, and in order that the polynomials should be approximations, the values of x must be considered to be sufficiently small. We are, then, implicitly in the neighbourhood of a point of the form $(0, y_0)$ on the curve with equation $P(x, y) = 0$ (in other words, y_0 is a solution of $P(0, y) = 0$).

As we shall see, it turns out that the exponents of x in these polynomials and series are not necessarily integers: they can be fractions. Here again, Newton explains his method through the use of examples.

I. Newton

De Methodus Fluxionum et Serierum infinitorum, written 1664–1671. Methods of Series and fluxions, in D. T. Whiteside, *The Mathematical papers of Isaac Newton*, Cambridge: Cambridge University Press, vol. III (1670–1673), 1969, pp. 51–57.

However, to make this rule still more evident, I thought it fitting to expound it in addition with the aid of the following diagram. Describing the right angle BAC, I divide its sides BA, AC into equal segments and from these raise normals distributing the space between the angle into equal squares or rectangles: these I conceive to be denominated by the powers of the variables x and y, as you see them entered in figure 1. Next, when some equation is proposed, I mark the rectangles corresponding to each of its terms with some sign and apply a ruler to two or maybe several of the rectangles so marked, one of which is to be the lowest in the left-hand column alongside AB, a second to the right touching the ruler, and all the rest not in contact with the ruler should lie above it. I then choose the terms of the equation which are marked out by the rectangles in contact with the ruler and thence seek the quantity to be added to the quotient.

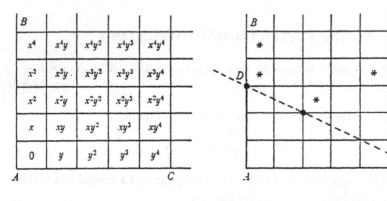

Fig. 1   Fig. 2

So to extract the root y from

$$y^6 - 5xy^5 + (x^3/a)y^4 - 7a^2x^2y^2 + 6a^3x^3 + b^2x^4 = 0,$$

I mark the rectangles answering to its terms with some sign *, as you see done in the second illustration. I then apply the ruler DE to the lower corner of the places marked out in the left-hand column and make it swing to the right from bottom to top until in like fashion it begins to touch a second or maybe several together of the other marked places. Those so touched I see to be x^3, x^2y^3 and y^6. Hence from the terms $y^6 - 7a^2x^2y^2 + 6a^3x^3$ as though set equal to nothing (and in addition, if it pleases, reduced to $v^6 - 7v^2 + 6 = 0$ by supposing $y = v\sqrt{ax}$). I seek the value of y and find it to be fourfold, $+ \sqrt{ax}$, $- \sqrt{ax}$, $+ \sqrt{2ax}$ and $- \sqrt{2ax}$. Any of these may be acceptable as an initial term in the quotient depending on whether the decision is made to extract one or other of the roots.

[...]

After these premisses it remains to illustrate the practical solution. Let, therefore, $y^3 + a^2y +axy - 2a^3 - x^3 = 0$ be the equation to be resolved. From its terms $y^3 + a^2y - 2a^3 = 0$ as fictitious equation by the third of the premisses I elicit $y - a = 0$ and write $+ a$ in the quotient. Next, since $+ a$ is not accurately the value of y, I set $a + p = y$ and in place of y in the terms of the equation written in the margin I substitute $a + p$, writing the resulting terms $(p^3 + 3ap^2 + axp ...)$ again in the margin. From these, once more by the third of the premisses, I select the terms $4a^2p + a^2x = 0$ for a fictitious equation and, since it yields $p \equiv -\frac{1}{4}x$, I write $- \frac{1}{4}x$ in the quotient. Further, since $-\frac{1}{4}x$ is not accurately the value of p, I set $- \frac{1}{4}x + q = p$ and in place of p

$$a - \frac{x}{4} + \frac{x^2}{64a} + \frac{131x^3}{512a^2} + \frac{509x^4}{16384a^3} ... \quad [= y]$$

$a + p = y.$	y^3	$a^3 + 3a^2p + 3ap^2 + p^3$
	$+ axy$	$+ a^2x + axp$
	$+ a^2y$	$+ a^3 + a^2p$
	$- x^3$	$- x^3$
	$- 2a^3$	$- 2a^3$
$-\frac{1}{4}x + q = p.$	p^3	$-\frac{1}{64}x^3 + \frac{3}{16}x^2q * -\frac{3}{4}xq^2 + q^3$
	$+ 3ap^2$	$+\frac{3}{16}ax^2 - \frac{3}{2}axq + 3aq^2$
	$+ axp$	$-\frac{1}{4}ax^2 + axq$
	$+ 4a^2p$	$- a^2x + 4a^2q$
	$+ a^2x$	$+ a^2x$
	$- x^3$	$- x^3$
$+\frac{x^2}{64a} + r = q$	q^3	*
	$-\frac{3}{4}xq^2$	*
	$+ 3aq^2$	$+\frac{3x^4}{4096a} * + \frac{3}{32}x^2r + 3ar^2$
	$+\frac{3}{16}x^2q$	$+\frac{3x^4}{1024a} * + \frac{3}{16}x^2r$
	$-\frac{1}{2}axq$	$-\frac{1}{128}x^3 - \frac{1}{2}axr$
	$+ 4a^2q$	$+\frac{1}{16}ax^2 + 4a^2r$
	$-\frac{65}{64}x^3$	$-\frac{65}{64}x^3$
	$-\frac{1}{16}ax^2$	$-\frac{1}{16}ax^2$

$$4a^2 - \frac{1}{2}ax \overline{)\frac{131}{128}x^3 - \frac{15x^4}{4096a} \left(\frac{131x^3}{512a^2} + \frac{509x^4}{16384a^3} ... \right.}$$

in the marginal terms I substitute $- \frac{1}{4}x + q$, writing the resulting terms $(q^3 - \frac{3}{4}xq^2 + 3aq^2 \ldots)$ once more in the margin. From these in turn by the above-stated rule I choose the terms

$$4a^2q - \frac{1}{16}ax^2 = 0$$

for a fictitious equation and, since this yields $q = \frac{x^2}{64a}$, I write $+\frac{x^2}{64a}$ in the quotient. Furthermore, since $\frac{x^2}{64a}$ is not accurately the value of q, I set $\frac{x^2}{64a} + r = q$ and in place of q in the marginal terms I substitute $\frac{x^2}{64a} + r$. And so I repeat the operation at pleasure, as the appended diagram accordingly indicates.

In substituting the series $\sum_{k \geq 0} b_k x^k$ for y in the polynomial $P(x, y)$, we are hoping to obtain 0: therefore, in particular, it is essential that the terms with small exponents, α, should cancel each other out.

Now, after substitution, the exponents of x are of the form $i + kj$. In order that the terms of lower degree cancel each other out, there must be at least two of them, therefore there exist at least two pairs (i_1, j_1) and (i_2, j_2) in the expression $P(x, y)$ such that:

$$i_1 + k j_1 = i_2 + k j_2.$$

The small squares considered by Newton, allow the terms $x^i y^j$ appearing in $P(x, y)$ to be arranged in order, which we do now by thinking of them as Cartesian coordinates. Using Newton's example, the procedure is set out as follows:

$$y^6 - 5xy^5 + (1/a)x^3y^4 - 7a^2x^2y^2 + 6a^3x^3 + b^2x^4 = 0$$

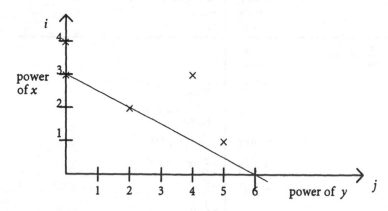

The straight line joining the points (i_1, j_1) and (i_2, j_2) has a slope of $- k$ and cuts the vertical axis at the point $(0, \alpha)$. Consider all the straight lines passing through at least two points of the graph. In order that the value $i + kj$ associated with such a line should correspond to the exponent of terms of lower degree, then there must be no point (i, j) below that line; hence the procedure explained by Newton.

The points – two or more – situated on a line satisfying this assertion, provide the terms to be considered in the expression $P(x, y)$; using these terms we have a simpler expression $Q(x, y)$, and the solution of $Q(x, y) = 0$ will give the first term bx^k of the expansion of y as a function of x.

We see in the example, following the first approximation, we have the fractional exponent $k = 1/2$ and several possible choices for the coefficient b, corresponding to several branches of the curve at the point $(0, 0)$.

In the second example, the graph appears as shown here. Which straight line must be considered? The vertical gives $x = \sqrt[3]{2}$, we are not in the neighbourhood of $x = 0$ and, moreover, k would be infinite. The horizontal gives $y = a$ and $k = 0$: the first approximation for y is the constant a.

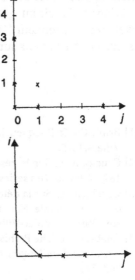

Let us now put $y = a + p$, and we come to the next graph. Again we have to stop to consider. Do we choose the horizontal or the line with slope – 1? The horizontal leads to a constant value p (here 0). This time we need to use one of the powers of x in order to carry out the approximation.

At each stage, the choices we make for the exponent k and for the coefficient b_k allow us to eliminate in $P(x, y)$ the terms in x of increasingly higher degree, and so to improve the approximation.

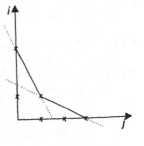

We also notice that, according to Newton, if x is not small but differs by only a small amount from some quantity a, we can substitute $z = x - a$ in the equation, and if x is 'indefinitely large' we can substitute $z = 1/x$, which will then be indefinitely small.

In the examples given by Newton, the distribution of the coordinate points (i, j) is such that the method always leads to joining a point on the horizontal axis to a point on the vertical axis. It can happen, however, that the configuration could be as shown here and so, in using one or

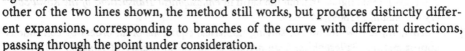

other of the two lines shown, the method still works, but produces distinctly different expansions, corresponding to branches of the curve with different directions, passing through the point under consideration.

Puiseux studied this in detail in his *Mémoire sur les fonctions algébriques* [18], published in 1850. By studying the integrals of functions u of a complex variable z, he was led to consider the different branches of a curve defined implicitly by the algebraic equation:

$$f(u, z) = 0.$$

To this extent, he demonstrated a method that was, in effect, Newton's parallelogram method, but in a more complete form, since he did not just consider a straight line sweeping across the plane to reach a certain position, but a succession of segments which made up Newton's polygon.

Puiseux's study follows, to some extent, the work by Cauchy on functions of a complex variable. The latter did not, however, carry out a systematic study of these implicit functions, functions which are multiformed whose values, except at certain critical points, are determined by continuity along the arc of a curve in the complex plane.

Puiseux's method is clearly explained in the second edition of the *Théorie des fonctions elliptiques* by Briot and Bouquet [1]. Puiseux's work, related to multiform functions and their integrals, is the starting point for the theory of Abelian functions developed by Riemann [20], where difficulties in dealing with branch-points were resolved by his famous surfaces.

Bibliography

[1] Briot, Ch. & Bouquet, J.-C., *Théorie des fonctions elliptiques*, 2nd ed., Paris: Gauthier-Villars, 1875.

[2] Cauchy, A.-L. Sur la résolution numérique des équations, Note III of *Cours d'Analyse*, 1821. *Oeuvres*, 2nd series, Paris: Gauthier-Villars, vol. III, 1897, pp. 378–425.

[3] Cauchy, A.-L., Sur la détermination approximative des racines d'une équation algébrique ou transcendante, Note to *Leçons sur le Calcul différentiel*, Paris: de Bure, 1829, 2nd series, Paris: Gauthier-Villars, vol. IV, 1899, pp.573–609.

[4] Chabert, J.-L. The Prehistory of Fractals, in I. Grattan-Guiness (ed.) *Encyclopaedia of the History and Philosophy of the Mathematical Sciences*, London: Routledge, 1993, vol. I, pp. 367–374.

[5] Douady, A., *L'étude dynamique des polynômes quadratiques complexes et ses réinvestissements*, Journée annuelle de la Société Mathématique de France du 26 janvier 1985, 21–42.

[6] Fatou, A., Sur les équations fonctionnelles, *Bulletin de la Société Mathématique de France*, 47 (1919), 161–271; 48 (1920), 33–94 and 208–314.

[7] Fourier, J., Question d'analyse algébrique, *Bulletin des Sciences par la Société philomatique* (April 1818), 61–67 in *Oeuvres*, Paris: Gauthier-Villars, vol. II, 1890, pp. 241–253.

[8] Fourier, J., *Analyse des équations déterminées*, Paris: Didot, 1831.

[9] Halley, E., A new, exact and easy Method of finding the Roots of any Equations generally, and that without any previous Reduction, *Philosophical Transactions of the Royal Society*, 210 (1694), 136; English tr. from the Latin in *Philosophical Transactions of the Royal Society of London*, 3 (1809), 640–649.

[10] Horner, W., A new method of solving numerical equations of all orders by continuous approximation, *Philosophical Transactions of the Royal Society of London*, 109 (1819), 308–335.

[11] Julia, G., Mémoire sur l'itération des fractions rationnelles, *Journal de Mathématiques Pures et Appliquées*, 4 (1918), 47–245.

[12] Lagrange, J.-L., *Traité de la résolution des équations numériques*, Paris, 1798.

[13] Mandelbrot, B., *The fractal Geometry of Nature*, New York: Freeman and Co., 1977.

[14] Mourraille, J.-R., *Traité de la Résolution des Equations en général*, London & Marseille, 1768.

[15] Newton, I., *Arithmetica universalis*, Cambridge, 1707, in D.T. Whiteside (ed.), *The Mathematical Papers of Isaac Newton*, vol. V (1683–1684), Cambridge: Cambridge University Press, 1972.

[16] Newton, I. *De Analysi per Aequationes numero terminorum infinitas*, London, 1711, in D.T. Whiteside (ed.), *The Mathematical Papers of Isaac Newton*, vol. II (1667–1670), Cambridge: Cambridge University Press, 1968.

[17] Newton, I., *Methodus Fluxionum et Serierum infinitarum*, written between 1664 and 1671, English tr. in in D.T. Whiteside (ed.), *The Mathematical Papers of Isaac Newton*, vol. III (1670–1673), Cambridge: Cambridge University Press, 1969.

[18] Puiseux, V., Mémoire sur les fonctions algébriques, *Journal de Mathématiques Pures et Appliquées*, 15 (1850), 365–480.

[19] Raphson, J., *Analysis Aequationum universalis*, London, 1690.

[20] Riemann, B., Theorie der Abelschen Functionen, *Journal für die reine und angewandte Mathematik*, 54 (1857), 115–155.

[21] Simpson, T., *Essays on Mathematics*, London, 1740.

[22] Wallis, T., *Algebra*, 1685, in *Opera mathematica*, Oxford, vol. II, 1693, pp. 1–482.

[23] Waring, E., *Meditationes algebraicae*, 1770.

7 Solving Equations by Successive Approximations

> Physicists insist, with reason, on the fact that for them a theorem has no interest unless it yields the possibility of numerically calculating numbers or functions which are under consideration [...], a real number is only 'known' when one is given a procedure by which it can be approximated – with an error that the mathematician wishes to be arbitrarily small, while the user in practice is content with rather less.
>
> Jean Dieudonné
> *Calcul infinitésimal*, Paris: Hermann, 1968, p. 8.

A good introduction to this chapter is the following entry on approximation by d'Alembert which appeared in the *Encyclopédie*:

APPROXIMATION, approximation, *f.f.* (in Mathematics) is an operation by which one approaches ever more closely to the value of a required quantity, without however ever finding the exact value. See ROOT.

Wallis, Raphson, Halley, & others have given us different methods of approximation: all these methods consist in finding convergent series, by the use of which one approaches the exact value of the required quantity as close as is wished; & this more or less rapidly according to the nature of the series. See CONVERGENT & SERIES.

If a number is not a perfect square, it is not possible to obtain the exact root in terms of rational numbers, whether integers or fractions; for these cases it is necessary to have recourse to methods of approximation, & to be satisfied with a value which only differs by a very small quantity from the exact value of the desired square root. It is the same for the cube root of a number which is not a perfect cube, & so on for other powers ...

Like d'Alembert, we shall start with methods for extracting square roots. Then we shall go on to consider various numerical methods for the solving equations by successive approximations. Finally we shall look at technical improvements to these iterative methods.

The idea of approximation has been touched on earlier: in Chapter 4 we showed how irrational numbers can be approximated using continued fractions, and in Chapter 5 we looked at approximations to π. Here we shall not be considering approximations to given numbers, but approximations to the roots of a given equation.

We considered linear equations at some length in Chapter 3, using methods of false position. We shall start here by considering the next most simple type of equation, which is an equation corresponding to the square root of a given number. We shall see that there are several approaches to solving the equation, as shown in the first three texts.

Heron of Alexandria, (first century AD) seems to have been the first to explicitly propose an iterative method of approximation in which each new value is used to obtain the next value (Section 7.1). But it was only with the standard method for calculating square roots, or the 'root, not a root' method described here by Ibn al-Bannā (Section 7.3), that we can begin to see an analogy with the methods of solving algebraic equations which proceed by successive determination of the digits of the roots, like those proposed by al-Ṭūsī (Section 7.4) or those of Viète (Section 7.5). To a certain extent, these methods have their culmination in the Newton-Raphson method, a method which is so efficient and simple that we have just dealt with it separately in Chapter 6.

At the beginning of the 19th century, Ruffini, Budan and Horner established technical improvements that made it simple to obtain the transformed polynomial equations that the use of these methods required (Section 7.9). The techniques involved were not previously known, but traces of their origins can be found in 12th century Arabic mathematical texts (by al-Ṭūsī) and in 13th century Chinese texts.

The idea of the method of successive approximations, that is the use of iterative algorithms which generate a convergent sequence whose limit is the solution of the proposed problem, described as a 'fixed point method', only developed gradually from a consideration of several equations of the type $x = g(x)$, whose solutions are fixed points of the function g. For use in astronomy, at the beginning of the 15th century, al-Kāshī used this approach to calculate the value of $\sin 1°$, given the value of $\sin 3°$, making use of the relation: $x = (\sin 3° + 4x^3)/3$ (see end of Section 7.5). Again, for use in astronomy, Kepler used this process to solve a transcendental equation, now known as Kepler's equation (Section 7.6).

In general, with this type of method, we have to start with an initial value x_0, chosen so as to guarantee the convergence of the sequence of successive terms defined by the recurrence relation $x_{k+1} = g(x_k)$. At the beginning of the 20th century, the general way forward in developing this method was shown by Picard in his work on the construction and existence of approximate solutions to differential equations (see Section 12.2 below). In this way, the different methods for determining approximate solutions to numerical equations (linear interpolation and linearisation by differential calculus) were brought together under the same schema.

There are also other more specific methods for finding approximate solutions. We describe, for example, Daniel Bernoulli's method of recurrent series (Section 7.7) and Lagrange's use of continued fractions (Section 7.8). But we do not claim to deal with all methods in this chapter.

Extraction of Square Roots

Babylonian tablets provide us with a number of examples of approximations to square roots. For example, in tablet VAT 6598 (Berlin Museum), dating from the early Babylonian period (2000–1700 BC) problem 6 ia about finding the diagonal of a rectangular door of height 40 and width 10 which calls for an approximation to $\sqrt{40^2 + 10^2} = 10\sqrt{17}$. The Babylonians, as we know, used a sexagesimal place value notation (see Chapter 1) and they knew the 'Pythagoras' relation. In this tablet, the Babylonian scribe provides the answer 41;15 (that is 41 + 15/60), which agrees with the well known standard approximation of $h + \ell^2/2h$ for the value of $\sqrt{h^2 + \ell^2}$, provided ℓ is small compared with h (here h and ℓ are the height and width of the door).

In the absence of any other information, it is difficult to know how the Babylonians arrived at such an approximation since, as we shall see, there are a number of ways of arriving at the same formula. We also note that there is no hint here of an iteration process, to find the answer, that is approaching a more exact value through stages of approximations, whereas this is typical of the cases which we shall consider below.

In order to approximate to the square root r of a number A which is not a square, the most natural approach is to start from a number a whose square is close to A, whether greater than it or less than it.

From this, a first way of reasoning is to notice that both a and A/a are approximations to r, and that r lies between them. We can then take, as a better approximation to the root, the arithmetic mean of these two numbers:

$$\frac{1}{2}\left(a + \frac{A}{a}\right).$$

This value is always greater than the value r of the root. Notice that if, in the Babylonian problem, h is taken as the first approximation for the diagonal, then a better approximation is given by:

$$\frac{1}{2}\left(h + \frac{h^2 + \ell^2}{h}\right) = h + \frac{\ell^2}{2h}.$$

This calculation of the arithmetic mean of the pair of approximations can be repeated again, to provide an even better approximation. This was the method that was described by Heron of Alexandria in the first century (7.1 below).

An alternative approach is to look at the difference x between the approximate value a and the exact value r.

Suppose first, that a^2 is less than A so that $a + x = r$. We then have:

$$(a + x)^2 = A, \quad \text{from which} \quad x^2 + 2ax = A - a^2.$$

If a is close to r, then x is small and, neglecting the x^2 term, we have:

$$2ax = A - a^2 \quad \text{and so} \quad x = \frac{A - a^2}{2a}.$$

We now have a better approximation to r: $a + \dfrac{A - a^2}{2a}$, which value is greater than the value r of the root. It is interesting to notice that this reasoning gives the same result as the previous one, since:

$$a + \frac{A - a^2}{2a} = \frac{1}{2}\left(a + \frac{A}{a}\right).$$

If we start with a value that is greater than the root r, identical calculations will lead us to an approximation of the root equal to:

$$a - \frac{a^2 - A}{2a} = \frac{1}{2}\left(a + \frac{A}{a}\right).$$

This type of reasoning was presented in a geometrical way by Theon of Alexandria in around 370 AD. By repeating the calculations, Theon set out a procedure for calculating approximations to square roots (see Section 7.2).

Finally, the problem can be considered by starting from a first approximation a that is less than the root, and to use the formula:

$$x^2 + 2ax = A - a^2$$

in order to look for a value of x to make $x(x + 2a)$ as close as possible to $A - a^2$, but less than it. These calculations can then be repeated in order to get closer to r. This reasoning, based on the expansion of the binomial expression $(x + a)^2$, turns out to be particularly effective when the numbers are set out using a place value system. It is the basis of the method for the extraction of square root taught to countless children right up to the 1960s. The method was described in the 13th Century by the mathematician Ibn al-Bannā (Section 7.3).

As we shall see from the set of texts presented in this section, the particular procedure proposed is a consequence of the numeration system being used.

7.1 The Method of Heron of Alexandria

The method of successive approximations, which has turned out to have been so profitable, would appear, from the documents available to us, to have been used first by Heron of Alexandria ($c.$ 50? AD).

Heron of Alexandria
Heron, *Metrica* i, 8, ed. H. Schöne (Heron iii.) 18. 12–24. 21,
in Ivor Thomas, *Greek Mathematical Works*, vol. II, p. 471.

Since 720 has not a rational square root, we shall make a close approximation to the root in this manner. Since the square root nearest to 720 is 729, having a root of 27, divide 27 into 720; the result is $26\frac{2}{3}$; add 27; the result is $53\frac{2}{3}$. Take half of this; the result is $26\frac{2}{3} + \frac{1}{3}$ ($= 26\frac{5}{6}$). Therefore the square root of 720 will be very nearly $26\frac{5}{6}$. For $26\frac{5}{6}$ multiplied by itself gives $720\frac{1}{36}$; so that the difference is $\frac{1}{36}$. If we wish to make the difference less than $\frac{1}{36}$, instead of 729 we shall take the number now found, $720\frac{1}{36}$, and by the same method we shall find an approximation differing by much less than $\frac{1}{36}$.

In this text, discovered by Schöne in 1896, Heron sets out an algorithm for approximating a square root. From knowing a first approximate value a of $\sqrt{A}$, Heron finds a second approximation $\frac{1}{2}\left(a+\frac{A}{a}\right)$ and then proposes repeating the method.

The explanation of the method, is based on an example, the need to find the 'side', in the original Greek, of 720, that is $\sqrt{720}$ and Heron uses unit fractions, of the type $1/n$, and the 'special' fraction 2/3 (see Chapter 1).

The approximation to the square root is firmly in the tradition of using mean values, beloved of the Pythagoreans, since what we have here is the arithmetic mean of two approximations, the one greater than, the other less than, the required value. The application of this formula is, as we have already pointed out, equivalent to the formula:

$$a+\frac{A-a^2}{2a}\,.$$

Nevertheless, the calculations that lead to these two formulas are quite different. The numerical context of using unit fractions may, perhaps, have lead Heron to prefer to use the arithmetic mean, and in actual fact, it is difficult to subtract unit fractions so as to calculate $(A - a^2)$.

We note that the iterative character of the procedure is clearly identified in the last phrase of the text. A more thorough examination of Heron's works shows that he did not always use this process.

7.2 The Method of Theon of Alexandria

Theon of Alexandria, who lived around 370 AD, and his daughter Hypatia, provide us with the evidence about the nature of mathematical activity during the later period of Ancient Greece. In his *Commentary on Ptolemy's Almagest*, Theon explains how to find the square root of a sexagesimal number to the nearest second. The procedure is entirely geometrical.

We may recall that the expansion of the binomial $(a + b)^2$ is given in a geometrical form in Euclid's *Elements*, II. 4. The proposition reads:

If a straight line be cut at random, the square on the whole is equal to the squares on the segments and twice the rectangle contained by the segments.

This corresponds to the formula

$(a + b)^2 = a^2 + b^2 + 2ab.$

The example used by Theon is to find the value of $\sqrt{4500}$. This particular example was suggested to Theon by his reading of Ptolemy, who gives the length of the side of a square of area 4500, as almost exactly 67° 4' 55", but with no further explanation.

Theon of Alexandria

Commentary on Ptolemy's Syntaxis, i 10, ed. Rome, *Studi e Testi*, lxxii (1936), pp. 469–473
in Ivor Thomas, *Greek Mathematical Works*, vol. 1, 57–61.
[with some small changes to the notation]

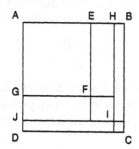

In order to show visually for one of the numbers in the *Syntaxis*, this extraction of the root by taking away the parts, we shall construct the proof for the number 4500°, whose side he [Ptolemy] made 67° 4' 55". Let *ABCD* be a square area, the square alone being rational, and let its contents be 4500°, and let it be required to calculate the side of a square approximating to it. Since the square which approximates to 4500° but has a rational side and consists of a whole number of units is 4489° on a side of 67°, let the square *AF*, with area 4489° and side 67°, be taken away from the square *ABCD*. The remainder, the gnomon *DFFB*, will therefore be 11°, which we reduce to 660' and set out. Then we double *GF*, because the rectangle on *GF* has to be taken twice, as though we regarded *FE* as on the straight line *GF*, divide the result 134° into 660', and by the division get 4', which gives us each of *GJ*, *EH*. Completing the parallelograms *JF*, *FH*, we have for their sum 536', or 268' each. Continuing, we reduce the remainder, 124', into 7440", and subtract from it also the complement *FI*, which is 16', in order that by adding a gnomon to the original square *AF* we may have the square *AI* on a side of 67° 4' and consisting of 4497° 56' 16". The remainder, the gnomon *DIIB*, consists of 2° 3' 44", that is 7424". Continuing the process, we double *JI*, as though *IH* were on a straight line with *JI* and equal to it, divide the product 134° 8' into 7424", and the result is approximately 55", which gives us an approximation to *JD*, *HB*. Completing the parallelograms *DI*, *IB*, we shall have for their joint area 7377" 20''', or 3688" 40''' each. The remainder is 46" 40''', which approximates to the square *IC* on a side of 55", and so we obtain for the side of the square *ABCD*, consisting of 4500°, the approximation 67° 4' 55".

In general, if we seek the square root of any number, we take first the side of the nearest square number, double it, divide the product into the remainder reduced to minutes, and subtract the square of the quotient; proceeding in this way we reduce the remainder to seconds, divide it by twice the quotient in degrees and minutes, and we shall have the required approximation to the side of the square area.

Theon starts from a first approximate value that is less than the root. Let $A = 4500$ be the area of the square $ABCD$ and $a^2 = 67^2$ the area of the square $AEFG$, the remaining gnomon $EBCDGF$, that is the area between the two squares, has an area $A - a^2 = 660'$. If we let x stand for the length GD, the gnomon consists of two rectangles of area ax, together with a square of area x^2. Neglecting the area of this square, the area of the gnomon is approximated by $2ax$, and so the side x by:

$$\frac{A - a^2}{2a}$$

This explanation corresponds to the argument set out by Theon. But he always takes the integer part of the ratio, and this gives him another approximation that is less than the root. Theon then repeats the process and calculations to obtain a new approximation to the root. To calculate the area of the gnomon $HBCDJI$, he does not use the area of the square AH, but subtracts the area of the gnomon $EHIJGF$ from that of the gnomon $EBCDGF$.

7.3 Mediaeval Binomial Algorithms

The method for the extraction of square roots, sometimes known as 'root, not a root' was taught in schools right up to the 1960s. The process depends on the expansion of the binomial $(a + b)^2$ and on writing numbers as decimals. It can easily be extended to the extraction of cube roots. The method can be found, in different guises, in third century China, in India and, throughout the Mediterranean region, in the Middle Ages.

The earliest general Chinese procedures for the extraction of square and cube roots are to be found in Chapter 4 of the *Jiuzhang suanshu* (Mathematical Prescriptions in Nine Chapters). The commentary on the work by Liu Hui at the end of the third century shows us that the method depended on dividing up the number whose root is sought into blocks of 2 digits (or 3 digits for the cube root). These numbers are expressed in the base ten positional notation of the counting sticks, and the calculation of the root is determined place by place. At each stage, the calculations depend, effectively, on the expansion of the binomial $(a + b)^n$, with $n = 2$ for the square root or $n = 3$ for the cube root. At the end of the process, after the root has been found, there were two techniques for handling the 'remainder', that is the non-integer part of the root, to give a suitable approximation.

In the same way that the traditional Chinese method using counting sticks was widespread, we also find the 13th century method of Ibn al-Bannā used in all the arithmetic books of the Middle Ages, irrespective of the cultural context.

Ibn al-Bannā

Talkhīṣ aʿmal al-ḥisāb d'Ibn al-Bannā,
Critical edition by M. Souissi, Tunis, 1968.
From the French translation of the Arabic by A. Djebbar.

The procedure for extracting the square root [of a number] is to count off its digits 'root', 'not a root', up to the end of the line. Then, you go to the last of the digits which is labelled 'root' and you put below it a number [such that if] you multiply it by itself, you will exhaust the number above it or there remains [a number such] that it is not possible to have an integer which is smaller than it. Then you put the number back, [after having] doubled it, under the digit 'not a root', and you look for a number to place under the digit 'root' to go in front of it [which is such] that if you multiply it by the previous doubled number, then by itself, it will exhaust what is above them, or there remains a [number such] that it is not possible to have any number smaller than it. Then do not stop, proceeding thus: doubling the previous [number] and setting [it] out, until you have exhausted all the line. That which results in the second line before the doubling is the root.

In order to understand Ibn al-Bannā's text, we need to remember that he wrote the numbers from right to left. We need, therefore, from our point of view, to replace "last of these digits" in his text by "first of these digits", and "previous" by "following".

We shall set out Ibn al-Bannā's calculations, using the number 189574. Ibn al-Bannā starts by pairing off the digits, from right to left for us, and labels them 'root' (r), 'not a root' (n):

$$18 \quad 95 \quad 74$$
$$nr \quad nr \quad nr$$

We have to imagine that the calculations are carried out on a grid where the spaces are clearly evident. The square root will be found, digit by digit. We shall write the digits above the digits labelled 'root' (r), as al-Ṭūsī did in his solution of numerical equations (see Section 7.4). In our example the root will have three digits. The intermediary calculations will appear below the number, in a third line.

Ibn al-Bannā first looks for the greatest number whose square is contained in 18: this is 4 (Fig. 1). The remainder, $18 - 16 = 2$ replaces the 18 in the grid (Fig. 2). He then doubles the 4 and writes it on the third line, moving on, that is placing 8 under the 9 of 95 (Fig. 3).

4	4	4
18 95 74	2 95 74	2 95 74
		8
Fig. 1	Fig. 2	Fig. 3

Since we can now read this as 80 is under 295, we have to find the largest number n such that $(80 + n)n < 295$. This number is 3, and it is put on the first line above the

5 of 95. The remainder, 295 – 249 = 46 replaces 295 (Fig. 4). At the next stage, we must double the 43 to give 86 and write this on the third line, again moving on (Fig. 5). This now gives 860 below 4674, and we need to find the largest number n such that $(860 + n)n < 4674$. This number is 5, and this is put on the first line, above the 4 of 74. And so 435 turns out to be the integer part of the square root of 189574.

4 3		4 3		4 3 5
46 74		46 74		46 74
		8 6		8 6

| **Fig. 4** | **Fig. 5** | **Fig. 6** |

We shall now set out the principle underlying the method 'root, not a root' for extracting the square root of an integer N together with the details of the calculations.

Initialisation	**Example**
We write N in the form:	$N = 18\ 95\ 74$

$$N = Q\ 10^{2k} + R, \text{with } Q < 100 \text{ and } R < 10^{2k}$$

	$k = 2, Q = 18, R = 9574$

then n, the integer part of the root of N, must necessarily have $k + 1$ digits.

Let q be the first digit of n, then n can be written as:

$$n = q\ 10^k + r, \text{with } r < 10^k .$$

Consequently, q is the greatest integer such that

$$q^2 < Q$$

	$q^2 < 18, q = 4$

Recurrence	**Example**

Suppose there remain only the last h digits of n to be found, then n can be written as:

$$n = a\ 10^h + b, \text{with } b < 10^h \text{ and } a \text{ known.}$$

	$a = 4$

We write N in the form:

$$N = A\ 10^{2h} + B, \text{with } B < 10^{2h}.$$

	$A = 18, B = 9574$

Since $n^2 = N$ then b must satisfy:

$$(a\ 10^h + b)^2 = A\ 10^{2h} + B.$$

and, by expanding the binomial and rearranging terms:

$$b(b + 2a\ 10^h) = (A - a^2)10^{2h} + B.$$

Let c be the first digit of b and C the number given by the first two digits of B, then c is the greatest integer such that

	$C = 95$

$$c\ 10^{h-1}(c\ 10^{h-1} + 2a\ 10^h) < (A - a^2)10^{2h} + C\ 10^{2h-2},$$

that is:

$$c(c + 20a) < (A - a^2)10^2 + C.$$

	$c(c + 80) < (18 - 16)10^2 + 95$
	$c(c + 80) < 295, c = 3.$

The calculations are set out like this:

First step:

N		18 95 74		4		a
$(A - a^2)10^2 + C$		2 95		$c(c + 80) < 295$		

we find $c = 3$:

Second step:

N		18 95 74		43		a
		2 95		$3(3 + 80) < 295$		
$(A - a^2)10^2 + C$		46 74		$c(c + 860) < 4674$		

we find $c = 5$, and therefore 435 is the integer part of the square root of 189574. The calculation can then be continued to find the decimal part of the root.

Numerical Solutions of Equations

7.4 Al-Ṭūsī's Tables

Al-Ṭūsī's *Treatise on Equations* contains the most thorough discussion of the theory of cubic equations of all the known mediaeval mathematical texts. The study by al-Ṭūsī, in the tradition of the work of al-Khayyām, both completed and extended the earlier work.

Omar al-Khayyām wrote a *Treatise on al-Jabr and the Muqābala* in which he gave a classification of the 25 types of equations of degree equal to or less than 3, and whose coefficients are positive. He distinguished between those equations for which the existence of a positive solution can be found from Euclid's *Elements*, and those whose solutions are obtained from the intersection of a conic with a circle [17]. Al-Ṭūsī proposed a different classification from al-Khayyām's, which took account of whether or not there were positive solutions. For eight cases, the equation always has a positive root and for five others the impossibility of their solution can be shown. Nonetheless, al-Ṭūsī, just as al-Khayyām, did not indicate that, for certain given values, the equation "a cube and its roots equal to squares and a number", which we would write as

$$x^3 + 3bx = 3ax^2 + N,$$

could have three positive roots.

In the example given below, al-Ṭūsī sets out the solution method for the equation "a cube plus some squares in number of the form 12 and some roots in numbers of the form 102 equal to a number of the form 34 345 395", which we would write as:

$$x^3 + 12x^2 + 102x = 34\,345\,395.$$

This particular equation is used as a basic example to show how to solve cubic equations of the type $x^3 + 3ax^2 + 3bx = N$, an equation that will always have a positive root. In his equation, al-Ṭūsī refers to 12 as the 'number of squares', 102 as 'the number of roots' and 34 345 395 as the 'number'. Al-Ṭūsī's procedure has three parts:

- a geometrical justification for the existence of a root. For this, he uses al-Khayyām's method of considering the intersection of a semi-circle with a branch of a rectangular hyperbola, and proves that these two curves have a common point which corresponds to a root of the cubic.
- a numerical solution, by a method which derives from the digit by digit method for the extraction of roots of a cubic already in use by Arab mathematicians since the eleventh century, and modified by using intermediary steps to reduce the number of arithmetic operations.
- a geometrical justification of the algorithm, which is similar to the one used by Theon of Alexandria.

The principle behind al-Ṭūsī's method is to use a table to find the root of the equation one digit at a time. The columns of the table determine the orders of magnitude of the numbers used in the calculations, which avoids having to write the powers of ten in exponential form. The contents of the rows of the table change as the calculations take place, and play the role of a memory register of certain intermediate results. There are four lines. The 'upper line' contains the successive approximations of the required root. The line below initially contains the number, and the two lower lines, called the 'intermediary line' and the 'lower line' contain initially, and respectively, one-third of the 'number of roots' and the 'number of squares'.

In order to make the text easier to read, we have set it out in two columns, the right hand column containing the corresponding calculations.

---

al-Ṭūsī

Sharaf al-Dīn al-Ṭūsī oeuvres mathématiques, algèbre et géométrie au XIIème siècle,
Critical edition by R. Rashed, 2 vol., Paris: Les Belles Lettres, 1986, vol. I, pp. 78–79.

We say that: a cube with some squares – 12 in this case – and some roots – 102 in this case – is equal to a number – 34 345 395 in this case. We count off the number by *root, not a root*, and we determine by how many the order of the last [digit] of the number of squares is less than the same order of the last cube, and we move the last order of the number of the squares as far as the order below the last cube. And we move the last order of the number of the roots as far as the order which is below the same root of the last cube, of that value. Then we reduce the number of the squares and the number of the roots to thirds. Then we have this figure:

```
          O               O               O
      3 4 3 4 5 3 9 5
```

```
          O               O               O
      3 4 3 4 5 3 9 5
              1 0 2
          1 2
```

```
          O         O         O
      3 4 3 4 5 3 9 5
              3 4
          4
```

Then, we determine the number which is sought for the cube – and it is three – ; and we put it [in the place of] the last cube.

```
        3         O         O
    3 4 3 4 5 3 9 5
              3 4
          4
```

We subtract its cube from the number

$34\,345\,395 - 27\,000\,000 = 7\,345\,395$

and we multiply it by a third of the number of the squares

$300 \times 4 = 1200$

and we add the result to the intermediary line – which is the one containing the third of the number of the roots –;

$1200 + 34 = 1234$

we multiply it by the intermediary line,

$300 \times 1234 = 370\,200$

we subtract three times each product from the number

$7\,345\,395 - 3 \times 370\,200 = 6\,234\,795$

and we add the square of [the number] sought to the intermediary line in the order which is to the side of it.

$(300)^2 = 90\,000$

$1234 + 90\,000 = 91\,234$

We multiply it again by a third of the number of the squares and we add the result to the [intermediary] line. It will then be of this form:

$91\,234 + 1200 = 92\,434$

```
3        O        O
  6  2  3  4  7  9  5
  9  2  4  3  4
              4
```

Then we move the upper line and the lower line by two ranks and the intermediary [line] by one rank.

```
    3  O          O
6  2  3  4  7  9  5
   9  2  4  3  4
             4
```

Then we consider another sought [number] – and it is two –; we subtract its cube from the number.

we multiply it by the first sought [number] and the third of the number of squares,

$(20)^3 = 8000$
$6\,234\,795 - 8000 = 6\,226\,795$

we add the result of the product to the intermediary [line]

we multiply it by the intermediary [line]

$20 \times (300 + 4) = 6080$

and we subtract three times each product from the number.

$6080 + 92\,434 = 98\,514$

Then we add its square to the intermediary line

and we multiply it by the first sought [number] and by the third of the number of squares, and we add the result to the intermediary line. It will now be of this form:

$98\,514 \times 20 = 1\,970\,280$
$6\,226\,795 - 3 \times 1\,970\,280 = 315\,955$

$(20)^2 = 400; \quad 98\,514 + 400 = 98\,914$

```
        3  2     O
    3  1  5  9  5  5
 1  0  4  9  9  4
             4
```

$98\,914 + 6080 = 104\,994$

Then we move the upper and lower [lines] two ranks and the intermediary [line] one rank, we consider the third sought [number] – and it is one –; we subtract its cube from the number, we multiply it by the sum of the first and the second sought [numbers] and by the third of the number of squares, we add the result to the intermediary [line], we multiply it by the intermediary [line] and we subtract three times each product by the number. The number disappears and so the result is that the upper line is like this: 321

```
          3  2  O
    3  1  5  9  5  5
 1  0  4  9  9  4
              4
```

The method used by al-Ṭūsī certainly similar to what we today call an algorithm. The case studied by al-Ṭūsī: "a cube plus some squares in number of the form 12 and some roots in numbers of the form 102 equal to a number of the form 34 345 395", would now be written as:

$$x^3 + 12x^2 + 102x = 34\ 345\ 395,$$

a particular case of the cubic equation: $x^3 + 3ax^2 + 3bx = N$, where $a = 4$, $b = 34$ and $N = 34\ 345\ 395$.

Al-Ṭūsī begins by configuring the table. He first determines the order of magnitude of the root, that is the number of its digits. To do this, he inspects the number N, from right to left, and places a mark 'o' over the digits that correspond to the cube root and nothing over the next two digits where there is no root. The root will, therefore, have as many digits as there are marks 'o' in the upper line. Here the root will contain three digits since $(100)^3 < N < (1000)^3$.

Next, al-Ṭūsī works on the first digit of the root. Since this root is of the order 10^2, its square is of the order 10^4. Therefore, al-Ṭūsī places $b10^2$ in the 'intermediary line' and $a10^4$ in the lower line. The initial configuration of the table is therefore like this:

	o			o			o		
3	4	3	4	5	3	9	5		N
				3	4				b
		4							a

The principle of the algorithm is to determine the root, digit by digit. The first loop of the algorithm contains two parts:

- first stage: estimate the first digit of the root x, and so provide a first approximation x_0 for the root.
- second stage: calculate an equation which has as a root $s = x - x_0$.

Following this, since the number of digits of s must be less than those of x, we can start again, applying the first two steps to the new equation. And so on.

Al-Ṭūsī does not say how he obtains the value 3 for the first digit of the root. But, a and b being small in comparison with N, it is sufficient to look for the greatest number whose cube is less than 34: it is therefore 3.

The number s is the solution to the equation

$$(x_0 + s)^3 + 3a\,(x_0 + s)^2 + 3b\,(x_0 + s) = N$$

or, more simply, of the equation: $s^3 + 3a_1s^2 + 3b_1s = N_1$
with:
$$N_1 = N - x_0^3 - 3x_0\,(ax_0 + b)$$
$$a_1 = x_0 + a$$
$$b_1 = x_0^2 + 2ax_0 + b$$

This equation can be found in al-Ṭūsī's calculations, where he uses the columns of the table to establish the exponential order of the digits. He calculates, in order:

$$N - x_0^{\ 3} = 7\,345\,395$$
$$ax_0 = 1200$$
$$ax_0 + b = 1234$$
$$x_0(ax_0 + b) = 370\,200$$
$$N_1 = N - x_0^{\ 3} - 3x_0(ax_0 + b) = 6\,234\,795$$
$$b_1 = x_0^{\ 2} + (ax_0 + b) + ax_0 = 92\,434$$

The table then becomes:

3		O			O		x_0
6	2	3	4	7	9	5	N_1
9	2	4	3	4			b_1
4							a

The number a_1 is found by summing the numbers in the first and last lines. Al-Ṭūsī then works on finding the first digit of s. Since s is of order 10, he moves the line b_1 to the right by one rank, so b_1 10 appears on that line, and he moves the first and last lines two ranks, so that a_1 10^2 is the sum of these two lines. He obtains the following table:

	3	O			O		x_0
6	2	3	4	7	9	5	N_1
	9	2	4	3	4		b_1
		4					a

Now al-Ṭūsī considers that the first digit of s is 2, but he does not say why. This time the coefficient b_2 is large, it is therefore sufficient to look for the greatest integer such that multiplying it by $3b_1$ makes it less that 6234795, and this turns out to be 2. The second stage of the algorithm provides him with a new table:

	3	2			O		x_1
	3	1	5	9	5	5	N_2
1	0	4	9	9	4		b_2
		4					a

and after adjusting the positions:

		3	2	O			x_1
3	1	5	9	5	5		N_2
1	0	4	9	9	4		b_2
		4					a

The number a_2 is then equal to the sum of the numbers of the first and last lines: $a_2 = s + a_1 = x_1 + a$, that is 324. Al-Ṭūsī takes 1 to be the last digit of the root. It can be seen, in the same way as before, that 1 is the largest integer which can multiply $3b_2$ with the product less than 315955. Then, in calculating N_3, he states that the 'number disappears'. This means that the equation obtained at the second stage has 0 for a root, and so 321 is the solution of the equation we started with.

7.5 Viète's Method

The treatise *De numerosa potestatum ad exegesim resolutione* (On the solution of numerical powers) is the final work of *The Analytical Art or mathematical Analysis restored*, by Viète, to be published before his death in February 1603.

Reading the works of Viète on the numerical solution of equations is hardly easy: we have archaic language, the absence of a clear symbolism and a conciseness of presentation, all of which make the comprehension difficult. Viète's project was an extremely ambitious one: to describe, with the use of some forty examples, a general method for finding the positive solutions of equations, from the second degree to the sixth degree, in the form of decimal fractions (Viète had made a plea for the use of decimal fractions in his earlier *Canon*). This method is an extension of the methods for extracting roots of the same degree which he had explained in the first part.

The numerical examples of the work divide themselves into:

- a first part dealing with the extraction of pure roots, that is finding nth roots for $n \le 6$;
- a second part concerning 'affected' powers, which may be
  affirmative, as for example: $x^6 + 6000x = 191\,246\,976$, or
  negative, as for example: $x^5 - 5x^3 + 500x = 7\,905\,504$, or
  pulled apart, as for example: $65x^3 - x^4 = 1\,481\,544$.

Viète's method illustrates the principles explained in the *Introduction to the Analytical Art* and relies on certain 'synthetic theorems'. These are stated rhetorically, that is in words and without the use of algebraic symbols. Examples, in our notation are:

$$(x_1 + x_2)^5 = x_1^5 + (5x_1^4)x_2 + (10x_1^3)x_2^2 + (10x_1^2)x_2^3 + (5x_1)x_1^4 + x_2^5$$
$$(x_1 + x_2)^3 = x_1^3 + (3x_1^2)x_2 + (3x_1)x_2^3 + x_2^3$$

We shall present here Viète's approach to the solution of his problem XV:

$$x^5 - 5x^3 + 500x = 7\,905\,504.$$

In order to respect the 'law of homogeneity' (only quantities of the same magnitude can be added), Viète associates the number 5 with a *number-plane*, that is the product of two lengths, and the coefficient 500 with a *plane-plane*, that is the product of two areas. The adherence to this law shows us that Viète is continuing in the geometrical tradition of Greek mathematics, as does his use of 'side' to denote a root and its approximations.

F. Viète

De numerosa potestatum ad exegesim resolutione, Paris, 1600. pp. 24–27
from an unpublished French translation by F. Ritter, Bibliothèque de l'Institut.

Problem XV

To extract the side of a square-cube given in affected numbers by the addition of a plane-solid product of the side and a given plane-plane coefficient, and by the subtraction of a plane-solid product of the cube of the side and a given length coefficient.

A number multiplied by its square-square and by 500 and diminished by the product of 5 times its cube makes 7,905,504. What is that number?

or, to use our notation, 1QC – 5C + 500N equals 7 905 504. How many units does 1N contain?

7,905,504 is a square-cube affected by the adjoining of a plane-solid product of the side and of a given plane-plane 500 and diminished by a plane-solid product of the cube and a plane 5.

[...]

Example of the analysis of a square-cube doubly affected by affirmation for the side and by negation for the cube.

1. Extraction of the first partial side

– plane coefficient		5 • •			to be multiplied by the cube	
+ plane-plane coefficient		5 0 0 • •			to be multiplied by the side	
affected square-cube to be solved	7 9 • QCj	0 5 5 0 4 QQ C Q N • QCij		0 N 2 Q 4 C 8 QQ 16 QC 32	0 4 16 64 256 1024	
plane-solids	subtractive	3 2		square-cube of the first side		
		1 0 0 0		first side by the plane-coefficient		
	additive	4 0 •		cube of the first side by the plane coefficient		
excess of the subtractive plane-solids	3 1	7 0 0 0				
remainder of the affected square-cube to be solved	4 7	3 5 5 0 4				

2. Extraction of the second partial side

upper part of the divisors	plane-plane of the expletion, plane coefficient by the triple square of the first side		6 0
	– solid of the expletion, plane coefficient by the triple of the first side		3 0
	– plane coefficient		5
	+ plane-plane coefficient		5 0 0 •
remainder of the affected square-cube to be solved		4 7	3 5 5 0 4 •
lower part of the divisors	quintuple square-square of the first side ten-fold cube of the same ten-fold square of the same quintuple tuple of the first side	8	0
			8 0
			4 0
			1 0
sum of the divisors of the additive affection		8	8 4 6 0
sum of the divisors of the subtractive affection			6 3 0 5
excess of the divisors of the additive affection		8	7 8 2 9 5
		3 2	0
		1 2	8 0
plane-solids to be taken away arising from the division	uppers	2	5 6 0
			2 5 6 0
	lower		1 0 2 4
			2 0 0 0
sum of the subtractive plane solids		4 7	6 6 6 2 4

second side by the quintuple square-square of the first side, square of the second side by the ten-fold cube of the first, cube of the second side by the ten-fold square of the first, square-square of the second side by the quintuple of the first, square-cube of the second side, second side by the plane-plane coefficient

		2 4 0	second side by the plane-plane of the expletion
additive plane-solids		4 8 0	square of the second side by the solid of the expletion
		3 2 0	cube of the second side by
sum of the additive plane-solids		2 9 1 2 0	the plane of the coefficient
excess of the subtractive plane-solids equal to the affected square-cube to be solved	4 7	3 5 0 4	

Therefore, if 1QC – 5C + 500N equals 7,905,504, then 1N is 24, by following exactly the same path as that of the composition, but in the opposite direction.

This is how Viète sets out to solve equation of Problem XV:

$$x^5 - 5x^3 + 500x = 7\ 905\ 504$$

The procedure requires an initialisation which consists in determining the order of magnitude of the positive root being sought, and its first decimal digit. Viète only deals with those terms of the equation that affect that order of magnitude. Here, taking account of the size of the coefficients, he only deals with the constant term. He puts a mark under the digit of the units of the constant term, as well as under the digit of the fifth powers whose rank is of the form $1 + 5k$, $k = 1, 2, \dots$ Counting the number of blocks of 5 digits gives the number of digits of the root x, and here x will have two digits.

If n is the first digit of x then $10n$ is a first approximation to x. Therefore n has to be the greatest integer such that:

$$10^5 n^5 - 5.10^3 n^3 + 500.10n \leq 7\ 905\ 504.$$

The greatest integer for which $n^5 < 79$ is $n = 2$. Therefore, taking account of the size of the coefficients of the equation, we can say that 20 is a first approximation to the root.

Let $x = 20 + s$ in the equation and we get:

$$(20 + s)^5 - 5(20 + s)^3 + 500(20 + s) = 7\ 905\ 504.$$

In order to determine the different terms of this quintic equation, Viète uses the 'synthetic theorems' which provide the expansions of the binomials. He distinguishes between the 'upper part of the divisors' which come from the terms $- 5(20 + s)^3 + 500(20 + s)$, and the 'lower part of the divisors' which come from the expansion of $(20 + s)^5$ and which occur in all extractions of fifth roots. The adjectives 'upper' and 'lower' refer to the way the table is laid out, the upper part occupying the lines situated above the remainder of the 'affected square-cube to be solved', the lower part being written under it. Also, in the intermediate operations, Viète separates out the terms that are to be added, and the terms that are to be subtracted.

Viète starts by calculating the constant term of the polynomial

$$(20 + s)^5 - 5(20 + s)^3 + 500(20 + s)$$

He obtains:

$+ (20)^5$	$+ 3\,2$	$0\,0\,0\,0\,0$
$+ 500 \times 20$	$+$	$1\,0\,0\,0\,0$
$- 5 \times (20)^3$	$-$	$4\,0\,0\,0\,0$
so the constant term is	$+ 3\,1$	$7\,0\,0\,0\,0$
This term taken from:	$7\,9$	$0\,5\,5\,0\,4$
leaves a remainder:	$4\,7$	$3\,5\,5\,0\,4$

Therefore, by using the binomial expansions, the equation to be solved is now:

$$5. (20)^4 s + 10. (20)^3 s^2 + 10. (20)^2 s^3 + 5. 20 s^4 + s^5 - 5. 3. (20)^2 s - 5. 3. 20 s^2 - 5 s^3 + 500 s = 4\,735\,504.$$

The number s is an integer between 0 and 9. To reduce the number of possible values for s, we can write the above equation in the form:

$$s^5 + sP(s) = 4\,735\,504$$

where $P(s) = 5. (20)^4 + 10. (20)^3 s + 10. (20)^2 s^2 + 5. 20 s^3 - 5. 3. (20)^2 - 5. 3. 20 s - 5 s^2 + 500$, and we notice that $sP(1) < s\,P(s) < s^5 + sP(s) = 4\,735\,504$. Therefore s must necessarily be less than 4 735 504 divided by $P(1)$.

This reasoning corresponds to Viète's approach. He calculates the value of $P(1)$ and obtains:

$- 5. 3. (20)^2$	$-$	$6\,0\,0\,0$
$- 5. 3. (20)$	$-$	$3\,0\,0$
$- 5$	$-$	5
to be subtracted:	$-$	$6\,3\,0\,5$
$+ 500$	$+$	$5\,0\,0$
$+ 5. (20)^4$	$+ 8$	$0\,0\,0\,0\,0$
$+ 10. (20)^3$	$+$	$8\,0\,0\,0\,0$
$+ 10. (20)^2 s$	$+$	$4\,0\,0\,0$
$+ 5. 20$	$+$	$1\,0\,0$
to be added:	$+ 8$	$8\,4\,6\,0\,0$
and subtracting	$-$	$6\,3\,0\,5$
$P(1)$ equals:	$+ 8$	$7\,8\,2\,9\,5$

On dividing 4 735 504 by 878 295 we obtain a number greater than 5, so this is the first value to test for s. The calculations start from this value and show that 5 is too large: these calculations use the values of the coefficients already found. Therefore, the next value to test is 4. Viète obtains:

$+ 800\,000 s$	$+ 3\,2$	$0\,0\,0\,0\,0$
$+ 80\,000 s^2$	$+ 1\,2$	$8\,0\,0\,0\,0$
$+ 4\,000 s^3$	$+ 2$	$5\,6\,0\,0\,0$
$+ 100 s^4$	$+$	$2\,5\,6\,0\,0$
$+ s^5$	$+$	$1\,0\,2\,4$
$+ 500 s$	$+$	$2\,0\,0\,0$
to be added:	$+ 4\,7$	$6\,4\,6\,2\,4$

$-6000s$	$-$	2 4 0 0 0
$-300s^2$	$-$	4 8 0 0
$-5s^3$	$-$	3 2 0
to be subtracted:	$-$	2 9 1 2 0
which gives the value	$+47$	3 5 5 0 4

which shows that 24 is the exact value for the root of the equation.

In the case of a number of three digits, the same calculations can be carried out, by means of shifting the numbers displayed in the table. Of course, once the second digit has been found, the same algorithm can be applied to find the third digit. This is done by Viète in problems I, II, III. The process can be continued as far as is needed to obtain the required. Viète himself chose coefficients so that the roots turned out to be integers, and the most common of these are 8, 12, 19, 24, 27, 34, 45, 57 or, exceptionally, 243 or 432.

Al-Kāshī and Sin 1°

The digit by digit calculation of roots by Ibn al-Bannā (Section 7.3) and by Viète depends on numbers being represented with positional notation, so that at each stage of the process we can move on to the next lower order of magnitude. With Viète, of course, numbers were represented in base ten, but the principle still holds true in any other base. The sexagesimal number system, which was positional, continued to have widespread use, particularly for astronomy, right up to the Middle Ages. The mathematician and astronomer al-Kāshī used a 'digit by digit' method for obtaining accurate approximations, though the 'digit' here is the value of the different orders of magnitude of the number in sexagesimal form.

Astronomers have always had a need for accurate 'trigonometric tables' and that was why Ptolemy, for example, included tables of chords in his *Almagest*. If a chord b is subtended by an angle b in a circle of radius $R = 60$, then we have the relation: $\mathrm{crd}.b = 2R\sin b/2 = 120\sin b/2$. Ptolemy's calculations, based on geometrical reasoning, enabled him to obtain very accurate values for crd.12°, crd.6°, crd.3° and crd.1° 30'. However, despite the level of the sophistication of his method, the best sexagesimal values that Ptolemy was able to obtain for crd.2° and crd.1° were the approximations 2;5,40 and 1;2,50 respectively (see Section 10.1).

Indian mathematicians introduced the idea of the half chord, later taken up by the Arabs, and this corresponds closely to our use of sine since, if we use Sin a for the half chord of a circle of radius 60, then we have Sin a = 60 sin a. In his *Epistle of the chord and the sine* (Rīsalat al-watar wa-l-Jayb) (c. 1400) al-Kāshī describes an algorithm that allows him to determine Sin 1°, with considerable accuracy, from Sin 3°. This was already known to be 3;8,24,33,59,34,28,15, to an accuracy of '60⁻⁷'.

The original work has not been found, but al-Kāshī's method is described in a commmentary on astronomical tables written by his grandson Mīram Shalabī (c. 1500). The Sine is taken as a 'thing', that is an unknown, and the problem comes down to this: 45 units of the first order multiplied by the unknown are equal to the sum of the cube and a number (see [1], [21], [22]).

If we let $x = \text{Sin } 1°$, and use the relation $\sin 3a = 3 \sin a - 4 \sin^3 a$, then al-Kāshī has to solve the cubic equation $3x = 4x^3/60^2 + \text{Sin } 3°$, which is of the form:

$$qx = x^3 + p$$

where p and q, in sexagesimal form, are $p = 47,6;8,29,53,37,3,45$ and $q = 45,0$. (This is the equation for the *trisection of the angle*, see Section 6.2).

Al-Kāshī solves the equation by considering x as the sum of its different sexagesimal orders of magnitude: $x = a + b + c + \ldots$. Since x^3 is small compared with p, we have an approximation for a by neglecting the x^3 term: $a \approx p/q \approx 47/45 \approx 1$. Now b can be found by substituting in the equation the approximation $a^3 = 1$ for x^3, and letting $a + b = 1 + b$. This gives:

$$b = x - a \approx (p + a^3)/q - a = (p - aq + a^3)/q = (p - q + 1)/q \approx 0;2.$$

Now c can be found in a similar way:

$$\text{Let } x = (a + b) + c = 1;2 + c \text{ and } x^3 = (a + b)^3 = (1;2)^3.$$

We now obtain:

$$c \approx [p + (a + b)^3]/q - (a + b) = [p - (a + b)q + (a + b)^3]/q \approx 0;0,49$$

And so on ....

Al-Kāshī carried out nine iterations of this type and obtained the value:

$$\text{Sin } 1° = 1;2,49,43,11,14,44,16,26,17$$

which in decimal form is equivalent to

$$\sin 1° = 0.\,017\,452\,406\,437\,283\,571.$$

The rate of convergence of the process is remarkable. To put it symbollically, al-Kāshī calculates $\text{Sin } 1°$ using ten successive approximations x_k starting with $x_0 = 1$, using the relation $x_{k+1} = f(x_k)$ where $f(x) = (p + x^3)/q$.

7.6 Kepler's Equation

Johannes Kepler, in Book V of his abridgement of Copernicus' Astronomy, *Epitomes astronomiae Copernicanae* (1618) [9], needed to solve the following equation, now named after him:

$$x = t - e \sin x$$

where t and e are given values.

The equation arose from describing the motion of a heavenly body C which moves about a second heavenly body A. From 'Kepler's first law', C will describe an ellipse with A as one of its foci. The eccentricity of the ellipse, e is given by $e = AB/RB$ (Fig. 1). Let u be the angle PAC, which gives the position of C in its orbit, referred to its perihelion P.

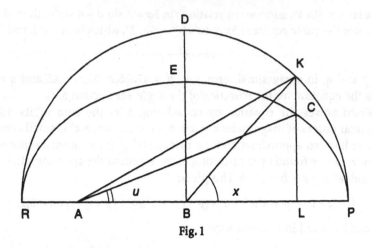

Fig. 1

From 'Kepler's second law', during its movement, the area *PAC*, swept out by the line *AC* connecting the two bodies, is proportional to the time taken. It is natural to introduce a fictitious angle *t*, which corresponds to the movement of a body under uniform circular motion, completing a revolution in the same time. Kepler called this angle *t* the *mean anomaly* and, by contrast, the angle *u* is called the *true anomaly*.

The mean anomaly *t* is therefore proportional to the elliptic area *PAC*. From the geometry of the ellipse and its auxiliary circle, this area is also proportional to the area *PAK*, where *K* is the point on the circle where the perpendicular through *C* to the diameter *PR* meets the circle. The point *K* is given by the angle *x = PBK*, called the *eccentric anomaly*.

The equation Kepler obtained as an equality of areas is:

$$\text{area } PAK = \text{area } PBK + \text{area } KBA$$

With *a* for the semi-major axis, we have:

> area *PAK* / *t* = constant = $a^2/2$,
> area *PBK* = *x*. $a^2/2$,
> area *KBA* = *AB*. *KL* / 2 = *ae*. *a* sin *x*/2.

From this we have: $1/2\ a^2t = 1/2\ a^2x + 1/2\ a^2e \sin x$
or: $t = x + e \sin x$.

Knowing the eccentric anomaly *x*, it is then easy to deduce the mean anomaly *t*, but it is the inverse problem that is of interest to astronomers. Now, the mean anomaly *t* is proportional to the time elapsed, and it is as a function of time that we wish to know the position of the heavenly body *C*. This is, in fact, perfectly determined by the eccentric anomaly *x*, that is by the arc *PK*, since *C* is at the intersection of the ellipse and the perpendicular to the major axis through *K*.

The relation being transcendental, its solution cannot be found directly but only by approximation methods. In fact, work had already been done on an equation of exactly the same type as Kepler's, in preparing tables needed for the theory of parallax. In the 9th century, Ḥabash al-Ḥāsib had solved it by an iterative method [8] similar to the one used by al-Kāshī to find sin 1° (see previous Section) and the one used by Kepler to solve the equation that has been called after him.

In each case we have what is called a *fixed point method*. To solve the equation $x = f(x)$, we start from some value x_0 close to the sought root, and consider the sequence of approximations defined by

$$x_{n+1} = f(x_n).$$

If the sequence x_n converges to a limit, then that limit must necessarily be a solution to the equation.

In order to produce new astronomical tables, his *Rudolphine Tables*, Kepler needed to solve the equation given above for different values of t. We should point out that Kepler expressed areas in terms of their ratio to the area of the circle. This ratio is expressed in sexagesimal units, the area of the circle being 360°, or 21 600′, or 1 296 000″. As with al-Kāshī, the sine is multiplied by the radius of the circle, in this case $r = 100\,000$, so that in Kepler's text, the sine of 50° 9′ 10″ is 76 775.

J. Kepler
Epitomes astronomiae Copernicanae, Linz, 1618–1622, Book V, Ch. IV, pp. 694–696.
From the French translation of the Latin by A. Michel-Pajus.

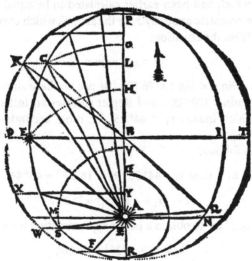

Can you teach through an example an easy method so that someone who is inexperienced may not err too far on account of badly determined positions?

Let us take the example above and let the mean anomaly, or the area *PKA*, be already given as equal to 50° 9′ 10″; it is clear that, if the area of the triangle *KBA* is known, the remaining area *KBP* can be obtained, with the same number of degrees as its arc *PK*, and so, after taking the value of *KBA* from *PKA* what is left is the eccentric anomaly *PK*. Therefore, since *PKA* is greater than PKB, the sine of the arc PK will be smaller than the sine of $50° 9' 10''$ and therefore less than 76 775. Let this sine be taken at the first position as 70 000 in order to make the multiplication easier.

Therefore, this multiplied once by the value of the triangle DBA, which was in the example above 11 910'', and suppressing the 5 [last digits], gives for BKA 8 337'', that is $2° 18' 57''$, which you add to $44° 25'$, the arc whose sine is 70 000, so the area PKA will be $46° 44'$. This value is too small by $3° 25'$, since it ought in fact to be the given value of $50° 9'$. Let now this sine be taken in the second position, obtained by adding $3° 25'$ to the arc first proposed equal to $44° 25'$, so that PC becomes close to $47° 50'$, whose sine is very nearly 74 000, which I have again chosen to make the multiplication easier. This figure, multiplied by 11 910 now adds $7' 56''$ to BKA, which becomes $2° 26' 53''$, which I add to the second position of PK, that is to the value $47° 44' 6''$ of PKB; now PKA comes out at $50° 10' 59''$, and we go beyond the value of $50° 9' 10''$ by $1' 49''$. That is why we now feel that this small excess must be taken away from the second position of PK, and the eccentric anomaly sought, that is PK, will be $47° 42' 17''$. This value is acceptable. The sine of the arc is in fact 73 979, which [multiplied] by 11 910'' requires the value of KBA to be $2° 26' 50''$, and this being added on will give $50° 9' 7''$, which differs from the correct value of $50° 9' 10''$ by an imperceptible amount.

Kepler calls his method the 'rule of position', an allusion probably to the long tradition of rules of false position (see Chapter 3). He explains it by means of an example, based on a value of the mean anomaly t, that is the area of PKA, equal to $50° 9' 10''$. The eccentricity e, or rather the quantity $ea^2/2$, corresponds to the area of the triangle ABD, which had been earlier calculated to be equal to 11 910''. The task is to determine the eccentric anomaly x, or the arc PK, which corresponds to the area PKB, and starting from the equation:

$$t = x + e \sin x.$$

Since $x < t$, we have Sin $x <$ Sin $t = 76\,775$ [we use Sin here since it refers to the sine multiplied by the radius 100 000], and Kepler chooses an initial value x_1 such that Sin $x_1 = 70\,000$, which makes $x_1 = 44° 25'$ (or, more accurately, Arc Sin 70 000 $= 44° 25' 37''$).
Kepler calculates as follows:

$$x_1 + e \sin x_1 \approx 44° 25' + 2° 18' 57'' \approx 46° 44'.$$

From this: $t - (x_1 + e \sin x_1) \approx 3° 25'.$

He uses this error of $3° 25'$ to obtain a better approximation for x:

$$x_2 = x_1 + 3° 25' = 47° 50'.$$

Now, since Sin x_2 is close to 74 000, he makes the calculations easier by taking:

$$x_2 = \text{Arc Sin } 74\,000 \approx 47° 44' 6''$$

(the correct value of Arc Sin 74 000 is $47° 43' 53''$).
Using this new value of x_2 Kepler repeats the calculations:

$$e \sin x_2 = e \sin x_1 + e (\sin x_2 - \sin x_1) = 2° 18' 57'' + 7' 56'' = 2° 26' 53''$$
and: $x_2 + e \sin x_2 \approx 47° 44' 6'' + 2° 26' 53'' \approx 50° 10' 59''.$

From this: $t - (x_2 + e \sin x_2) \approx 50° 9' 10'' - 50° 10' 59'' = - 1' 49''$

and we have a new approximation:

$$x_3 = x_2 - 1' 49'' = 47° 42' 17''.$$

Kepler stops here, since:

$$x_3 + e \sin x_3 = 50° 9' 7''.$$

What Kepler did can be summarised by saying that the successive approximations are defined by:

$$x_n = f(x_{n-1}) \quad \text{where} \quad f(x) = t - e \, \text{Sin} \, x.$$

In the *Principia* (1687) [13], Newton showed an interest in Kepler's equation. In Book I, Proposition XXXI he showed how "to find the place of a body moving in a given ellipse at any assigned time". He did this by means of an extremely elegant geometrical proof using a tronchoid, which is a sort of deformed cycloid (the nail on the wheel not situated on its rim). Newton recognised, however, that the construction of this curve is difficult and in a scholium he proposed the use of an approximate solution. This description was, according to Whiteside, the first publication by Newton of what we know as Newton's Method (see Chapter 6).

At a later time, approximate solutions to Kepler's equation were considered by Lagrange [12] in his work on analytic functions.

7.7 Bernoulli's Method of Recurrent Series

Daniel Bernoulli, son of Jean Bernoulli, was professor of mathematics at St Petersburg from 1725 to 1733. He applied himself to a number of problems drawn from physics, notably in hydrodynamics and elasticity. He attempted to solve partial differential equations, in particular through the use of series, both integer series and trigonometric series.

In 1728, he published an article with the title: 'Observations concerning series formed by adding or subtracting successive terms in any manner, where it is shown in particular the remarkable use of these same series to find the roots of all algebraic equations' [3]. The procedure which interests us here is therefore a corollary of a more general study of recurrent series. Bernoulli states it thus:

A new method for finding the roots of all equations, with numerical or algebraic coefficients. Write the proposed equation in the form:

$$1 = ax + bx^2 + cx^3 + ex^4 + etc.$$

Now form a sequence beginning with arbitrary terms, and in number equal to the degree of the equation, according to the following law: if A, B, C, D, E are consecutive terms taken in order, we must always have $E = aD + bC + cB + eA + etc.$ Let there be in this sequence, continued sufficiently far, two neighbouring terms M and N. Then we shall find that the term M divided by the following term N is nearly equal to a sought root.

Bernoulli does not give a satisfactory justification for the method. But he underlines the difficulties connected with the existence of several distinct roots in the same small interval and shows, by examples, how these may be overcome.

Daniel Bernoulli invited Euler to come to St. Petersburg and took him on as his assistant in 1733. He relinquished his post to him the same year when he left for Basle to study medicine, and Euler took up a number of problems that Bernoulli had begun to research. This is how Euler came to publish Bernoulli's method in 1748, in his *Introductio in analysin infinitorum* [5]. He gives a proof of it using recurrent series, that is series of the form $A + Bz + Cz^2 + ...$ whose coefficients $A, B, C, ...$ are related to each other by a linear recurrence relation, just as in the series studied by Bernoulli. In order to solve an algebraic equation $T(z) = 0$, where the constant term of T is 1, the idea is to consider the polynomial $T(z)$ as the denominator of a rational fraction in z, $S(z)/T(z)$ with arbitrarily chosen numerator $S(z)$. This fraction can be expanded as a series in z, with integer coefficients. In practice, we put $T(z) = 1 - zU(z)$ where $U(z)$ is a polynomial; then $S(z)/T(z) = S(z)/(1 - zU(z)) = S(z)[1 + zU(z) + z^2U(z)^2 + ...]$. All that is needed is to expand the terms $U(z)^n$, multiply them by $S(z)$ and then to rearrange in ascending powers of z to obtain the recurrent series $A + Bz + Cz^2 + ...+ Pz^n + Qz^{n+1} + ...$ We use the fact that, as we shall see later in the case where all the roots are real and distinct, when n is sufficiently large, the ratio P/Q is a good approximation to the root with the least absolute value.

L. Euler

Introductio in analysin infinitorum, Lausanne, 1748, Book I, Ch. XVII.
On the Use of Recurrent Series to find Roots of Equations

333. Since all recurrent series can be thought of as the expansion of some rational fraction, let us take the fraction

$$= \frac{a + bz + cz^2 + dz^3 + ez^4 + \&c.}{1 - \alpha z - \beta z^2 - \gamma z^3 - \delta z^4 - \&c.}$$

which gives rise to the following recurrent series:

$$A + Bz + Cz^2 + Dz^3 + Ez^4 + Fz^5 + \&c.$$

whose coefficients A, B, C, D, &c. are determined in the following way:

$$A = a$$
$$B = \alpha A + b$$
$$C = \alpha B + \beta A + c$$
$$D = \alpha C + \beta B + \gamma A + d$$
$$E = \alpha D + \beta C + \gamma B + \delta A + c$$
$$\&c.$$

Now the general term, or the coefficient of the power z^n, is found by decomposing the proposed fraction into simple fractions, whose denominators are the factors of the denominator

$$1 - \alpha z - \beta zz - \gamma z^3 - \&c.,$$

as we have seen (Chap. XIII.)

334. As for the form of the general term, it depends principally on the nature of the simple factors of the denominator, according as they are real or imaginary, and according as they are not equal to each other, or two or several being equal to each other. To go through these different cases in order, suppose firstly that all the simple factors of the denominator are both real and distinct. Let then all the simple factors of the denominator be

$$(1 - pz)(1 - qz)(1 - rz)(1 - sz) \ \&c.;$$

from which the proposed fraction may be decomposed into the following simple fractions

$$\frac{A}{1-pz} + \frac{B}{1-qz} + \frac{C}{1-rz} + \frac{D}{1-sz} + \&c.$$

These being known, the general term of the recurrent series will be

$$= z^n(Ap^n + Bq^n + Cr^n + Ds^n + \&c.)$$

which we shall put $= Pz^n$; that is that P will be the coefficient of the power z^n, those of the following powers will be Q, R, &c., so that the recurrent series becomes

$$A + Bz + Cz^2 + Dz^3 + \ldots + Pz^n + Qz^{n+1} + Rz^{n+2} + \&c.$$

335. Let us suppose now that n is a very large number, or the recurrent series continued as far as a very large number of terms. Since the powers of unequal numbers are themselves even more unequal, as they are raised to a higher degree, there will be so great a difference between the powers Ap^n, Bq^n, Cr^n, than that which arises from the greatest of the numbers p, q, r, &c., that the former will considerably surpass the latter, so that the latter will become absolutely nothing in comparison with the former, if n is infinitely large. Since the numbers p, q, r, &c. are not equal to each other, suppose that p is the largest of them. Consequently, if n is an infinite number, we have $P = Ap^n$; but if n is only a very large number, we have very nearly $P = Ap^n$. Similarly $Q = Ap^{n+1}$, and so $\frac{Q}{P} = p$. From this it is evident that, if now the recurrent series is prolonged very far, the coefficient of a term divided by that of the one before it, will provide the approximate value of the greatest letter p.

336. Therefore, if in the proposed fraction

$$\frac{a + bz + cz^2 + dz^3 + \&c.}{1 - az - \beta z^2 - \gamma z^3 - \delta z^4 - \&c.}$$

the denominator is only composed of simple factors which are real and different, from the recurrent series which arises from it, we will be able to know one simple factor, namely the factor $1 - pz$, where the letter p has the greatest value of all of them.

$$[\ldots]$$

337. It is now fairly easy to see how to apply this study to finding the roots of any algebraic equation. Indeed, with the factors of the denominator $1 - \alpha z - \beta z^2 - \gamma z^3 - \delta z^4 - \&c.$ being known, we can easily produce the roots of the equation

$$1 - \alpha z - \beta z^2 - \gamma z^3 - \delta z^4 - \&c. = 0$$

such that if one factor is $1 - pz$, one root of that equation will be $z = \frac{1}{p}$. Therefore, since the recurrent series will provide us with the greatest number p, at the same time we shall obtain the least root of the equation

$$1 - \alpha z - \beta z^2 - \gamma z^3 - \&c. = 0.$$

Or, if we let $z = \dfrac{1}{x}$, to obtain

$$x^m - \alpha x^{m-1} - \beta x^{m-2} - \gamma x^{m-3} - \&c. = 0,$$

we shall have, by the same method, the greatest root of this equation, $x = p$.

[...]

EXAMPLE II.

Let the proposed equation be $3x - 4x^3 = \frac{1}{2}$, whose roots represent the sines of three arcs, the triples of which have a sine $= \frac{1}{2}$.

After rearranging the equation to the form $0 = 1 - 6x* + 8x^3$, let us find, so that we can have it in terms of integers, the smallest root, so that it will not be necessary to put $\dfrac{1}{z}$ for x. We therefore form the fraction

$$\frac{a + bx + cxx}{1 - 6x* + 8x^3}$$

and from the first three terms, that we take according to our wish to be 0, 0, 1, in order to make the calculations very easy, we obtain, leaving out the powers of x, since we only need the coefficients, the recurrent series

$$0;\ 0;\ 1;\ 6;\ 36;\ 208;\ 1200;\ 6912;\ 39808;\ 229248.$$

The approximate value of the smallest root of the equation will therefore be $= \dfrac{39808}{229248} = \dfrac{311}{1791} = 0.1736515$, which on that account has to be the sine of 10°; but, from tables, this one is 0.1736482; which the found root surpasses by $\dfrac{33}{10000000}$.

Euler makes an error in his final calculation, since $\dfrac{311}{1791} = 0.1736460$, and so the calculated value of sin 10° is in fact less than the value given in the tables, by $\dfrac{22}{10000000}$.

In his article, Bernoulli, considering in particular geometric sequences satisfying recurrence relations, introduced what we now refer to as the *characteristic equation* of the recurrence relation. He gives, by induction, but without a valid proof, the general form for the solution sequence, which we would write as $\sum\limits_{1 \le i \le s} P_i(n) q_i^n$ where the $q_1, \ldots, q_s$ are the roots of the characteristic equation and P_i is a polynomial of degree equal to 1 less than the multiplicity of q_i. It is apparently this result that allows him to state his rule.

Euler establishes this result by a different approach. He uses the expansion of a rational fraction which he writes in two different ways: the first way shows that the series is a recurrent series, the coefficients of the recurrence relation being those of the polynomial $T(z)$, that is of the equation to be solved; the second way uses the decomposition of the rational fraction into simple elements which allow him to obtain a general term for the series.

Euler remarks, as did Bernoulli, that if there are two roots with the least absolute value, but of opposite sign, then taking the roots in pairs will give us the square of these roots. He also notes that if there are multiple roots, then the convergence may, perhaps, be very slow. He suggest a way of getting round this difficulty by iterating the process: after having found an approximation z_0 to the root z, the initial equation can be transformed by putting $y = z - z_0$ so that the transformed equation in y will have a root that is closer to 0 than all the others, and so this will give a good convergence. He also proposed a method for dealing with the case where there are two conjugate complex roots, but this takes us into areas that go beyond the scope of this chapter.

In Note VI of his *Traité de la résolution des équations numériques* [11], Lagrange revisits Bernoulli's method. Also, seeing that the convergence of the process can be very slow in the case of multiple roots, Lagrange took advantage of the fact that the numerator of the fraction $S(z)/T(z)$ was arbitrary, and proposed that $S(z)$ should be chosen to be the derivative of the polynomial $T(z)$. This ensures that the denominator of the fraction $T'(z)/T(z)$ will not have multiple roots.

Aitken [2] perfected this method to obtain all the roots by using determinants. In particular, he generalised Bernoulli's result in the following way: let $q_1, q_2, \ldots, q_s$ be the roots of an equation to be solved, arranged in order of increasing absolute values, and let (u_n) be a sequence satisfying the recurrence relation associated with the equation; if $|q_m| < |q_{m+1}|$, then:

$$q_1 q_2 q_3 \cdots q_m = \lim_{n \to \infty} U_n / U_{n+1}$$

where
$$U_n = \begin{vmatrix} u_n & u_{n+1} & \cdots & u_{n+m-1} \\ u_{n-1} & u_n & \cdots & u_{n+m-2} \\ \cdots & \cdots & \cdots & \cdots \\ u_{n-m+1} & u_{n-m+2} & \cdots & u_n \end{vmatrix}$$

When $m = 1$, the formula gives q_1 and we have Bernoulli's result. When the roots have distinct absolute values, we can successively calculate q_1, $q_1 q_2$, then q_2, $q_1 q_2 q_3$, then q_3, etc.... In this same article, Aitken describes his well known convergence acceleration process for sequences (see Section 14.4).

7.8 Approximation by Continued Fractions

The first works by Lagrange on approximating the roots of an equation were published in the *Memoirs of the Berlin Royal Academy of Sciences* in 1769. Lagrange developed there a new approach through the use of continued fractions (see Sections 4.5 and 8.11). The principle of the method is as follows.

Consider an algebraic equation that has to be solved:

$$Ax^m + Bx^{m-1} + Cx^{m-2} + \ldots + K = 0.$$

Suppose that an integer approximation p to the root is known, such that the root lies in the interval $[p, p + 1[$. If we change the unknown using the substitution $x = p + 1/y$ $(0 < 1/y < 1)$, we obtain a new equation in y:

$$A'y^m + B'y^{m-1} + C'y^{m-2} + \ldots + K' = 0.$$

for which we are assured that it has at least one root greater than 1. If we let q be the integer part of this root, the substitution $y = q + 1/z$ leads to a new equation in z which has at least one root greater than 1. Let r be its integer part ... and so on. The continued fraction

$$p + \cfrac{1}{q + \cfrac{1}{r + \cfrac{1}{\ldots}}}$$

will provide the value of a root of the initial equation, and its successive convergents provide alternately upper and lower approximations for the root. Of course, if the root is rational, the process will end, and provide the root, after a finite number of iterations. For irrational roots, the process also gives us an estimation of the error of approximation at each stage, something that the Newton-Raphson method cannot do. The text by Lagrange that follows needs no further commentary.

J. -L. Lagrange

Sur la résolution des équations numériques,
Mémoires de l'Académie royale des Sciences et Belles-Lettres de Berlin,
vol. XXIII (1769)
Œuvres, vol. II, Paris: Gauthier-Villars, 1868, pp. 539–578.

§III – A new method for approximating the roots of numerical equations

18. Consider the equation

(a) $$Ax^m + Bx^{m-1} + Cx^{m-2} + \ldots + K = 0$$

and suppose that we have already found, by the preceding method or otherwise, the integer value approximating to one of its roots, being real and positive: let this first value be p, such that we have

$$x > p \quad \text{and} \quad x < p + 1;$$

we make

$$x = p + \frac{1}{y},$$

and, substituting this value in the proposed equation, in the place of x, we shall have, after multiplying the whole equation by y^m and arranging the terms in order of powers of y, an equation of the form

(b) $$A'y^m + B'y^{m-1} + C'y^{m-2} + \ldots + K' = 0.$$

Now, since, by hypothesis, $\dfrac{1}{y} > 0$ and < 1, we have $y > 0$; therefore the equation (b) must necessarily have one real root greater than unity.

We therefore, by the methods of §1, look for the integer value approximating this root, and, since this root has to be positive, we only need to consider positive [values of] y (4).

Having found the integer value that approximates to y, which I shall call q, we now make

$$y = q + \frac{1}{z},$$

and, substituting this value of y in equation (b), we shall have a third equation of the form

(c) $$A''z^m + B''z^{m-1} + C''z^{m-2} + \ldots + K'' = 0,$$

which must necessarily have at least one real root greater than unity, for which we can find an approximate integer value in the same way.

This approximation to z being called r, we now have

$$z = r + \frac{1}{u},$$

and on substituting we shall have an equation in u which will have at least one real root greater than unity, and so on.

Continuing in the same manner, we shall approach closer and closer to the value of the required root; but, should it happen that some one of the numbers $p, q, \ldots$ should be an exact root, then $x = p$ or $y = q, \ldots$, and the operation will terminate; therefore in this case, we shall find for x a commensurable value.

In all other cases the value of the root must necessarily be incommensurable, and can only be approximated, as close as is wished.

[...]

§IV. – The application of the preceding methods to some examples.

25. I shall take as my first example the equation that Newton solved by his method, namely

$$x^3 - 2x - 5 = 0$$

[...]

I shall now, following the method of §III, put $x = 2 + \dfrac{1}{y}$; on substituting and arranging the terms as powers of y, we have the equation

$$y^3 - 10y^2 - 6y - 1 = 0,$$

in which I have changed the signs in order to make the first term positive.

This equation will therefore necessarily have a single root greater than unity (19), so that, in order to find an approximate value we need only substitute the numbers $0, 1, 2, 3, \ldots$, until we find two consecutive substitutions that give results of contrary sign.

In order not to carry out many unnecessary substitutions, I note that on putting $y = 0$ I get a negative result, and in putting $y = 10$ the result is still negative; I start therefore with the number 10, and I make successively $y = 10, 11, \ldots$. I find straight away the results $- 61, 54, \ldots$; from which I conclude that the approximate value for y is 10; therefore $q = 10$.

I therefore make $y = 10 + \dfrac{1}{z}$, and I have the equation

$$61z^3 - 94z^2 - 20z - 1 = 0,$$

and successively letting $z = 1, 2, \ldots$, I have the results $- 54, 71, \ldots$; therefore $r = 1$.

Again, I let $z = 1 + \dfrac{1}{u}$, and I shall now have the equation

$$54u^3 + 25u^2 - 89u - 61 = 0,$$

and, letting $u = 1, 2, \ldots$, I shall have the results $- 71, 293, \ldots$; therefore $s = 1$ and so on.

By continuing in this way, we find the numbers

$$2, 10, 1, 1, 2, 1, 3, 1, 1, 12, \ldots ,$$

so that the required root can be expressed by this continued fraction

$$x = 2 + \cfrac{1}{10 + \cfrac{1}{1 + \cfrac{1}{1 + \cfrac{1}{2 + \cdots}}}}$$

from which we can obtain the fractions (23)

$$\frac{2}{1}, \frac{21}{10}, \frac{23}{11}, \frac{44}{21}, \frac{111}{53}, \frac{155}{74}, \frac{576}{275}, \frac{731}{349}, \frac{1307}{624}, \frac{16415}{7837}, \ldots,$$

which are alternately smaller and greater than the value of x.

The last fraction $\dfrac{16415}{7837}$ is greater than the required root; but the error will be less than $\dfrac{1}{(7837)^2}$, (23, 2°), that is less than 0.000 000 016 3; therefore, if the fraction $\dfrac{16415}{7837}$ is reduced to decimals, it will be exact up to the seventh decimal; now, in carrying out the division, we obtain 2.094 551 486 5...; therefore, the required root will be between the numbers 2.094 551 49 and 2.094 551 47.

Newton, by his method, found the fraction 2.094 551 47 (*see his Method of infinite series*), from which we can see that that method gives, in this case, an extremely exact result; but we would be wrong to assume we would always have such accuracy.

Horner like Transformations of Polynomial Equations

7.9 The Ruffini-Budan Schema

At the beginning of the 19th century ingenious techniques for the transformation of polynomials were developed, apparently independently, by three mathematicians: Ruffini (1804) [18], Budan (1807 and 1813) [4] and Horner (1819) [7]. These techniques, combined with the results on locating the roots of a polynomial, allowed the decimals of the required root to be determined by successive approximations, often digit by digit, with considerable saving in operations.

Traces of this method of evaluating a polynomial using this approach, are already to be found in a text by Newton (see Section 5.1), however, a systematic method for finding, not only the constant term, but also all the other coefficients of a transformed polynomial $Q(x) = P(x + u)$ did not properly appear before the beginning of the 19th century. Mathematicians in the 18th century had not, in effect, succeeded in finding a practical and simple way of transforming one equation into a second equation, the roots of which were to be simply a constant added to each of the roots of the first equation.

It would appear that Paolo Ruffini [18], at least in Europe, was the first to formulate an algorithm for the transformation of the coefficients of an equation. He was followed, a little later, by François Budan [4]. Paolo Ruffini, a medical doctor and professor of mathematics at Medina, is principally known for his work on the impossibility of solving algebraically equations of degree 5. Budan, who was a doctor of medicine and professor of mathematics at Nantes, worked on the solution of polynomials.

What follows are extracts from the work of both authors. To make the reading of Ruffini's extract easier, we make some remarks about the notation used by Ruffini. The equation under consideration is

(A) $$Ax^m + Bx^{m-1} + Cx^{m-2} + \ldots + Sx^2 + Tx + V = 0.$$

The following quantities are introduced:

$$P' = A, \; P'' = P'p + B = A\,p + B, \; P''' = P''p + C = A\,p^2 + B\,p + C, \ldots$$
$$P^{(m+1)} = P^{(m)}p + V = Ap^m + Bp^{m-1} + \ldots + Tp + V.$$

The notation $P^{(k)}$ does not therefore correspond with the derivative. Ruffini also introduces:

$$Q^{(k)} = \frac{d\,P^{(k+1)}}{dp}, \quad R^{(k)} = \frac{dQ^{(k+1)}}{2d\,p}, \quad S^{(k)} = \frac{dR^{(k+1)}}{3dp}, \quad \text{etc.}$$

and shows that:

$$Q^{(k)} = Q^{(k-1)}p + P^{(k)}, \; R^{(k)} = R^{(k-1)}p + P^{(k)}, \; S^{(k)} = S^{(k-1)}p + P^{(k)}, \text{ etc.}$$

The first formula in the text is equivalent to the Taylor series for a polynomial Z of degree m:

$$Z(p+y) = Z(p) + Z'(p)y + \frac{Z''(p)}{2}y^2 + \frac{Z'''(p)}{2.3}y^3 + \ldots + \frac{Z^{(m)}(p)}{m!}y^m$$

where the derivative $Z^{(k)}$ is given as $\dfrac{d^k Z}{dx^k}$.

P. Ruffini
Sopra la determinazione delle radici nelle equazioni numeriche di qualunque grado, 1804, Modena.
Opere matematiche, vol.II, Edizioni Cremonese, Rome, 1953, pp. 300–302.

Problem 3: – *Given that $x = p + y$, to determine a quick method for transforming the equation (A) into another with unknown y.*

From the properties of transformations we know, if Z is the left hand side of the equation (A), the desired transformation will be obtained by putting p in the place of x in the equation

$$Z + \frac{dZ}{dx}y + \frac{d^2 Z}{2dx^2}y^2 + \frac{d^3 Z}{2.3.dx^3}y^3 + ... + Ay^m = 0.$$

Then, with the hypothesis $x = p$, we have $Z = P^{(m+1)}$, and so

$$\frac{dZ}{dx} = \frac{dP^{(m+1)}}{dp} = Q^{(m)}, \quad \frac{d^2 Z}{2dx^2} = \frac{dQ^{(m)}}{2dp} = R^{(m-1)}, \quad \frac{d^3 Z}{2.3.dx^3} = \frac{dR^{(m-1)}}{3dp} = S^{(m-2)}$$

Therefore these quantities $P^{(m+1)}$, $Q^{(m)}$, $R^{(m-1)}$, $S^{(m-2)}$, ... are simply the coefficients of the equation in y and for that reason the equation becomes:

(F) $P^{(m+1)} + Q^{(m)}y + R^{(m-1)}y^2 + S^{(m-2)}y^3 + ... + Ay^m = 0.$

This being so, all that is needed to solve our problem is to quickly determine the quantities $P^{(m+1)}$, $Q^{(m)}$, $R^{(m-1)}$, $S^{(m-2)}$, ...

To do this, the coefficients of equation (A) being set out in (IV) in the same way as in (I), and all the quantities A, P'', P''', ..., $P^{(m+1)}$ being determined in the same way, I put in a third line under P'', the coefficient A, I multiply this by p, I add the product $Ap = Q'p$ to P'' which is above, and the result $Q'p + P''$ being equal to Q'', is put in the third line under P''. This Q'' is now multiplied by p, the term P''' is added to it and we put the result $Q''p + P''' = Q'''$ under P^{iv}. Continuing in this way we shall obtain under $P^{(m+1)}$ the result $Q^{(m)}$.

In a fourth line we write under Q'' the term A, this being multiplied by p, we add to it Q'' which is above, and the result $Ap + Q'' = R'p + Q'' = R''$ is put in the same line underneath Q'''. R'' being multiplied by p and having Q''' added to it, I write the result $R''p + Q''' = R'''$ below Q^{iv}. I continue in the same way, and underneath $Q^{(m)}$ there will appear in the fourth line the quantity $R^{(m-1)}$.

By continuing the process in the same way, we shall have in the fifth line the sequence of quantities $A = S'$, S'', S''', ..., $S^{(m-2)}$, in the sixth line the quantities $A = T'$, T'', T''', ..., $T^{(m-3)}$, and so on.

	A,	B,	C,	D,	E,	F,	...,	V	
		A,	P'',	P''',	P^{iv},	P^v,	...,	$P^{(m)}$,	$P^{(m+1)}$
IV			A,	Q'',	Q''',	Q^{iv},	...,	$Q^{(m-1)}$,	$Q^{(m)}$
				A,	R'',	R''',	...,	$R^{(m-2)}$,	$R^{(m-1)}$
					A,	S'',	...,	$S^{(m-3)}$,	$S^{(m-2)}$
						A,	...,	$T^{(m-4)}$,	$T^{(m-3)}$

In the last $(m+1)^{th}$ vertical column, are contained the values of the coefficients of equation (F), I substitute them into equation (F) itself, and in this way we obtain the transformed equation by means of operations which are simple and easy to carry out, as everyone can easily see for himself, and as is shown in the following example:

Being given the equation $4x^5 - 6x^4 + 3x^3 - 5x^2 - 4x + 8 = 0$, it is wished to transform it into another in which $y = x - 6$. I carry out the operations in the manner which is set out below:

$$
\begin{array}{rrrrrrr}
 & & & 6 & & & \\
4 & -6 & 3 & -5 & -4 & 8 & \\
 & 4 & 18 & 111 & 661 & 3962 & 23780 \\
 & & 4 & 42 & 363 & 2389 & 20996 \\
 & & & 4 & 66 & 759 & 7393 \\
 & & & & 4 & 90 & 1299 \\
 & & & & & 4 & 114 \\
 & & & & & & 4
\end{array}
$$

and we obtain the transformed equation:

$$4y^5 + 114y^4 + 1299y^3 + 7393y^2 + 20996y + 23780 = 0$$

If we wish to transform this equation into another in which we have $x = y + 6$, then since we have $y = x - 6$, I carry out the operations shown below in the usual manner.

$$
\begin{array}{rrrrrrr}
 & & -6 & & & & \\
4 & 114 & 1299 & 7393 & 20996 & 23780 & \\
 & 4 & 90 & 759 & 2839 & 3962 & 8 \\
 & & 4 & 66 & 363 & 661 & -4 \\
 & & & 4 & 42 & 111 & -5 \\
 & & & & 4 & 18 & 3 \\
 & & & & & 4 & -6 \\
 & & & & & & 4
\end{array}
$$

and the transformed equation will be

$$4x^5 - 6x^4 + 3x^3 - 5x^2 - 4x + 8 = 0$$

which was the proposed equation to start with, as in fact must be apparent; therefore this second transformation can serve as a rule, by which means we can see if the transformation has been correctly done.

The extract that follows is from the second edition of Budan's *New Method for the Solution of Numerical equations of whatever degree* which appeared in 1822. In it he explains his theory of 'Syntagmatic sequences' which he had originally presented to the Paris Academy of Sciences in 1813.

F. -D. Budan

Nouvelle Méthode pour la résolution des équations numériques d'un degré quelconque (A New Method for the solution of numerical equations of whatever degree), Paris: Dondey-Dupré, 1822, pp. 99–100.

ARTICLE IV. - An algorithm for the transformation of a Polynomial in x of degree n, into an equivalent Polynomial, of the same degree, in (x - u), u being positive or negative, integer or fraction.

Being given a sequence $a_0 \ldots a_1 \ldots \ldots a_n$ etc.

We calculate a second sequence $a_0 \ldots a_1^{(1)} \ldots \ldots a_n^{(1)}$ etc.

in such a way that we have (whatever the value of n)

$$a_n = a_n^{(1)} - u\,a_{n-1}^{(1)}; u \text{ being a constant factor,}$$

or

$$a_n^{(1)} = a_n + u\,a_{n-1}^{(1)}.$$

I call the first sequence the *anti-syntagmatic* of the second; and a term $a_n^{(1)}$ of the second sequence will be called the *syntagma* of the first, from a_0 up to a_n.

In calculating in the same way, we obtain the syntagmatic of the syntagmatic, or the second syntagmatic of the given sequence, by which we can similarly obtain the 3rd, 4th, etc. syntagmatics; the general terms of the syntagmatics of the different order will be represented by $a_n^{(1)}$, $a_n^{(2)}$, $a_n^{(3)}$, etc.

This being given, in order to obtain the coefficients of a polynomial in $(x - u)$ equivalent to a polynomial of the same degree n, $a_0 x^n + \ldots$, whose given coefficients are $a_0 \ldots a_1 \ldots a_n$, write it, as if forming a sequence, writing 0 for a term, when 0 is the coefficient of a power of x; that is, when that power is missing in the polynomial: afterwards you will be able to obtain successively the first syntagmatic of that sequence; then the 2nd, 3rd, 4th, etc, syntagmatics, by calculating at each stage one fewer term; the last terms, being represented by, respectively, $a_n^{(1)}$, $a_{n-1}^{(2)}$, $a_{n-2}^{(3)}$, $\ldots a_1^{(n)}$ will be the coefficients of $(x - u)^0$, $(x - u)^1$, $\ldots (x - u)^{n-1}$; as for the coefficient of $(x - u)^n$, it will be the same as that of x^n; that is a_0.

The justification for this process, is that if you take the syntagmatic, using any constant module u, of the coefficients of a polynomial, the *syntagma* of these coefficients will be the remainder on division of the polynomial by $(x - u)$, and the terms which precede this syntagma will be the coefficients of the quotient polynomial; that is, denoting a given polynomial by P_n and the quotient polynomial by P_{n-1} we have

$$\frac{1}{x-u}P_n = P_{n-1} + \frac{a_n^{(1)}}{x-u}$$

then, on dividing the two sides of the equation by $x - u$, we have

$$\frac{1}{(x-u)^2}P_n = P_{n-2} + \frac{a_{n-1}^{(2)}}{(x-u)} + \frac{a_n^{(1)}}{(x-u)^2};$$

and continuing in this way with the divisions, we arrive at

$$\frac{1}{(x-u)^n}P_n = a_0 + \frac{a_1^{(n)}}{(x-u)} + \ldots + \frac{a_n^{(1)}}{(x-u)^n};$$

from which we have:

$$P_n = a_0(x-u)^n + a_1^{(n)}(x-u)^{n-1} + \ldots + a_{n-1}^{(2)}(x-u)^1 + a_n^{(1)}.$$

Let us consider the polynomial $2x^3 - 3x^2 + 5x - 3$ and calculate, in this way, the coefficients of the equivalent polynomial in $(x + 1)$, noting that here the module u is -1 in the relation equation $a_n^{(1)} = a_n + u\,a_{n-1}^{(1)}$.

Coefficients	....	2	– 3	+ 5	– 3
First Syntagmatic	....	2	– 5	+ 10	– 13
Second Syntagmatic	....	2	– 7	+ 17	
Third Syntagmatic	....	2	– 9		
Fourth Syntagmatic	....	2			

from which the equivalent polynomial is

$$2(x + 1)^3 - 9(x + 1)^2 + 17(x + 1)^1 - 13(x + 1)^0.$$

The algorithm defined by Ruffini allows us to use the $m + 1$ coefficients of a polynomial $P(x)$ to obtain the $m + 1$ coefficients of the polynomial $Q(x) = P(x + u)$. A purely algebraic justification can be given for these successive divisions of $P(x)$ by the term $(x - u)$.

Consider the polynomial $P(x) = a_0 x^m + a_1 x^{m-1} + \ldots + a_m$. The coefficients of the quotient polynomial $P^{(1)}$, such that $P(x) = P^{(1)}(x)(x - u) + r_0$ are the first m terms of the sequence defined by the relations:

$$a_0^{(1)} = a_0 \quad \text{and} \quad a_n^{(1)} = a_n + u a_{n-1}^{(1)} \qquad \text{for } n = 1, \ldots, m - 1.$$

Also, the remainder term $r_0 = a_m^{(1)}$, equal to $P(u)$, is the term of that sequence with suffix m.

Consider the sequence of polynomials obtained by repeated division by $(x - u)$:

$$P^{(k)}(x) = P^{(k+1)}(x)(x - u) + r_k \quad \text{for } k = 1, \ldots, m.$$

Here again, the sequence defined by the relations:

$$a_0^{(k+1)} = a_0 \quad \text{and} \quad a_j^{(k+1)} = a_j^{(k)} + u a_{j-1}^{(k+1)} \text{ for } j = 1, \ldots, m - k,$$

will give the $m - k$ coefficients of the polynomial $P^{(k+1)}$ and the value of the remainder term $r_k = a$.

Now we have: $P(x) = r_m(x - u)^m + r_{m-1}(x - u)^{m-1} + \ldots + r_1(x - u) + r_0$.

The algorithm described by Budan, which first appeared in 1807, shows how to calculate the coefficients of a polynomial $Q(x) = P(x + 1)$, from the $m + 1$ coefficients of a given polynomial $P(x)$, using only additions. Budan generalised the method in a memoir presented to the Institute in 1813, under the title *suites syntagmatiques*, which memoir appeared as an appendix to the 1822 edition. Except for the way that the intermediate calculations are set out, this procedure corresponds to that described by Ruffini. Horner, who published his algorithm in 1819, did not know of Ruffini's work, but he did draw on Budan's work, published in 1807.

We should point out, that this calculation procedure, which provides both the values of $P(a)$ and $P'(a)$ at the same time, and with considerable saving in the number of operations and demands on memorising intermediate values, considerably facilitates the use of Newton's method when applied to algebraic equations.

Bibliography

[1] Åboe, A., Al-Kāshī's Iteration Method for the determination of sin (1°), *Scripta mathematica*, 20 (1954), 24–29.

[2] Aitken, A.C., On Bernoulli's Numerical Solution of Algebraic Equations, *Proceedings of the Royal Society of Edinburgh*, vol. 46 (1926), 289–305.

[3] Bernoulli, D., Observationes de Seriebus quae formantur ex additione vel substractione quacunque terminorum se mutuo consequentium, *Commentarii Academiae Scientiarum Imperialis Petropolitanae*, vol. III (1728), 85–100. *Die Werke von Daniel Bernoulli*, vol. II, pp. 49–80, Basel: Birkhäuser, 1982.

[4] Budan, F.D., *Nouvelle Méthode pour la résolution des équations numériques d'un degré quelconque*, Paris: Courcier, 1807; repub. with additions, Paris: Dondey-Dupré, 1822.

[5] Euler, L., *Introductio in Analysin infinitorum*, Lausanne, 1748, in *Opera*, I viii & ix; English tr. J. D. Blanton, *Introduction to Analysis of the Infinite*, New York: Springer, 1988, 1990.

[6] Heron of Alexandria, *Metrica*, in Opera quae supersunt omnia, vol. III, 1903; Rationes Dimentiendi et Commentatio Dioptrica recensivit Hermannus Schoenne, Leipzig: B.G. Teubner, 1899-1914; extracts in Thomas, I., *Greek Mathematical Works*, Loeb Classical Library, Cambridge, Massachusetts & London; Harvard University Press & Heinemann, vol. 2, 1941.

[7] Horner, W., A new method of solving numerical equations of all orders by continuous approximation, *Philosophical Transactions of the Royal Society*, vol.109 (1819), 308–335.

[8] Kennedy, E.S., & Transue, W.R., A medieval iterative algorism, *American Mathematical Monthly*, vol. 63, 2 (1956), 513–516.

[9] Kepler, J., *Epitomes astronomiae Copernicanae*, Linz, 1618–1622. *Joannis Kepleri astronomi Opera omnia*, ed. Ch. Frisch, Frankfurt, 1858. *Gesammelte Werke*, vol. VII, ed. W. von Dyck & M. Caspar, Munich, 1953.

[10] Lagrange, J.-L., Sur la résolution des équations numériques, *Mémoires de l'Académie royale des Sciences et Belles-Lettres de Berlin*, vol. XXIII (1769). *Oeuvres*, vol. II, Paris: Gauthier-Villars, 1868.

[11] Lagrange, J.-L., *Traité de la résolution des équations numériques de tous les degrés*, 1798; 2nd ed., 1808. *Oeuvres*, vol. VIII, Paris: Gauthier-Villars, 1879.

[12] Lagrange, J.-L., Sur le problème de Kepler, *Mémoires de l'Académie royale des Sciences et Belles-Lettres de Berlin*, vol. XXV (1771). *Oeuvres*, vol. III, pp. 113–138, Paris: Gauthier-Villars, 1869.

[13] Newton, I., *Philosophiae naturalis Principia mathematica*, London, 1687. English tr. A. Motte, revised by Florian Cajori, Berkeley: University of California Press, 1947.

[14] Parker. R., *Demotic Mathematical Papyri*, Providence, Rhode Island: Brown University Press, 1972.

[15] Rashed, R., *Sharaf al-Dīn al-Ṭūsī, Oeuvres mathématiques, Algèbre et Géométrie au XIIe siècle*, text determined and tr. by R. Rashed, 2 vols, Paris: Les Belles Lettres, 1986.

[16] Rashed, R., Résolution des équations numériques et algèbre: Sharaf al-Dīn al-Ṭūsī, Viète, *Archive for History of Exact Sciences*, vol. 12 (1974), 243–290.

[17] Rashed, R. & Djebbar, A., *L'oeuvre algébrique d'al-Khāyyam*, Alep: I.H.A.S., Université d'Alep, 1981.

[18] Ruffini, P., *Sopra la determinazione delle radici nelle equazioni numeriche di qualunque grado*, Modena, 1804. *Opere matematiche*, vol. II, Rome: Edizioni Cremonese, 1953.

[19] Souissi, M., *Talkhīṣ aᶜmāl al-ḥisāb d'Ibn al-Bannā*, Tunis: Publication de l'Université de Tunis, critical ed. with French tr., 1968.

[20] Viète, F., *Opera mathematica*, Abrabim Elzeviriorum, 1646 ; reprnt. with Introduction by Joseph E. Hofmann, Hildesheim: Georg Olms Verlag, 1970.

[21] Youschkevitch, A.P. & Rosenfeld, B.A., article Al-Kāshī in *Dictionary of Scientific Biography*, Gillepsie (ed.), New York: Charles Scubmer's & sons, 1970-1980.

[22] Youschkevitch, A.P., *Les mathématiques arabes (VIIIe-XVe siècles)*, French tr. by M. Cazenave & K. Jaouiche, Paris: Vrin, 1976.

8 Algorithms in Arithmetic

The problem of distinguishing between prime numbers and composite numbers, and the decomposition of the latter into their prime factors, is known to be one of the most important and most useful in the whole of Arithmetic [...]. The dignity of the science seems to demand that we research carefully with all necessary tools to find the solution to so elegant and famous a problem.

Gauss

Disquisitiones Arithmeticae, 1801

Does a procedure exist for finding all the prime numbers? This question is, without doubt, as old as the science of Arithmetic itself and, as we shall see, leads on to other algorithmic problems. In this chapter we shall look at several famous algorithms in the history of Arithmetic.

We shall start with the Sieve of Eratosthenes, who provides a theoretical answer to the problem (Section 8.1). But, in practice, is it possible to determine whether a given number N is prime? As Pascal shows in a memoir of 1665, it is easy enough to establish if a number is divisible by 3, 5, 7, or 11 (Section 8.2). However, to test for all possible factors of N when N is fairly large requires an enormous number of calculations. In fact, finding efficient tests for primality proves to be difficult and has generated much research. In this chapter we present a number of tests for primality: tests produced by Lucas and Lehmer that arise from the reciprocal of Fermat's 'little' theorem, the test using so called Lucas sequences, and Pépin's test (Sections 8.4–8.6). Except for the last of these, the tests require the factorising of $N + 1$ or $N - 1$ when determining the primality of N. And this leads us to a new problem, namely, the prime factorisation of integers.

When the number to be factorised is large, this problem turns out to be even more difficult than the first problem. Three algorithms are offered here. Fermat's is neat and simple, but is only effective in particular cases (Section 8.7). Algorithms that have wider application, derived a century later by Gauss and Legendre, rely on more elaborate mathematics using quadratic residues and continued fractions (Sections 8.8 and 8.9). Besides using the idea of congruence, the algorithms given in this chapter are based on these two ideas. We shall remind the reader of such definitions and results as are needed to follow the texts and commentaries. The calculus of quadratic residues has itself led to an algorithm, due to Legendre, and we give it in this chapter (Section 8.3).

At the end of the chapter we include algorithms related to the Pell-Fermat equation. This Diophantine equation has played an important role in the history of mathematics. It was Fermat's challenge that opened up a rich area of research for mathematics, and the elegant response given by Lagrange offers a whole range of solutions (Sections 8.10 and 8.11).

The search for algorithms for divisibility, primality and factorisation doubtless found its origin in the fascination of numbers. The challenge this presented is linked to the history of the theory of numbers. Fascination about numbers, and research into number theory, are both very much alive, and a search for powerful algorithms continues to motivate mathematicians – and, today, computer scientists. But interest in all these questions has been recently renewed, in particular by the science of cryptography. In 1978, three American mathematicians Rivest, Shamir and Adleman invented a method for encrypting messages that uses a 'public key' which uses prime numbers [25].

Encrypting using a public key

This system allows all the members of a network to code their messages using a rule which is public knowledge, and to transmit their messages quite openly, since only the head of the network who has provided the code will be able to decode the messages.

The principle is as follows: the head of the network chooses two large prime numbers p and q and calculates the products $n = p.q$ and $\varphi(n) = (p - 1)(q - 1)$. He then chooses a number d, less than $\varphi(n)$ and prime to it, and finds a number e such that $d.e \equiv 1 \pmod{\varphi(n)}$, that is such that $d.e - 1$ is a multiple of $\varphi(n)$. Finally, he sends out the values of the numbers n and e to the members of his network, which need not be kept secret, and only he himself knows the values of $p, q, \varphi(n)$ and d.

Encoding is done by breaking the message down into blocks of numbers N, which are less than n. Then each of these numbers N is replaced by a number C defined by $C \equiv N^e \pmod n$. It is the number C which is transmitted.

In order to decode the message, that is to find the number N from C, we need to be able to calculate C^d, since by Fermat's little theorem,

$$C^d \equiv N^{ed} \equiv N^{1 + \lambda\varphi(n)} \equiv N \pmod n.$$

This decoding requires a knowledge of d, and so of $\varphi(n)$. Now, to find $\varphi(n)$ we need to be able to factorise n, which is impossible in practice, even with the most powerful computers, provided the numbers chosen are sufficiently large (of the order of a hundred digits).

Factors and Multiples

The starting point of Arithmetic is the study of a certain relationship between two numbers, namely whether one might be a factor or a multiple of the other. We have already dedicated a chapter of this book to the famous Euclidean algorithm by which the greatest common divisor of two numbers can be found. In this section we shall consider three other algorithms concerning the multiple/factor relation.

This relation puts the prime numbers, those numbers which cannot be divided except by themselves, as a set apart. In proposition 33 of Book VII of the *Elements* Euclid shows that every composite number, that is every number that is not prime, is divisible by a prime number. He also proves in proposition 20 of Book IX that "Prime numbers are more than any assigned multitude of prime numbers", which we would express today by saying that there is an infinity of primes. But how could we establish the list of all the primes less than some given 'magnitude'? Nicomachus of Gerasa (1st century AD) in his *Introductio Arithmeticae* proposed the use of a sieve by which the prime numbers could be separated from the other numbers. This sieve, attributed to Eratosthenes, isolates the primes by 'crossing off' the multiples of the primes (see 8.1 below).

We can find out if a number is divisible by any given number by using a division algorithm. But the calculations quickly become tedious when the number contains many digits. From the 10th century onwards, there started to appear tests of criteria for divisibility based on simple calculations based on the digits of the number being tested. The simplest and oldest of these tests is for divisibility by 9: any number written in our base ten system will be divisible by 9 if the sum of its digits is also divisible by 9. All the division criteria depend on congruence properties which were explained by Pascal in a memoir in 1654 (Section 8.2). Pascal himself did not use the language of congruences, this being first systematically set forth by Gauss in his *Disquisitiones Arithmeticae* in 1801.

Euler, in the 18th century, raised another, apparently quite simple, problem: given two numbers n and m with $n < m$, when might the remainder of the division of m by n be a square number? This relates to the problem of quadratic residues and a number of applications of the theory will be looked at in the two following sections of this chapter. Legendre proposed a simple to use algorithm, which depends on a fundamental result in the theory of numbers, namely the law of quadratic reciprocity (Section 8.3).

8.1 The Sieve of Eratosthenes

The sieve of Eratosthenes is explained by the first century neo-Pythagorean Nichomachus of Gerasa in his *Introductio Arithmeticae*. This work provides a systematic, clear and ordered presentation of the properties of numbers. Nichomachus deals with even and odd numbers, squares, rectangular and polygonal numbers, and gives a number of properties that were known from the Pythagoreans. He also studied different types of proportions. All this is done in the spirit of the Pythagoreans, whose concerns also included aesthetics, mysticism and morality related to numbers. The *Introductio Arithmeticae* became a work of reference down the centuries. Translated by Boetius in the 5th century, it was used in monastic schools in the Middle Ages. Eratosthenes considerably pre-dates Nichomachus, having lived at about 250 BC. He was the first great geographer in Alexandria, and he is credited with having determined the radius of the Earth.

The sieve allows us to set up a table that determines all the prime numbers less than some number N, fixed in advance. Leonardo of Pisa in his *Liber Abaci* of 1202 gives such a table for $N = 97$. Mathematicians through the centuries have constructed such tables for greater and greater values of N by perfecting the method of Eratosthenes.

What purpose can these tables serve? They can, for example, allow us to study the distribution of prime numbers, or the longest sequences of consecutive composite numbers. In this way, the tables allow mathematicians to make conjectures about prime numbers, conjectures which continue to be worked on today [7].

A century ago, tables of prime numbers filled up numerous pages. Today they are stored in computers in a more compact form. A very simple method is to code each odd number by 0 or 1 according to whether it is composite or prime. For example, the sequence 01110 means: 1 not prime, 3 prime, 5 prime, 7 prime, 9 composite. Thus, a five figure binary number can be used to represent the odd numbers in blocks of five, from 1 to 9, from 11 to 19, etc., showing which of them are prime:

0 1 1 1 0	1 1 0 1 1	0 1 0 0 1	1 0 0 1 0	1 1 0 1 0
1 3 5 7 9	11 13 15 17 19	21 23 25 27 29	31 33 35 37 39	41 43 45 47 49

In this way, a table of the prime numbers up to 50 can be given by the sequence: 01110 11011 01001 10010 11010. If we also exclude multiples of 5 and 3, an even more compact table can be produced: 0110 1111 0101 1010 1110. Each of the numbers 0110, 1111, 0101, 1010, 1110 is the binary representation of a number less than 15. Hence, each can be represented by a single symbol using hexadecimals (base 16). With A = ten, B = eleven, C = 12, D = 13, E = 14, F = 15, the table above now becomes: 6F5AE [24].

Nichomachus of Gerasa
Introductio Arithmeticae, Book 1, Chap. 13, 3–8, tr. Martin Luther D'Ooge *Introduction to Arithmetic*, New York: Macmillan, 1926, pp. 204–205.

The method of the 'sieve' is as follows. I set forth all the odd numbers in order, beginning with 3, in as long a series as possible, and then starting with the first I observe what ones it can measure, and I find that it can measure the terms two places apart, as far as we care to proceed. And I find that it measures not as it chances and at random, but that it will measure the first one, that is, the one two places removed, by the quantity of the one that stands first in the series, that is, by its own quantity, for it measures it three times; and the one two places from this by the quantity of the second in order, for this it will measure 5 times; and again the one two places further on by the quantity of the third in order, or 7 times, and the one two places still farther on by the quantity of the fourth in order, or 9 times, and so *ad infinitum* in the same way.

Then taking a fresh start I come to the second number and observe what it can measure, and find that it measures all the terms four places apart, and the first by the quantity of the first in order, or 3 times; the second by that of the second, 5 times; the third by that of the third, or 7 times; and in this order *ad infinitum*.

Again, as before, the third term 7, taking over the measuring function, will measure terms six places apart, and the first by the quantity of 3, the first of the series, the second by that of 5, for this is the second number, and the third by that of 7, for this has the third position in the series.

And analogously throughout, this process will go on without interruption so that the numbers will succeed to the measuring function in accordance with their fixed position in the series; the interval separating terms measured is determined by the orderly progress of the even numbers from 2 to infinity, or by the doubling of the position in the series occupied by the measuring term, and the number of times a term is measured is fixed by the orderly advance of the odd numbers in series from 3.

Now if you mark the numbers with certain signs, you will find that the terms which succeed one another in the measuring function neither measure all the same number – and sometimes not even two will measure the same one – nor do absolutely all of the numbers set forth submit themselves to a measure, but some entirely avoid being measured by any number whatsoever, some are measured by one only, and some by two or even more.

Now these that are not measured at all, but avoid it, are primes and incomposites, sifted out as it were by a sieve ...

Nichomachus says that a number measures another number if it is a factor of it, and he calls the 'quantity' of a number, the number of units it contains. Let us write out the odd numbers in a line, starting from the third, that is 3:

$$3\ 5\ 7\ 9\ 11\ 13\ 15\ 17\ 19\ 21\ 23\ 25\ 27\ 29\ 31\ 33\ 35\ 37\ 39\ \dots$$

The algorithm consists in marking off the multiples of 3: these numbers being separated by two numbers in between:

$$3\ 5\ 7\ 9^*\ 11\ 13\ 15^*\ 17\ 19\ 21^*\ 23\ 25\ 27^*\ 29\ 31\ 33^*\ 35\ 37\ 39^*\ \dots$$

Then mark off the multiples of 5: these numbers will be separated by an interval of four numbers. We obtain:

$$3\ 5\ 7\ 9^*\ 11\ 13\ 15^{**}\ 17\ 19\ 21^*\ 23\ 25^*\ 27^*\ 29\ 31\ 33^*\ 35^*\ 37\ 39^*\ \dots$$

Then mark off the multiples of 7: these numbers will be separated by an interval of six numbers. We now obtain:

$$3\ 5\ 7\ 9^*\ 11\ 13\ 15^{**}\ 17\ 19\ 21^{**}\ 23\ 25^*\ 27^*\ 29\ 31\ 33^*\ 35^{**}\ 37\ 39^*\ \dots$$

And so on. In our example, going as far as using 5 is enough since $[\sqrt{39}] = 6$. The numbers which have not been marked are prime and all the others are measured by one or two numbers. The remark that for a table of numbers up to N it is only necessary to test as far as $\sqrt{N}$ was made by Ibn al-Bannā in the 13th century in his *Talkhīṣ*.

8.2 Criteria for Divisibility

How can we know whether a number can be divided by a given number without having to carry out the division? According to the *Talmud* a number of the form $100a + b$ is divisible by 7 provided $2a + b$ is divisible by 7. This divisibility criterion is typical: it depends on a simple calculation to be carried out on the digits of the number being tested. Another criterion is even older, namely the test for divisibility by 9: any number can be divided by 9 if the sum of its digits is also divisible by 9. 'Casting out nines', based on this divisibility criterion, appears to have its origins in Indian mathematics.

Fibonacci's *Liber Abaci* (1202) gives criteria for divisibility by 7 and 11 and Ibn al-Bannā's *Talkhīṣ* (c. 1250) those for 7, 8 and 9. The use of these criteria for divisibility by 7 or by 9 are found as checks to check arithmetic operations in the 15th century, for example in Chuquet's *Triparty en la Science des nombres* and in Luca Pacioli's *Summa*. Pierre Forcadel of Béziers gives this explanation of testing for divisibility by 7 in his *Arithmétique* (1556): multiply the first digit by 3, subtract from it the largest multiple of 7, then add on the next digit and multiply by 3, and so on; if the last number obtained is a multiple of 7 then so is the number to be tested. Thus 5845 is divisible by 7 because:

$$5 \times 3 - 2 \times 7 + 8 = 9, 9 \times 3 - 3 \times 7 + 4 = 10, \text{ and } 10 \times 3 - 4 \times 7 + 5 = 7.$$

In a memoir addressed to the Paris Academy in 1654, Pascal collects all these tests together to give a general criterion for divisibility, which he proves. We give here an extract from this memoir. The idea is based on the following: in order to know whether a number is divisible by p, we need to use the remainders of dividing different powers of 10 by p.

In the years that followed, other mathematicians and amateur number enthusiasts produced criteria for divisibility [3], which can all be interpreted according to the general principles set forth by Pascal. Consider two tests for divisibility by 7. In 1728, Fontenelle stated this test: multiply the first digit by 3, add on the second digit, replace the first two digits of the number with this sum; continue in this way and the last number to be reached must be 7. For example, 7 will divide into 5845 since the sequence of calculations produces: 5845, 2345, 945, 315, 105, 35, 14, 7. In 1889, Tucker gave the test: split the number into two parts, of which the second part is the last digit of the number; take twice the second number from the first, and repeat this process; the final number to be obtained must be divisible by 7. For example, using 5845 we have the sequence of calculations: $584 - 2 \times 5 = 574, 57 - 2 \times 4 = 49$, and 49 is divisible by 7. It is easy to verify that these criteria are valid by using the theory of congruences applied to Pascal's method, the principles of which are given in the subsequent commentary.

Blaise Pascal

Des caractères de divisibilité des nombres déduits de la somme de leurs chiffres (On the divisibility properties of numbers deduced from the sum of their digits) (1665).
Oeuvres complètes, Paris: Ed. du Seuil, 1963, pp. 84–86.

Preliminary Remark

Nothing is better known in arithmetic than the proposition by which any multiple of 9 consists of digits whose sum is itself a multiple of 9. If, for example, we add together the digits which make up 18, the double of 9, we find $1 + 8 = 9$. In the same way, on adding the digits of any number whatsoever, we can recognise if that number is divisible by 9. Thus, 1719 is a multiple of 9, because the sum of all its digits, $1 + 7 + 1 + 9$, or 18, is itself divisible by 9. Although this rule is commonly used, I do not believe that anyone up till now has offered a proof, nor sought to generalise the principle. In this small treatise, I shall justify the property of divisibility by 9 and several other analogous properties; I shall also present a general method, by which it can be recognised, by the simple inspection of the sum of its digits, whether a given number is divisible by any number; this method is not applicable only to our decimal system of numeration (a system which is based on convention and not, as is commonly supposed, upon natural necessity), but also, without default, to all systems of numeration having for their base whatever number one wishes, as will be seen in the pages that follow.

A Single Proposition

To recognise, solely by inspection of the sum of its digits, if a given number is divisible by another given number.

For the sake of greater generality, we shall replace numbers by letters. Let there be any divisor whatsoever which we shall represent by the letter A, and let there be a dividend TVNM, the letters of which, M, N, V, T represent respectively the units, tens, hundreds, and thousands digits, and so on: in such a way that to change from literal quantities to numerical quantities, all that is needed is to replace each of the letters by one of the first 9 numbers, for example M by 4, N by 3, V by 5, T by 6, which gives 6534 for the dividend, the number A being any number such as 7. But we shall leave aside particular examples in order to understand all the possible cases in a single general solution. Being given, then, the dividend TVNM and any divisor A, it is required to recognise, solely by inspection of the sum of its digits, if the dividend is exactly divisible by A.

We write on the same line, and in decreasing order, the sequence of natural numbers, then underneath another sequence of numbers, as shown in this table:

10	9	8	7	6	5	4	3	2	1
K	I	H	G	F	E	D	C	B	1

In the table, the numbers of the second line are found as follows:

Underneath the 1, put 1.

From this number taken ten times, that is from the number 10, subtract the divisor A as many times as is possible, and write the remainder B below the number 2.

From B taken ten times, subtract the same divisor A as many times as is possible, and write the remainder C under the number 3.

From 10 C subtract again the divisor A as many times as is possible, and write the remainder D under the number 4.

And so on.

Now take the last digit of the dividend, M, being the first starting from the right, and multiply it by the 1 (which in our table is found below the digit 1).

Then take the second digit N, and multiply it by the number B, which in our table is found below the digit 2: then write the product below M.

Now take the third digit V, multiply it by C (the number placed underneath the digit 3), and write the product under the preceding products.

And do the same for T, and so on.

$$
\begin{array}{l}
M \\
N \times B \\
V \times C \\
T \times D
\end{array}
$$

I say that, for the number TVNM to be divisible by A, it is necessary and sufficient that the sum M + N×B + V×C + T×D, etc. should itself be divisible by A.

[...]

Examples

Suppose we want to find numbers that are divisible by 7. I write out the sequence of the first ten numbers and make up the table

$$10 \quad 9 \quad 8 \quad 7 \quad 6 \quad 5 \quad 4 \quad 3 \quad 2 \quad 1$$
$$6 \quad 2 \quad 3 \quad 1 \quad 5 \quad 4 \quad 6 \quad 2 \quad 3 \quad 1$$

by proceeding in the following way:

I write 1 underneath the 1.
From the 1, taken 10 times, I subtract 7 as many times as possible, and I place the digit 3 under the digit 2.
I multiply the remainder 3 by 10, and from the product 30, I subtract 7 as many times as possible; I put the new remainder 2 under the digit 3.
From 20 I subtract 7 as many times as possible; 6 is left over which I put under the 4.
From 60 I subtract 7 as many times as possible; 4 is left over which I put under the 5.
From 40 I subtract 7 as many times as possible; 5 is left over which I put under the 6.
From 50 I subtract 7 as many times as possible; 1 is left over which I put under the 7.
From 10 I subtract 7 as many times as possible, which brings me back to the first remainder found, that is 3, which I put under the 8.
From 30 I subtract 7 as many times as possible; I come back to the second remainder found, namely 2, which I put under the 9.
The remainders that have been now found, namely 1, 3, 2, 6, 4, 5, are thus found again in the same order, and so on indefinitely.
Let us now see if we can recognise whether some number 287 542 178 is a multiple of 7.
I take the first digit of the number starting from the right, and I multiply it by 1 (which in our table is found under the number 1).

I write down the product of 8 by 1, that is	8
I then write the product of 7 by the number 3, which appears under the 2 in our table, that is	21
Then the product of 1 by 2	2
the product of 2 by 6	12
the product of 4 by 4	16
the product of 5 by 5	25
the product of 7 by 1	7
the product of 8 by 3	24
the product of 2 by 2	4
and I find the sum	119

If 119 is divisible by 7, the proposed number 287 542 178 will be as well.

The theory of congruences derives directly from Pascal's method so it is entirely appropriate to use the language and symbolism of congruences to analyse this text. Let $A > 1$, we say that two integers a and b, with $a > b$, are congruent modulo A when $(a - b)$ is a multiple of A. In this case we write:

$$a \equiv b \ (\text{mod} \ A).$$

For example, $48 \equiv 20 \ (\text{mod} \ 7)$. If $a = Aq + r$ is the Euclidean division of a by A, then we have $a \equiv r \ (\text{mod} \ A)$ and r is the least number congruent to a modulo A. For example, $100 \equiv 2 \ (\text{mod} \ 7)$. Congruences have immediate and very useful properties:

if $a \equiv b \ (\text{mod} \ A)$ and $b \equiv c \ (\text{mod} \ A)$, then $a \equiv c \ (\text{mod} \ A)$,

if $a \equiv b \ (\text{mod} \ A)$ and $c \equiv d \ (\text{mod} \ A)$, then $a + c \equiv b + d \ (\text{mod} \ A)$ and $ac = bd \ (\text{mod} \ A)$.

From the first property we derive the result that two numbers are congruent to each other modulo A if and only if they have the same remainder when devided by A.

In the 18th century, the works of Euler, Lagrange and Legendre drew on the idea of congruences. Gauss developed the theory of congruences and introduced the symbolism of congruences in his *Disquisitiones Arithmeticae* (1801).

Pascal begins by finding the remainders of the division of powers of 10 and he arranges them in a table such that the remainder r_i in $10^i = Aq + r_i$ is placed underneath the index i:

10	9	8	7	6	5	4	3	2	1
r_{10}	r_9	r_8	r_7	r_6	r_5	r_4	r_3	r_2	r_1

We have $10^i \equiv r_i \ (\text{mod} \ A)$ and we can calculate the remainders r_i by recurrence, which is what Pascal does. In fact, since $10^i \equiv r_i \ (\text{mod} \ A)$, we have $10^{i+1} \equiv 10r_i \ (\text{mod} \ A)$ and so $r_{i+1} \equiv 10r_i \ (\text{mod} \ A)$. Consequently the terms of the second row of the table are found from right to left, the first term being 1 and the successive terms being found by multiplying the preceding term by 10 and calculating the remainder on division by A.

Suppose N is a number we wish to test for divisibility by A. We can write

$$N = a_n 10^n + a_{n-1} 10^{n-1} + \ldots + a_2 10^2 + a_1 10 + a_0, \text{ with } 0 \leq a_i \leq 9.$$

For all $i = n, \ldots, 1$, we have $a_i 10^i \equiv a_i r_i \ (\text{mod} \ A)$, so

$$N \equiv a_n r_n + a_{n-1} r_{n-1} + \ldots + a_2 r_2 + a_1 r_1 + a_0 \ (\text{mod} \ A).$$

Therefore, if the right hand side is divisible by A then so must be N: this is Pascal's rule.

In the particular case where $A = 9$, all the powers of 10 being congruent to 1 modulo 9, a number is congruent modulo 9 to the sum of its digits. Hence Pascal's rule is an extension of the fortunate result for 9.

In the particular case where $A = 7$, we are able to justify the criteria that were set out above. Those of Forcadel and Fontenelle are both based on the fact that

$$a_n 10^n + a_{n-1} 10^{n-1} + \ldots + a_0 \equiv (3 \times a_n + a_{n-1}) 10^{n-1} + \ldots + a_0 \ (\text{mod} \ 7).$$

As for Tucker's test, it is based on the fact that $10b + a$ is divisible by 7 if and only if $b - 2a$ is divisible by 7.

8.3 Quadratic Residues

The concept of quadratic residues appeared in a memoir by Euler (1754–1755) and is a very important idea in the theory of numbers. We shall see several applications of this in tests of primality, such as those of Lucas and Pépin, and in factorisation algorithms, such as Gauss's method.

Being given an odd prime p and an integer $a > 0$, finding those numbers for which the residue modulo p is a, that is solving the equation $x \equiv a \pmod{p}$, presents no particular difficulty. There is always an infinity of solutions and we can take $x = a + kp$. On the other hand, finding the quadratic residues of a modulo p, that is solving the equation $x^2 \equiv a \pmod{p}$, is less simple. Consider, for example, the equation $x^2 \equiv 32 \pmod{5}$. First we can note that this reduces to $x^2 \equiv 2 \pmod{5}$, and then to calculate the remainders modulo 5 all we need to do is to consider each of the numbers from 1 to 4. Since $1^2 \equiv 1$, $2^2 \equiv 4$, $3^2 \equiv 4$ and $4^2 \equiv 1 \pmod{5}$, we can see that the equation has no solution. We shall say, with Euler, that a is a *quadratic residue* modulo p if the equation $x^2 \equiv a \pmod{p}$ has a solution, and that a is a *non-residue* (mod p) otherwise. Thus in our example, 1 and 4 are quadratic residues modulo 5 while 2 and 3 are non-residues modulo 5. Half the numbers from 1 to $p - 1$ are quadratic residues modulo p.

How can we tell whether or not a is a quadratic residue modulo p? When p is large, the calculations given above are tedious. In his *Théorie des nombres*, Legendre provides a very simple algorithm for testing this. This algorithm which we reproduce here, rests on a very powerful result in the theory of numbers: *the law of quadratic reciprocity*.

The law of quadratic reciprocity is stated by Euler in a memoir of 1783 but without proof. Legendre also stated the result independently in 1785, but his proof is incomplete. The proof given in the *Théorie des nombres* is also incomplete. A first rigorous proof was proposed by Gauss in *Disquisitiones Arithmeticae* (1801). Since then more than fifty proofs have been provided [9]. Gauss himself considered this law to be the jewel of arithmetic.

To express this law, we shall make use of the symbolism developed by Legendre. From Euler's theorem, if p does not divide a then $a^{(p-1)/2} \equiv \pm 1 \pmod{p}$. We can deduce Euler's criterion from this, which states that a is a quadratic residue modulo p if and only if $a^{(p-1)/2} \equiv 1 \pmod{p}$. Legendre lets $\left(\dfrac{a}{p} \right)$ stand for the remainder of the division of $a^{(p-1)/2}$ by p, so $\left(\dfrac{a}{p} \right) = 1$ if a is a quadratic residue modulo p and $\left(\dfrac{a}{p} \right) = -1$ otherwise.

Legendre states the law of quadratic reciprocity as follows: let p and q be two odd numbers, then if p and q are not both of the form $4n + 3$, we have $\left(\dfrac{p}{q} \right) = \left(\dfrac{q}{p} \right)$, and if p and q are both of the form $4n + 3$ then $\left(\dfrac{p}{q}\right) = -\left(\dfrac{q}{p}\right)$. He remarks that the proposition can be summarised by the formula:

$$\left(\frac{p}{q}\right)\left(\frac{q}{p}\right) = (-1)^{(p-1)(q-1)/4}.$$

The law encompasses all possible cases since the remainder on dividing an odd number by 4 is necessarily 1 or 3.

A. -M. Legendre

Théorie des nombres (1798)
4th edition, repub. Blanchard, Paris, 1955, vol.1, 244–245

Use of the preceding theorem to determine whether a prime number c will divide the expression $x^2 + a$. Cases where the number x can be determined a priori.

When c is a rather large number, and we need to know whether c is a divisor of $x^2 + a$, it can be done by the rather long process of raising a to the power of $(c-1)/2$, while shortening the process as much as is possible, taking care to omit the multiples of c whenever they occur. Here is a procedure which is provided by the preceding theorem, and which leads very quickly to the desired value of $\left(\dfrac{a}{c}\right)$.

$1°$. If a is greater than c, then in place of a put the remainder of the division of a by c; thus we can always assume that a is smaller than c. In fact, it can easily be seen that $(mc + a)^{(c-1)/2}$ divided by c leaves the same remainder as $\alpha^{(c-1)/2}$.

$2°$. If the number a thus reduced is a prime number, the expression $\left(\dfrac{a}{c}\right)$ will change, according to the theorem, either to $\left(\dfrac{c}{a}\right)$ or to $-\left(\dfrac{c}{a}\right)$, the second case only when both a and c are of the form $4n + 3$. But since $c > a$, we can put in place of c the remainder of the division of c by a; let this remainder be c', then we have $\left(\dfrac{c}{a}\right) = \left(\dfrac{c'}{a}\right)$; hence finding $\left(\dfrac{c}{a}\right)$ is reduced to finding $\left(\dfrac{c'}{a}\right)$, which involves smaller numbers; the result will be found eventually, by using what we have already said, and by what we are going to add.

$3°$. If a is not prime, decompose a into its prime factors α, β, γ, ... which may include 2, you will have $\left(\dfrac{a}{c}\right) =$ to the product of expressions $\left(\dfrac{\alpha}{c}\right)\left(\dfrac{\beta}{c}\right)\left(\dfrac{\gamma}{c}\right)$, etc. Leave out factors which are squares, since in general $\left(\dfrac{\alpha^2}{c}\right) = \left(\dfrac{\alpha}{c}\right)\left(\dfrac{\alpha}{c}\right) = +1$; notice further, following n° 150, that $\left(\dfrac{2}{c}\right) = +1$, if c is of the form $8n \pm 1$, and $\left(\dfrac{2}{c}\right) = -1$, if c is of the form $8n \pm 3$.

By means of these precepts and the inversions given by the theorem in the preceding paragraph, we can soon find the value of the proposed expression $\left(\dfrac{a}{c}\right)$. And the operation, just like the one used to find the greatest common divisor of two numbers, will be just as quick.

Example 1

To find the value of the expression $\left(\dfrac{601}{1013}\right)$ I observe that these two numbers are prime, and so, by virtue of the theorem, $\left(\dfrac{601}{1013}\right)=\left(\dfrac{1013}{601}\right)$; dividing 1013 by 601 gives a remainder of 412, and 412 being the product of 4 and 103, we can omit the square factor four, which gives $\left(\dfrac{601}{1013}\right)=\left(\dfrac{103}{601}\right)$. But 103 being again a prime number, we have by the theorem, $\left(\dfrac{103}{601}\right)=\left(\dfrac{601}{103}\right)$ (dividing 601 by 103 and keeping only the remainder) $\left(\dfrac{86}{103}\right)=\left(\dfrac{-17}{103}\right)=-\left(\dfrac{17}{103}\right)$
$$=-\left(\dfrac{103}{17}\right)=-\left(\dfrac{1}{17}\right)=-1.$$
Therefore $\left(\dfrac{601}{1013}\right)=-1$. Therefore 1013 is not a quadratic residue of 601.

Before this extract, Legendre had proved two properties of quadratic residues:

1. If c is of the form $4n+3$, then $\left(\dfrac{-a}{c}\right)=\left(\dfrac{a}{c}\right)$, if not then $\left(\dfrac{-a}{c}\right)=-\left(\dfrac{a}{c}\right)$,

2. $\left(\dfrac{ab}{c}\right)=\left(\dfrac{a}{c}\right)\left(\dfrac{b}{c}\right)$.

At the beginning of this extract he also proves, using Euler's criterion, that if $a=mc+r$ is the Euclidean division of a by c, then $\left(\dfrac{a}{c}\right)=\left(\dfrac{r}{c}\right)$. These properties, together with the law of quadratic reciprocity, allow him to construct his algorithm.

As Legendre remarks, this algorithm is comparable to Euclid's algorithm for calculating the greatest common divisor of two numbers. In fact the inverse of the law of quadratic reciprocity allows us to replace calculating the value of $\left(\dfrac{a}{c}\right)$ with that of $\left(\dfrac{c}{b}\right)$ with $c<a$, so by starting again a number of times, we end up with numbers sufficiently small to enable us to easily come to an answer.

Thus $\left(\dfrac{601}{1013}\right)=\left(\dfrac{103}{601}\right)$ since $601=4\times125+1$. Since $1013=601+412$ is the Euclidean division of 1013 by 601, we have: $\left(\dfrac{103}{601}\right)=\left(\dfrac{412}{601}\right)$. The decomposition of

412 into prime factors is $2^2 \times 103$, therefore $\left(\dfrac{412}{601}\right) = \left(\dfrac{2^2}{601}\right)\left(\dfrac{103}{601}\right) = \left(\dfrac{103}{601}\right)$, since $\left(\dfrac{2^2}{601}\right) = \left(\dfrac{2}{601}\right)\left(\dfrac{2}{601}\right) = 1$. We therefore end up with having to calculate $\left(\dfrac{103}{601}\right)$. We simply have to use the law of quadratic reciprocity again, etc. Finally we arrive at $\left(\dfrac{1}{17}\right)$, which is equal to 1, since the equation $x^2 \equiv 1 \pmod{17}$ has the trivial solution 1.

Tests for Primality

There exists a very simple algorithm for testing the primality of an odd integer N, that is determining whether it is prime or composite. All that is required is to divide N successively by all integers d which are smaller than $\sqrt{N}$ (see Section 8.1). Unfortunately, when N is large the number of operations becomes enormous. As we shall see in the next section, factorisation algorithms for N require even more operations. Also, before setting out to produce the factorisation of any number, it would be well worth knowing whether or not the number is in fact composite, and therefore it would be worth carrying out a preliminary test for primality.

Thus, finding useful tests for primality, that is ones where the number of operations required would not be significantly important, has been a research area for mathematicians for a long time. This research has gone on in parallel with the search for the 'largest prime number'. When they were carried out by hand, the calculations had to be humanly feasible. With the invention of computers, it is now possible to tackle numbers consisting of more than 60000 digits, and the limiting factor is a function of machine time. The ideal would be to find an algorithm which was of polynomial type, that is where the number of operations needed increased as a polynomial in n, where n is the number of digits to be tested. Unfortunately, we do not know if such an algorithm exists [23].

We can classify the tests of primality according to a different criteria. First there are tests which can be applied to any integer whatever, which is the case for the two tests we shall give here, and those which only apply to integers of a particular given form, like Pépin's test. Also, some tests are deterministic, like these three, and others are probabilistic. We shall give an idea of how these latter function by looking at Fermat's theorem.

The three tests we present are based on elementary results of the theory of congruences and the theory of quadratic residues, of which we have given the principles and some fundamental results in the earlier part of this chapter. We have chosen these three tests to present here because of their historical roots; being based on ancient theories, perfected through time, and still today the object of continued research.

The first two tests require that the factors of $N + 1$ or $N - 1$ should be known. This particularity explains why they have been used so much to test numbers of the form $2^n \pm 1$ and $10^n \pm 1$. The Fermat numbers, of the form $2^{2^n} + 1$, which Pépin's test exclusively deals with, and Mersenne numbers, of the form $2^n - 1$, have thus been particularly studied. The 'greatest known prime number' at any time is often a Mersenne number. This was the case again in 1998 with the number $2^{3021377} - 1$ which possesses 909526 digits.

8.4 The Converse of Fermat's Theorem

Pierre de Fermat began his interest in the theory of numbers after reading the *Arithmetic* by Diophantus, published by Bachet in 1621. From 1636, he entered into a correspondence on this subject with Père Mersenne and Frénicle de Bessy. In a letter to the latter of October 1640 [5], he announced what we today call 'Fermat's little theorem': if a and p are two integers prime to one another, then if p is prime, $a^{p-1} \equiv 1$ (mod p). This appeared to be known in the particular case of $a = 2$ to the Chinese in about 500 BC who knew that $2^p - 2$ was divisible by prime p [3].

Fermat's letter

Every prime number is always a factor [*mesure infailliblement*] of one of the powers of any progression – 1, and the exponent [*exposant*] of this power is a divisor of the prime number – 1. After one has found the first power that satisfies the proposition, all those powers of which the exponents are multiples of the exponent of the first power also satisfy the proposition.

Example: Let the given progression be

	1	2	3	4	5	6	
	3	9	27	81	24?	729	etc.

with its exponents written on top.

Take, for instance, the number 13. It is a factor of the third power minus 1, of which 3 is the exponent and a divisor of 12 which is one less than the number 13, and because the exponent of 729 which is 6 is a multiple of the first exponent which is 3, it follows that 13 will also be a factor of this power 729 minus 1.

And this proposition is generally true for all progressions and for all prime numbers, of which I would send you the proof if I were not afraid of being too long.

Fermat never gave a proof of his theorem. One has been found among the manuscripts of Leibniz but it was Euler who first published a proof in 1736. That proof relied on the arithmetic properties of the binomial coefficients.

Fermat's theorem allows us to establish if a number N is composite, but it cannot serve as a test for primality, since the converse of Fermat's theorem is false. In fact, for $a = 2$ and $N = 341$, we have $2^{340} \equiv 1$ (mod 341) however, N is not prime since $341 = 11 \times 31$. But if we take $a = 3$, then $3^{340} \equiv 56$ (mod 341) which indicates that 341 is composite. A number which satisfies Fermat's result for an integer a but is not prime is said to be a 'pseudo-prime to base a'. Thus 341 is a pseudo-prime to base 2.

Pseudo-primes are rare, even though they are infinite in number, in fact there are only 1770 numbers less than 25×10^9 which are pseudo-prime to all the bases 2, 3, 5 and 7 [24].

But Fermat's theorem can also be used to provide probabilistic tests for primality, like that provided by Solovay and Strassen in 1974. These tests consist of choosing 'witness' numbers $a_1, a_2, \ldots, a_k$. If no witness recognises N as being composite, that is $a_i^{N-1} \equiv 1 \pmod{N}$ for $i = 1, 2, \ldots, k$, then there is a high probability that N is a prime number [9].

Other probabilistic tests make use of Euler's theorem that, if a and p are two numbers prime to each other, then if p is prime, $a^{(p-1)/2} \equiv \pm 1 \pmod{p}$. The converse is false. A number that satisfies Euler's result for a without itself being prime (in fact, a stricter condition), is said to be a 'strong pseudo-prime to base a'. These strong pseudo-primes are extremely rare. The most recent tests for primality, based on the theory of cyclotomic fields, consist of testing the possibility of N being a strong pseudo-prime. The test of Adelman, Pomerance and Rumely of 1983 is based on this principle: it is almost of a polynomial order [1].

In order to obtain a deterministic test of primality, we need to use a partial converse of Fermat's theorem, that is to find a supplementary condition which allows us to assert that N is prime. Lucas established such a condition in 1876. But the supplementary condition introduced by Lucas is costly. Lehmer also proposed a new form of the converse of Fermat's theorem in 1927. Over a week-end in 1949, he had the opportunity of testing out the feasibility of his method on the first electronic calculator, the US Army's *Electronic Numerical Integrator and Computer* (ENIAC). The invention of calculators has stimulated the search for other forms for the converse of Fermat's theorem, like that by Selfridge in 1975 [24].

We reproduce here extracts from the works of Lucas and Lehmer. Lucas uses the term 'aliquot part' for a factor of a number, excluding the number itself, but including 1.

E. Lucas
Théorie des nombres, (1891), repub. Paris: Blanchard, 1961, p. 441.

Reciprocal of Fermat's Theorem. If $a^x - 1$ is divisible by n, for x equal to $(n-1)$, and is not divisible by n for x equal to an aliquot part of $(n-1)$, the number n is prime.

We stated this theorem for the first time in 1876 at the Congress of the French Association for the Advancement of Science at Clermont-Ferrand, in a note entitled: *Sur la recherche des grands nombres premiers* (On searching for very large prime numbers). We shall give, in the second volume, a large number of corollaries.

Example 1. Let $a = 3$, $n = 65537 = 2^{16} + 1$;

the divisors of $(n-1)$ are $1, 2, 2^2, 2^3, 2^4, 2^5, \ldots 2^{16}$.

If we calculate the remainders on dividing n into the powers of 3 whose exponents are equal to the terms of the preceding sequence, noting that each remainder on dividing by n is equal to the square of its predecessor, we obtain $3, 9, 81, 6561, -11088, \ldots, -1$, and, since there is no remainder equal to 1 before the last of the sequence, we conclude that $2^{16} + 1 = 65537$ is a prime number.

The difficulty of the algorithm is that it requires the factorisation of $n - 1$ and a verification for each of the factors. This difficulty is reduced if the test is applied to numbers of the form $n = p^k + 1$ with p prime. Lucas considered the case where $p = 2$, which greatly simplifies the verifications, and thereby established the primality of a large number, the Fermat number $2^{16} + 1$.

Lucas also took $a = 3$ and considered the sequence of powers 3^x for $x = 1, 2, 2^2, 2^3,$ $2^4, 2^5, \ldots 2^{16}$. Each of the terms of this sequence is the square of the preceding one, and so its remainder modulo n is the remainder of the division by n of the square of the preceding term of the sequence. In fact, if $3^x \equiv r \pmod{n}$, then $3^x = nq + r$ giving $(3^x)^2 = n^2 q^2 + 2nqr + r^2$ and so $(3^x)^2 \equiv r^2 \pmod{n}$. Even so, calculations carried out by hand remain tiresome.

Lehmer began his article of 1927 by stating defects in the theorem by Lucas, and proposed improvements.

D. H. Lehmer

Tests for Primality by the Converse of Fermat's Theorem,
Bulletin of the American Mathematical Society, 33 (1927), 327–340.

Theorem 1. *If $a^x \equiv 1 \pmod{N}$ for $x = N - 1$, but not for x a proper divisor of $N - 1$, then N is prime.*

When applied to a particular N, this theorem exhibits three defects. In the first place, the complete factorization of $N - 1$ must be known. Secondly, the number of values of x which must be tried in order to show that the second part of the hypothesis is fulfilled, may be impossibly large. Thirdly, the condition for primality is sufficient but not necessary. If, however, N is of the form $2^n + 1$, the first two defects vanish; for in this case all the divisors of $N - 1$ are powers of 2, so that in testing for the first part of the hypothesis, the second part is automatically taken care of in the successive squarings of the residues modulo N. Unfortunately the only numbers of the form $2^n + 1$ that have any chance to be primes are the Fermat numbers in which n is a power of 2. The numbers $2^{128} + 1$ and $2^{256} + 1$ have been tested in this way by Morehead and A. E. Western, and both numbers were found to be composite. The next such number awaiting investigation is $2^{1024} + 1$, a number of 309 digits. A skilful computer could test this number in about ten years. As far as is known to the present author, no prime above the range of ordinary methods of factorization has ever been identified by the converse of Fermat's theorem. It is clear that Theorem 1 must be improved before further results are possible.

[...]

The next improvement in Theorem 1 that suggests itself is the reduction of the number of values of x to be tried. In the following theorem this number is reduced from the divisors of $N - 1$ to the number of its prime factors.

Theorem 2. *If $a^x \equiv 1 \pmod{N}$ for $x = N - 1$, but not for x a quotient of $N - 1$ on division by any of its prime factors, then N is a prime.*

[...]

Theorem 3. *If $a^x \equiv 1 \pmod{N}$ for $x = N - 1$ and if $a^x \equiv r > 1$ for $x = (N - 1)/p$, and if $r - 1$ is prime to N, then all the prime factors of N are of the form $np^\alpha + 1$, where α is the highest power to which the prime p occurs as a divisor of $N - 1$.*

[...]

Example 3. $N = 9\,999\,999\,900\,000\,001$

This number is $(10^{24} + 1)/(10^8 + 1)$. In this case we have

$$N - 1 = \frac{10^{24} + 1}{10^8 + 1} - 1 = \frac{10^8}{10^8 + 1}(10^{16} - 1) = 2^8 \cdot 5^8 \cdot 3^2 \cdot 11 \cdot 73 \cdot 101 \cdot 137 .$$

The divisors of 2^8 and 5^8 were chosen as values of p^α and tested in one operation as follows. It was first found that

$$7^{(N-1)/10} \equiv 7\,128\,121\,476\,353\,673 = r \pmod{N}.$$

Then

$$r^2 \equiv 7^{(N-1)/5} \equiv 428\,233\,546\,143\,224 \pmod{N},$$

and finally

$$r^5 \equiv -1 \quad \text{and} \quad r^{10} \equiv 1 \pmod{N}.$$

Since $r^2 - 1$ is prime to N, it follows that every factor of N is of the form $5^8 n + 1$ and also of the form $2^8 n + 1$. But $N^{1/2} < 10^8$, so that N is a prime and we have the complete factorization:

$$10^{24} + 1 = 17 \cdot 5\,882\,353 \cdot 9\,999\,999\,900\,000\,001.$$

Lehmer proves Theorems 2 and 3 with reasoning based on congruences. The algorithm Lehmer deduces from Theorem 3 is particularly suitable for numbers N for which $N - 1$ has few prime factors. This is the case for the seven examples dealt with by Lehmer, all of which concern divisors of the form $10^n \pm 1$.

To test the primality of $N = (10^{24} + 1)/(10^8 + 1)$, Lehmer shows that

$7^{N-1} \equiv 1 \pmod{N}$,

$7^{(N-1)/2} \equiv N - 1 > 1 \pmod{N}$ with $(N - 2)$ and N prime to each other,

$7^{(N-1)/5} \equiv r^2 > 1 \pmod{N}$ with $(r^2 - 1)$ and N prime to each other.

He deduces from this that all the factors of N are of the form $2^8 n + 1$, and also of the form $5^8 n + 1$. Therefore N will have factors of the form $10^8 n + 1$, which is impossible since $\sqrt{N}$ is less than this number.

To facilitate the calculations, Lehmer notes that if N is a divisor of a number of the form $10^n + 1$, then the remainder modulo N can be obtained immediately from the remainder modulo $10^n + 1$. Now, the remainder modulo $10^n + 1$ of a number of $2n$ digits can be easily found by dividing up the number into 2 blocks of n digits and subtracting the first from the second. For example, $256743 = 256 \times 10^3 + 743 = 256(10^3 + 1) + 743 - 256$, therefore $256743 \equiv 743 - 256 \equiv 487 \pmod{10^3 + 1}$. Before the invention of computers, examples had to be well chosen and required skilful calculations.

A result due to John Selfridge in 1975 simplifies Theorem 2 which permits changing the value of a for each divisor of $N - 1$. He states that if for each prime divisor p of $N - 1$ there exists an a such that

$$a^{(N-1)/p} \not\equiv 1 \pmod{N} \quad \text{while} \quad a^{(N-1)} \equiv 1 \pmod{N}$$

then N is prime. There are other simplifications of Theorem 2 which do not require $N - 1$ to be completely factorised [24].

8.5 The Lucas Test

The use of Lucas sequences to test the primality of a number N comes from research into the arithmetic properties of certain recurrent sequences, beginning initially with the Fibonacci sequence.

In the *Liber Abaci* of 1228 Fibonacci, or Leonardo di Pisa, uses the sequence 1, 2, 3, 5, 8, 13, ... in a problem about the number of rabbits produced by an original single pair of rabbits. Albert Girard, in *L'Arithmétique de Simon Stevin de Bruges* (1634), notes that the Fibonacci sequence can be obtained by means of a recurrence formula: $u_1 = 1$, $u_2 = 2$ and $u_{n+2} = u_{n+1} + u_n$. Then, in 1753, Robert Simson notes that it can be found from the continued fraction expansion of $(1 + \sqrt{5})/2$, the positive root of the quadratic equation $x^2 - x - 1 = 0$. From 1750 onwards a number of mathematicians, such as Lagrange, Legendre, Gauss, Dirichlet and others, became interested in the arithmetic properties of recurrent sequences.

In 1876, Lucas let a and b stand for the roots of the equation $x^2 - x - 1 = 0$ and considered the two sequences defined by:

$$U_n = (a^n - b^n)/(a - b) \text{ and } V_n = a^n + b^n = U_{n-1} + U_{n+1}.$$

He proved a number of arithmetic properties of these sequences. These properties led to effective tests for primality which were well suited to the study of Mersenne numbers, that is numbers of the form $2^m - 1$, and Lucas was able to establish that $2^{127} - 1$ is prime. This number remained the 'greatest known prime number' until fairly recently. With the advent of computers, the Lucas test has been used to break new records. In 1988, W. N. Colquitt and L. Welsh showed that the Mersenne number $2^{110503} - 1$ is prime [23].

The 'Lucas sequences' generalise the above sequences to quadratic equations $x^2 - Px + Q = 0$, where P and Q are integers and coprime. The terms of these sequences can be easily found by the use of the following formulas:

$$U_{2n} = U_n V_n, \ V_{2n} = (V_n)^2 - 2Q^n, \ U_{2n+1} = U_{n+1} V_n - Q^n \text{ and } V_{2n+1} = V_{n+1} V_n - PQ^n$$

In the extract that follows Lucas states a general theorem on the arithmetic properties of 'Lucas series' which immediately provides a test for primality. Then he states a corollary of this theorem which lets him test for the primality of the number $2^{31} - 1$.

E. Lucas
Théorie des fonctions numériques simplement périodiques,
American Journal of Mathematics Pure and Applied, 1 (1878), 302–307.

Fundamental theorem:

If in one of the recurrent series U_n the term U_{p-1} is divisible by p, without any of the terms of the series whose rank is a divisor of $p - 1$ being so, the number p is prime; in the same way, if U_{p+1} is divisible by p, without any of the terms of the series whose rank is a divisor of $p + 1$ being so, the number p is prime.

[...]

Example

Let $2^7 - 1 = 127$.
To know whether 127 is prime, we calculate U_{128} in the Fibonacci series; now we have the formulas:

$$V_{4n+2} = V_{2n+1}^2 + 2, \quad V_{4n} = V_{2n}^2 - 2 \, ;$$

and thus we form the table:

$$U_4 \ = U_2(V_1^2 + 2) \ = U_2 \times 3,$$
$$U_8 \ = U_4(V_2^2 - 2) \ = U_4 \times 7,$$
$$U_{16} = U_8(V_4^2 - 2) \ = U_8 \times 47,$$
$$U_{32} = U_{16}(V_8^2 - 2) = U_{16} \times 2207,$$
$$U_{64} = U_{32}(V_{16}^2 - 2) = U_{32} \times 48\,70847,$$
$$U_{128} = U_{64}(V_{32}^2 - 2) = U_{64} \times 2732\,51504\,97407.$$

Now, 127 divides the last factor and does not divide any of the preceding ones, so $2732\,51504\,97407 = 127 \times 18\,68122\,08641$, and consequently 127 is a prime number. The calculations can be considerably simplified by the use of congruences, continually replacing the numbers V_2, V_4, V_8, ... by their residues modulo 127. Taking account of this observation, the preceding table becomes:

$$V_4 \ = 3^2 - 2 \ = \ 7,$$
$$V_8 \ = 7^2 - 2 \ = 47,$$
$$V_{16} = 47^2 - 2 \equiv 48 \quad (\mathrm{mod}\ 127),$$
$$V_{32} \equiv 48^2 - 2 \equiv 16 \quad (\mathrm{mod}\ 127),$$
$$V_{64} \equiv 16^2 - 2 \equiv \ 0 \quad (\mathrm{mod}\ 127).$$

This method of testing for large prime numbers, which relies on the principle which we have just proved, is *the only direct and practical method*, presently known for solving the problem in question [...].

To test the final claim by P. Mersenne of the supposed prime nature of the number

$$2^{257} - 1,$$

a number of seventy eight digits, the whole of humanity, consisting of one thousand million individuals, calculating simultaneously and without interruption, would require a time greater than a number of centuries represented by a twenty digit number; by our method it would be enough to successively evaluate the squares of 250 numbers of 78 digits or more; this operation carried out by two skilful calculators, checking their work, would not require more than eight months of labour. We shall first of all apply the fundamental theorem to testing large prime numbers of the Fermat series which have the form

$$p = 2^{4q+3} - 1,$$

in which we suppose the exponent $4q + 3$ to be equal to a prime number such that $8q + 7$ is composite. [...]

Theorem II

Let $p = 2^{4q+3} - 1$ for which $4q + 3$ is prime, and $8q + 7$ is composite; we produce the series r_n

$$1, 3, 7, 47, 2207, \ldots$$

by means of the relation, for $n > 1$,

$$r_{n+1} = r_n^2 - 2;$$

the number p is prime whenever the rank of the first term divisible by p occupies a rank between $2q + 1$ and $4q + 2$; the number p is composite if none of the $4q + 2$ first terms of the series is divisible by p [...].

In practice, calculations are carried out using congruences, carrying forward only the residues modulo p, as we showed earlier for the number $p = 2^7 - 1$. We have indicated another calculation procedure, using the binary system of numeration, and which leads to the construction of a literally mechanistic method for testing large prime numbers.

In this number system, multiplication consists simply in the longitudinal displacement of the multiplicand; further, it is clear that the remainder of the division of 2^m by $2^n - 1$ is equal to 2^r, r designating the remainder of the division of m by n; thus in testing $2^{31} - 1$, for example, it

Table I: calculating the residue of $V_{2^{26}}$ modulo $2^{31} - 1$, from $V_{2^{25}}$.

is sufficient to operate on numbers having at most 31 digits. Table I gives the calculations of the residue of $V_{2^{26}}$ deduced from the residue of $V_{2^{25}}$ modulo $2^{31} - 1$, using the formula

$$V_{2^{26}} \equiv (V_{2^{25}})^2 - 2 \pmod{2^{31} - 1};$$

the black squares representing the units of the different orders in the binary system, and the white squares representing zeros. The first line is the residue of $V_{2^{25}}$; the 31 first lines, numbered 0 - 30 provide the square of $V_{2^{25}}$; the 4 lines numbered 0, 1, 2, 3 at the foot of the page show the addition of each column with carrying; one unit has been subtracted from the first column on the left; finally the last line is the residue of $V_{2^{26}}$.

Table II contains all the residues of $V_2, V_{2^2}, V_{2^3}, \ldots, V_{2^{29}}, V_{2^{30}}$ modulo $2^{31} - 1$. The last line, consisting entirely of zeros, shows us that $2^{31} - 1$ is prime.

Table II: diagram for the prime number $2^{31} - 1$.

The fundamental theorem applied to the Fibonacci sequence is particularly well suited for the Mersenne numbers, that is numbers of the form $p = 2^m - 1$. In fact, in this case, divisors of $p + 1 = 2^m$ are the powers of 2. The algorithm consists in calculating $U_2, U_{2^2}, U_{2^3}, U_{2^4}, ..., U_{2^m}$:

if $U_{2^m} \equiv 0 \pmod{p}$ *and* $U_{2^k} \not\equiv 0 \pmod{p}$ for $k = 1, 2, 3, 4, ..., m - 1$,

then $p = 2^m - 1$ is prime.

The calculation of the terms of the sequence is carried out easily with the use of the formulas $U_{2n} = U_n V_n$ and $V_{2n} = (V_n)^2 - 2(-1)^n$.

In the case of numbers of the form $p = 2^{2^n} - 1$, these formulas become

$$U_{2^{n+1}} = U_{2^n} V_{2^n} \text{ and } V_{2^{n+1}} = (V_{2^n})^2 - 2,$$

and Lucas provides a more rapid search algorithm by considering the sequence r_n such that $r_n = V_{2^n}$. Moreover, the calculations are simplified by writing the terms of the sequence as binary numbers. Since the calculations are carried out to modulo $2^{2^n} - 1$, it is sufficient to consider binary numbers of 2^{2^n} digits. These calculations are easy to carry out since, if

$$r = a_m 2^m + a_{m-1} 2^{m-1} + ... + a_1 2 + a_0$$

then

$$2r = a_m 2^{m+1} + a_{m-1} 2^m + ... + a_1 2^2 + a_0 2$$

and

$$2r \equiv a_{m-1} 2^m + ... + a_1 2^2 + a_0 2 + a_m \pmod{2^m + 1}.$$

Therefore, to double a number modulo $2^m + 1$ we just move the coefficients to the left, and since $r^2 = a_m(2^m r) + a_{m-1}(2^{m-1} r) + ... + a_1 2r + a_0 r$ all we need to do is to add up each of the terms of that sum in a table.

The fundamental theorem can be improved so as to provide a more effective test of primality. In 1932, A. E. Western gave a particularly simple test for the primality of Mersenne numbers: the number $N = 2^m - 1$, with m odd and prime, is a prime if and only if N divides the $(n - 1)$th term of the sequence defined by $V_1 = 4$ and $V_n = V_{n-1}^2 - 2$. Lehmer provided an elementary proof of this result in 1934 in an article: *Lucas's test for the primality of Mersenne's numbers* [14].

8.6 Pépin's Test

Pépin's theorem provides an effective algorithm for testing the primality of Fermat numbers. A Fermat number is a number of the form $F_n = 2^{2^n} + 1$. Fermat maintained that all numbers of that form were prime. He even claimed, in a letter to Carcavi in 1659, to have proved the result by the method on infinite descent. In 1729, Goldbach drew Euler's attention to the claim that Fermat had made. This famous correspondence proved, in 1732, that the claim was false, since $F_5 = 2^{32} + 1$ turns out to be composite: it can be written as $641 \times 6\,700\,417$. The date marks the beginning of a major period of investigation into the properties of Fermat numbers, and interest remains to this day. Interest grew through the works of Gauss from 1801, which showed that these numbers turn up in investigating the possibility of ruler and compass constructions of regular polygons.

Towards the end of the 19th century we find F_6 defying the attacks of mathematicians: is it prime or composite? In 1877, Lucas claimed that it would require just 30 hours of calculation (by a human!) to test its primality, using one of his sequences. The same year, Pépin proposed a simple test, whose proof uses the theory of quadratic residues, developed by mathematicians from Euler's work of 1754. Pépin ends his article by enumerating the 63 congruences which need to be calculated to determine the nature of F_6.

The introduction of computers has proved the value of Pépin's theorem for testing the primality of large Fermat numbers. It is particularly interesting because of the very simple binary representation of these numbers. The largest Fermat number that has been explored using Pépin's test is F_{22}. This number, of 1262612 digits, turns out to be composite.

It is only since 1990 that we know the prime divisors of F_9. As of 1994, the largest known prime Fermat number was F_4, the largest composite Fermat number whose decomposition was known was F_{11} (Brent and Morain, 1988), the largest Fermat number known to be composite was F_{23471} (Keller, 1983) and the first Fermat number whose nature is unknown was F_{24}. But with the development of algorithms and the improvement of computers, records are always falling!

J. -F. T. Pépin
Sur la formule $2^{2^n} + 1$, *Comptes rendus de l'Académie des Sciences de Paris*, 84 (July–December 1877), pp. 329–331.

Theorem

The necessary and sufficient condition for the number $a_n = 2^{2^n} + 1$ to be prime, when n is > 1, is that the number $5^{(a_n - 1)/2} + 1$ should be divisible by a_n.

1. This condition is necessary; for n being > 1, the number a_n is congruent to 2 modulo 5; a_n is therefore a quadratic non-residue of 5; and conversely 5 is a quadratic non-residue of a_n if this last is prime. Now, all the quadratic non-residues of a_n satisfy the congruence

(1) $$x^{(a_n-1)/2} \equiv -1 \pmod{a_n};$$

if then a_n is prime it must necessarily divide the number $5^{(a_n-1)/2} + 1$.

2. This condition is sufficient; for when it is satisfied, we can deduce that every prime divisor of a_n, other than unity, is equivalent to a_n. Suppose, in fact, that P is a prime divisor of a_n; we have, by the hypotheses, the congruences

(2) $$5^{(a_n-1)/2} + 1 \equiv 0, \quad 5^{(a_n-1)} \equiv 1 \pmod{P}$$

and, by Fermat's theorem,

(3) $$5^{P-1} \equiv 1 \pmod{P}.$$

The greatest common divisor of the two numbers $P - 1$, $a_n - 1$ is a power of 2; let this be 2^α. We can conclude immediately, from the congruences (2) and (3), that we also have

(4) $5^{2^{\alpha}} \equiv 1 \,(\mathrm{mod}\, P)$.

Now, 2^{α} cannot be less than $a_n - 1$; since, by that hypothesis, 2^{α} would be a divisor of $(a_n - 1)/2$, and we could deduce from formula (4)

$$5^{(a_n-1)/2} \equiv 1 \,(\mathrm{mod}\, P)$$

and we would therefore have, at the same time, the two congruences

$$5^{(a_n-1)/2} + 1 \equiv 0 \text{ and } 5^{(a_n-1)/2} - 1 \equiv 0$$

which is manifestly impossible. The greatest common divisor of the two numbers $P - 1$ and $a_n - 1$ is therefore $a_n - 1$, and, since P is a divisor of a_n, it must necessarily be that $P = a_n$, that is that the number a_n has no other divisor than itself and unity. The congruence (2) is sufficient therefore, when it is satisfied, to guarantee us that a_n is prime.

To recognise whether congruence (2) is satisfied, we form the sequence

(A) $5^2, 5^4, 5^8, \ldots, 5^{(a_n-1)/2}$,

composed of $2^n - 1$ terms, each of which is the square of its predecessor; but we should take care to reduce each term to its minimum residue modulo a_n, in order that all the necessary operations reduce to squaring terms whose number of digits never surpasses that of the number of digits of a_n. For example, let $n = 6$; the sequence (A) will consist of sixty three terms, the number a_6 is prime or composite according to whether or not the minimum positive residue of the sixty-third term reduces to the number $a_n - 1 = 2^{64}$.

The proof of the necessary condition uses the theory of quadratic residues. The number a_n is congruent to 2 modulo 5 because

$$2^4 = 16 \equiv 1 \,(\mathrm{mod}\, 5) \text{ and } 2^{2^n} = 4^{2^{n-1}} \equiv 1 \,(\mathrm{mod}\, 5)$$

Since 2 is not a quadratic residue of 5, the same is true for a_n, that is, using the Legendre symbol, $\left(\dfrac{a_n}{5}\right) = -1$. From the law of quadratic reciprocity (see §3), we have:

$$\left(\frac{a_n}{5}\right)\left(\frac{5}{a_n}\right) = (-1)^{(a_n-1)(5-1)/4} = 1$$

hence $\left(\dfrac{5}{a_n}\right) = -1$ and 5 is a quadratic non-residue of a_n. From Euler's criterion:

$$\left(\frac{5}{a_n}\right) \equiv 5^{(a_n-1)/2} \,(\mathrm{mod}\, a_n), \text{ hence } 5^{(a_n-1)/2} \equiv -1 \,(\mathrm{mod}\, a_n)$$

Therefore, if a_n is prime, it will divide $5^{(a_n-1)/2} + 1$.

The proof of the sufficient condition is based on a reasoning using divisibility and congruences and presents no difficulty.

Pépin's theorem remains true when 5 is replaced by any number which is a quadratic non-residue of a_n, for example 3. Riesel carried out the algorithm, with 3, in order to show that F_4 is prime [24]:

$$3^8 \equiv 6561, 3^{16} \equiv 54449, 3^{32} \equiv 61869, 3^{64} \equiv 19139, 3^{128} \equiv 15028, 3^{256} \equiv 282, 3^{512} \equiv 13987 \pmod{F_4}$$
$$3^{1024} \equiv 2^{13} + 2^5, 3^{2048} \equiv -2^3, 3^{4096} \equiv 2^{12}, 3^{16384} \equiv -2^8, 3^{32768} \equiv 2^{16} = -1 \pmod{F_4}.$$

A formula for Prime Numbers

As an alternative to testing for the primality of a number, we could try to obtain all the prime numbers by use of a formula. Euler proposed the formula $n^2 - n + 41$, which produces a prime number for $n = 1, 2, 3, ..., 40$ but not, obviously, for 41. In fact, it turns out that the set of prime numbers is recursively enumerable (see Section 15.4) and so, from work done in the 1960s, this set can be represented by a polynomial formula.

There remained the problem of finding explicitly a good polynomial. This was given in 1976 in an article by J. P. Jones, D. Sato, H. Wada and D. Wiens, who produced a polynomial of degree 25 with 26 variables [8]. The polynomial can be written:

$$
\begin{aligned}
(k + 2) \{ 1 &- [wz + h + j - q]^2 - [(gk + 2g + k + 1)(h + j) + h - z]^2 - [2n + p + q + z - e]^2 \\
&- [16(k + 1)^3 (k + 2)(n + 1)^2 + 1 - f^2]^2 - [e^3 (e + 2)(a + 1)^2 + 1 - o^2]^2 \\
&- [(a^2 - 1)y^2 + 1 - x^2]^2 - [16r^2 y^4 (a^2 - 1) + 1 - u^2]^2 \\
&- [((a + u^2(u^2 - a))^2 - 1)(n + 4dy)^2 + 1 - (x + cu)^2]^2 \\
&- [n + \ell + v - y]^2 - [(a^2 - 1)\ell^2 + 1 - m^2]^2 - [ai + k + 1 - \ell - i]^2 \\
&- [p + \ell(a - n - 1) + b(2an + 2a - n^2 - 2n - 2) - m]^2 \\
&- [q + y(a - p - 1) + s(2ap + 2a - p^2 - 2p - 2) - x]^2 \\
&- [z + p\ell(a - p) + t(2ap - p^2 - 1) - pm]^2 \}
\end{aligned}
$$

The set of prime numbers is exactly the set of the positive values taken by this polynomial when the variables are replaced by natural numbers. The authors of the article explain that it is possible to find a polynomial of degree 5, but with 42 variables, and that it is also possible to find a polynomial of only 12 variables, but to a greatly higher degree.

Factorisation Algorithms

How can we find the factorisation of an integer N? Before even starting to begin on this task, it would be as well to determine if the given number is in fact composite, since factorisation algorithms require many more steps than tests for primality. Moreover, as Knuth notes in *The Art of Computer Programming*, factorisation is much more difficult than finding the GCD of two numbers. Also, before launching into a factorisation algorithm, it would be worth starting by checking to see if N is divisible by small prime numbers. Much earlier, Gauss, in *Disquisitiones Arithmeticae* (1801), proposed two factorisation methods, but he advises the reader: "Before calling upon the following methods, it is always very useful to try to divide the given number by some of the smaller primes, say by 2, 3, 5, 7, etc. up to 19, or a little beyond, in order to avoid using subtle and artificial methods when division alone would be easier." Riesel gives the same advice in *Prime numbers and computer methods for factorization* (1985) but with the advent of more powerful machines, he advises testing for prime numbers up to 100 or even 1000. On the other hand, to test for all prime numbers less than $\sqrt{N}$, when N is large, requires calculations of gigantic proportions.

The majority of the algorithms in use are based on the methods which we present here, due to Fermat, Gauss and Legendre. In particular, Legendre's idea of using a continued fraction expansion of $\sqrt{N}$ to obtain a factorisation of N has been taken up again in the 20th century to construct high performance algorithms. These three methods are quite general, that is they search for factors, of any form, of a number of any form and all three are, of course, deterministic. More recently, in 1975, Pomerance constructed an algorithm based on statistical methods [24].

There exist, then, several factorisation algorithms, but they are costly. New improvements of any algorithms are therefore judged by their complexity. To date an algorithm has been found with a complexity of the order of $N^{\sqrt{1.5\ln\ln N/\ln N}}$ but theoretical results suggest that it should be possible to find one with an order of complexity of a polynomial in $\ln N$ [24]. Furthermore, although the search for factorisation algorithms has become much intensified during the last twenty years, it is very far from reaching a conclusion.

The fact is, that the arithmetician today has at his command a whole collection of algorithms, which currently offer the possibility of working on numbers up to 75 digits. The problem always is: which one to choose? In *Prime numbers and computer methods for factorization* Riesel goes as far as suggesting a strategy to choose a factorisation procedure which is an algorithm in itself! [24].

8.7 Factorisation by the Difference of Two Squares

In 1643 Père Mersenne set Fermat the challenge of factorising the number 100 895 598 169. His reply provided a systematic method for factorising a number. The starting point was to try to write the number N in the form of the difference of two square numbers. We would then have:

$$N = x^2 - y^2 = (x + y)(x - y).$$

The number x is necessarily greater than the square root of the integer N. It is sufficient therefore to test, one by one, the numbers x greater than that square root, to determine whether the difference $x^2 - N$ is, or is not, a square. Fermat provides a very simple algorithm for successfully carrying out the calculations.

Fermat's method requires a great many calculations. Successors to Fermat have tried to find other ways of writing N as the difference of two squares. Kausler, in 1805, and Collins, in 1840, made use of congruences modulo 4: N being odd, it must necessarily be congruent to 1 or 3 modulo 4. The method used by Kraïtchik in 1911 uses several congruences and also the theory of quadratic residues. If $N = x^2 - y^2 \equiv r$ (mod n) then $a = r + y^2$ is a quadratic residue modulo n, as well as $a - r$. This allows us to find possible values for r, x and y [3]. However, Fermat's method was taken up again in 1974 by Sherman Lerman. His idea is to eliminate unfavourable cases, that is those for which N has small prime factors [24].

P. de Fermat
Extract from *Œuvres de Fermat*, ed. Tannery & Henry, vol. II, 1894.
Letter of 1643, pp. 257–258

The question is, given a number, for example 2 027 651 281, I am asked whether it is prime or composite, and of what numbers it is composed, in the case where it is composite. I extract the root, to know the least of the said numbers, and I find 45 029 with 40 440 left over, which I subtract from double plus 1 of the root that was found, namely from 90 059: the result is 49 619, which is not a square, because no square can end with 19, and going on I add to it 90 061, namely 2 more than 90 059 which is the double plus 1 of the root 45 029. And because again the sum 139 680 is not a square, as can be seen from the last digits, I add to it again the same number augmented by 2, namely 90 063, and I continue in the same way to add so that the sum becomes a square, as one can see here. This does not happen until 1 040 400, which is the square of 1020, and from this the number is composite; for it is easy, from inspection of the said sums, to see that none can be a square number except the last one, for squares cannot have the final digits that they have, if not 499 944 would be squared, which however it is not. To know now the numbers which compose 2 027 651 281, I take away the number that I first added, namely 90 061, from the last one added 90 081. The result is 20, to half of which plus 2, namely 12, I add the previously found root 45 029. The sum is 45 041, to which number adding and subtracting 1020, the root of the last sum 1 040 400, we have 46 061 and 44 021, which are the two closest numbers which compose 2 027 651 281. These are also the only ones, since both are prime.

The copy of the letter which has come down to us does not mention the calculations which Fermat probably wrote in the margins.

Let $N = x^2 - y^2 = (x + y)(x - y)$. The first step in the algorithm is to write the number to be factorised in the form $N = x^2 - z$ with the least possible value for x, namely $x = [\sqrt{N}] + 1$.

Let $[\sqrt{N}] = n$ and $N = n^2 + r$. The smallest values of z and x to be tested are therefore:

$$z = x^2 - N = (n + 1)^2 - n^2 - r = 2n + 1 - r \quad \text{and} \quad x = n + 1$$

For the example dealt with by Fermat,

$$n = 45\,029, \quad r = 40\,440, \quad z = 2 \times 45\,029 + 1 - 40\,440 = 49\,619, \quad \text{and } x = 45\,030$$

The number z is not a square because the number 19, from the last two digits of the number, is not a square. Therefore we need to carry on.

The second step is to test the next x number, so we write N in the form $N = (x + 1)^2 - z'$. The next z' and x' to be tested are, therefore:

$$z' = (x + 1)^2 - N = (x + 1)^2 - x^2 + z = 2x + 1 + z \quad \text{and} \quad x' = x + 1$$

For the example dealt with by Fermat,

$$z' = 2 \times 45\,030 + 1 + 49\,619 = 139\,680, \quad \text{and } x' = 45\,031.$$

The number z' is not a square since 80, the number formed by its final two digits, is not a square.

We begin again with a new step. At each stage, the new values x' and z' to be tested are obtained from the preceding x and z by the formulas: $z' = 2x + 1 + z$ and $x' = x + 1$. The steps in the calculations can be set out in a table:

step	z	x	2x + 1
1	49619	45030	90061
2	139680	45031	90063
...	...	...	...
11	950319	45040	90081
12	1040400	45041	

Fermat does not explicitly find the successive values of x, as we have done. But he shows how he obtains the successive values of $2x + 1$. To find x, Fermat calculates the number of steps of the calculation, which is equal to

$$2 + (90\ 081 - 90\ 061)/2 = 12.$$

He therefore obtains $x = 45\ 029 + 12 = 45\ 041$. Fermat could more simply have noted that the last x value is $(90\ 081 + 1)/2$.

Fermat finds, in the end, that his number is equal to $46\ 061 \times 44\ 021$, after 12 calculation steps. We could say that this was a well chosen example, since the two factors are close to each other. In fact, if $N = x^2 - y^2 = ab$, with $a > b$, the number of steps needed to execute the algorithm is equal to

$$x - \sqrt{N} = (a+b)/2 - \sqrt{N} = (a + N/a)/2 - \sqrt{N} = (a - \sqrt{N})^2 / 2a$$

Carrying out the algorithm is therefore, in general, a long process.

What happens if N is prime? Lucas offers an answer to this question in his *Théorie des nombres* of 1891 by making the remark that an odd number is prime if and only if it can be written as the difference of two squares in only one way. In fact, if N is prime then we must necessarily have

$$x + y = N, \quad x - y = 1, \quad \text{and so } x = (N + 1)/2.$$

But if $N = ab$ is composite, with $a > b$, then

$$x + y = a, \quad x - y = b \quad \text{and } x = (a + b)/2 = (a + N/a)/2 < (N + 1)/2.$$

Hence, as Lucas notes, Fermat's method is a consequence of this property. The calculations have to be carried out until x reaches the value $(N + 1)/2$: if we have not by then found a smaller value of x for which $x^2 - N$ is a square, then N will be prime [17].

8.8 Factorisation by Quadratic Residues

In 1801, in his *Disquisitiones Arithmeticae*, Gauss complained that factorisation methods so far known were either restricted to particular cases or were long and tiresome. He proposed two new methods for factorisation, and we shall present the first of these here. The method is based on a search for small quadratic residues of the number to be factorised. We recall that a is a quadratic residue of N if there exists an x such that $x^2 \equiv a \pmod{N}$.

This method is important because it was used a great deal in theory and practice during the ensuing 150 years, a time when calculations were carried out by hand, or with the aid of small calculators.

The method used by Gauss is a method of exclusion. The basic idea is this: let N be a number to be factorised, the fact that we know that a number a is a quadratic residue of N allows us to exclude half of the set of possible prime factors of N. By successively using several quadratic residues, the possible prime factors of N rapidly diminishes. If none remain then the number N is prime, if more than one remains then we can carry out tests by division. This method presupposes that we know how to exclude prime numbers. Gauss had studied this question and had established a table for small quadratic residues. What we need to know, therefore, is how to calculate the small quadratic residues of the number to be factorised.

Carl Frederich Gauss, Disquisitiones Arithmeticae, Leipzig, 1801, tr. Arthur A. Clarke, S.J., New Haven, Connecticut: Yale University Press, 1966. §§ 330–331, pp. 397–401

330. The foundation of the FIRST METHOD is the theorem which states that *any positive or negative number which is a quadratic residue of another number M, is also a residue of any divisor of M.* We know in general that if M is divisible by no prime number below $\sqrt{M}$, M is certainly prime; but if all prime numbers below this limit that divide M are p, q, etc., the number M is composed of these *alone* (or their powers), or there is only *one* other prime factor greater than $\sqrt{M}$. It is found by dividing M by p, q, etc. as often as we can. Therefore, if we designate the complex of all prime numbers below $\sqrt{M}$ (excluding those which we already know do not divide the number) by ω, manifestly it will be sufficient to find all the prime divisors of M contained in ω. Now if we know in some manner that a number r, (nonquadratic) is a quadratic residue of M, certainly no prime number of which r is a nonresidue can be a divisor of M; therefore we can remove from ω all prime numbers of this type (they will usually compose about half the numbers in ω). And if it becomes clear that another nonquadratic number r' is a residue of M, we can exclude from the remaining prime numbers in ω those for which r' is a nonresidue. Again we will reduce these numbers by almost half, provided the residues r and r' are independent (i.e. unless one of them is necessarily a residue of all numbers of which the other is a residue; this happens when rr' is a square). If we know still other residues of M, r'', r''', etc. all of which are independent* of those remaining, we can institute similar exclusions with each of them. Thus the number of numbers in ω will diminish rapidly until they are all removed, in which case M will certainly be a prime number, or so few will remain (obviously all prime divisors of M will appear among them, if there are any such) that division by them can be tried without too much difficulty. For a number that does not exceed a million or so, six or seven exclusions will usually suffice; for a number with eight or nine digits, nine or ten exclusions will certainly suffice. There remain now two things to do, *first* to find suitable residues of M and a sufficient number of them, *then* to effect the exclusion in the most convenient way. But we will invert the order of the questions, especially since the second will show us which residues are the most suitable for this purpose.

331. We have shown at length in Section IV how to distinguish prime numbers whose residue is a given number r (we can suppose that it is not divisible by a square) from those for which it is a nonresidue; that is to say, how to distinguish divisors of the expression $x^2 - r$ from non-divisors. All the former are contained under formulae like $rz + a$, $rz + b$, etc. or like $4rz + a$, $4rz + b$, etc. and the latter under similar formulae. Whenever r is a very small number, we can evolve satisfactory exclusions with the help of these formulae; e.g. when $r = 1$ all numbers of the form $4z + 3$ will be excluded; when $r = 2$ all numbers of the form $8z + 3$ and $8z + 5$ etc. But since it is not always possible to find residues like this for a given number M, and the application of the formulae is not very convenient when the value of r is large, much will be gained and the work of exclusion will be greatly reduced if we have a table for a sufficiently large number of numbers (r) both positive and negative which are not divisible by squares. The table should distinguish prime numbers which have each (r) as residue from those for which they are nonresidues. Such a table can be arranged like the example at the end of this book which we have already described above; but in order that it be useful for our present purposes, the prime numbers (moduli) in the margin should be continued much farther, to 1000 or 10000. [...]

An example will illustrate this sufficiently well. If somehow we know that the numbers -6, $+13$, -14, $+17$, $+37$, -53 are residues of 997331, then we should join together a first column (which in this case should be continued as far as the number 997, i.e. up to the prime number next smaller than $\sqrt{997331}$) and the columns which have at the top the numbers -6, $+13$, etc. Here is a section of this scheme:

	-6	$+13$	-14	$+17$	$+37$	-53
3	-	-	-		-	-
5	-		-			
7	-		-		-	
11	-				-	
13		-	-	-		-
17		-		-		-
19			-	-		-
23		-	-			-
			etc.			
113		-	-			-
127	-	-	-	-	-	-
131	-	-	-			
			etc.			

From merely inspecting the numbers *contained in this part of the scheme* we see that after all the exclusions with the residues -6, 13, etc. only the number 127 remains in ω. And the whole scheme extended to the number 997 would show us that there would be no other number remaining in ω. When we try it, we find that 127 actually divides 997331. In this way we find that this number can be resolved into the prime factors 127 × 7853**.

From this example it is abundantly clear that this method is useful if the residues are not too large or at least if they can be decomposed into prime factors that are not too large, for the immediate use of the auxiliary table does not extend beyond the numbers at the head of the columns, and the immediate use includes only those numbers that can be resolved into factors contained in the table.

---

* If the product of any number of numbers r, r', r'', etc. is a square, each of them, e.g. r, will be a residue of any prime number (which does not divide any one of them) that is a residue of the others, r', r''. etc. Thus for the residues to be independent, no product of pairs, or triples, etc. of them can be a square.[Gauss's footnote]

**The author has constructed for his own use a large section of the table described here, and he would gladly have published it if the small number of those for whom it would be useful had sufficed to justify such an undertaking. If there is any devotee of arithmetic who understands the principles involved and desires to construct such a table on his own, the author would find great pleasure in communicating to him by letter all the procedures and devices that he used. [Gauss's footnote]

In Gauss's table, the first column contains all the prime numbers up to 997, a mark – in the other columns is used to indicate that the positive or negative number is a quadratic residue of the prime number of that line. Thus the numbers – 6, + 13, – 14, + 17, + 37, – 53 are all quadratic residues of 127.

The principle of Gauss's method and its application are very simple. But, as Gauss notes, we need to know the small quadratic residues of the number we wish to factorise. How should we proceed? How can we find the quadratic residues of, for example, 997331? We can begin by noting that the quadratic residues of a number N can be found by subtracting from N the squares of numbers close to $\sqrt{N}$, and then that the quadratic residues of multiples of N are also quadratic residues of N. Furthermore, the ratio and the product of two quadratic residues of N is also a quadratic residue of N. We can therefore try to combine the quadratic residues of N and of its multiples in order to find small residues. For example, for $N = 997331$, we have: $2N = (1412)^2 + 918$ and $918 = 2 \times 3^3 \times 17$ so 6×17 is a quadratic residue of N.

A more rapid and more elegant method for finding small quadratic residues of N is to use the continued fraction expansion of $\sqrt{N}$. We shall look at this method in the following section.

8.9 Factorisation by Continued Fractions

The idea of using a continued fraction expansion of $\sqrt{N}$ in order to find the factorisation of N is long established: it was proposed by Legendre, in the way of a remark, in his *Théorie des nombres* of 1798. The idea was taken up later by several mathematicians, but it achieved its place of honour in an article in 1931 by Lehmer and Powers, an extract from which we present here. With the advent of computers, their algorithm has attracted considerable interest, and in the twenty years that followed several improvements were made to it, so as to produce extremely powerful algorithms.

Let us recall (see Ch. 4) that the continued fraction expansion of $\sqrt{N}$ can be written as

$$\sqrt{N} = b_0 + \cfrac{1}{b_1 + \cfrac{1}{b_2 + \cfrac{1}{b_3 + \cfrac{1}{\cdots}}}}$$

which is obtained from the recurrence formulas:

$$x_0 = \sqrt{N}, \quad b_i = [x_i], \quad x_{i+1} = \frac{1}{x_i - b_i}.$$

If each x_i is written in the form: $x_i = \dfrac{\sqrt{N} + P_i}{Q_i}$ then it can be verified by recurrence that the P_i and Q_i are integers.

Let A_n/B_n be the nth convergent of the continued fraction expansion of $\sqrt{N}$, then there exists a relation between the integers defined thus:

$$A_{n-1}^2 - NB_{n-1}^2 = (-1)^n Q_n \text{ and so } A_{n-1}^2 \equiv (-1)^n Q_n \ (\text{mod } N).$$

In his *Théorie des nombres*, Legendre makes the remark that if $(-1)^n Q_n$ is a square then there exist two numbers x and y such that $x^2 \equiv y^2 \ (\text{mod } N)$ and so $(x + y)(x - y) \equiv 0 \ (\text{mod } N)$. Consequently, N and $x \pm y$ have common factors which can be determined by finding their GCD. The search for a $(-1)^n Q_n$ is limited because the continued fraction expansion of $\sqrt{N}$ is periodic. This last result had been proved by Lagrange in 1767.

The method proposed by Lehmer and Powers is based on a similar idea. They present a clear statement of it and illustrate its effectiveness.

D. H. Lehmer & R. E. Powers
On factorising Large Numbers,
Bulletin of the American Mathematical Society, 37 (1931), 772–773.

If A_n/B_n is the nth convergent to $N^{1/2}$, we have the well known relation

$$A_{n-1}^2 - NB_{n-1}^2 = (-1)^n Q_n = Q_n^*,$$

which can be written

$$Q_n^* \equiv A_{n-1}^2 \ (\text{mod } N).$$

Hence if Q_i^* and Q_j^* are equivalent, so that $x^2 Q_i^* = y^2 Q_j^*$, then

$$(xA_{i-1})^2 - (yA_{j-1})^2 \equiv 0 \ (\text{mod } N).$$

Unless N divides either $xA_{i-1} \pm yA_{j-1}$ it is possible by the greatest common divisor process to obtain a factorization of N. In the same way more than two Q^*'s may be used. For example, if $x^2 Q_i^* Q_j^* = y^2 Q_k^*$, then

$$(xA_{i-1}A_{j-1})^2 - (yA_{k-1})^2 \equiv 0 \pmod{N}.$$

The A's are of course calculated by the familiar recurrence

$$A_n = q_n A_{n-1} + A_{n-2}, \qquad (A_{-2} = 0, A_{-1} = 1)$$

and when necessary the A's may be reduced modulo N.

[...]

Consider the case $N = 13290059$. The elements for $N^{1/2}$ are as follows.

n	Q_n^*	q_n	$A_n \pmod{N}$
0	1	3645	3645
1	-2×2017	1	3646
2	3257	1	7291
3	-5×311	4	32810
4	1321	5	171341
5	$-2 \times 5^2 \times 41$	3	546833
...	...	...	...
24	5×877	1	39935
25	-5×571	1	11455708
26	$2 \times 31 \times 53$	1	11495643
27	-13×271	1	9661292
28	2381	2	4238109
29	-5×571	2	4847451
...	...	...	...

Lehmer and Powers provide a complete table as far as $n = 36$. We observe, with them, that $Q_{25}^* = Q_{29}^* = -5 \times 571$ and so we use Euclid's algorithm to look for the GCD of $A_{24} + A_{28}$ and N. This is found to be 4261, and so the factorisation is:

$$N = 3119 \times 4261.$$

We may note that, whereas Fermat's algorithm is based on the possibility of writing N in the form $x^2 - y^2$, that of Lehmer and Powers looks for x and y such that $x^2 \equiv y^2 \pmod{N}$, that is that a multiple of N could be written in the form $x^2 - y^2$.

The Lehmer and Powers algorithm was perfected in 1970 by Shanks, and by Morrison and Brillhart. These two had the idea of decomposing $(-1)^n Q_n$ into prime factors

$$(-1)^n Q_n = (-1)^{e_0} p_1^{e_1} p_2^{e_2} \cdots p_r^{e_r}.$$

The prime numbers form a base and the factorisations are then stored in the form of vectors $(e_0, e_1, e_2, \ldots, e_r)$. Partial products of the $(-1)^n Q_n$ can then be carried out by summing the components of the vectors. This is done to obtain a vector for which all the components are even, and so a product of $(-1)^n Q_n$ will be a square number [24]. This algorithm is very effective: it can deal with a number of 25 digits in about 30 seconds, and a number of 40 digits in 50 minutes, using an IBM 360/91. It achieved its first triumph in 1970 with the factorisation of the Fermat number $2^{128} + 1$ [9]. In 1983, Pomerance and Wagstaff proposed an 'Early Abort Strategy' which provided new improvements: this last algorithm has a cost factor of the order of $C \cdot N^{\sqrt{1.5 \ln \ln N / \ln N}}$.

The Pell-Fermat Equation

A Diophantine equation is an equation of the form $P(x_1, x_2, ..., x_n) = 0$, where P is a polynomial with integer coefficients and where the required solutions are to be integers or rational numbers. Why the description Diophantine? Because in the thirteen books of his *Arithmetica* Diophantus resolved problems which correspond to equations of the type $ay^2 + by + c = x^2$. We shall present an extract from the work of Diophantus.

Does there always exist a decision procedure, that is an algorithm, that allows us to determine whether a Diophantine equation has a solution? That question was the subject of Hilbert's 10th problem. As we shall see, at the end of this book, the mathematician Matijasevič was able to provide the answer no to this question in 1970.

Here, we shall restrict ourselves to those algorithms which concern the Pell-Fermat equation, since the general study of Diophantine equations is too vast for this work. On the other hand, the Pell-Fermat equation has played a very important role in the history of mathematics. A Pell-Fermat equation is a Diophantine equation with integer solutions of the form $x^2 - ay^2 = 1$, or its equivalent $ay^2 + 1 = x^2$, where a is a non-square integer.

The search for integer values for x and y satisfying the equation is equivalent to the search for a rational approximation to $\sqrt{a}$. In fact the equation can be written: $(x/y)^2 = a + 1/y^2$, and the greater the value of y the better the approximation to $\sqrt{a}$. Some of the approximations to $\sqrt{2}$ given by the Greeks and early Indian mathematicians correspond to solutions of $2y^2 + 1 = x^2$. This is the case with the approximation given by Theon of Smyrna, with side and diagonal numbers, and also with the approximations 17/12 and 577/408 given by Baudhayana [3].

This, perhaps, explains the interest of Indian mathematicians in solutions of the Pell-Fermat equation [2]. In the 6th century, Brahmagupta gave a general rule for finding solutions, which we would now express like this: let x and y be such that $ay^2 + R = x^2$, then:

$$m = 2xy \quad \text{and} \quad n = ay^2 + x^2 \qquad \text{satisfy} \quad am^2 + R^2 = n^2, \quad \text{and}$$
$$p = 2xy/R \quad \text{and} \quad q = (ay^2 + x^2)/R \qquad \text{satisfy} \quad ap^2 + 1 = q^2.$$

Bhāskara, in the 12th century, stated a rule of composition that allowed other solutions of an equation to be obtained starting from a single one. We shall give this rule, since we find Bhāskara formulas repeatedly in the history of mathematics, including in Lagrange's proof. Let x, y, and x', y' be such that $ay^2 + R = x^2$, and $ay'^2 + R' = x'^2$, then:

$$p = xy' + x'y \quad \text{and} \quad q = ayy' + xx' \quad \text{satisfy} \quad ap^2 + RR' = q^2.$$

Hence, if x and y are solutions of the Pell-Fermat equation then so also are $2xy$ and $ay^2 + x^2$.

The Pell-Fermat equation, sometimes called Pell's equation, ought really to be called Fermat's equation, and the fact that it isn't is simply due to a confusion on the part of Euler. For it was Fermat, in 1657, who set the challenge to other mathematicians to prove that this equation always possesses an infinite number of solutions. Although Fermat claimed to have a proof, it has never been found. But Fermat's challenge caused other mathematicians to become interested in these types of problems, and Lagrange furnished the first proof in 1766. It is interesting to note that this proof uses a sequence of rationals that converge to $\sqrt{a}$, being the convergents of the continued fraction expansion of $\sqrt{a}$ since, as we have already said, it was the search for approximations to square roots that first led to the study of equations of this type.

Lagrange thought that the solution to Fermat's problem would be the key to solving all questions of this type. Effectively, Legendre, in 1798, uses Lagrange's methods to give the conditions for the solubility of quadratic Diophantine equations in his *Théorie des nombres*. He proves, for example, that if a is a prime number of the form $4n + 1$, then the equation $x^2 - ay^2 = -1$ is possible. This was at the beginning of an intense period of study into Diophantine equations. Today, the methods used for solving these equations call upon a great variety of techniques in algebra, geometry and analysis, but approximation methods have lost nothing of their value. Such methods were used to prove Roth's theorem from which it can be proved that the equation $x^3 - 2y^3 = 1$ has only a finite number of solutions.

8.10 The Arithmetica of Diophantus

The books of the *Arithmetica* by Diophantus, written about AD 250, are set out, like the Egyptian papyri, in the form of a series of solved problems. Diophantus writes, in his dedication to the most esteemed Dionysus, that although the subject matter may appear to be difficult, "it will be easy to grasp, with your enthusiasm and my teaching".

Diophantus wrote that the number "which possesses in itself an indeterminate quantity of units is called arithmos." It is apparent from his consideration of powers of the arithmos as far as order 6, that Diophantus goes beyond the geometric idea of number possessed by the classical Greeks. The unknown quantities to be found are not line segments, but rational (positive) numbers. Whether this work is different, and marginal, or perhaps provides evidence of a current of Greek thinking about number is a matter that divides historians today.

The statements of problems by Diophantus are entirely general, for example Problem 1 of Book 1 requires that "a given number should be divided into two numbers whose difference is given", but they are solved for particular values; in this case the given number is taken to be 100 and the difference 40. However, this takes nothing away from the generality of the procedure: even though they have numerical values, the given numbers have the character of parameters.

In solving the problems, Diophantus deals with about forty equations of the type $ay^2 + by + c = x^2$. He considers different cases: often b is zero, if not, a or c is generally a square. For the most part, Diophantus contents himself with one solution. But in the lemma we present here, he offers a problem, indeed a whole family of problems, which can have an infinity of solutions. This lemma provides a method for solving certain of the Pell-Fermat equations.

Diophantus of Alexandria

The Arithmetica, Book VI, in Ver Eecke (tr. into French from the Greek) *Les six Livres d'Arithmétique et le Livre des nombres polygones*, Paris: Blanchard, 1959, pp. 251-252.

Lemma for Proposition 12.

Being given two numbers whose sum forms a square, we can find an infinity of squares which, multiplied by one of the given numbers, [and added to the other number], form a square.

Let the two given numbers be 3 and 6, and we must find a square which, multiplied by 3, and augmented by 6 units, forms a square.

Let the required square be 1 square arithmos plus 2 arithmos plus one unit. From this, 3 square arithmos plus 6 arithmos plus 9 units become equal to a square. Now the square can be found in an infinity of ways, because the quantity of units is quadratic. For this square, therefore, to be that which has as a root 3 units less 3 arithmos, the arithmos becomes 4 units; from which the root of the square is 5 units, and we can find an infinity of others.

This lemma precedes proposition XII of Book VI. Diophantus uses neither the vocabulary nor the algebraic symbolism we use today, but his solutions are easily translated into our algebraic form. And it is worth doing this, so that we can more easily appreciate his reasoning.

The question concerns solutions of, in our symbolism, $3z^2 + 6 = x^2$. Diophantus takes $z = y + 1$, with y as the arithmos, and the equation becomes $3y^2 + 6y + 9 = x^2$. Then he remarks that if we take, for example, $x = 3 - 3y$, the equation we obtain, $3y^2 + 6y + 9 = (3 - 3y)^2$, has the solution $y = 4$. Thus, $z = 5$ and $x = 9$ are solutions of the equation we started with. Furthermore, we can find an infinity of other values of x which will lead to solutions.

Why use $z = y + 1$ and $x = 3 - 3y$, and how can other values of x be found that will give solutions? A geometrical interpretation, called the 'chord method', suggests an explanation [10]. Consider the case of a Pell-Fermat equation $az^2 + 1 = x^2$ where $a + 1$ is a square b^2. By putting $z = y + 1$, Diophantus makes b^2 appear and the equation becomes:

$$ay^2 + 2ay + b^2 = x^2.$$

If we consider this as the equation of a conic, then solving the equation is equivalent to finding points of the conic with rational coordinates. The point P with coordinates $x = b, y = 0$ lies on the conic, and so every straight line through P, with equation $x = b - my$, with m rational, meets the conic at a point with rational coordinates. The calculations carried out by Diophantus are equivalent to finding the coordinates of this intersection point. If we put $x = b - my$, we obtain the equation

$$ay^2 + 2ay + b^2 = m^2y^2 - 2mby + b^2,$$

which has the solution $y = (2a + 2mb)/(m^2 - a)$. We can now derive the rational solutions of our initial equation.

This 'chord method' is very useful: its generalisation to hyper-surfaces allows us to solve Diophantine equations of any degree. But did Diophantus himself proceed in this way? It would seem that the first to explicitly give a geometrical interpretation of the method used by Diophantus was Newton. But Newton, thanks to Descartes, knew that an equation could be related to a curve, whereas Diophantus, who certainly would have been familiar with conics, had no way of relating them to an equation. For Diophantus, the explicit value of the 'chord method' lies in it being an algorithm.

8.11 The Lagrange Result

In a letter to Frénicle, in February 1657, Fermat stated the following theorem: "Every non-square number is of such a nature that there are infinitely many squares which, multiplied by the given number and one added, give a square." His correspondent is being asked to find a general rule which would provide a solution, in terms of integers, of the equation $ay^2 + 1 = x^2$, where a is not a square number. Fermat takes the example of $a = 3$, and he offers: $3 \times 1^2 + 1 = 2^2$ and $3 \times 4^2 + 1 = 7^2$. He adds: "There are infinitely many which, multiplying by 3 and adding a unit, become as a square number" [5].

There is no doubt that Fermat considered this a serious subject: the same month, he threw out the "Second challenge to mathematicians". He wrote:

There is hardly anyone who proposes purely arithmetic questions, there is hardly anyone who knows how to solve them. Is that because Arithmetic has been treated up to the present, rather by means of geometry than by its own means? [...] However, Arithmetic is a proper domain, the theory of whole numbers [...]; arithmeticians should therefore develop it or restore it. To make the way clearer for them, I propose that they should prove as a theorem the following statement: if they are able to do so, they will recognise at least that questions of this type are not inferior to the most famous ones of geometry, neither in subtlety, nor in difficulty, nor in the method of proof. Being given any non-squared number, there are an infinity of determined squares such that, adding a unit to the product of one of them by the given number, makes a square. [5]

Solutions arrived from England: those by Lord Brouncker and John Wallis between November 1657 and January 1658. Frénicle himself proposed tables of solutions as far as $a = 150$ and asked Wallis to continue as far as $a = 200$, or at least to solve $a = 151$ and $a = 313$ which he was unable to do. Brouncker replied that for $a = 313$, one solution is $x = 2 \times 7170685 \times 126862368$, a result which he had found after less than two hours. In a letter to Carcavi in August 1659, Fermat provides a "Description of new discoveries in the Science of numbers": he considers that Fréni-cle and Wallis have "given many particular solutions", but that "the general proof will be found by the method of descent duly and properly applied" [5]. This proof was never given by Fermat, but he had nonetheless succeeded in launching mathe-maticians into the 'proper domain' of mathematics which is called the theory of numbers.

Thus, in a letter of 10 August 1730 to his friend Goldbach, Euler notes that the solution of a Diophantine equation $ay^2 + by + c = x^2$ reduces to solving the equation $ay^2 + 1 = x^2$, and he calls this equation Pell's equation. Euler had made a confusion: he attributes Brouncker's method to Pell, which Wallis had presented, whereas Pell himself had never dealt with equations of this type. Following this letter, Euler con-tinued to interest himself in equations of this type. In 1732, he notes that if one solu-tion of $ay^2 + 1 = x^2$ is known, then we can exhibit an infinity of solutions. In letters to Goldbach in 1753 and 1755 he rediscovers Brahmagupta's formulas, and he gives the smallest solutions of $61y^2 + 1 = x^2$ and of $109y^2 + 1 = x^2$.

The relation between the solution of the equation $ay^2 + 1 = x^2$ and a sequence of fractions convergent to $\sqrt{a}$ was established in the 1750s. In 1753, R. Simson pro-duced a sequence such that each of the fractions p/q satisfied $aq^2 + 1 = p^2$. In 1759, Euler remarked that Brouncker's method, called Pell's method, could be more easily obtained if the continued fraction expansion of $\sqrt{a}$ was used. He analysed the prop-erties of this expansion and noted its periodicity in certain cases. The study of con-vergents led him to some interesting results, for example: let $v = [\sqrt{a}]$, if a period contains $2v$ and if p/q is the corresponding convergent, then $aq^2 \pm 1 = p^2$.[3]

The work done by Lagrange on Pell-Fermat equations was published in a memoir for 1766-1769. He starts by recalling the statement of the problem by Fermat, thereby crediting him with his due, and he notes the inadequacies of Wallis's arguments. Then he provides a full account of the properties of the continued fraction expan-sion of a square root. These properties, and in particular the property that the resi-dues are convergent, allowed him to prove three principal results "with all possible rigour and generality". Firstly the proof that every Pell equation has a solution, to-gether with the means of finding it. Secondly, that if a solution has been found, it is possible to find an infinity of solutions. Thirdly, the proof that all solutions of the equation $x^2 - ay^2 = 1$ can be found from the continued fraction expansion of $\sqrt{a}$, to-gether with an algorithm for finding them. We present here extracts from Lagrange's memoir, and an example of the application of the algorithm.

J. -L. Lagrange
Solution d'un problème d'arithmétique, *Miscellanea Taurinensia*, 1766–1769,
Oeuvres, vol. I, Paris: Gauthier-Villars, 1867, pp. 671–718.

The problem which I undertake to solve in this Memoir is this:

Being given any whole non-square number, to find a whole square number such that the product of these two numbers, augmented by a unit, shall be a square number.

This problem is one of those that M. Fermat had proposed, in the way of a challenge, to all English geometers, and particularly to M. Wallis, who was the only one, that I know of, who was able to solve it, or at least had published the solution (*see* Chapter XCVIII of his *Algebra* and the letters XVII and XIX of his *Commercium Epistolicum*); but the method used by this wise geometer consists only of just stepping into the problem, by which he arrives at the result, but only by a rather uncertain method, and without even knowing if he will got there; further, it must be proved that the solution to the problem is always possible, whatever the given number, a proposition that is generally regarded as being true, but which has never been established, as far as I know, in a solid and rigorous way; it is true that M. Wallis claimed to have proved it, but by an argument that Mathematicians hardly find satisfactory, and which is only basically, it seems to me, an appeal to principle (*see* Chapter XIX of his *Algebra*). It follows from this, that the problem in question has not yet been resolved in a satisfactory manner and in a way that leaves nothing to be desired; it is this that has fired my determination to make it the object of my researches, the more so in that the solution to this problem acts as a key to all the problems of this type.

1. Let a be the given non-square number, y^2 the required square, and x^2 any other square whatever, the problem reduces to satisfying this equation: $ay^2 + 1 = x^2$, with x and y only taking integer values; thus it is a matter of finding two whole numbers x and y such that

$$x^2 - ay^2 = 1.$$

If we extract the square root of a by approximation, we shall have a decimal fraction which can be changed, by well known methods, into a continuous fraction, which will necessarily be infinite, because $\sqrt{a}$ is an irrational quantity, by hypothesis.

In order to do this, we only need to divide, first the numerator of the found fraction by its denominator, then the denominator by the remainder, and so on, carrying out the same operation that we use to find the greatest common measure of two numbers, and letting q, q', q'', q''', ... stand for the quotients that arise from the different divisions, we shall have:

$$\sqrt{a} = q + \cfrac{1}{q' + \cfrac{1}{q'' + \cfrac{1}{q''' + \cfrac{1}{\cdots}}}}$$

Now, this continuous fraction, being successively interrupted at the first term, the second term, the third term, etc., will give an infinity of particular fractions which I shall designate by m/n, M/N, m'/n', M'/N', ..., to which, adding the fraction $1/0$, we shall have the infinite sequence of fractions:

$$1/0, \, m/n, \, M/N, \, m'/n', \, M'/N', \, m''/n'', \, M''/N'', \, m'''/n''', \, M'''/N''', \, \ldots,$$

which are such that:

$$m = q \qquad\qquad n = 1$$
$$M = q'm + 1 \qquad\qquad N = q'n$$
$$m' = q''M + m \qquad\qquad n' = q''N + n$$
$$M' = q'''m' + M \qquad\qquad N' = q'''n' + N$$
$$m'' = q^{iv}M' + m' \qquad\qquad n'' = q^{iv}N' + n'$$
$$M'' = q^{v} m'' + M' \qquad\qquad N'' = q^{v}n'' + N'$$

....

These types of fractions have many properties which have been known for a long time from geometers, but which we believe must be reviewed here in a few words, because we shall make considerable use of them in what follows.

[...]

18. I say now that all numbers x and y which satisfy the equation

$$x^2 - ay^2 = 1.$$

will necessarily be found among the numbers $M, M', M'', ...,$ and $N, N', N'', ...$ which make up the fractions $M/N, M'/N', M''/N'', ...$ which converge towards the root of a, but are always greater than that root (n° 1); that is to say that each of the numbers x is equal to one of the terms of the series $M, M', M'', ...,$ and that the corresponding number y is equal to the corresponding term of the series $N, N', N'', ...,$ in such a way that the fraction x/y will always be one of those about which we have just been talking.

[...]

22. Example 1.

– *Let it be proposed that we are to find two numbers x and y such that $x^2 - 13y^2 = 1$.*

I begin by extracting the square root of 13 as a decimal fraction, and this I find, in taking the approximation as far as nine characters, which can be done easily using Vlacq's large Tables of logarithms; I find, as I say:

$$\sqrt{13} = 3.60551950 = \frac{36055195}{10000000}$$

I divide the numerator of this fraction by its denominator, then the denominator by the remainder, and so on, as I would do to find the greatest common measure between the numerator and the denominator, and these different divisions give me these quotients: 3, 1, 1, 1, 1, 6, 1, 1, 1, 1, 6, 1, 1, ..., from which I form, beginning with 1/0, the following fractions:

$$\begin{array}{ccccccccc} & 1 & 1 & 1 & 1 & 6 & 1 & 1 & 1 \; ... \\ \frac{1}{0}, & \frac{3}{1}, & \frac{4}{1}, & \frac{7}{2}, & \frac{11}{3}, & \frac{18}{5}, & \frac{119}{33}, & \frac{137}{38}, & \frac{256}{71}, ... \end{array}$$

where it can be seen that the numerator of each fraction is equal to the sum of the numerator of the preceding fraction, multiplied by the number written above it (these numbers being none other than the quotients which need to be written out in sequence, and following the order in which they were found), and the numerator of the fraction that is before that; and it is the same for the denominators, which agrees with what was said in n°1.

I now substitute the numerators of these different fractions for x, and the corresponding denominators for y in the formula $x^2 - ay^2 = R$, and I have

x	y	R
1	0	1
3	1	- 4
4	1	3
7	2	- 3
11	3	4
18	5	- 1
119	33	4
137	38	- 3
256	71	3
...	...	...

I notice here two values for x and y, namely: $x = 4, y = 1$ and $x' = 256, y' = 71$, which both give $R = 3$ which is a prime number; hence I can make use of n°6.

I thus have: $a = 13, R = 3, x = 4, y = 1, x' = 256, y' = 71$;

hence $xy' + yx' = 540$ which is divisible by 3; from which I would have had to start with $q = \dfrac{540}{3} = 180$; and then $xx' + ayy' = 1947$, which is also divisible by 3; from which I deduce $p = 1947/3 = 649$; thus the required numbers will be $x = 649$ and $y = 180$; in fact the square of 649 is 421201, and the square of 180 is 32400, which multiplied by 13 gives 421200; from which we have

$$(649)^2 - 13(180)^2 = 1.$$

[...]

Moreover, in continuing the series $\dfrac{1}{0}, \dfrac{3}{1}, ...$, we shall come across these: $\dfrac{393}{109}, \dfrac{649}{180}, ...$, from which we have

x	y	R
393	109	- 4
649	180	1
...	...	...

from which we can see that 649 and 180 are the least numbers satisfying the equation $x^2 - 13y^2 = 1$ (n°18); from which, by substituting these numbers for p and q in formulas given in n°16 or 17, we shall find all the possible values of x and y; thus using $x, x', x'', ...$ and $y, y', y'', ...$ for their values, we shall have

$x = 649$	$y = 180,$
$x' = 842401$	$y' = 253640,$
$x'' = 1093435849$	$y'' = 303264540,$
....	....

and we can be assured that there are no other numbers smaller than these which solve the problem (n°17).

Lagrange proved that, if x and y are solutions of $x^2 - 13y^2 = 1$, then x/y is a convergent of the continued fraction expansion of $\sqrt{13}$. Consequently, having calculated these convergents, he sets out a first column of the numerators of the convergents x, a second column of the corresponding denominators y, and a third column of the numbers R such that $x^2 - 13y^2 = R$.

He also proved certain results that avoided the need to calculate convergents up until $R = 1$ appears. From the presence of two different convergents having the same value for R he was able to deduce a solution. In fact, in part 6 of his memoir, Lagrange notes that if $x^2 - ay^2 = R$ and $x'^2 - ay'^2 = R$, then

(1) $(xx' + ayy')^2 - a(xy' + yx')^2 = (x^2 - ay^2)(x'^2 - ay'^2) = R^2$,
(2) $x^2y'^2 - y^2x'^2 = R(y'^2 - y^2)$.

From (2), R divides $xy' + yx'$, and so from (1) R divides $xx' + ayy'$. Consequently, if $p = (xx' + ayy')/R$ and $q = (xy' + yx')/R$, then $p^2 - aq^2 = 1$. This allows Lagrange to deduce a solution to $x^2 - 13y^2 = 1$ from the two convergents giving the same $R = 3$.

In part 15 of his memoir, Lagrange proved that, if $p^2 - aq^2 = 1$, then for all integer $m > 1$,

$$x_m = \frac{(p+q\sqrt{a})^m + (p-q\sqrt{a})^m}{2} \quad \text{and} \quad y_m = \frac{(p+q\sqrt{a})^m - (p-q\sqrt{a})^m}{2}$$

also satisfy $x_m^2 - ay_m^2 = 1$. He further showed, in part 17, that, if p and q are the least integers for which $p^2 - aq^2 = 1$, then all solutions of $x^2 - ay^2 = 1$ are of the form given above. We therefore have an algorithm here that gives all solutions of the Pell-Fermat equation.

We should add that these solutions (x_m, y_m) can also be calculated by use of the recurrence formula:

$$\begin{pmatrix} x_{m+1} \\ y_{m+1} \end{pmatrix} = \begin{pmatrix} p & aq \\ q & p \end{pmatrix} \begin{pmatrix} x_m \\ y_m \end{pmatrix}.$$

Bibliography

[1] Adleman, L. M., Pomerance, C. & Rumely, R. S., On distinguishing prime numbers from composite numbers, *Annals of Matematics*, 117 (1983), 173–206.

[2] Colebrooke, H.T., *Algebra with Arithmetic and Mensuration from the Sanscrit of Brahmegupta and Bhascara*, London: Murray, 1817.

[3] Dickson, L.E., *History of the Theory of numbers*, 2 vols., New York: Chelsea Publishing Company, 1952.

[4] Diophantus of Alexandria: *Les six Livres d'Arithmétique et le Livre des nombres polygones*, French tr. by Ver Eecke, Paris: Blanchard, 1959. See also *A Study in the History of Greek Algebra*, ed. & tr. by T. L. Heath, 2nd ed. Cambridge: Cambridge University Press, 1910; reprnt. New York: Dover, 1964.

[5] Fermat, P. de, *Oeuvres*, ed. Tannery & Henry, vols II & III, Paris: Gauthier-Villars, 1894 & 1896.

[6] Gauss, C. F., *Disquisitiones Arithmeticae*, Leipzig, 1801; English tr. by A. A. Clarke, SJ, New Haven: Yale University Press, 1966.

[7] Hardy, G. H. & Wright, E. M., *An introduction to the Theory of numbers*, Oxford: Clarendon Press, 1938, 5th edn. 1979.

[8] Jones, J., Sato, D., Wada, H. & Wiens, D, Diophantine representation of the set of prime numbers, *American Mathematical Monthly*, 83 (June–July 1976), 449–474.

[9] Knuth, D.E., *The art of computer programming*, vol. 2, *Seminumerical Algorithm*, 2nd edn, Reading, Massachussetts: Addison-Wesley, 1981.

[10] Lachaud, G, Exactitude et approximation en analyse diophantienne, *L'à-peu-près*, Paris: Editions de l'Ecole des Hautes Etudes en Sciences Sociales, 1988, 28–45.

[11] Lagrange, L.-A., Solution d'un problème d'arithmétique, *Miscellanea Taurinensia*, 1766-1769, *Oeuvres*, vol. I, Paris: Gauthier-Villars, 1867.

[12] Legendre, A.-M., *Théorie des nombres* (1798), 4th edn, repub. Paris: Blanchard, 1955.

[13] Lehmer, D.H. & Powers R.E., On factoring large numbers, *Bulletin of the American Mathematical Society*, vol. 37 (1931), 770–776. Reproduced in *Selected papers of D.H. Lehmer*, ed. Mc Carthy, The Charles Babbage Research Centre, vol 1, Winnipeg, 1981.

[14] Lehmer, D.H., Lucas's test for the primality of Mersenne's numbers, *Journal of the London Mathemetical Society*, vol.10 (1935), 162–165. Reproduced in *Selected papers of D.H. Lehmer*, ed. Mc Carthy, The Charles Babbage Research Centre, vol 1, Winnipeg, 1981.

[15] Lehmer, D.H., Tests for primality by the converse of Fermat's theorem, *Bulletin of the American Mathematical Society*, vol.33 (1927), 327–340. Reproduced in *Selected papers of D.H. Lehmer*, ed. Mc Carthy, The Charles Babbage Research Centre, vol 1, Winnipeg, 1981.

[16] Lucas, E., Théorie des fonctions numériques simplement périodiques, *American Journal of Mathematics Pure and Applied*, vol. 1, 1878.

[17] Lucas, E., *Théorie des nombres*, 1891; repub. Paris: Blanchard, 1961.

[18] Molk, J., *Encyclopédie des Sciences Mathématiques pures et appliquées*, vol. I, Chap. 15, Paris: Gauthier-Villars, 1906.

[19] Nicolas, J.L., Tests de primalité, *Expositiones Mathematicae*, vol.2, 1984, 223–234.

[20] Nicomachus of Gerasa, *Introductio Arithmeticae*, ed. Hoche, Leipzig, 1866. Translated into English by Martin Luther D'Ooge as *Introduction to Arithmetic*, New York: Macmillan Company, 1926.

[21] Pascal, B., *Des caractères de divisibilité des nombres déduits de la somme de leurs chiffres*, 1665, Oeuvres complètes, Paris: Ed. du Seuil.

[22] Pépin, P., Sur la formule $2^{2^n} + 1$, *Comptes rendus de l'Académie des Sciences*, vol. 84 (July–December, 1877).

[23] Ribenboim, P., *The New Book of Prime Number Records*, New York: Springer-Verlag, 1996.

[24] Riesel, H., *Prime Numbers and Computer Methods for Factorization*, Boston: Birkhäuser, 1985, 2nd edn. 1994.

[25] Rivest R., Shamir, A. & Adleman, L. M., A method for obtaining digital signatures and public key cryptosystems, *Communications ACM*, 21 (1978), 120–128.

9 Solving Systems of Linear Equations

> ... Elimination is the most difficult and longest
> part of the work, so much so that one is reluc-
> tantly led to wish that science might discover the
> means for doing it that would be as useful as the
> invention of logarithms has been for multiplica-
> tion ...
>
> Christian Ludwig Gerling
> *Die Ausgleichungs-Rechnungen*, 1843

The solution of some ancient problems can be considered today as the solution of
systems of linear equations. We come across such problems frequently in Babylonian
and Egyptian mathematics, and also in Indian mathematics of the Middle Ages, as
well in Islamic countries and in Europe [29]. It becomes quite difficult however to
decide in which branch of mathematics we should place the corresponding algorithms for solving these problems. Their solution is given as a sequence of instructions, followed by a validation of the results, presented in ways that make their use
quite general. However, in order to assess the validity of these algorithms, and to
help our understanding, we shall consider them as belonging to the domain of algebra, as we understand the term today. Of course, describing these old problems as
being problems about solving systems of linear equations involves an anachronism,
since the idea of systems of equations is very much later. We have already seen such
examples in Chapter 3, particularly with the text by Clavius, where we saw that he
used the method of repeated double false position to solve, what we would now call a
'system of order 3' (see also [23]).

All the same, the Chinese methods deserve a special mention. In fact, the description of some of their methods bears a remarkable resemblance to our modern
matrix algorithmic methods. The method called *fangcheng*, described for the first
time in Chapter 8 of the *Jiuzhang Suanshu* displays the coefficients of the system of
equations in tabular form. Certain operations can then be carried out directly on the
numbers in the table (see the illustration following Section 9.3). These include multiplying columns by a given number, and adding and subtracting all the terms of
columns in order to eliminate certain coefficients, and these methods are perfectly
general and correspond to our matrix methods.

It was not until the end of the 17th century that systems of linear equations with
literal coefficients appear. Leibniz even uses double indexation ([20], see also [14]).
In about 1730 Maclaurin [21] calculates the solutions of systems of 2 and 3 equations
explicitly, and gives by induction the formulas for the solution of 4 equations. Fi-

nally, in 1750, Cramer [6] provides general formulas, known as Cramer's Rule, for any number of equations, though without offering a proof (Section 9.1).

Cramer's Rule should have brought the matter to a close. But advances in astronomy and geodesics required the solution of systems of a large number of equations, and the number of operations needed for their solution quickly becomes enormous (of the order of 300 million multiplications for solving a system of ten equations). It was for this reason that other methods had to be sought: for example the Gauss 'pivot' method (Section 9.3) which reduces the number of operations, and also methods which provide solutions by successive approximations.

Another consideration was that the coefficients of these equations came from measurements, of greater or lesser accuracy, and so as to obtain the greatest possible reliability, these measurements would usually be more numerous than the unknowns.The first job for mathematicians was to try to find a way of replacing the original equations by another system which has the same number of equations as unknowns, and which gives the 'best possible' approximations to the required values. Among a number of methods proposed, the method of least squares used by Legendre [19] and Gauss [11] is designed to minimise errors by giving values for which 'the likely mean error is the least possible' (Section 9.2).

The second task was to find solution methods that were more practical than Cramer's Rule or the standard elimination procedure offered by the Gauss pivot method, when needing to handle large systems of equations, in particular those that were derived from the least squares method. This led to the use of iterative methods, developed in the 19th century, which allowed for solutions to be obtained to any given degree of accuracy, among which were those of Gauss [10], Jacobi [13] and Seidel [26] (Sections 9.4, 9.5 and 9.6). The rate of convergence was also important and Nekrasov gave results concerning Seidel's method (Section 9.7). Finally, we report on Cholesky's method [1], this time an exact method, but one which comes from using the method of least squares (Section 9.8).

In all we shall consider three 'exact methods': Cramer's Rule, the Gauss pivot method and Cholesky's method, and three 'approximate methods': the iterative processes due to Gauss, Jacobi and Seidel. As we shall see, Cramer's Rule apart, all the other methods are inspired to a certain extent by the method of least squares.

9.1 Cramer's Rule

It should be borne in mind that systems of equations with literal coefficients started to appear towards the end of the 17th century and that Maclaurin [21] solved them explicitly in the case where there are a small number of equations. In the text by Cramer, which we give below, formulas are obtained for any n. These are described by using induction from the particular cases of $n = 1, 2$ and 3. The solutions are given in the form of a quotient of two homogeneous polynomials of degree n. The use of a determinant was not yet available (see [24]), but we can recognise expressions for determinant expansions, notably in the presence of signs linked to the signature of permutations.

The passage is taken from the *Introduction à l'analyse des lignes courbes* of 1750. Cramer had, in fact, to deal with formulas of this type in his work on determining conics passing through five given points. His notation, which was better than Maclaurin's, was not however as good as Leibniz's.

The notation used by Leibniz, which was little known since it had not been published, anticipates our index notation if we interpret his double numbers as the double index notation for coefficients, and his single numbers as the index of the unknowns. Thus, in a manuscript of 1678 (see [14], pp. 5–6), Leibniz considers a general system of four linear equations in four unknowns, referred to by 2, 3, 4 and 5, and he represents the system in the form:

$$12,2 + 13,3 + 14,4 + 15,5 - A \text{ equals } 0$$
$$22,2 + 23,3 + 24,4 + 25,5 - B \text{ equals } 0$$
$$32,2 + 33,3 + 34,4 + 35,5 - C \text{ equals } 0$$
$$42,2 + 43,3 + 44,4 + 45,5 - D \text{ equals } 0$$

In this same manuscript, Leibniz gives 'Cramer's Rule', 75 years before Cramer himself, the value of each unknown being a fraction whose numerator and denominator are explicitly given as 'literal' expansions in the case of four unknowns and, at the same time, the general rule is explicitly stated (see [23]). This text was only discovered later and so could not have had any influence on the development of methods for solving systems of linear equations.

G. Cramer
Introduction à l'analyse des lignes courbes algébriques, Geneva, 1750 (pp. 657–658).

Let there be several unknowns z, y, x, v, &c. and as many equations

$$A^1 = Z^1z + Y^1y + X^1x + V^1v + \&c.$$
$$A^2 = Z^2z + Y^2y + X^2x + V^2v + \&c.$$
$$A^3 = Z^3z + Y^3y + X^3x + V^3v + \&c.$$
$$A^4 = Z^4z + Y^4y + X^4x + V^4v + \&c.$$
$$\&c.$$

where the letters A^1, A^2, A^3, A^4, &c. do not indicate, as usual, the powers of A, but the left-hand side, supposed known, of the first, second, third, fourth &c. equation. Similarly Z^1, Z^2, &c. are the coefficients of z; Y^1, Y^2, &c. are those of y; X^1, X^2, &c. are those of x; V^1, V^2, &c. are those of v; &c. in the first, second, &c. equation.

With this notation, if there is only one equation & only one unknown z; we shall have $z = \dfrac{A^1}{Z^1}$.

If there are two equations and two unknowns z & y; we shall find that $z = \dfrac{A^1Y^2 - A^2Y^1}{Z^1Y^2 - Z^2Y^1}$, & $y = \dfrac{Z^1A^2 - Z^2A^1}{Z^1Y^2 - Z^2Y^1}$. If there are three equations & three unknowns $z, y,$ & x; we shall find

$$z = \frac{A^1Y^2X^3 - A^1Y^3X^2 - A^2Y^1X^3 + A^2Y^3X^1 + A^3Y^1X^2 - A^3Y^2X^1}{Z^1Y^2X^3 - Z^1Y^3X^2 - Z^2Y^1X^3 + Z^2Y^3X^1 + Z^3Y^1X^2 - Z^3Y^2X^1}$$

$$y = \frac{Z^1A^2X^3 - Z^1A^3X^2 - Z^2A^1X^3 + Z^2A^3X^1 + Z^3A^1X^2 - Z^3A^2X^1}{Z^1Y^2X^3 - Z^1Y^3X^2 - Z^2Y^1X^3 + Z^2Y^3X^1 + Z^3Y^1X^2 - Z^3Y^2X^1}$$

$$x = \frac{Z^1Y^2A^3 - Z^1Y^3A^2 - Z^2Y^1A^3 + Z^2Y^3A^1 + Z^3Y^1A^2 - Z^3Y^2A^1}{Z^1Y^2X^3 - Z^1Y^3X^2 - Z^2Y^1X^3 + Z^2Y^3X^1 + Z^3Y^1X^2 - Z^3Y^2X^1}$$

Examining these formulas gives us a general Rule. The number of equations and unknowns being n, we shall find the value of each unknown by forming n fractions whose common denominator has as many terms as there are different arrangements of n different things. Each term is composed of the letters $ZYXV$ &c. always written in the same order, but to which are distributed, as exponents, the first n digits arranged in all possible ways. Thus, when we have three unknowns, the denominator has $[1 \times 2 \times 3 =]$ 6 terms, composed of the three letters ZYX, which receive successively the exponents 123, 132, 213, 231, 312, 321. To these terms we attach the signs + or –, according to the following Rule. When an exponent is followed in the same term, immediately or later, by an exponent that is smaller, I call this a *derangement*. When we count the number of derangements in each term: if it is an even number or null, the term will have the sign +; if it is odd, the term will have the sign –. For example in the term $Z^1Y^2X^3$ there is no derangement: this term will therefore have the sign +. The term $Z^3Y^1X^2$ will also have the sign +, because there are two derangements, 3 before 1 & 3 before 2. But the term $Z^3Y^2X^1$, which has three derangements, 3 before 2, 3 before 1, & 2 before 1, will have the sign –.

The common denominator being thus formed, we shall find the value of z by giving a numerator to this denominator which is formed by changing Z into A in all the terms. And the value of y is the fraction that has the same denominator & for numerator the quantity that results when we change Y for A, in all the terms of the denominator. And the values of the other unknowns are found in a similar way.

In the text we can recognise what is now called Cramer's Rule, which we write as:

$$z = \frac{\begin{vmatrix} A^1 & Y^1 & X^1 & \cdots \\ A^2 & Y^2 & X^2 & \cdots \\ A^3 & Y^3 & X^3 & \cdots \\ \cdot & \cdot & \cdot & \cdots \end{vmatrix}}{\begin{vmatrix} Z^1 & Y^1 & X^1 & \cdots \\ Z^2 & Y^2 & X^2 & \cdots \\ Z^3 & Y^3 & X^3 & \cdots \\ \cdot & \cdot & \cdot & \cdots \end{vmatrix}}, \quad y = \frac{\begin{vmatrix} Z^1 & A^1 & X^1 & \cdots \\ Z^2 & A^2 & X^2 & \cdots \\ Z^3 & A^3 & X^3 & \cdots \\ \cdot & \cdot & \cdot & \cdots \end{vmatrix}}{\begin{vmatrix} Z^1 & Y^1 & X^1 & \cdots \\ Z^2 & Y^2 & X^2 & \cdots \\ Z^3 & Y^3 & X^3 & \cdots \\ \cdot & \cdot & \cdot & \cdots \end{vmatrix}}, \quad x = \frac{\begin{vmatrix} Z^1 & Y^1 & A^1 & \cdots \\ Z^2 & Y^2 & A^2 & \cdots \\ Z^3 & Y^3 & A^3 & \cdots \\ \cdot & \cdot & \cdot & \cdots \end{vmatrix}}{\begin{vmatrix} Z^1 & Y^1 & X^1 & \cdots \\ Z^2 & Y^2 & X^2 & \cdots \\ Z^3 & Y^3 & X^3 & \cdots \\ \cdot & \cdot & \cdot & \cdots \end{vmatrix}}$$

We use the term *rearrangement* for what Cramer calls *derangement*. Note that, for Cramer, the proof of the formulas for any number n of equations, which includes the proof of what we now refer to as the expansion of a determinant, results from induction on the cases $n = 1, 2, 3$.

Cramer also saw that systems might be contradictory or indeterminate. Bézout [2] went on to show that the existence of non-null solutions for a homogeneous system of equations was equivalent to a zero value for the determinant of the coefficients.

Further work on formulas that were easier to apply in practice, and then proofs for these formulas, gave birth to the theory of determinants. While we may recognise the work of Vandermonde [30], Laplace [17] and Lagrange [15] in this field, it was Cauchy [3] who gave the first proof of Cramer's Rule in 1815, establishing the way we write them now and, above all, beginning a systematic study of determinants. It was Sylvester [28] who was responsible for introducing the term *matrix* in 1850. As Cayley [5] notes, although, from a mathematical point of view, the idea of the matrix precedes that of the determinant, the theory of matrices was only developed after their fundamental properties had been clearly established, just as the theory of determinants followed their use in dealing with linear systems of equations.

9.2 The Method of Least Squares

Suppose we wish to determine the values of a certain number of unknowns $x, y, z, \ldots$ which cannot be measured directly, but we are able to find the values $v, v', v'', \ldots$ of functions $V, V', V'', \ldots$ of these unknowns. Then we need to be able to solve the system of equations:

$$
\begin{aligned}
V(x, y, z, \ldots) &= v \\
V'(x, y, z, \ldots) &= v' \\
V''(x, y, z, \ldots) &= v''
\end{aligned}
$$

$$\cdots\cdots\cdots\cdots\cdots\cdots$$

Since errors can creep into measurements, the number of measurements taken is increased, in the hope that accuracy can be improved, and so we need to be able to solve a system of n equations in k unknowns where $n > k$. In general this system will be contradictory. A method for deducing approximate values for the unknowns $x, y, z, \ldots$ is the following. We determine the differences or deviations $e, e', e'', \ldots$ where

$$
\begin{aligned}
e &= V(x, y, z, \ldots) - v \\
e' &= V'(x, y, z, \ldots) - v' \\
e'' &= V''(x, y, z, \ldots) - v''
\end{aligned}
$$

$$\cdots\cdots\cdots\cdots\cdots\cdots$$

and we then say that the values $x, y, z, \ldots$ should satisfy the equations as well as possible when the sum of the squares of the deviations is required to be a minimum.

An alternative method could be to try to minimise the maximum of the absolute values of the deviations, as Euler [7] does, or to try to minimise the sum of the absolute values of the deviations which was the approach used by Laplace [18]. But the method of least squares is the method that leads to the most convenient calculations.

Legendre explains his method in an appendix to *Nouvelles Méthodes pour la Détermination des Orbites des Comètes* in the case where $V, V', V'', \ldots$ are linear functions of the unknowns $x, y, z, \ldots$.

A.-M. Legendre

From the Appendix to *Nouvelles Méthodes pour la détermination des Orbites des Comètes,* Paris, 1805.

In the majority of questions where it is a matter of deriving given measures from observations that are as exact as they can be, we almost always have to deal with a system of equations of the form:

$$E = a + bx + cy + fz + \text{etc.}$$

in which a, b, c, f, etc. are known coefficients that vary from one equation to another, and x, y, z, etc. are unknowns which have to be determined with the condition that the value of E should, for each equation, be reduced to zero or a very small quantity. If there are as many equations as there are unknowns x, y, z, etc. there will be no difficulty in determining these unknowns, and the errors E can be made absolutely zero. But most often, the number of equations is greater than the number of unknowns, and it is impossible to eliminate all the errors.

In these circumstances, which is the case for the majority of the problems of physics and astronomy, where one tries to determine some important elements, something of the arbitrary necessarily enters into the distribution of the errors, and we cannot expect that all hypotheses would lead to the same results; but above all we should ensure that the extreme errors, without regard to their signs, should be held within the strictest limits possible.

Of all the principles that could be proposed for this purpose, I consider that there is no principle that is more general, more exact, or more easy to apply, than that which we have used earlier, and which consists of making the sum of the squares of the errors a *minimum*. By this means a sort of equilibrium is established between the errors, preventing the extremes from dominating the others, and is very suitable for finding the state of the system of equations that is closest to the truth.

The sum of the squares of the errors $E^2 + E'^2 + E''^2 + \text{etc.}$ being

$$(a + bx + cy + fz + \text{etc.})^2$$
$$+ (a' + b'x + c'y + f'z + \text{etc.})^2$$
$$+ (a'' + b''x + c''y + f''z + \text{etc.})^2$$
$$+ \text{etc.};$$

if we want to find its *minimum* as x alone varies, we shall have the equation

$$0 = \int ab + x \int b^2 + y \int bc + z \int bf + \text{etc.}$$

in which $\int ab$ is understood to mean the sum of similar products $ab + a'b' + a''b'' + \text{etc.}$, $\int b^2$ means the sum of the squares of the coefficients of x, that is $b^2 + b'^2 + b''^2 + \text{etc.}$, and so on.

The *minimum* with respect to y will similarly produce the equation

$$0 = \int ac + x \int bc + y \int c^2 + z \int fc + \text{etc.}$$

and the *minimum* with respect to z, the equation

$$0 = \int af + x \int bf + y \int cf + z \int f^2 + \text{etc.}$$

where we see the same coefficients $\int bc$, $\int bf$, etc., appearing in two equations, which helps to make the calculations easier.

In general, *to form the equation of the* minimum *with respect to one of the unknowns, all the terms of each proposed equation has to be multiplied by the coefficient of the unknown in that equation, taken with its sign, and a sum must be made of all these products.*

In this way, we obtain as many equations of the *minimum* as there are unknowns, and these equations have to be solved by the usual methods.

[...]

The rule by which the mid-value of the results of a number of observations (for a single element) is to be taken, is but a simple consequence of our general method, which we call the *method of least squares*. In fact, an experiment has given different values a', a'', a''' etc. for a certain quantity x, the sum of the errors will be $(a' - x)^2 + (a'' - x)^2 + (a''' - x)^2 +$ etc.; and in equating this sum to a *minimum*, we have

$$0 = (a' - x) + (a'' - x) + (a''' - x) + \text{etc.}$$

from which the result $x = \dfrac{a' + a'' + a''' + ... + \text{etc.}}{n}$, n being the number of observations.

What Legendre says can be restated in the following way. Given a system of equations:

$$ax + by + cz + ... + n = 0$$
$$a'x + b'y + c'z + ... + n' = 0$$
$$a''x + b''y + c''z + ... + n'' = 0$$

$$\cdots\cdots\cdots\cdots\cdots\cdots\cdots\cdots\cdots\cdots\cdots\cdots$$

the sum of the squares of the deviations will be

$$\Omega = (ax + by + cz + ... + n)^2 + (a'x + b'y + c'z + ... + n')^2 + (a''x + b''y + c''z + ... + n'')^2 + ...$$

that is :

$$\Omega = [aa]x^2 + [bb]y^2 + [cc]z^2 + ... + [nn]$$
$$+ 2[ab]xy + 2[ac]xz + 2[bc]yz + ...$$
$$+ 2[an]x + 2[bn]y + 2[cn]z + ...$$

where we use the notation introduced by Gauss (see Section 9.3 below):

$$[ab] = ab + a'b' + a''b'' + ...$$

This sum Ω is a minimum when its partial derivatives with respect to $x, y, z, ...$ are all zero, that is when

$$[aa]x + [ab]y + [ac]z + ... + [an] = 0$$
$$[ab]x + [bb]y + [bc]z + ... + [bn] = 0$$
$$[ac]x + [bc]y + [cc]z + ... + [cn] = 0$$

$$\cdots\cdots\cdots\cdots\cdots\cdots\cdots\cdots\cdots\cdots\cdots\cdots$$

This new system consists of k equations in the k unknowns $x, y, z, ...$ and is now called the system of *normal equations*.

If we were to restate this using modern matrix notation, we would say, if the initial system is written:

$$MX + N = 0, \text{ where } M = \begin{bmatrix} a & b & c & \cdots \\ a' & b' & c' & \cdots \\ a'' & b'' & c'' & \cdots \\ . & . & . & \cdots \end{bmatrix}, X = \begin{bmatrix} x \\ y \\ z \\ \cdots \end{bmatrix} \text{ and } N = \begin{bmatrix} n \\ n' \\ n'' \\ \cdots \end{bmatrix},$$

then the system of normal equations is simply : $SX + P = 0$, where $S = M^T.M$ is a symmetric matrix and $P = M^T.N$.

The method therefore consists in replacing a system of n initial equations by a system of only k equations, these being obtained from linear combinations of the former (which come from left multiplication by the matrix M^T). But this choice of method is not arbitrary: while for Legendre it was the most suitable method because it was the easiest to apply, for Gauss it was the most suitable because it could be justified - in fact *a posteriori* - by Probability Theory (see for example [27] or [31], chapter IX).

The method described by Legendre in 1805 had been used by Gauss, since at least 1801, since we know he used it to determine, to general astonishment, the orbit of the minor planet Ceres although only brief observations had been made. Gauss justified his method in1809 by appeal to the axiom that the arithmetical mean of a number of observed values of a quantity is the most probable value of that quantity [11]. In fact, in 1821, he showed that this choice of linear combinations corresponds exactly to the one for which 'the likely mean error' for each variable is the least possible [10] or, according to Seidel [26]:

The probability that the true values of the unknowns should lie within the given length intervals about the calculated values is greater with this system than with any other.

The method of least squares can be interpreted like this. We have to solve $MX + N = 0$, but N does not necessarily belong to the image of M; also we want to find X_0 such that $MX_0 + N$ should be 'the least possible'. The method of least squares corresponds to the structure of Euclidean space where we seek, in effect, a vector X_0 which minimises $\|MX + N\|^2$, that is the sum of the squares of the components.

Consider the quantity $\Phi_Y(t) = \|M(X_0 + tY) + N\|^2 = \|(MX_0 + N) + tMY\|^2$
$$= \|MX_0 + N\|^2 + t^2\|MY\|^2 + 2t<MX_0 + N, MY>.$$

For X_0 to be a minimum requires that, for all Y, $\Phi_Y(t)$ should be a minimum at $t = 0$, that is, for all Y:

$$\Phi_Y'(0)/2 = <MX_0 + N, MY> = <M^T(MX_0 + N, Y)> = 0.$$

from which: $M^T(MX_0 + N) = 0$, that is $M^TMX_0 + M^TN = 0$,

which we gave before as: $SX_0 + P = 0$.

We could also say that $\|MX_0 + N\|$ is the distance from $- N$ to the image of M and thus that X_0 is such that $- N_0 = MX_0$ is the orthogonal projection of $- N$ onto the image of M.

9.3 The Gauss Pivot Method

Gauss's main preoccupation was with astronomy. At the age of thirty, in 1807, he became director of the Observatory at Göttingen, a position that he held for the rest of his life. We have already mentioned the fact that Gauss used the method of least squares to determine the orbit of Ceres in 1801 just after the discovery of the planet by Piazzi. When a second minor planet, Pallas, was later discovered by Olbers, Gauss again set about determining its precise orbit, and he explained his approach in a memoir dated 1810.

> Starting from six observations made when the planet was in opposition - and so close to the Earth -, Gauss obtained 12 equations between 6 unknowns (the mean anomaly, the mean diurnal movement, the longitude of the perihelion, the eccentricity, the longitude of the node, the inclination). After obtaining an approximate solution, he determined 12 linear equations that must satisfy corrections to be made to the 6 unknowns. Rejecting the tenth because it was too imprecise on account of the observation, he used 11 equations from which he derived 6 normal equations and finally 6 corrections

In order to make it easier to deal with the algebraic solution of the system formed by the normal equations, Gauss made certain remarks at the end of the *Theory of the Movement of Celestial Bodies* (1809) which he developed in his Memoir on the minor planet Pallas (1810), and then in the second part of the *Theory of the Combination of Observations* which appeared in 1823. These remarks remind us of what is now called, in linear algebra, Gaussian elimination, or the Gauss pivot method, and the transformation of a quadratic into quadratic form, that is into a sum of squares, which produces a diagonal matrix.

The method of least squares requires us to minimise the quantity:

$$\Omega = w^2 + w'^2 + w''^2 + \ldots$$

where the w, w', w'', ... are linear combinations of unknowns p, q, r, ... Gauss rewrites Ω as another sum of squares, in such a way that at each stage an unknown disappears. (This is what is now called Gaussian reduction, applied here to positively defined quadratic forms). By elimination of each of the new squares, Gauss arrives at a 'triangular' system of equations.

C. F. Gauss
Disquisitio de Elementis Ellipticis Palladis ex oppositionibus annorum 1803, 1804, 1805, 1807, 1808, 1809. Memoir presented, 25 November 1810, to the Royal Society of Sciences, Göttingen; *Works*, vol. VI (1874), pp. 3-24.

As it is not possible to determine the six unknowns so as to satisfy all eleven proposed equations, that is to make the functions of the unknowns together on the left equal to zero, we try to find values that make the sum of the squares of these functions as small as possible. And it can easily be seen that if, generally, the linear functions of the unknowns p, q, r, s, etc. are set out as:

$$n \; + ap \; + bq \; + cr \; + ds \; + \text{etc.} = w$$
$$n' \; + a'p \; + b'q \; + c'r \; + d's \; + \text{etc.} = w'$$
$$n'' \; + a''p \; + b''q \; + c''r \; + d''s \; + \text{etc.} = w''$$
$$n''' \; + a'''p \; + b'''q \; + c'''r \; + d'''s + \text{etc.} = w'''$$

etc., then the conditional equations for $w^2 + w'^2 + w''^2 + $ etc. $= \Omega$ to be a minimum are the following:

$$aw \; + a'w' + a''w'' + a'''w''' + \text{etc.} = 0$$
$$bw \; + b'w' + b''w'' + b'''w''' + \text{etc.} = 0$$
$$cw \; + c'w' + c''w'' + c'''w''' + \text{etc.} = 0$$
$$dw \; + d'w' + d''w'' + d'''w''' + \text{etc.} = 0$$
$$\text{etc.}$$

or, to shorten the expressions, by representing

$$an + a'n' + a''n'' + a'''n''' + \text{etc. by } [an]$$
$$aa + a'a' + a''a'' + a'''a''' + \text{etc. by } [aa]$$
$$ab + a'b' + a''b'' + a'''b''' + \text{etc. by } [ab]$$
$$\text{etc.}$$
$$bb + b'b' + b''b'' + b'''b''' + \text{etc. by } [bb]$$
$$bc + b'c' + b''c'' + b'''c''' + \text{etc. by } [bc]$$
$$\text{etc. etc.}$$

p, q, r, s, etc. have to be determined by elimination from the equations:

$$[an] + [aa]p + [ab]q + [ac]r + [ad]s + \text{etc.} = 0$$
$$[bn] + [ab]p + [bb]q + [bc]r + [bd]s + \text{etc.} = 0$$
$$[cn] + [ac]p + [bc]q + [cc]r + [cd]s + \text{etc.} = 0$$
$$[dn] + [ad]p + [bd]q + [cd]r + [dd]s + \text{etc.} = 0$$
$$\text{etc.}$$

However, when the number of unknowns p, q, r, s etc. is somewhat large the elimination requires lengthy and tedious labour, which can be significantly shortened in the following way. As well as the coefficients $[an], [aa], [ab]$, etc. (whose number is $= \frac{1}{2}(ii + 3i)$ if the number of unknowns $= i$), suppose that one has also calculated the sum $nn + n'n' + n''n'' + n'''n''' + $ etc. $= [nn]$, it can easily be seen that we have

$$\Omega = [nn] + 2[an]p + 2[bn]q + 2[cn]r + 2[dn]s + \text{etc.}$$
$$+ [aa]pp + 2[ab]pq + 2[ac]pr + 2[ad]ps + \text{etc.}$$
$$+ [bb]qq + 2[bc]qr + 2[bd]qs + \text{etc.}$$
$$+ [cc] \; rr + 2[cd]rs + \text{etc.}$$
$$+ [dd]ss + \text{etc.}$$
$$\text{etc.}$$

and representing

$$[an] + [aa]p + [ab]q + [ac]r + [ad]s + \text{etc. by } A$$

it is clear that all the terms of $\dfrac{A^2}{[aa]}$ which contain the factor p are found in the expression Ω

and, therefore, $\Omega - \dfrac{A^2}{[aa]}$ is a function independent of p. That is why, on putting

$$[nn] - \frac{[an]^2}{[aa]} \qquad = [nn,1]$$

$$[bn] - \frac{[an].[ab]}{[aa]} \qquad = [bn,1]$$

$$[cn] - \frac{[an].[ac]}{[aa]} \qquad = [cn,1]$$

$$[dn] - \frac{[an].[ad]}{[aa]} \qquad = [dn,1] \text{ etc.}$$

$$[bb] - \frac{[ab]^2}{[aa]} \qquad = [bb,1]$$

$$[bc] - \frac{[ab].[ac]}{[aa]} \qquad = [bb,1]$$

$$[bd] - \frac{[an].[ad]}{[aa]} \qquad = [bd,1] \text{ etc. etc.}$$

we shall have

$$\Omega - \frac{A^2}{[aa]} = [nn,1] + 2[bn,1]q + 2[cn,1]r + 2[dn,1]s + \text{etc.}$$

$$+ [bb,1]qq + 2[bc,1]qr + 2[bd,1]qs + \text{etc.}$$
$$+ [cc,1]rr + 2[cd,1]rs + \text{etc.}$$
$$[dd,1]ss + \text{etc.}$$
$$+ \text{etc.}$$

which function we designate by Ω'.
Similarly, putting

$$[bn,1] + [bb,1]q + [bc,1]r + [bd,1]s + \text{etc.} = B,$$

the function

$$\Omega' - \frac{B^2}{[bb,1]}$$

will be independent of q; which we shall represent by Ω''.
In the same way, by putting

$$[nn,1] - \frac{[bn,1]^2}{[bb,1]} \qquad = [nn,2]$$

$$[cn,1] - \frac{[bn,1].[bc,1]}{[bb,1]} \qquad = [cn,2]$$

$$[cc,1] - \frac{[bc,1]^2}{[bb,1]} \qquad = [cc,2]$$

etc. etc. and

$$[cn,2] + [cc,2]r + [cd,2]s + \text{etc.} = C,$$

thus $\Omega'' - \frac{C^2}{[cc,2]}$ will be a function independent of r.

Continuing in this way, we shall come in the sequence Ω, Ω', Ω'', etc. to a last term which is independent of all the unknowns, which will be $[nn, \mu]$, if we use μ for the number of unknowns p, q, r, s etc. We have therefore:

$$\Omega = \frac{A^2}{[aa]} + \frac{B^2}{[bb,1]} + \frac{C^2}{[cc,2]} + \frac{D^2}{[dd,3]} + \text{ etc. } + [nn,\mu].$$

Now, since $\Omega = w^2 + w'^2 + w''^2 + $ etc. cannot by its nature take a negative value, it is easy to prove that the divisors $[aa]$, $[bb, 1]$, $[cc, 2]$, $[dd, 3]$, etc. must necessarily be positive (however, in the interests of brevity, I shall not set out the proof here). From this, it naturally follows that the minimum value of Ω is produced, if we make $A = 0, B = 0, C = 0, D = 0$, etc. From this therefore, the unknowns p, q, r, s, etc. can be determined from the μ equations, which can also be easily done in the reverse order, since clearly the last equation only contains one unknown, the one before it two, and so on. At the same time this method which recommends itself, gives the minimum value of the sum Ω, which in fact is clearly equal to $[nn, \mu]$.

In matrix terms, the system can be written:

$$N + MX = 0, \text{ where } M = \begin{bmatrix} a & b & c & \cdots \\ a' & b' & c' & \cdots \\ a'' & b'' & c'' & \cdots \\ \cdot & \cdot & \cdot & \cdots \end{bmatrix}, X = \begin{bmatrix} p \\ q \\ r \\ \cdots \end{bmatrix} \text{ and } N = \begin{bmatrix} n \\ n' \\ n'' \\ \cdots \end{bmatrix},$$

If we put:

$$W = N + MX,$$

then the question is to minimise:

$$\Omega = \|W\|^2 = \langle N + MX, N + MX \rangle$$

where $\langle ... \rangle$ stands for the Euclidean scalar product and $\|...\|$ the corresponding modulus. The expansion of Ω can be read in the following way:

$$\Omega = (N + MX)^T(N + MX) = N^TN + 2N^TMX + X^TM^TMX$$

or as

$$\Omega = \|N\|^2 + 2\langle M^TN, X \rangle + \langle M^TMX, X \rangle.$$

The formulas then found by Gauss remind us of what is called the pivot method. We can also notice here that the system of equations is symmetric since it concerns à system of normal equations. The pivots are $[aa]$, $[bb, 1]$, $[cc, 2]$, ...; but this metaphor is not made explicit in the text.

The pivots that appear as denominators can be written in terms of the principal diagonal minors of the matrix M^TM:

$$[aa], \quad \begin{pmatrix} [aa] & [ab] \\ [ab] & [bb] \end{pmatrix}, \quad \begin{pmatrix} [aa] & [ab] & [ac] \\ [ab] & [bb] & [bc] \\ [ac] & [bc] & [cc] \end{pmatrix}, \quad \cdots$$

These determinants are strictly positive since the quadratic form associated with M^TM is defined positive.

Chinese techniques ... and the Gauss pivot method

The Chinese techniques rely upon the arrangement of the numbers arising from some arithmetic problem into parallel columns (*hang*), each column translating a linear condition imposed on a set of unknowns, called the 'things' (*wu*). These columns of numbers (represented in concrete form by groups of counting-rods marked with various decimal units of numbers) are placed, in the proper sense of the word, upon a 'calculating surface' and so there appear configurations of numbers corresponding to what we would call the 'augmented matrix'. Since the whole arrangement is like a square, or more accurately a rectangle, the Chinese methods are called by the general name of *fangcheng* [arrangement of numbers in a square]. Starting from this arrangement, the methods of solution use a whole arsenal of manipulations such as 'multiplication throughout' (*bian cheng*), multiplication of a column of numbers by the same factor, and even 'direct reduction' (*zhi chu*), that is subtracting the terms of two columns term by term with the purpose of eliminating the coefficient of one of the unknowns. One of the most remarkable of the Chinese methods is to reduce the 'matrix' of the system to the triangular form, and then to calculate the unknowns by successive substitution, just as in the Gauss pivot method ([16], [22]).

Extract from *Sangaku keimō genkai taisei* by Takebe Katahiro, Vol. 3, Ch. 4, p. 32, verso.

The *Sangaku keimō genkai taisei* [Sum of the commentaries on the *Suanxue qimeng*] is a work of the Japanese mathematician Takebe Katahiro (1664-1739). The *Suanxue qimeng* [Introduction to Mathematics] under consideration here is the work of the Chinese mathematician Zhu Shijie (dates unknown) which appeared in 1299. After being reprinted in Korea, the text eventually arrived in Japan and was studied there as a manual of algebra in the second half of the 17th Century. At the base of the page, Takebe has drawn three pictures of 'numbers on rods', arranged as in a matrix, and representing three successive stages of the solution of a system of three equations in three unknowns. Each equation corresponds to a column. The Chinese characters appearing at the top of the diagram signify respectively 'right column', 'central column', 'left column'; those at the right signify respectively (silk) 'gauze', (silk) 'muslin', 'lustre' and 'sapeks' (monetary coinage). What we have are respectively the three unknowns of a concrete problem from which we get the 'matrix' of the system and sums of money put in the 'right hand side' of the equations. Finally, in each of the twelve cells of the diagram we can read the numbers expressed by means of the counting-rod system of numeration. Note that an oblique bar across a number indicates that the numeral is to be taken to be negative, and where a number is more than a single numeral, it is written in a compact way, as in a monogram; the corresponding symbols have to be entered into the appropriate space in each case. The reader will be able to identify the different numbers in the three 'matrices' in Takebe's work with the aid of these translations:

0	0	4		8	0	4		8	4	4
−14	−1	5		−4	−1	5		−4	4	5
−16	−14	6		−4	−14	6		−4	−8	6
−2262	−1023	1219		176	−1023	1219		176	196	1219

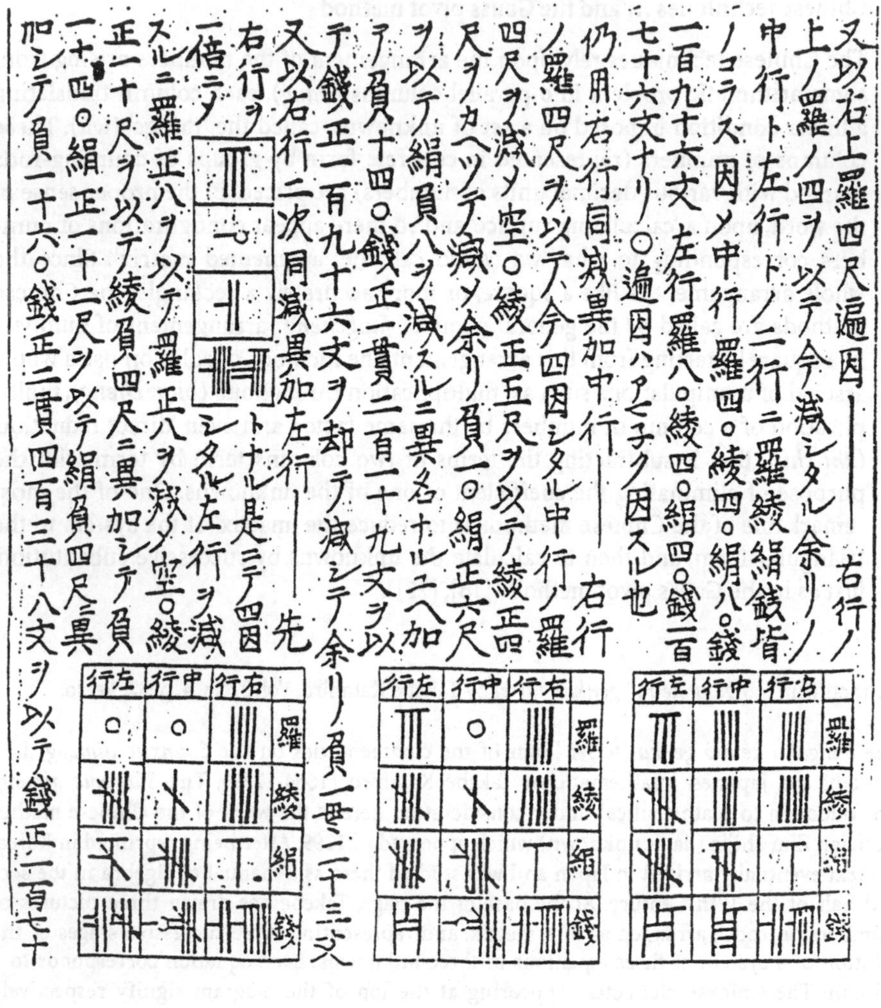

又以右上羅四尺ヲ遍因
上ノ羅ノ四ヲ以テ今減シタル余リノ
中行ト左行トノ二行ニ羅綾絹銭皆
ノコラズ因ノ中行ノ羅四〇綾四〇絹八〇銭
一百九十六文。左行ノ羅八〇綾四〇絹四〇銭一百
七十六文ナリ〇遍因ハヒチク因スルナリ

右
行
ノ

仍用右行同減異加中行
ノ羅四尺ヲ以テ今四因シタル中行ノ羅
四尺ヲ減メ空〇綾正五尺ヲ以テ綾正四
尺ヲカヘツテ減メ余リ負一〇絹正六尺
ヲ以テ絹正八ヲ減スルニ異名ナルニヘ加
フ負十四〇銭正一貫二百一十九文ヲ以
テ銭正二百九十六文ヲ却テ減シテ余リ負一每二十三文

又以右行ヲ
二次同減異加左行ニ

トナし是ヲ三ヶ四囘
スルニ羅正八ヲ以テ羅正八ヲ殺メ空〇綾
正二十尺ヲ以テ綾負四尺二異加シテ負
一十四〇絹正二十二尺ヲ以テ絹負四尺ニ異
加シテ負二十六〇銭正二貫四百三十八文ヲ以テ銭正二百七十

羅
綾
絹
銭

一倍ニヲ

先

行左	行中	行右	
〇	〇	三	羅
			綾
			絹
			銭

行左	行中	行右	
三	〇	三	羅
			綾
			絹
			銭

行左	行中	行右	
			羅
			綾
			絹
			銭

9.4 A Gauss Iterative Method

The basic idea behind the method that we know as the Gauss-Seidel method can be read in a letter Gauss wrote to Gerling in 1823, in which he explains what he calls an 'indirect' method - we would say 'iterative' method - for solving the system of normal equations arising from geodesic measurements.

The first triangulations, carried out to provide an accurate map of the Earth's surface, were started in the 16th century by Tycho-Brahe. They consist of constructing, from some side assumed to be known, a canvas of triangles all of whose angles are measurable on the Earth. In the 17th Century, Picard measured the arc of the meridian in the neighbourhood of Paris in order to fix the value of the radius of the Earth. In the 18th century, the work carried out by the French expeditions to Lapland by Maupertuis and to Peru by Condamine were intended to determine the extent of the flattening of the Earth, and the measuring of the French meridian by Delambre and Méchain between 1792 and 1799 was carried out in order to define the length of the metre. Triangulations were used in all cases and the data was then handled by the method of least squares.

At about 1817 Gauss completed his theoretical work on Astronomy and then turned to Geodesics. His plan for the triangulation of Hanover was officially approved in 1820 and he occupied himself with this work almost entirely up to 1825. He used a chain of 26 triangles and, naturally, the method of least squares. (For the particular application of the method to geodesics, see Section 9.8 below.)

Now, a very large number of measurements was taken and the systems of equations to be solved were enormous, including those of the normal equations; exact calculations were impracticable even by the 'pivot' method. For this reason, Gauss proposed iterative methods in order to speed up the calculations. Thus, in the *Supplement* of 1826, he suggested dividing up the equations into two or more groups, to look for corrections that would satisfy the first group, then to consider the second group starting from these corrected values, and so on, and then to return to consider the first group until the corrections became negligible.

What follows shows how Gauss handled a numerical example involving four unknowns, which appears in a letter to Gerling in 1823. The problem concerns the size of four angles which Gerling has given him, together with the number of repetitions of the respective measurements, thus providing weights to the measures. Gauss takes account of these weights to obtain a first set of corrected approximate values which he then sets about changing with the use of quantities a, b, c and d which are to be determined. Hence he calculates, for all the measured angles, the difference between the value obtained by Gerling and the value changed by the use of a, b, c and d; he then multiplies this by the corresponding number of measurements. Finally, from the method of least squares, since the sum of the squares of all the deviations has to be a minimum, equating the partial derivatives to zero produces the normal equations given below.

The extract of the letter given here concerns solving this system of normal equations, which Gauss does by successive approximations. The method can be summarised as follows: starting with a set of approximate values, he alters them one by one, choosing each time the unknown for which the change will be the greatest.

C. F. Gauss
Letter to Gerling, 26 December, 1823, *Brief wechsel zwischen Carl Friedriech Gauss und Christian Ludwig Gerling*, Berlin, 1927.

Göttingen, 26 December, 1823

My letter was too long in the post and it has come back to me. Also, I have reopened it in order to add some practical advice about elimination. There are really many particular little advantages to it that can only be appreciated through use.

I take your measurements of Orber-Reisig, by way of example:

[...]

The conditional equations are:

$$0 = + \quad 6 + 67a - 13b - 28c - 26d$$
$$0 = - 7558 - 13a + 69b - 50c - 6d$$
$$0 = - 14604 - 28a - 50b + 156c - 78d$$
$$\underline{0 = + 22156 - 26a - 6b - 78c + 110d}$$
$$\text{sum} = 0$$

To eliminate indirectly now, I notice that, if three of the quantities a, b, c, d are set equal to 0, the fourth will take the greatest value if d is chosen as the fourth quantity. Naturally each quantity has to be determined from its own equation, and so d from the fourth one. So I put d = – 201 and substitute this value. The constant terms then become: + 5232, – 6352, + 1074, +46; the rest remaining unchanged.

Now I let b be next, and I find b = + 92, I substitute it, and I find the constant terms: + 4036, – 4, – 3526, – 506. I continue in this way until there is nothing left to correct. But in actual fact, for the whole of this calculation, I merely write out the following table:

$d = -201$	$b = +92$	$a = -60$	$c = +12$	$a = +5$	$b = -2$	$a = -1$
+ 6	+ 5232	+ 4036	+ 16	– 320	+ 15	+ 41	– 26
– 7558	– 6352	– 4	+ 776	+ 176	+ 111	– 27	– 14
– 14604	+ 1074	– 3526	– 1846	+ 26	– 114	– 14	+ 14
+ 22156	+ 46	– 506	+ 1054	+ 118	– 12	0	+ 26

In that I am only taking the calculation to the next 1/2000th of a second, I see that there is now nothing more to correct. I collect up the terms:

$a = -60$	$b = +92$	$c = +12$	$d = -201$
+ 5	– 2		
– 1			
– 56	+ 90	+ 12	– 201

[...]

Almost every evening, I make up a new version of the table, which is always easy to improve. Compared with the monotony of the work of taking measurements, it is always a pleasant distraction: you can also see immediately if something doubtful has crept in, what remains to be found, etc. I recommend this method to you as a model. You will hardly ever again have to carry out a direct elimination, at least not when you have more than two unknowns. The indirect procedure can be carried when you are half asleep, or you can think about other things while doing it.

To solve the system of normal equations, Gauss uses successive approximations, changing one unknown at a time. That is, to use matrix notation, to solve the system of equations:

$$SX + P = 0$$

where the components of the vector P are the absolute (constant) terms of the equations, and where S is the matrix of the system of equations, he considers a vector Δ_1 having just one non-zero component and satisfying 'at best':

$$D\Delta_1 + P = 0$$

where D is the diagonal matrix formed from the diagonal of the matrix S, that is a vector Δ_1 whose non-zero component has the greatest possible absolute value. If we now define P_1 by:

$$S\Delta_1 + P = P_1$$

we can carry out an iteration on the new system $SX + P_1 = 0$ and so obtain a new vector Δ_2 having just one non-zero component and satisfying 'at best':

$$D\Delta_2 + P_1 = 0;$$

from which we can obtain P_2 such that: $S\Delta_2 + P_1 = P_2$.

And we can continue in this way to produce an iteration with the recurrence relations:

$$P_0 = P$$
$$D\Delta_k + P_{k-1} = 0$$
$$S\Delta_k + P_{k-1} = P_k$$

We stop at the step n when the absolute value of the components of Δ_n are less than 0.5 (which corresponds to the approximation 1/2000 given by Gauss) and we then take the solution:

$$X = \Delta_1 + \Delta_2 + \ldots + \Delta_n$$

since: $SX + P = S\Delta_1 + S\Delta_2 + \ldots + S\Delta_n + P = S\Delta_2 + \ldots + S\Delta_n + P_1 = \ldots$
$$= S\Delta_n + P_{n-1} = P_n \approx 0.$$

Gauss works here with integers, both for the coefficients of the equations and for the solution values, and so he knows that the process will end after a finite number of steps. This method was described by Gerling in 1843 in *Die Ausgleichungs-Rechnungen* [12]. For an overview of the iterative methods by Gauss, Seidel and Jacobi, see the commentary below at the end of Section 9.6.

We can also see that in this example the normal equations are not independent, since their sum is zero. This is because Gauss was not working on the measured angles, but on the directions that contain the angles. And what may appear to be an inconvenience is in fact according to Gauss an advantage, since he is easily able to verify his corrections at each stage of the procedure. He notes:

with the indirect method, it is very advantageous to make a variation to each direction. You can easily convince yourself of this by carrying out the calculations in this same example without this procedure, you will then lose the great advantage of always being able to verify that the sum of the constants is = 0.

This refers to the sum of the constant terms in the system of equations at the start, as well as in all the systems of equations occurring during the calculations. This requirement reminds us of the loop invariant that we find in modern algorithms.

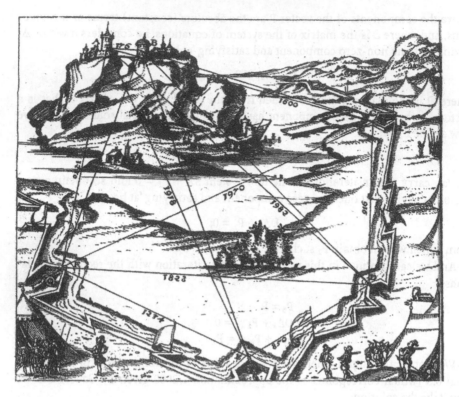

Casparus *Waserus. Novum Instrumentum Geometricum*, 1607, p. 16. (B. N.)

9.5 Jacobi's Method

Carl Gustav Jacobi needed to deal with linear systems of equations when considering physical systems subject to small oscillations. The brief extract of the text we reproduce below
shows an iterative method which he thought effective when the
diagonal coefficients were in the preponderance. If this were
not the case, he used the rotation of axes - so as to eliminate the
larger coefficients that were not lying on the diagonal - and
which, in particular, gave the eigenvalues and eigenvectors of a
symmetric matrix.

We only give the beginning of the text, which describes the method for the solution of linear equations which is known as Jacobi's method. Note that the matrix under discussion does not have to be symmetric.

C. G. J. Jacobi

Über eine neue Auflösungsart der bei der Methode der kleinsten Quadrate vorkommenden linearen Gleichungen, *Astronomische Nachrichten*, vol. 22 (1845), 297–306. From the French tr. of the German by C. Bloch.

The difficulty of solving exactly a large number of linear equations (as is often the case) when the method of least squares is used, has led to the consideration of approximate methods. One of these occurs when, in the various equations, it is never the same variable which has a coefficient clearly greater than the others. Suppose we have:

$$(00)x + (01)x_1 + (02)x_2 + \ldots = (0m)$$
$$(10)x + (11)x_1 + (12)x_2 + \ldots = (1m)$$
$$(20)x + (21)x_1 + (22)x_2 + \ldots = (2m)$$
$$\text{etc. etc. etc.}$$

in which all the coefficients (ik) are very small compared with the (ii) of the diagonal; an approximate value for the unknowns x, x_1, x_2, etc. is found from:

$$(00)x = (0m), \ (11)x_1 = (1m), \ (22)x_2 = (2m), \text{etc.}$$

If these values are represented by a, a_1, a_2, etc., their first corrections $\Delta, \Delta_1, \Delta_2$, etc. are obtained from:

$$(00)\Delta = -\{(01)a_1 + (02)a_2 + \ldots\},$$
$$(11)\Delta_1 = -\{(10)a + (12)a_2 + \ldots\}, \text{etc.}$$

and if we put, in general:

$$x = a + \Delta + \Delta^2 + \Delta^3 + \ldots,$$
$$x_1 = a_1 + \Delta_1 + \Delta_1^2 + \Delta_1^3 + \ldots,$$
$$x_2 = a_2 + \Delta_2 + \Delta_2^2 + \Delta_2^3 + \ldots, \text{etc.}$$

where the upper indices indicate the successive corrections, successively smaller, we can deduce the Δ^{i+1} from the Δ^i by the equalities:

$$(00)\Delta^{i+1} = -\{(01)\Delta_1^i + (02)\Delta_2^i + \ldots\},$$
$$(11)\Delta_1^{i+1} = -\{(10)\Delta^i + (12)\Delta_2^i + \ldots\},$$
$$(22)\Delta_2^{i+1} = -\{(20)\Delta^i + (21)\Delta_1^i + (23)\Delta_3^i + \ldots\}, \text{etc.}$$

In the equations produced by the method of least squares, the diagonal coefficients are effectively preponderant because they are the sums of squares, whereas the others come from the addition of positive and negative numbers which to an extent cancel each other out [...]

We need to solve the system:

$$AX = B$$

where A is a square matrix whose diagonal elements are non-zero.

Decompose A into the form:

$$A = D - C$$

where D is the matrix formed from the diagonal of A. Jacobi's method is to introduce a first approximate solution X_0 satisfying:

$$DX_0 = B,$$

and then to improve it by:

$$X_1 = X_0 + \Delta \text{ where } D\Delta = CX_0.$$

More generally, he defines successive approximations by:

$$X_i = X_0 + \Delta + \Delta^2 + \ldots + \Delta^i \text{ where } D\Delta^{i+1} = C\Delta^i.$$

It can be shown, by recurrence on i, that:

$$DX_{i+1} = CX_i + B.$$

If the sequence X_i has a limit X, then it will satisfy:

$$DX = CX + B$$

that is: $$AX = B.$$

It should be noted that this method requires the values of both X_i and X_{i+1} to be retained at each stage of the calculations, since the components of X_{i+1} are calculated from those of X_i (see the commentary at the end of Section 9.6).

We know now that, when the matrix A is 'diagonally dominant', that is when, for all i:

$$|a_{ii}| > \sum_{j \neq i} |a_{ij}|,$$

the method converges to a unique solution. Jacobi does not give a theoretical justification of the convergence here, but the remainder of the article is devoted to transformations of the system which are intended to make the matrix one in which the diagonal is sufficiently preponderant, so that "starting from a certain point, left to the discretion of the person carrying out the calculations, it will be advantageous to introduce the method of approximation. If it is done too early, this method itself will throw up coefficients which will compromise its success and which must then be eliminated by new transformations."

9.6 Seidel's Method

Ludwig Seidel, a pupil of Jacobi's, carried out a great many calculations for him. In particular, he had to solve a system of equations with 72 unknowns in a study of the luminosity of stars. Seidel, in his turn, also proposed an iterative technique for finding the solution to a system of normal equations. His inspiration came both from Jacobi's idea of successive approximations and from the Gauss 'pivot' method, and turns out to correspond in part to the method described by Gauss in his letter to Gerling given above. It appears, however, that Seidel would not have known about that method, since it seems that it was disseminated among geodesists but not among astronomers.

The text that follows starts by recapitulating the method of least squares, the system of normal equations, and then the Gauss 'pivot' method and Jacobi's method. We find here the notations introduced by Gauss and some analogy with a beginning of a triangulation of the system (see Section 9.3). We also find a justification for the convergence of the method, and that by a strict decrease at each stage of the sum of the squares of the deviations. Also, the optimal choice for the component to be altered differs somewhat, as we shall see, from that recommended by Gauss (see also the commentary at the end of this section).

P. L. von Seidel

Ueber ein Verfahren die Gleichungen, auf welche die Methode der kleinsten Quadrate führt, sowie lineäre Gleichungen überhaupt, durch successive Annäherung aufzulösen. Communication to the Mathematics-Physics Section of the Royal Academy of Berlin, meeting of 7 February, 1874. From the French tr. of the German by C. Bloch.

[...]

Suppose, first of all, that we have taken for the unknowns $x, y, z, \ldots$ some set of values which does not yet satisfy the most probable system of "normal equations" (B) but which give:

$$[aa]x + [ab]y + \ldots + [an] = N_1$$
$$[ab]x + [bb]y + \ldots + [bn] = N_2, \text{etc.} \ldots$$

For identification purposes, the sum of the squares of the deviations can be written:

$$Q = \{[aa]x + [ab]y + [ac]z + \ldots + [an]\}^2/[aa] + [bb.1]y^2 + [cc.1]z^2 + \ldots + 2[bc.1]yz + \ldots$$
$$+2[bn.1]y + 2[cn.1]z + \ldots + [nn.1]]$$
$$= N_1^2/[aa] + [bb.1]y^2 + \ldots + [nn.1].$$

In this form the unknown x only appears in the first term (that is in N_1); from which it can be seen straight away that Q will be diminished by the quantity

$$N_1^2/[aa]$$

if, without changing the values of $y, z, \ldots$, the value of x is changed in such a way that the expression for N_1 becomes zero. This will happen by correcting x by:

$$\Delta x = - N_1/[aa]$$

and the thus improved value of $x + \Delta x$ is that which, for the first unknown, *agrees the best* with the chosen values of the other unknowns and will thus be the best if the provisional values of the other variables are already the required true values.

The change in x, which replaces N_1 by $N_1' = 0$, at the same time changes the values of N_2, $N_3, \ldots$, to:

$$N_2' = N_2 + [ab]\Delta x,$$
$$N_3' = N_3 + [ac]\Delta x, \text{etc.} \ldots$$

If, instead of keeping $y, z, \ldots$ and correcting x by $- N_1/[aa]$, we had kept $x, z, \ldots$ and corrected y by $- N_2/[bb]$, the sum Q would have been reduced by $N_2^2/[bb]$; it would have been reduced by $N_3^2/[cc]$ if we had corrected just z (by $- N_3/[cc]$).

A second step is then to correct y by:

$$\Delta y = - N_2'/[bb];$$

$y + \Delta y$ will be the best value that agrees with the system of values $x + \Delta x, z, \ldots$ and reduces Q by $N_2'^2/[bb]$, while Q had already been reduced by $N_1^2/[aa]$.

This change in y replaces the system of values $N_1' = 0, N_2', N_3', \ldots$ by $N_1'', N_2'' = 0, N_3'', \ldots$, where:

$$N_1'' = N_1' + [ab]\Delta y = [ab]\Delta y,$$
$$N_2'' = N_2' + [bb]\Delta y = 0,$$
$$N_3'' = N_3' + [bc]\Delta y, \text{ etc.}$$

If now we correct z so that its new value $z + \Delta z$ makes the best possible agreement with the values $x + \Delta x, y + \Delta y$, and with the other initial values of the following unknowns, Q would be reduced by $N_3''^2/[cc]$; on the other hand, Q would be reduced by $N_1''^2/[aa]$ if we were to go back to x (because $x + \Delta x$ is not the best value agreeing with $y + \Delta y, z, \ldots$) and if we were to give it second correction of $- N_1''/[aa]$.

If then, starting from any system of initial values, they are changed successively taking them in any arbitrary order (and it is not necessary to go through the whole cycle before coming back to an unknown whose value has already been changed), always taking care to determine each correction so as to satisfy those of the normal equations where the variable concerned occupies the 'diagonal' position, then the sum of the squares of the deviations will be reduced at each step (and each time by an immediately known quantity, of the form $N^2/[aa]$ or $[aa]\Delta x^2$), and this for as long as it can be reduced.

Since the reductions in the size of Q, and the corrections made to the unknowns (these being of the form $- N/[aa]$), do not become negligible until all the N have simultaneously been reduced to values which are practically zero. When this has been achieved, all the normal equations will be satisfied and, by successive approximations, the unknowns will have taken the most likely values.

It must be pointed out that the proof of the decrease of Q and also of the convergence of this approximation process rests entirely on the condition that for each variable the successive improvements are calculated with respect to the equation where it occurs as the diagonal term; for if, in any system of n linear equations with n unknowns, one attempts to begin from a set of arbitrary values for the unknowns and make successive improvements by determining the correction for x starting from any one of the equations, then the correction for y from another one, arbitrarily chosen, etc. ..... - it would not be possible to prove in general that the solution system would be continuously approached - it might happen that (as one can easily convince oneself) the successive values of the unknowns tend towards infinity or oscillate continually between two distinct finite values.

But all systems of n linear equations in n unknowns can be put in the normal form [...].

Seidel notes that the guaranteed convergence of the method comes from the particular properties of the coefficients of the system of normal equations $[S = M^T.M]$, but that all systems of k equations with k unknowns $[A.X = B]$ can be changed into a system of normal equations by associating them with the system of its normal equations $[A^T.A.X = A^T.B]$.

Note that at each stage of the process Seidel uses Gauss's method of reduction to a quadratic form of the sum of squares and, more specifically, each time to the first stage of this reduction. In fact, it is because $Q = \sum_i N_i^2$ can be written in the form:

$$Q = N_1^2 / [aa] + Q_1(y, z, \ldots),$$

where Q_1 does not contain the variable x, that Q can be reduced by the amount $N_1^2 / [aa]$ when x is replaced by $x + \Delta x$, where $\Delta x = N_1/[aa]$, even though all the other N_i may be changed and may happen to be increased. This method may therefore be considered as an optimisation method.

Furthermore, "at each stage, the most rational thing to do", according to Seidel, "is to choose to change the variable whose correction will have the maximum effect on the sum of the squares of the deviations". And so, "it is not absolutely necessary to improve all the variables one after the other in order". Also, to use Seidel's notation, we can say that, at each stage, the optimum is to choose i, so as to maximise the corresponding reduction of $N_i^2 / [a_i a_i]$ in the value of Q, whereas for Gauss the optimum is to choose i so as to maximise the change $N_i / [a_i a_i]$ in x_i. In both cases, what we have is what, since the 1940s with Southwell and Temple, is called a 'relaxation method', in which at each stage just one element is modified, here so as to have an optimal effect on the deviations.

If, on the other hand, we forget about the optimal choices indicated by Gauss and Seidel, and we proceed to carry out changes 'cyclically', that is to take each variable in order, and only coming back to the first after having changed the last one, the method becomes what we now call the Gauss-Seidel method.

Indirect methods for solving linear equations

It would be worth summarising here the different indirect methods of solving linear equations that we have considered. These are relaxation methods and iterative methods.

Suppose we have to solve the system $MX + B = 0$,

$$\text{where } M = \begin{bmatrix} m_1^1 & m_1^2 & \ldots & m_1^n \\ m_2^1 & m_2^2 & \ldots & m_2^n \\ \ldots & \ldots & \ldots & \ldots \\ m_n^1 & m_n^2 & \ldots & m_n^n \end{bmatrix}, \quad X = \begin{bmatrix} x_1 \\ x_2 \\ \ldots \\ x_n \end{bmatrix} \text{ and } B = \begin{bmatrix} b_1 \\ b_2 \\ \ldots \\ b_n \end{bmatrix}.$$

The vector $W = MX + B$ is called the *residue* relative to the vector X. The indirect methods of solution are those which, starting from some vector $X^{(0)}$, construct a sequence of vectors $X^{(p)}$ by recurrence, such that the sequence of corresponding residues: $W^{(p)} = MX^{(p)} + B$ tend to the vector 0. When the matrix M is invertible, the sequence $X^{(p)}$ has a limit, and this limit is the required solution of the system of equations.

Relaxation methods

Here just one component is changed at a time. The vector $X^{(p+1)}$ is found from $X^{(p)}$ by choosing just one component, say the kth, and just one equation, say the ith: $\sum_{1 \le j \le n} m_i^j x_j + b_i = 0$, and changing it so that the residue $W^{(p+1)}$ becomes zero -

or nearly zero. Then the components of $X^{(p+1)}$ will be:

$$x_k^{(p+1)} = -\frac{b_i + \sum_{j \ne k} m_i^j x_j^{(p)}}{m_i^k} = x_k^{(p)} - \frac{w_i^{(p)}}{m_i^k},$$

and for $j \ne k$, $\qquad\qquad\qquad x_j^{(p+1)} = x_j^{(p)}.$

For the methods considered here, the choice has always been to take $k = i$, which requires non-zero diagonal coefficients. The methods differ by their choice of

criteria for i: For Gauss the choice is make $\dfrac{|w_i^{(p)}|}{m_i^i}$ a maximum, for Seidel it is

$\dfrac{|w_i^{(p)}|^2}{m_i^i}$ and , for Southwell, it is $|w_i^{(p)}|$.

Iterative methods

These are relaxation methods that are cyclical: at each stage, i is increased by one (modulo n). We can also express $X^{(p+1)}$ as a function of $X^{(p)}$ in matrix form (which is clearly the case in the text by Nekrasov which follows).

If $M = T + A$ where T is invertible, a recurrence can be defined by the formula:

$$X^{(p+1)} = - T^{-1}[AX^{(p)} + B].$$

If the sequence $X^{(p)}$ converges, then its limit X is the solution of the system of equations since $TX^{(p+1)} = - AX^{(p)} - B$ and so $(T + A)X + B = 0$.

The methods differ in their choice of T: it corresponds to the diagonal of M in Jacobi's method, and to the lower triangle of M in the cyclical version of Seidel's method.

9.7 Nekrasov and the Rate of Convergence

Alexander Ivanovich Nekrasov investigated Seidel's method at the request of the astronomer Tzeraki. In an article of 1885, he raises the question of its rate of convergence, which he explores in the case where the successive changes are carried out in a 'cyclic' order. This allows him to consider, in a general way, the change of quantities $x_n, y_n, ..., t_n$ to $x_{n+1}, y_{n+1}, ..., t_{n+1}$, and the formulas expressing this change illustrate the role of the lower triangular matrix T of the matrix S of the system of equations. Using matrices, we can summarise these formulas by:

$$TX_{n+1} + (S - T)X_n - N = 0.$$

By using the technique of solving recurrence sequences that looks for particular solutions of a 'geometric sequence' type, Nekrasov shows that the rate of convergence is linked to what are now called the eigenvalues of the matrix defining the iteration, that is the matrix $T^{-1}(T - S)$.

The article begins with a description of Seidel's method for solving the system of normal equations:

(1)
$$
\begin{aligned}
(aa)x + (ab)y + (ac)z + \ldots + (ap)t - (aq) &= 0 \\
(ba)x + (bb)y + (bc)z + \ldots + (bp)t - (bq) &= 0 \\
(ca)x + (cb)y + (cc)z + \ldots + (cp)t - (cq) &= 0 \\
&\ldots\ldots\ldots\ldots\ldots\ldots\ldots\ldots\ldots\ldots\ldots \\
(pa)x + (pb)y + (pc)z + \ldots + (pp)t - (pq) &= 0
\end{aligned}
$$

and recalls the argument for convergence put forward by Seidel (see Section 9.6).

A. I. Nekrasov

Determination of unknowns by the method of least squares for a large number of unknowns, *Matematicheskii Sbornik*, vol. X 22 (1885), 189–204. From the French tr. of the Russian by J. Simon.

§ 3 Expression of the error in the approximate values obtained by Seidel's method

Seidel's method necessarily converges towards the required solutions of the system (1), but to evaluate the rate of convergence we need to have a general expression for the errors in the approximate magnitudes produced by this method. This expression, which we shall investigate, is missing from Seidel's work.

So as to evaluate this expression, we suppose that the successive approximations are made in the following order: first $x_1, \ldots, t_1$ successively, and then the magnitudes $x_2, \ldots, t_2$, etc. ...

Let $x_n, \ldots t_n$ be a system of approximate magnitudes generated by Seidel's method after the n operations described. After this, and by the same method, we shall obtain the system: $x_{n+1}, y_{n+1}, z_{n+1}, \ldots, t_{n+1}$. These magnitudes will satisfy the equations:

(4)
$$
\begin{aligned}
(aa)x_{n+1} + (ab)y_n + (ac)z_n + \ldots + (ap)t_n - (aq) &= 0 \\
(ba)x_{n+1} + (bb)y_{n+1} + (bc)z_n + \ldots + (bp)t_n - (bq) &= 0 \\
(ca)x_{n+1} + (cb)y_{n+1} + (cc)z_{n+1} + \ldots + (cp)t_n - (cq) &= 0 \\
&\ldots\ldots\ldots\ldots\ldots\ldots\ldots\ldots\ldots\ldots\ldots \\
(pa)x_{n+1} + (pb)y_{n+1} + (pc)z_{n+1} + \ldots + (pp)t_{n+1} - (pq) &= 0
\end{aligned}
$$

These equations can be regarded as difference equations with finite differences with respect to the variable n. [...]

Let $x', y', z', \ldots, t'$ be the exact values satisfying equations (1). The errors corresponding to the approximation $x_n, \ldots t_n$ are:

(5) $\qquad \xi_n = x_n - x', \ \eta_n = y_n - y', \ \zeta_n = z_n - z', \ \ldots, \ \tau_n = t_n - t'.$

These values will satisfy the following equalities:

$$(aa)\xi_{n+1} + (ab)\eta_n \quad + (ac)\zeta_n \quad + \ldots + (ap)\tau_n \quad = 0$$
$$(ba)\xi_{n+1} + (bb)\eta_{n+1} + (bc)\zeta_n \quad + \ldots + (bp)\tau_n \quad = 0$$
$$(4') \qquad (ca)\xi_{n+1} + (cb)\eta_{n+1} + (cc)\zeta_{n+1} + \ldots + (cp)\tau_n \quad = 0$$
$$\ldots\ldots\ldots\ldots\ldots\ldots\ldots\ldots\ldots$$
$$(pa)\xi_{n+1} + (pb)\eta_{n+1} + (pc)\zeta_{n+1} + \ldots + (pp)\tau_{n+1} = 0$$

A particular solution for this system can be sought by putting:

$$\xi_n = A\alpha^n, \eta_n = AB\alpha^n, \zeta_n = AC\alpha^n, \ldots, \tau_n = AP\alpha^n.$$

Substituting these values in the equalities (4') we obtain, after simplifying:

$$(aa)\alpha + (ab)B \quad + (ac)C \quad + \ldots + (ap)P = 0$$
$$(ba)\alpha + (bb)\alpha B + (bc)C \quad + \ldots + (bp)P = 0$$
$$(6) \qquad (ca)\alpha + (cb)\alpha B + (cc)\alpha C + \ldots + (cp)P = 0$$
$$\ldots\ldots\ldots\ldots\ldots\ldots\ldots\ldots\ldots$$
$$(pa) \quad + (pb)B \quad + (pc)C \quad + \ldots + (pp)P = 0$$

On eliminating the magnitudes B, C, ..., P, we obtain for a given α the equality:

$$(7) \qquad \begin{vmatrix} (aa)\alpha & (ab) & (ac) & \ldots & (ap) \\ (ba)\alpha & (bb)\alpha & (bc) & \ldots & (bp) \\ (ca)\alpha & (cb)\alpha & (cc)\alpha & \ldots & (cp) \\ \ldots & \ldots & \ldots & \ldots & \ldots \\ (pa) & (pb) & (pc) & \ldots & (pp) \end{vmatrix} = 0$$

This equation, of degree $\mu - 1$ in α will have $\mu - 1$ roots $\alpha_1, \ldots, \alpha_{\mu-1}$. Associated with each of these is a system of magnitudes B, C, ..., P which can be determined by the equalities (6). There are $\mu - 1$ particular solutions of the form (5) with an arbitrary constant A. From these particular solutions it is easy to produce the general solution which will be:

$$\xi_n = A_1\alpha_1^n \quad + A_2\alpha_2^n + \ldots + \quad A_{\mu-1}\alpha_{\mu-1}^n$$
$$\eta_n = A_1 B_1 \alpha_1^n + A_2 B_2 \alpha_2^n + \ldots + A_{\mu-1} B_{\mu-1} \alpha_{\mu-1}^n$$
$$(8) \qquad \zeta_n = A_1 C_1 \alpha_1^n + A_2 C_2 \alpha_2^n + \ldots + A_{\mu-1} C_{\mu-1} \alpha_{\mu-1}^n$$
$$\ldots\ldots\ldots\ldots\ldots\ldots\ldots\ldots\ldots$$
$$\tau_n = A_1 P_1 \alpha_1^n + A_2 P_2 \alpha_2^n + \ldots + A_{\mu-1} P_{\mu-1} \alpha_{\mu-1}^n$$

where $A_1, \ldots, A_{\mu-1}$ are arbitrary constants, $\alpha_i, B_i, C_i, \ldots, P_i$ are the magnitudes defined by the equalities (6) and α_i satisfies the equality (7). The arbitrary constants $A_1, \ldots, A_{\mu-1}$ are related to the initial errors $\eta_0, \zeta_0, \ldots, \tau_0$ and to the arbitrarily chosen values $y_0, z_0, \ldots, t_0$ by the equalities:

$$\eta_0 = y_0 - y' = A_1 B_1 + A_2 B_2 + \ldots + A_{\mu-1} B_{\mu-1}$$
$$(9) \qquad \zeta_0 = z_0 - z' = A_1 C_1 + A_2 C_2 + \ldots + A_{\mu-1} C_{\mu-1}$$
$$\ldots\ldots\ldots\ldots\ldots\ldots\ldots\ldots\ldots$$
$$\tau_0 = t_0 - t' = A_1 P_1 + A_2 P_2 + \ldots + A_{\mu-1} P_{\mu-1}.$$

If the roots of equation (6) are not all distinct, the right hand side of the equalities (8) need to be changed in an obvious way, but the conclusions stated above remain valid.

§ 4. Rate of convergence by Seidel's method.

Since the errors of the magnitudes $x_n, y_n, z_n, \ldots$ must tend to 0 as n tends to $+\infty$, then as has been proved in §2, the roots of equation (7) must have a modulus less than 1. By this condition alone, the right-hand side of the equalities (8) will tend towards 0 as n increases without restriction.

Let α be the root of (7) with the greatest modulus, which we shall call the principal root. With respect to the root α, the errors $\xi_n, \eta_n, \zeta_n, \ldots$, determined by the equalities (8) will be $A\alpha^n, Ab\alpha^n, Ac\alpha^n, \ldots$. The magnitude A, dependant on $y_0 - y', \ldots, t_0 - t'$, as is shown in equations (9), cannot be given in advance since $y', \ldots, t'$ are supposed unknown. Therefore, for an arbitrary choice of $y_0, \ldots, t_0$, A will not be zero in general and this is why the values of the errors $A\alpha^n, Ab\alpha^n, Ac\alpha^n, \ldots$ will diminish slowly as n increases, compared with the other errors. In Seidel's method, convergence will be rapid if the principal root has a small modulus. But if the principal root has a modulus close to 1, which is not rare, then the convergence towards the solutions of (1) will be slow unless, but this is hardly probable, the choice of initial values $y_0, \ldots, t_0$ lead to $A = 0$.

In general, the rate of convergence in Seidel's method is entirely dependent on the value of the principal root of the equation (7). If all the solutions of (7), including the principal root, are small, the method converges quickly, otherwise it will converge slowly.

Proceeding in this way, treating the 'system of unknowns' as a whole rather than considering changes in the unknowns one at a time, Nekrasov's approach to dealing with the unknowns can be described as 'vectorial', although he only uses determinants and not matrices. The different systems of equations he considers can in fact be expressed in matrix form as follows:

(1) $\qquad SX - N = 0$ (the normal equations)

(4) $\qquad TX_{n+1} + (S - T)X_n - N = 0$

or $\qquad X_{n+1} = T^{-1}(T - S)X_n + T^{-1}N$ (recurrence formula)

Using X' for a solution of (1), the error at step n becomes

(5) $\qquad \Xi_n = X_n - X'$,

satisfying

(4') $\qquad T\Xi_{n+1} + (S - T)\Xi_n = 0$.

The solutions of (4') of the form $\Xi_n = \alpha^n B$ correspond to the vectors B satisfying

(6) $\qquad [\alpha T + (S - T)]B = 0$,

and so to the values of α satisfying

(7) $\qquad \det[\alpha T + (S - T)] = 0$.

Now, $\qquad \alpha T + (S - T) = T[\alpha I + T^{-1}(S - T)] = -T[T^{-1}(T - S) - \alpha I]$,

in other words, α is an eigenvalue of the matrix $T^{-1}(T - S)$ of the iteration (4).

After giving examples where the convergence is very slow, Nekrasov shows that the optimum choice indicated by Seidel does not significantly increase the rate of convergence.

9.8 Cholesky's Method

"The Commandant of Artillery Cholesky, of the Army Geographic Service, killed during the Great War had, during his research into corrections of geodesic curves, conceived of a very ingenious procedure for the solution of the so-called normal equations, obtained by the application of the method of least squares to linear equations fewer in number than the number of unknowns. From this he had concluded a general method for the solution of linear equations." (Commandant Benoit [1]).

In contrast to the three iterative methods just described Cholesky's technique leads, as a rule, to an exact solution. It was developed for use in geodesics and, to understand how it came about it, we shall say something about the particular use of the method least squares applied to geodesics. Here angles play the role of both the unknowns and the measured values and these unknowns need to satisfy certain necessary relations. For example, the sum of the angles of each triangle has to be 180°, plus a certain amount – which can be calculated to a high degree of accuracy – to take account of the curvature of the Earth. In general there are more unknowns than there are relations and it is therefore always possible to adjust the measured values of the angles in order for the 'conditional relations' to be exactly satisfied, what Gauss called the 'compensation of the observations', and this can be done in an infinite number of ways. Now, the choice of 'the most plausible compensations', in the sense of Gauss's second theory, leads to the method of least squares and this leads here to minimising the sum of the squares of the changes.

Suppose we have n 'compensations' $X = (x_1, ..., x_n)$ to make satisfying p equations ($p < n$) which we can write using vectors as:

(1) $MX + K = 0$ (conditional equations)

where M is a matrix of p rows and n columns. The method, which requires the minimisation of the sum of the squares of the compensations:

(2) $x_1^2 + x_2^2 + ... x_n^2$

can be handled in the following way.

We introduce p new variables $\Lambda = (\lambda_1, ..., \lambda_p)$ related to $X = (x_1, ..., x_n)$ by the p equations

(3) $X = M^T\Lambda$ (correlative equations)

These new variables will satisfy the system of equations obtained by the substitution of (3) in (1):

(4) $MM^T\Lambda + K = 0$(normal equations)

that is

(5) $S\Lambda + K = 0$

where $S = MM^T$.

Cholesky's procedure is to investigate another system of conditional equations:

(6) $TY + H = 0$

where T is a triangular matrix with the correlative equations:

(7) $Y = T^T \Lambda$

and leading to the same system of normal equations:

(8) $TT^T \Lambda + H = 0,$

where we let:

(9) $S = TT^T$ and $H = K.$

The question therefore becomes one of finding a triangular matrix T such that $TT^T = S$. In fact, if we know how to deduce T from S by (9), then we can obtain Y from (6), Λ from (7) (the systems (6) and (7) being triangular) and finally X from (3). Cholesky obtains the elements of T by equating the coefficients of TT^T and S. If T $= (\alpha_{i,j})$ and $S = (a_{i,j})$, then the way in which the $\alpha_{i,j}$ can be found from the $a_{i,j}$ can be written as the formulas:

$$\alpha_{ii} = \sqrt{a_{i,i} - \sum_{1 \le k \le i-1} \alpha_{k,i}^2}$$

$$\alpha_{i,i+r} = \left[a_{i,i+r} - \sum_{1 \le k \le i-1} \alpha_{k,i}\alpha_{k,i+r} \right] / a_{i,i} \qquad (r > 0).$$

Commandant Benoit

Note sur une méthode de résolution des équations normales provenant de l'application de la méthode des moindres carrés à un système d'équations linéaires en nombre inférieur à celui des inconnues, (Procédé du Commandant Cholesky), *Bulletin géodésique*, 2 (1923), 67–77.

[...] Let us write the system of p normal equations more simply:

$$a_1^1 \lambda_1 + a_1^2 \lambda_2 + ... + a_1^p \lambda_p + K_1 = 0$$

$$a_2^1 \lambda_1 + a_2^2 \lambda_2 + ... + a_2^p \lambda_p + K_2 = 0$$

(5)

$$\cdots\cdots\cdots\cdots\cdots\cdots\cdots\cdots\cdots\cdots\cdots\cdots$$

$$a_p^1 \lambda_1 + a_p^2 \lambda_2 + ... + a_p^p \lambda_p + K_p = 0$$

where $a_1^2 = a_2^1$, $a_1^3 = a_3^1$ etc. and in general $a_i^k = a_k^i$, since the system is symmetric about the diagonal. The terms of this diagonal a_1^1, a_2^2, etc. are always positive.

Commandant Cholesky's method consists in comparing these normal equations with other normal equations derived from a system of p linear equations that have *already been solved*. From the known solutions of this system, the solutions of the normal equations can be derived (through use of their correlative equations as intermediary) and so, after comparing coefficients, those of the primitive system (5).

Let:

$$\alpha_1^1 y_1 \qquad\qquad\qquad\qquad\qquad +H_1 = 0$$
$$\alpha_2^1 y_1 + \alpha_2^2 y_2 \qquad\qquad\qquad\quad +H_2 = 0$$

(6)
$$\alpha_3^1 y_1 + \alpha_3^2 y_2 + \alpha_3^3 y_3 \qquad\quad +H_3 = 0$$

$$\cdots\cdots\cdots\cdots\cdots\cdots\cdots\cdots\cdots\cdots$$

$$\alpha_p^1 y_1 + \alpha_p^2 y_2 + \alpha_p^3 y_3 + \ldots + \alpha_p^p y_1 + H_p = 0$$

be the system of the p linear equations; it is immediately solvable since we can find y_1 from the 1st, y_2 from the 2nd, etc....

Let us now apply the method of least squares to it; we should obtain the minimum solution of the system (6), but this will coincide with the former, since the system has only one solution.

Its correlative equations are:

$$y_1 = \alpha_1^1 \lambda_1 + \alpha_2^1 \lambda_2 + \ldots + \alpha_p^1 \lambda_p$$

(7)
$$y_2 = \qquad\quad \alpha_2^2 \lambda_2 + \ldots + \alpha_p^2 \lambda_p$$

$$\cdots\cdots\cdots\cdots\cdots\cdots\cdots\cdots$$

$$y_p = \qquad\qquad\qquad\quad \alpha_p^p \lambda_p$$

which will let us derive the values of the λ from those of the unknowns y drawn from the equations (6). We therefore know the solutions of the derived normal equations.
Let us write this system as:

$$\alpha_1^{1^2}.\lambda_1 + \alpha_1^1 \alpha_2^1.\lambda_2 + \alpha_1^1 \alpha_3^1.\lambda_3 + \ldots + \alpha_1^1 \alpha_p^1.\lambda_p + H_1 = 0$$

$$\alpha_1^1 \alpha_2^1.\lambda_1 + (\alpha_2^{1^2} + \alpha_2^{2^2}).\lambda_2 + (\alpha_2^1 \alpha_3^1 + \alpha_2^2 \alpha_3^2).\lambda_3 + \ldots + (\alpha_2^1 \alpha_p^1 + \alpha_2^2 \alpha_p^2).\lambda_p + H_2 = 0$$

(8) $\alpha_1^1 \alpha_3^1.\lambda_1 + (\alpha_2^1 \alpha_3^1 + \alpha_2^2 \alpha_3^2).\lambda_2 + (\alpha_3^{1^2} + \alpha_3^{2^2} + \alpha_3^{3^2}).\lambda_3 + \ldots + (\alpha_3^1 \alpha_p^1 + \alpha_3^2 \alpha_p^2 + \alpha_3^3 \alpha_p^3).\lambda_p + H_3 = 0$

$$\cdots\cdots\cdots\cdots\cdots\cdots\cdots\cdots\cdots\cdots\cdots\cdots$$

$$\alpha_1^1 \alpha_p^1.\lambda_1 + (\ldots).\lambda_2 + (\ldots)\lambda_3 + \ldots + (\alpha_p^{1^2} + \alpha_p^{2^2} + \ldots + \alpha_p^{p^2}).\lambda_p + H_p = 0$$

Let us identify these equations with the normal equations (5) of the primitive system [...]

Now, on comparing with equations (6), $y_1 = -\dfrac{H_1}{\alpha_1^1} = -\dfrac{K_1}{\sqrt{a_1^1}}$.

We then have, finally, for the first correlative equation:

$$\sqrt{a_1^1}.\lambda_1 + \frac{a_1^2}{\sqrt{a_1^1}}.\lambda_2 + \ldots + \frac{a_1^p}{\sqrt{a_1^1}}.\lambda_p + \frac{K_1}{\sqrt{a_1^1}} = 0 .$$

By comparing the second of the normal equations, in the same way, we compare the y_2 term with its value derived from the equations (6), and can write the second correlative equation with no longer a term in λ_1:

$$\sqrt{a_2^2 - \alpha_2^{1^2}}.\lambda_2 + \frac{a_2^3 - \alpha_2^1 \alpha_3^1}{\alpha_2^2}.\lambda_3 + \ldots + \frac{a_2^p - \alpha_2^1 \alpha_p^1}{\alpha_2^2}.\lambda_p + \frac{K_2 - \alpha_2^1 K_1'}{\alpha_2^2} = 0 ;$$

the α have already been calculated and K' is used to designate the constant term of the preceding equation. In general, for the correlative equation of rank i, we shall obtain, after comparing terms of the corresponding normal equations, and finding the constant term by this system:

$$\alpha_i^i = \sqrt{a_i^i - \alpha_i^{1^2} - \alpha_i^{2^2} - \ldots - \alpha_i^{(i-1)^2}}$$

$$\alpha_{i+1}^i = \frac{a_i^{i+1} - \alpha_i^1 \alpha_{i+1}^1 - \ldots - \alpha_i^{i-1} \alpha_{i+1}^{i-1}}{\alpha_i^i}$$

$$\ldots\ldots\ldots\ldots\ldots\ldots\ldots\ldots\ldots\ldots\ldots\ldots\ldots$$

$$\alpha_{i+r}^i = \frac{a_i^{i+r} - \alpha_i^1 \alpha_{i+r}^1 - \ldots - \alpha_i^{i-1} \alpha_{i+r}^{i-1}}{\alpha_i^i}$$

$$\ldots\ldots\ldots\ldots\ldots\ldots\ldots\ldots\ldots\ldots\ldots\ldots\ldots$$

$$K_i' = \frac{K_i - \alpha_i^1 K_1' - \alpha_i^2 K_2' \ldots - \alpha_i^{i-1} K_{i-1}'}{\alpha_i^i}$$

the α and the K' are coefficients already calculated from the previous equations.

The general rule for forming the coefficients can be seen:

The first coefficient α_i^i of a correlative equation is the square root of the corresponding coefficient a_i^i less the sum of the squares of the coefficients α_i already calculated;

Any coefficient α_{i+r}^i is equal to the corresponding term a_i^{i+r} less the sum of the products two by two of the coefficients of λ_i by those of λ_{i+r}, of the correlative equations already written, the whole divided by first coefficient α_i^i which has already been calculated..

$$[\ldots]$$

Checking: – These long calculations would be practically inextricable if frequent checks were not carried out. The Commandant Cholesky obtained these checks by use of the terms Σ, Σ', Σ''. The Σ terms are the sums of the coefficients of the same variable in the conditional equations. *Treated as ordinary coefficients*, they provide the Σ' terms, respectively equal to the sum of the coefficients of the corresponding normal equation; from which the forming of that normal equation can be verified. The coefficients Σ', treated as ordinary coefficients of their normal equation, give the Σ'' terms, equal to the sum of the coefficients of the corresponding equation. We do not therefore proceed to the calculation of a column without first making a sure verification of the preceding one.

In what is written here, we can see the presence of an invariant in the procedure: a new coefficient which must, at each stage, remain equal to the sum of the coefficients of the corresponding equation.

Cholesky's method is applied more generally for solving a system of equations AX = B where A is written in the form A = LU, and where L and U are respectively lower and upper triangular matrices. The problem is then changed to one of solving two systems of triangular matrices.

9.9 Epilogue

We have confined ourselves so far to considering systems of linear equations. But, the method of least squares is, in fact, useful for functions of any kind and Gauss [11] explains they can be dealt with, by considering approximate values of the unknowns (obtained, for example, from k equations), and by taking the coefficients to be the partial derivatives of the functions at a point (in other words, by a linearisation of the functions in the neighbourhood of the approximate values). If the values obtained are too different from the initial approximate values, the process is repeated, beginning from the new values.

In a brief note in 1847, which we reproduce here, Cauchy [4] indicates a general method for systems of equations, not necessarily linear. At the end of the note the method of least squares is apparent. It can therefore be considered as the precursor of what are called 'gradient' methods or 'greatest descent' methods, which, in order to determine a value x which minimises a function $f(x)$, proceed by successive approximations $x^{(k)}$ by choosing at each stage $x^{(k+1)}$ on the normal at $x^{(k)}$ to the contour surface:

$$f(x) = f(x^{(k)}).$$

Cauchy's note

Méthode générale pour la résolution des systèmes d'équations simultanées. *Comptes Rendus de l'Académie des Sciences de Paris*, vol. 25 (1847), 536–538.

Let $u = f(x, y, z)$

be a function of several variables $x, y, z, \ldots$, which never becomes negative and which remains continuous, at least within certain limits. In order to find the values of $x, y, z, \ldots$, which will satisfy the equation

(1) $u = 0$

it will be sufficient to decrease the function u indefinitely, to the point where it vanishes. Now, let

$$x, y, z, \ldots$$

be particular values taken by the variables $x, y, z, \ldots$; u the corresponding value of u: X, Y, Z, $\ldots$ the corresponding values of $D_x u, D_y u, D_z u, \ldots$, and $\alpha, \beta, \gamma, \ldots$ small increases given to the particular values x, y, z, $\ldots$. When we put

$$x = x + \alpha, \quad y = y + \beta, \quad z = z + \gamma, \quad \ldots,$$

we shall have, approximately,

(2) $u = f(x + \alpha, y + \beta, \ldots) = u + \alpha X + \beta Y + \gamma Z + \ldots$

Suppose now, θ being a positive quantity, that we take

$$\alpha = -\theta X, \quad \beta = -\theta Y, \quad \gamma = -\theta Z, \ldots,$$

The formula (2) will give, approximately

(3) $f(x - \theta X, y - \theta Y, z - \theta Z, \ldots) = u - \theta(X^2 + Y^2 + Z^2 + \ldots).$

From this it is easy to conclude that the value Θ of u, given by the formula

(4) $$\Theta = f(x - \theta X, y - \theta Y, z - \theta Z, \ldots),$$

will be less than u if θ is sufficiently small. If now θ increases and if, as we have supposed, the function $f(x, y, z, \ldots)$ is continuous, the value Θ of u will decrease until it vanishes, or at least it will decrease to the point where it coincides with a *minimum* value, determined by the equation with just one unknown

(5) $$D_\theta \Theta = 0.$$

It is sufficient therefore, either to solve this last equation, or at least to give θ a sufficiently small value, to obtain a new value of u less than u. If the new value of u is not a minimum, it is possible to derive a third even smaller value, operating in just the same way; and, continuing thus, successive smaller and smaller values of u will be found, which will converge towards a minimum value of u. [...]

Suppose now that the unknowns $x, y, z, \ldots$ have to satisfy, not just a single equation but a system of simultaneous equations

(7) $$u = 0, \quad v = 0. \quad w = 0, \quad \ldots,$$

the number of which may even be greater than the number of unknowns. To bring this case back to the previous one, it is sufficient to substitute for the system of equations (7) the single equation

(8) $$u^2 + v^2 + w^2 + \ldots = 0$$

Bibliography

[1] Benoit (Commander), Note sur une méthode de résolution des équations normales provenant de l'application de la méthode des moindres carrés à un système d'équations linéaires en nombre inférieur à celui des inconnues, (Procédé du Commandant Cholesky), *Bulletin géodésique*, vol. 2 (1923), 67–77.

[2] Bézout, E., *Histoire de l'Académie Royale des Sciences de Paris*, year 1764, p. 288.

[3] Cauchy, L. A., Mémoire sur les fonctions qui ne peuvent obtenir que deux valeurs égales et de signes contraires par suite des transpositions opérées entre les variables qu'elles renferment, *Journal de l'Ecole Polytechnique*, vol. 10 (1815), 29–112. *Oeuvres* 2nd series, vol. 1, pp. 91–169, Paris: Gauthier-Villars, 1905.

[4] Cauchy, L. A., Méthode générale pour la résolution des systèmes d'équations simultanées, *Comptes Rendus de l'Académie des Sciences de Paris*, vol. 25 (1847), 536–538. *Oeuvres*, series I, vol. 10, pp. 399–402, Paris: Gauthier-Villars, 1897.

[5] Cayley, A., Remarques sur la notation des fonctions algébriques, *Journal für reine und angewandte Mathematik*, vol. 50 (1855), 282–285. *Collected Mathematical Papers*, vol. 2, pp. 185–188.

[6] Cramer, G., *Introduction à l'analyse des lignes courbes algébrique*, Geneva, 1750.

[7] Euler, L., *Recherches sur la question des inégalités du mouvement de Saturne et de Jupiter*, 1749. *Opera omnia*, 2nd series, vol. 25, Turin, 1960, pp. 45–157.

[8] Gauss, C. F., Disquisitio de elementis ellipticis Palladis, Memoir presented to Royal Society of Sciences, Göttingen, 25 November, 1810. *Werke*, vol. VI, 1874, pp. 3–24. French tr. in J. Bertrand, *Méthodes des moindres carrés*, Paris, 1855.

[9] Gauss, C. F., Letter of 26 December 1823, *Briefwechsel zwischen Carl Friedrich Gauss und Christian Ludwig Gerling*, Berlin: Elsmer, 1927, rprnt. Hildesheim: Ohms-Verlag, 1975.

[10] Gauss, C. F., Theoria Combinationis observationum erroribus minimis obnoxiae, presented to Royal Society of Sciences, Göttingen, 1st part 1821, 2nd part 1823, Supplement 1826. *Werke*, vol. IV, 1873, pp. 1–108. English by G. W. Stewart, Philadelphia: Siam, 1955.

[11] Gauss, C. F., *Theoria Motus Corporum coelestium in sectionibus solem ambientiem*, 1809, Hamburg. *Werke*, vol. VII, 1871. English tr. C. H. Davis, 1857, rprnt. New York: Dover, 1963.

[12] Gerling, C. L., *Die Ausgleichungs-Rechnungen der praktischen Geometrie oder die Methode der kleinsten Quadrate mit ihrer Anwendungen für geodätische Aufgaben*, Hamburg, 1843.

[13] Jacobi, C. G., Über eine neue Auflösungsart der bei der Methode der kleinsten Quadrate vorkommenden linearen Gleichungen, *Astronomische Nachrichten*, vol. 22 (1845), 297–306. *Gesammelte Werke*, vol. 3, 1884, pp. 469–478.

[14] Knobloch, E., *Der Begin der Determinantentheorie, Leibnizensnachgelassens Studien zum Determinantenkalkül; im Zuzammenhang mit dem gleichnamigen Abhandlungsband fast ausschlieblich zum ersten Mal nach den Originalschriften*, Hildesheim: Gerstenberg Verlag, 1980.

[15] Lagrange, J. L., Nouvelle solution du problème du mouvement de rotation d'un corps de figure quelconque qui n'est animé par aucune force accélératrice, *Nouveaux Mémoires de l'Académie royale des Sciences et Belles-Lettres de Berlin*, 1773, pp. 85–128. *Oeuvres*, vol. 3, pp. 577–616.

[16] Lam Lay-Yong & Ang Tian-Se, The Earliest Negative Numbers : How they Emerged from a Solution of Simultaneous Linear Equations, *Archives Internationales d'Histoire des Sciences*, vol. 37 (1987), 222–262.

[17] Laplace, P. S., Recherches sur le calcul intégral et sur le système du monde, *Mémoires de l'Académie des Sciences de Paris*, 1772. *Oeuvres*, vol. VIII, pp. 395–406.

[18] Laplace, P. S., *Traité de Mécanique céleste*, Paris, 1799, First Part, Book III, Chap. 5, § 39. *Oeuvres*, vol. II, Paris, 1843.

[19] Legendre, A. M., *Nouvelles Méthodes pour la détermination des Orbites des Comètes*, Paris, 1805.

[20] Leibniz, G. W., Letter to de l'Hospital of 28 April 1693. *Mathematische Schriften*, vol. 2, pp. 236–241, Hildesheim: Ohms-Verlag, 1962.

[21] Maclaurin, C., *Treatise of Algebra*, London, 1748.

[22] Martzloff, J.-C., *Histoire des mathématiques chinoises*, Paris: Masson, 1987, English tr. by S. S. Wilson, *A History of Chinese Mathematics*, New York: Springer, 1997.

[23] Michel-Pajus, A., Fragments d'une histoire des systèmes linéaires, *Mnémosyne*, 3 (1993), IREM Paris VII.

[24] Muir, T., *The theory of determinants in the historical order of development*, 4 vols., London, 1906–1923.

[25] Nekrasov, P.A., Determination of unknowns by the method of least squares for a large number of unknowns, *Matematicheskii Sbornik*, vol. X 22 (1885), 189–204 (in Russian).

[26] Seidel, L., Ueber ein Verfahren die Gleichungen, auf welche die Methode der kleinsten Quadrate führt, sowie lineäre Gleichungen überhaupt, durch successive Annäherung aufzulösen. Communication to the Mathematics-Physics Section of the Royal Academy of Berlin, meeting of 7 February, 1874.

[27] Sheynin, O. B., C. F. Gauss and the Theory of Errors, *Archive for the History of the Exact Sciences*, vol. 1, 20 (1979), 21–72.

[28] Sylvester, J. J., Additions to the articles, "On a new class of theorems", and "On Pascal's theorem", *Philosophical Magazine*, vol. 37 (1850), 363–370. *Collected Mathematical Papers*, vol. 1, New-York: Chelsea Publishing Co., 1973, pp. 145–151.

[29] Tropfke, J., *Geschichte der Elementarmathematik*, 4th ed., vol 1, Berlin and New York: Walter de Gruyter, 1980.

[30] Vandermonde, A.T., Mémoire sur l'élimination, Histoire de l'Académie Royale des Sciences de Paris, year 1772, 2nd part, Paris, pp. 516–532, 1776.

[31] Whittaker, E. T. & Robinson, G., *The Calculus of Observations, A Treatise on Numerical Mathematics*, 2nd ed., London & Glasgow: Blackie & Son Ltd., 1932.

10 Tables and Interpolation

> The theory of Interpolation ... the science of
> reading between the lines of a mathematical table.
>
> E. T. Whittaker
> *The calculus of observations*, (Chap. 1, p. 2)

The construction of tables turned out to be of vital importance for facilitating calculations and for avoiding the need to carry out the same operations many times. We have seen how tables have been constructed from the earliest times, for example, the Babylonians produced tables for calculating 'inverses' (Section 1.2 above). In Astronomy, trigonometric tables have played a major role, and these correspond to the tables of chords that Ptolemy produced in the 2nd century (Section 10.1). A table of logarithms is another type of table, specifically introduced to make calculations easier, by transforming multiplications into additions; we shall look at the decimal tables established by Briggs in the 17th century following on from the work of Napier (Section 10.2).

What happens usually with these tables is that, after a certain number of values have been calculated, by more or less ingenious methods, the other values are entered by use of *interpolation* from the previously calculated values. As tables of greater sophistication became necessary, so interpolation methods of greater effectiveness were developed. In this respect we shall look at the so-called Gregory-Newton interpolation formulas and the polynomial interpolation formulas of Newton and Lagrange (Sections 10.3, 10.4 and 10.5). With Cauchy and his 'interpolation functions' we come to appreciate the errors introduced with interpolation (Section 10.6). All these different interpolation formulas were established between the end of the 17th century and the beginning of the 19th century. In a more contemporary setting we shall see with Neville how the iterative calculations for the Lagrange polynomial can be made easier (Section 10.7), and how the CORDIC algorithm is used in electronic calculators to create their tables of standard functions.

But what exactly does the term 'interpolation' mean? The original meaning comes from the Latin 'interpolare', meaning to 'polish in-between', that is to put a gloss on what is missing. In this sense the word implies an unfair or biased reading of a text in which the original is corrupted by the insertion of spurious words or passages. This certainly happens with interpretations of historical mathematical material, and we have seen examples of this in earlier chapters. As for the scientific use of the terms 'interpolate' and 'interpolation', these made their appearance at the beginning of the 19th century and refer to the process of inserting intermediary values between known terms of a sequence.

Whenever we think of interpolation, we also think of the associated process of *extrapolation*. In current usage, this word has negative connotations; its use in a scientific context also corresponds to a procedure that is not always easy, nor certain, which is to calculate the values of a function outside the range of its known values. We shall consider later how extrapolation can be handled when we look at Richardson's extrapolation procedure (Section 14.5).

We can never think about interpolation without also considering the associated idea of *approximation*, if only because the value of a function obtained by interpolation is always – or almost always – an approximation to its true value. We shall make the following distinction here which, however, is not entirely free of ambiguity: the interpolation of a function over an interval corresponds to the calculated approximation of the value of a function at a point from other given known values; the approximation of a function conveys a more global idea, and concerns replacing a badly formulated or complicated expression by a known function or a simpler expression, which is close to the initial function over the whole interval and not just at a single point. The term 'close to' remains to be defined of course, and different degrees of 'closeness' are possible.

In practice, even where the interpolation of a function consists in calculating its approximate values at certain points from other known values of the function, we most often use another function of a given type which has these same values as the function to be interpolated, and take the interpolated values to be the values of the new function. All the same, this idea of an interpolatory function of a given type is not necessarily specific; this is particularly the case with tables. The most common interpolatory functions are polynomials, linear combinations of trigonometric and exponential functions or rational fractions. We shall restrict ourselves here to polynomial interpolation.

We should note that whatever the choice of type of interpolatory function, it will certainly be an approximation to the initial function. Will not the polynomial then be itself an approximation to the function? Here we come back to the subtle distinction between interpolation and approximation. These different points will be touched on as we work through this chapter, and also in the chapter on approximation to functions (Chap. 13).

The most elementary interpolation is *linear interpolation*. It is sometimes called *Lagrange's method* or the *secant method*. This involves replacing a function f, for which we know the values $f(a)$ and $f(b)$ for two values a and b of the variable, by the linear function g which takes these same values at a and b, namely:

$$g(x) = \frac{f(a) - f(b)}{a - b} x - \frac{b \cdot f(a) - a \cdot f(b)}{a - b}$$

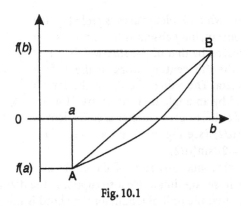

Fig. 10.1

In the Cartesian plane, this means replacing the curve $y = f(x)$ by the straight line $y = g(x)$ (see Fig. 10.1). Hence, if we want to solve $f(x) = 0$, that is to find where the curve $y = f(x)$ cuts the x-axis, interpolation involves replacing the curve by the secant passing through the points A and B whose coordinates are $(a, f(a))$ and $(b, f(b))$ respectively.

In the case where $f(a).f(b) < 0$, the solution of $g(x) = 0$ gives:

$$x = \frac{b.f(a) - a.f(b)}{f(a) - f(b)}$$

and here we have again the famous rule of double false position, if $f(a)$ and $f(b)$ are taken to be the errors corresponding to the false values a and b of the solution of the linear equation (Chap. 3).

10.1 Ptolemy's Chord Tables

Ptolemy wrote his *Mathematical Syntaxis* about 150 AD, which has since been called the *Almagest* – from the Arabic, meaning the 'greater' – and it became the essential reference work for astronomy, just as Euclid's *Elements* was for geometry. Ptolemy explained, in particular, how to construct a table of the lengths of chords of a circle as a function of the values of the corresponding arcs. These tables are the oldest of this type known to us, though it is almost certain that both Hipparchus (2nd century BC) and Menelaus (1st century AD) had used tables of this type.

We have already remarked that the tradition of numerical tables is an ancient one, which is to be found as far back as Babylonian mathematics. But such tables as those relied on operations that are simple and easy to put into practice (tables of additions, multiplications, squares, cubes, inverses, etc.). In contrast, other types of tables can only be produced with the use of a whole arsenal of mathematical theory. Ptolemy's tables belong to this latter category, since their construction requires a theory of plane trigonometry, which is very clearly presented in the *Almagest* itself.

The idea of the chord which Ptolemy uses is probably due to Hipparchus. This concept, long absent from our books on trigonometry, was replaced by that of the *sine*, which is met for the first time in the 6th century work of the Indian mathematician Āryabhata. The value of crd.α is the length of the chord subtended by an angle α at the centre of a circle, so that the ratio of crd.α to the diameter is equivalent to what we now call $\sin(\alpha/2)$ (see Fig. 10.2): for a circle of radius R we have: $\mathrm{crd}.\alpha = 2R\sin(\alpha/2)$.

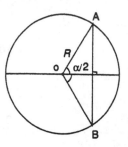

Fig. 10.2

Ptolemy uses the Babylonian division of the circle into 360° and he calculates in sexagesimals. He also supposes the diameter to be divided into 120 equal parts, so that the unit of length for the chord is the part (^{P}), divided in its turn into minutes (') and seconds ("). The *Almagest* gives the values of the chords in half-degree steps and to an accuracy corresponding to close to six decimal places.

The schema adopted by Ptolemy is as follows:

1) *Determining the values of crd.72° and crd.36°.* Ptolemy starts from the geometric construction of the sides of the regular pentagon and the regular decagon, which subtend angles at the centre of 72° and 36° respectively. By using the Pythagoras theorem, he then derives the values:

$$\mathrm{crd}.72° = 70^P\ 32'\ 3''\quad \text{and}\quad \mathrm{crd}.36° = 37^P\ 4'\ 55''$$

2) *Ptolemy's Theorem.* He derives the result known by his name, that for any cyclic quadrilateral the product of the diagonals is equal to the sum of the products of the opposite sides.

3) *The chord of the difference of two arcs.* Using the previous result, Ptolemy shows how, given the chords of two arcs α and β, the chord of the arc of the difference α − β can be calculated. Hence, in particular, crd.12° can be found from the previously known values of crd.72° and the easily derived crd.60°.

4) *The chord of half an arc.* Ptolemy explains how to derive crd.α/2 from crd.α. From this he can find crd.6°, crd.3°, crd.3/2° and crd.3/4°.

5) *The chord of* 1°. The method used so far does not enable Ptolemy to obtain crd.1° and certainly not crd.1/2°. Ptolemy uses a subtle method for finding boundary values for crd.1°, and then crd.1/2°, which are sufficiently accurate for his purposes.

6) *The table of chords in half-degree steps.* Ptolemy then establishes his table, using the techniques described and also, thanks to Ptolemy's theorem, using the fact that whenever we know the chords of two angles α and β, we can find crd(α − β) and we can also find crd(α + β).

The extract below gives parts 2, 3 and 4 of Ptolemy's schema.

Ptolemy
Syntaxis, i 10, ed. Heiberg i, §§ i. 31. 7–32. 9. Translation: Ivor Thomas, *Greek Mathematical Works*, 1941, 1993, Cambridge Mass. and London: Loeb Classical Library, Harvard University Press, pp. 423–431, 444–445.

Let ABCD be any quadrilateral inscribed in a circle, and let AC and BD be joined. It is required to prove that the rectangle contained by AC and BD is equal to the sum of the rectangles contained by AB, DC and AD, BC.

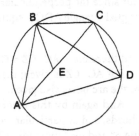

For let the angle ABE be placed equal to the angle DBC. Then if we add the angle EBD to both, the angle ABD = the angle EBC. But the angle BDA = the angle BCE, for they subtend the same segment; therefore the triangle ABD is equiangular with the triangle BCE.

∴   BC : CE = BD : DA;
∴   BC . AD = BD . CE.

Again, since the angle ABE is equal to the angle DBC, while the angle BAE is equal to the angle BDC, therefore the triangle ABE is equi-angular with the triangle BCD;

∴   BA : AE = BD : DC;
∴   BA . DC = BD . AE.

But it was shown that

BC . AD = BD . CE;
and ∴   AC . BD = AB . DC + AD . BC;

which was to be proved.

This having been first proved, let ABCD be a semicircle having AD for its diameter, and from A let the two [chords] AB, AC be drawn, and let each of them be given in length, in terms of the 120^P in the diameter, and let BC be joined. I say that this also is given.

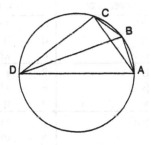

For let BD, CD be joined; then clearly these also are given because they are the chords subtending the remainder of the semicircle. Then since ABCD is a quadrilateral in a circle,

AB . CD + AD . BC = AC . BD.

And AC . BD is given, and also AB . CD; therefore the remaining term AD . BC is also given. And AD is the diameter; therefore the straight line BC is given.

And it has become clear to us that, if two arcs are given and the chords subtending them, the chord subtending the difference of the arcs will also be given. It is obvious that, by this theorem we can inscribe many other chords subtending the difference between given chords; and in particular we may obtain the chord subtending 12°, since we have that subtending 60° and that subtending 72°.

Again, given any chord in a circle, let it be required to find the chord subtending half the arc subtended by the given chord. Let ABC be a semicircle upon the diameter AC and let the chord CB be given, and let the arc CB be bisected at D, and let AB, AD, BD, DC be joined, and from D let DZ be drawn perpendicular to AC. I say that ZC is half of the difference between AB and AC.

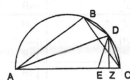

For let AE be placed equal to AB, and let DE be joined. Since AB = AE and AD is common, [in the triangles ABD, AED] the two [sides] AB, AD are equal to AE, AD each to each; and the angle BAD is equal to the angle EAD; and therefore the base BD is equal to the base DE. But BD = DC; and therefore DC = DE. Then since the triangle DEC is isosceles and DZ has been drawn from the vertex perpendicular to the base, EZ = ZC. But the whole EC is the difference between the chords AB and AC; therefore ZC is half of the difference. Thus, since the chord subtending the arc BC is given, the chord AB subtending the remainder of the semicircle is immediately given, and ZC will also be given, being half of the difference between AC and AB. But since the perpendicular DZ has been drawn in the right-angled triangle ACD, the right-angled triangle ADC is equiangular with DCZ, and

$$AC : CD = CD : CZ,$$

and therefore $AC . CZ = CD^2$.

But AC . CZ is given; therefore CD^2 is also given. Therefore the chord CD, subtending half of the arc BC, is also given.

And again by this theorem many other chords can be obtained as the halves of known chords, and in particular from the chord subtending 12° can be obtained the chord subtending 6° and that subtending 3° and that subtending 1½° and that subtending ½° + ¼° (= ¾°). We shall find, when we come to make the calculation, that the chord subtending 1½° is approximately $1^P 34' 15''$ (the diameter being 120^P) and that subtending ¾° is $0^P 47' 8''$.

[...]

Ptolemy's table of chords in Ivor Thomas, *Greek Mathematical Works II*, pp. 444–445 (with correction to the English translation).

κανονιον των εν κυκλω ευθειων

περεφερειῶν	ενῳειῶν			εξηκοστῶν			
∠'	ο	λα	κε	ο	α	β	ν
α	α	β	ν	ο	α	β	ν
α ∠'	α	λδ	ιε	ο	α	β	ν
β	β	ε	μ	ο	α	β	ν
β ∠'	β	λζ	δ	ο	α	β	μη
γ	γ	η	κη	ο	α	β	μη
γ ∠'	γ	λθ	νβ	ο	α	β	μη
δ	δ	ια	ις	ο	α	β	μζ
δ ∠'	δ	μβ	μ	ο	α	β	μζ
...	...	...	...	...	...	...	...
ξ	ξ	ο	ο	ο	ο	νδ	κα
...	...	...	...	...	...	...	...
ρος	ριθ	νε	λη	ο	ο	β	γ
ρος ∠'	ριθ	νς	λθ	ο	ο	α	μζ
ροζ	ριθ	νζ	λβ	ο	ο	α	λ
ροζ ∠'	ριθ	νη	ιη	ο	ο	α	ιδ
ροη	ριθ	νη	νε	ο	ο	ο	νζ
ροη ∠'	ριθ	νθ	κδ	ο	ο	ο	μα
ροθ	ριθ	νθ	μδ	ο	ο	ο	κε
ροθ ∠'	ριθ	νθ	νς	ο	ο	ο	θ
ρπ	ρκ	ο	ο	ο	ο	ο	ο

Table of the chords in a circle

Arcs	Chords			Sixtieths			
½°	0^P	31'	25"	0^P	1'	2"	50'''
1	1	2	50	0	1	2	50
1½	1	34	15	0	1	2	50
2	2	5	40	0	1	2	50
2½	2	37	4	0	1	2	48
3	3	8	28	0	1	2	48
3½	3	39	52	0	1	2	48
4	4	11	16	0	1	2	47
4½	4	42	40	0	1	2	47
...	...	...	...	...	...	...	...
60	60	0	0	0	0	54	21
...	...	...	...	...	...	...	...
176	119	55	38	0	0	2	3
176½	119	56	39	0	0	1	47
177	119	57	32	0	0	1	30
177½	119	58	18	0	0	1	14
178	119	58	55	0	0	0	57
178½	119	59	24	0	0	0	41
179	119	59	44	0	0	0	25
179½	119	59	56	0	0	0	9
180	120	0	0	0	0	0	0

The chord of the difference of two arcs.
Ptolemy's theorem states that, for ABCD a cyclic quadrilateral, we have:

$$AC.BD = AB.CD + AD.BC$$

Hence, in the case where AD is a diameter, when $AC = \mathrm{crd}.\alpha$ and $AB = \mathrm{crd}.\beta$ are known, then $BC = \mathrm{crd}(\alpha - \beta)$ is also known.

From Ptolemy's theorem, we have:

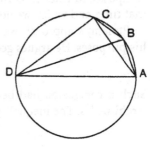

Fig. 10.3

$$\mathrm{crd}(\alpha - \beta) = BC = \frac{AC.BD - AB.CD}{AD}$$
$$= \frac{\mathrm{crd}(\alpha).\mathrm{crd}(180° - \beta) - \mathrm{crd}(\beta).\mathrm{crd}(180° - \alpha)}{120}$$

And by Pythagoras, we can find the values:

$$\mathrm{crd}^2(180° - \alpha) = CD^2 = AD^2 - AC^2 = 120^2 - \mathrm{crd}^2.\alpha$$
$$\mathrm{crd}^2(180° - \beta) = BD^2 = AD^2 - AB^2 = 120^2 - \mathrm{crd}^2.\beta$$

The chord expression above is equivalent to:

$$\sin\{(\alpha - \beta)/2\} = \sin(\alpha/2)\sin(90° - \beta/2) - \sin(\beta/2)\sin(90° - \alpha/2)$$

From which we can recognise 'our' trigonometric formula:

$$\sin(\theta - \phi) = \sin\theta \cos\phi - \cos\theta \sin\phi, \text{ using } \cos\gamma \text{ for } \sin(90° - \gamma).$$

Using the formula, and the known values for crd.72° = 70^P 32' 3" and crd.60° = 60^P (see above), we have the result:

$$\text{crd.}12° = 12^P \ 32' \ 36".$$

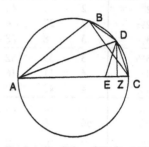

The chord of half an arc. Again, from another geometric argument, Ptolemy shows that when BC = crd.α is known then we also know CD = crd.α/2.

From the diagram, we have:

$CD^2 = AC.CZ$, with AC =120, CZ = (AC – AB)/2 and
AB = crd(180° – α).

In other words: $\text{crd}^2.\alpha/2 = 60[120 - \text{crd}(180° - \alpha)]$
and this gives us the standard formula:

Fig. 10.4

$$2\sin^2(\alpha/2) = 1 - \cos\alpha.$$

Using this, and starting from crd.12° = 12^P 32' 36", Ptolemy is able to calculate the values of

$$\text{crd.}6°, \text{crd.}3°, \text{crd.}3/2° = 1^P 34' \ 15" \text{ and crd.}3/4° = 0^P \ 47' \ 8".$$

The chord of 1°. These procedures cannot provide the values for crd.1° nor for crd.1/2°. In fact there is no algebraic expression using only square roots for expressing crd.α/3 in terms of crd.α. This is equivalent to the problem of finding a compass and ruler construction for the trisection of an angle which was unsolved at that time, and which we now know is not possible, except for particular values of α.

Ptolemy's approach was to find an approximation that was sufficiently good for his purposes. He argues geometrically that, given two arcs α and β, then:

$$\text{if } \alpha < \beta, \text{ then } [\text{crd.}\alpha \ / \ \text{crd.}\beta] > \alpha/\beta.$$

Such a conclusion had been used previously by Aristarchus of Samos, Euclid and Archimedes. The inequality can be restated as:

$$\text{crd.}\alpha/\alpha > \text{crd.}\beta/\beta$$

and is intuitively reasonable, as we can see that the greater the arc increases in size, the greater is the difference between the length of the arc and that of the chord (see diagram).

From this observation, we have:

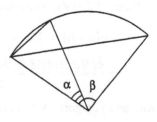

$$\frac{\text{crd.}3/4°}{3/4} < \frac{\text{crd.}1°}{1} < \frac{\text{crd.}3/2°}{3/2}$$

Fig. 10.5

From the already calculated values of crd.3/4° and crd.3/2° we find that the outer terms of this inequality are almost exactly equal, and so for all practical purposes crd.1° = (crd.3/2°)/(3/2) = 1^P 2' 50". And we can then deduce that crd.1/2° = 0^P 31' 25".

The table of chords. Ptolemy is now able to create his table of chords of angles, increasing in half angle steps, using to their best advantage the formulas for halving arcs, subtracting arcs and adding arcs, this last also derived from Ptolemy's theorem.

Interpolation as such has not appeared as part of this process, except perhaps for the derivation of crd.1° by interpolation of the function crd.α/α (a constant!) in the interval [3/4, 3/2]. It does appear however in the table drawn up by Ptolemy (see above): the first column gives the values of the arcs in half-degree intervals, the second has the corresponding values of the arcs and in a third column, headed sixtieths, the corresponding increase in value for a minute of angle over the half-degree interval. These values are calculated by taking a thirtieth of the half-degree increase, and so Ptolemy is inviting the reader to use linear interpolation for calculating the chords of smaller sub-divisions of angles down to minutes of arc.

We shall say something here briefly about tables that were established after Ptolemy's time. Indian mathematicians used the half chord, whose ratio to the radius corresponds to our sine. A number of different relations between multiples and sub-multiples of arcs were used to make up a table of 'sines' of arcs from 0 to 90°, tables which were also used for the purposes of astronomy. It is not known whether these tables were inspired by Ptolemy's table. What appears probable is that these Indian tables, as well as Ptolemy's, drew on a common source, namely that of Babylonian astronomy and its use of sexagesimal numbers.

In his treatise on astronomy *Khaṇḍa Khādyaka*, written in 665, Brahmagupta explains how to interpolate for sines of angles from values given in a table. In modern terms, he uses the finite differences $\Delta\sin pt$ and $\Delta\sin (p+1)t$ to approximate for $\sin p(t+\Delta t)$ by use of

$$\sin pt + \Delta t\{[\Delta\sin (p+1)t + \Delta\sin pt]/2 + \Delta t[\Delta\sin (p+1)t - \Delta\sin pt]/2\},$$

and this is equivalent to the order 2 Newton-Stirling difference formula (see Sections 10.3 and 10.4 below).

The Arab mathematicians were of course acquainted with the *Almagest*, and also some of the works of Indian mathematicians. The first table of sines was drawn up in a work on astronomy by al-Khwārizmī. The Indians had introduced the magnitudes corresponding to our sine and cosine and the Arabs drew up tables for the values of tangents and cotangents. These latter could be used to find the altitude α of the sun from the length ℓ of the shadow of a vertical side of a gnomon:

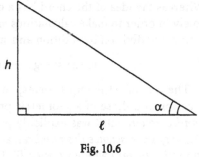

Fig. 10.6

$\ell = h \cot \alpha$, where h is the height of the side.

The Arabs raised the study of trigonometry, both plane and spherical, to the level of a true discipline. We can cite as evidence the *Revision of the Almagest* by al-Battānī (c. 920), the *Complete Book* by Abū-l-Wafā (c. 980) and the *Qānūn al-Mas'ūdī* by al-Bīrūnī (*c.* 1000) (see [33]). In his tables of sines, al-Bīrūnī used a quadratic interpolation which expressed in terms of differences would approximate sin $p(t + \Delta t)$ by:

$$\sin pt + \Delta t\, \Delta\sin (p - 1)t + (\Delta t)^2[\Delta\sin pt - \Delta\sin (p - 1)t].$$

Ibn Yūnis (*c.* 1000) establishes a formula that was essentially:

$$\cos\alpha \cos\beta = [\cos(\alpha + \beta) + \cos(\alpha - \beta)]/2.$$

Before the invention of logarithms, this relation was used in the 16th century, by Tycho Brahe in particular, for replacing multiplications by additions.

As for the problem faced by Ptolemy of determining a true value for crd.1°, we may recall the iterative method
used by al-Kāshī in the 15th century for solving the cubic equation $\sin 3° = 3x - 4x^3$, where he determines an approximate value for sin 1° from the known value of sin 3° (see Section 7.5).

Finally, we should mention the astronomer Rheticus, who used a team of workers over a period of many years to calculate the values for a table of sines in steps of 10" for the angles and to fifteen decimals for the sines.

10.2 Briggs and Decimal Logarithms

> The word algorithm itself is quite interesting; at first glance it may look as though someone intended to write logarithm but jumbled up the first four letters.
>
> Donald D. Knuth
> *The Art of Computer Programming* (vol. I, p.1)

Whereas the idea of the chord has a geometrical origin, logarithms were introduced solely in order to make calculations easier. Logarithms enable us to replace multiplication and division by addition and subtraction:

$$\log xy = \log x + \log y, \quad \log x/y = \log x - \log y.$$

The idea of setting up a correspondence between the terms of an arithmetic progression and those of a geometric progression occurs much earlier in Archimedes' *Sandreckoner*, but it was explicitly presented by Chuquet in the 15th century in his *Triparty en la science des nombres*, and then in the 16th century by Stiefel in his *Arithmetica integra* where we even find negative exponents in evidence. In such a correspondence, additions and subtractions in the one system correspond to multiplications and divisions in the other. Once tables of logarithms became available, the need for laborious calculations was finally at an end.

It was the Scotsman John Napier who developed the theory and practice of logarithms, or 'ratio numbers'. To obtain a sufficiently accurate approximation, he introduced intermediate progressions generated by a continuous movement, thereby linking continuity to approximation.

The Logarithme therefore of any sine is a number very neerely expressing the line, which increased equally in the meane time, whiles the line of the whole sine decreased proportionally into that sine, both motions being equal-timed, and the beginning equally swift. ([28], p. 515)

The reason why the calculation of the sine lay at the origin of the discovery of logarithms was because they were intended to make astronomical calculations easier. Napier, in fact, considered a circle with a radius $R = 10^7$ so that his logarithm turns out to be what we would now write as $R \ln (R/x)$, that is $10^7 \ln 10^7 - 10^7 \ln x$, using $\ln x$ for the natural or *Naperian* logarithm, as it was called by the French mathematician Lacroix. In 1614, Napier [19] published a table of logarithms, obtained by successive extractions of square roots, in his *Mirifici logarithmorum canonis Descriptio*, which appeared soon after in an English translation by Edward Wright under the title: *A Description of the Admirable Table of Logarithmes* (1616).

These tables enjoyed an immediate success and many other versions were soon published. In particular we should mention the tables of *Antilogarithms* published in Prague in 1620 by the Swiss astronomer Bürgi [3], calculated somewhat earlier and independently of Napier. We should also mention, of course, Henry Briggs who, concerned with navigation, was looking for quick and reliable calculation methods and so naturally became interested in Napier's logarithms. His first table *Logarithmorum Chilias Prima* appeared in 1617 but his most important work was the *Arithmetica Logarithmica* [2] which appeared in 1624. These new logarithms differed from those of Napier in that they used 0 and 1 for the logarithms of 1 and 10 respectively and so were decimal logarithms. The table of the logarithms of integers from 1 to 20 000 and from 90 000 to 101 000 was calculated to 14 decimal places.

While the practical advantages of logarithms were immediately apparent, theoretical interest in logarithms only developed later with the study of the logarithmic spiral and then the logarithmic curve itself, and with the problem of determining the area under the hyperbola. Preparing a table of logarithms presented the mathematician with certain technical difficulties: the tables of logarithms had to contain sufficient decimal places to allow calculations to be carried out to a sufficient degree of accuracy, yet the calculations of the logarithms themselves was very laborious. In fact, proceeding only from the basic definition of the logarithm, new ratios had to be inserted between known ones, and therefore successive square roots had to be calculated and this in a way that had to give the desired degree of accuracy. In order to reduce the considerable amount of work, and so only calculate a limited number of specific values, methods of interpolation were introduced.

In the extract that follows, Briggs proposes a first method for calculating the logarithm of a number, here log 2. It depends on the fact that, if the decimal form of 2^n contains k digits, that is if $10^{k-1} \leq 2^n \leq 10^k$, then $(k-1)/n \leq \log 2 \leq k/n$, and greater precision is achieved as n increases in value.

Henry Briggs

Arithmetica Logarithmica, London 1624.
From the French translation of the Latin by Adrien Vlacq, *Arithmétique logarithmétique*, Gouda, 1628, pp. 9–10.

Thus the two being multiplied by itself makes four, whose Index is 2, the four similarly makes 16, & its Index is 4, ditto 16 gives in like fashion 256, whose Index is 8, & then this last product being multiplied by 4 gives 1024, whose Index 10 is equal to the Indices of the factors: And the Characters of the last product are four in number. All that is given above is evident by the fourth and fifth Lemmas preceding. And here is the construction.

1	0	
2	1	
4	2	1
16	4	2 1st
256	8	3 Quaternion
1024	10	4
10,48576	20	7
109,95116,27776	40	13 2nd
12089,25819,61463	80	25 Quaternion
12676,50600,22823	100	31
16069,38044,25899	200	61
25822,49878,08685	400	121 3rd
66680,14432,87940	800	241 Quaternion
10715,08607,18618	1000	302
11481,30695,27407	2000	603
13182,04093,43051	4000	1205 4th
17376,62031,93695	8000	2409 Quaternion
19950,63116,87912	10000	3011
	Indices	Number of Characters

Now these four numbers 4, 16, 256, 1024, make the first Quaternion: after which yet another must be arranged; & for the start, the first number will be found by multiplying the last of the preceding Quaternion by itself; the second will be the square of the first, & the third the square of the second; the fourth is the product of the first multiplied by the third of this Quaternion: & these four have for Indices 20, 40 , 80, 100, the first containing seven Characters, the second 13, the third 25, & the fourth 31. Thus I achieve similarly the other Quaternions until the fourth number of the last one has for Index 1,00000,00000,0000 and contains 30102,99956,6399 Characters in number. And so about fifty and seven principle numbers have to be placed in the second arrangement*. Of which the first which is the side & the root of the others, is Two: & the last has the same Index, which has been given, in the first infinite progression by Ten.

For the third progression, it has almost been entirely described by the second Lemma, if the last product of the second should be known: which has to be by the first Lemma equal to the last product of this third progression. Of which are already known the number of Characters for the last products, which will not go beyond the Index of that product to the third rank, & that of Two in the first infinite progression, except for a single unit. For to find this

Index required at the third rank, it is only necessary to divide continually the last product by Ten the number given, whose Index has also similarly been given. And in dividing thus some number by Ten, the Quotient will be the Tenth of the dividend, that is, the same number only cut off at the right by a Character (like 4357 being divided by 10 gives the quotient 435) so that the number of the quotients or the divisions will only be one Character less than the number being divided itself, but the number of the quotients will show how far the number being divided is from the Unit, where the Index of this same number divided in this third arrangement, being the same as the Index of the first infinite progression, which has to be added to the side of the second rank, namely to the Two, & is the Logarithm of Two: the same as we are required to find, namely 30102,99956,6398.

*Not that they need to be written down in their entirety, for it is not necessary or possible; But it is sufficient for each one to write down some of the Characters on the left, in order to indicate the number of these Characters. Nonetheless, we have more precisely described the first Characters for each of these Quaternions, because otherwise we would not be able to find the true number of the Characters of the last product, as is shown in the numbers marked B, being reduced by a Unit from the last number of the first Quaternion 1024, it makes the last number of the third Quaternion a number of 301 Characters, which ought to be, as has been found above, 302.

B		
1023	10	4
1046529	20	7
1095223	40	13 1st
1199513	80	25 Quat.
1255332	100	31
1575870	200	61
2483363	400	121 2nd
6167075	800	241 Quat.
9718508	1000	301
	Index	Number of Characters

Briggs went on to propose another method for calculating logarithms. This was based on the following observations:

1. for $a > 1$, $\sqrt[2^n]{a}$ tends to 1 as n tends to infinity,
2. if $a = 1 + x$, where $x \ll 1$, then $\sqrt{a} \approx 1 + x/2$,
3. for $x \ll 1$ (say $x < 10^{-6}$), then $\log(1 + x) \approx kx$ where $k = 0.43429...$

Hence, in order to compute a table of decimal logarithms, it is sufficient to first calculate the logarithm of every prime number, since that will then give the logarithm of every integer, and so of every decimal number. For n sufficiently large,

$$\sqrt[2^n]{p} = 1 + x, \text{ where } x < 10^{-6}$$

whence:

$$\log p = 2^n \log(1 + x) \approx 2^n kx.$$

Essentially then, all that is needed is to extract square roots successively (see Chapter 7). Briggs was able to speed up his calculations by the use of a number of astute observations. For example, in the case of log 2, he notes that $2^{10}/1000 = 1.024$; it is therefore better to start extracting square roots from 1.024 which is closer to 1 than the number 2.

But all this is tedious and time consuming and Briggs developed other ingenious techniques, in particular calculating from finite differences, which is explored in the following sections. This technique is based on the fact that for practically all tabulated functions the finite differences of a certain order are zero or very small. This interpolation with the aid of finite differences was adapted in this case to the extraction of square roots. We shall not deal with this explicitly here, except to mention that this is not a case, as with Gregory later, of interpolation by functions.

Tables of logarithms, just as tables of trigonometric functions, have of course now fallen into disuse with the advent of electronic computers.

2 LOGARITHMS

	0	1	2	3	4	5	6	7	8	9	1	2	3	4	5	6	7	8	9
10	·0000	0043	0086	0128	0170	0212	0253	0294	0334	0374	4	8	12	17	21	25	29	33	37
11	·0414	0453	0492	0531	0569	0607	0645	0682	0719	0755	4	8	11	15	19	23	26	30	34
12	·0792	0828	0864	0899	0934	0969	1004	1038	1072	1106	3	7	10	14	17	21	24	28	31
13	·1139	1173	1206	1239	1271	1303	1335	1367	1399	1430	3	6	10	13	16	19	23	26	29
14	·1461	1492	1523	1553	1584	1614	1644	1673	1703	1732	3	6	9	12	15	18	21	24	27
15	·1761	1790	1818	1847	1875	1903	1931	1959	1987	2014	3	6	8	11	14	17	20	22	25
16	·2041	2068	2095	2122	2148	2175	2201	2227	2253	2279	3	5	8	11	13	16	18	21	24
17	·2304	2330	2355	2380	2405	2430	2455	2480	2504	2529	2	5	7	10	12	15	17	20	22
18	·2553	2577	2601	2625	2648	2672	2695	2718	2742	2765	2	5	7	9	12	14	16	19	21
19	·2788	2810	2833	2856	2878	2900	2923	2945	2967	2989	2	4	7	9	11	13	16	18	20
20	·3010	3032	3054	3075	3096	3118	3139	3160	3181	3201	2	4	6	8	11	13	15	17	19
21	·3222	3243	3263	3284	3304	3324	3345	3365	3385	3404	2	4	6	8	10	12	14	16	18
22	·3424	3444	3464	3483	3502	3522	3541	3560	3579	3598	2	4	6	8	10	12	14	15	17
23	·3617	3636	3655	3674	3692	3711	3729	3747	3766	3784	2	4	6	7	9	11	13	15	17
24	·3802	3820	3838	3856	3874	3892	3909	3927	3945	3962	2	4	5	7	9	11	12	14	16
25	·3979	3997	4014	4031	4048	4065	4082	4099	4116	4133	2	3	5	7	9	10	12	14	15
26	·4150	4166	4183	4200	4216	4232	4249	4265	4281	4298	2	3	5	7	8	10	11	13	15
27	·4314	4330	4346	4362	4378	4393	4409	4425	4440	4456	2	3	5	6	8	9	11	13	14
28	·4472	4487	4502	4518	4533	4548	4564	4579	4594	4609	2	3	5	6	8	9	11	12	14
29	·4624	4639	4654	4669	4683	4698	4713	4728	4742	4757	1	3	4	6	7	9	10	12	13
30	·4771	4786	4800	4814	4829	4843	4857	4871	4886	4900	1	3	4	6	7	9	10	11	13
31	·4914	4928	4942	4955	4969	4983	4997	5011	5024	5038	1	3	4	6	7	8	10	11	12
32	·5051	5065	5079	5092	5105	5119	5132	5145	5159	5172	1	3	4	5	7	8	9	11	12
33	·5185	5198	5211	5224	5237	5250	5263	5276	5289	5302	1	3	4	5	6	8	9	10	12
34	·5315	5328	5340	5353	5366	5378	5391	5403	5416	5428	1	3	4	5	6	8	9	10	11
35	·5441	5453	5465	5478	5490	5502	5514	5527	5539	5551	1	2	4	5	6	7	9	10	11
36	·5563	5575	5587	5599	5611	5623	5635	5647	5658	5670	1	2	4	5	6	7	8	10	11
37	·5682	5694	5705	5717	5729	5740	5752	5763	5775	5786	1	2	3	5	6	7	8	9	10
38	·5798	5809	5821	5832	5843	5855	5866	5877	5888	5899	1	2	3	5	6	7	8	9	10
39	·5911	5922	5933	5944	5955	5966	5977	5988	5999	6010	1	2	3	4	5	7	8	9	10
40	·6021	6031	6042	6053	6064	6075	6085	6096	6107	6117	1	2	3	4	5	6	8	9	10
41	·6128	6138	6149	6160	6170	6180	6191	6201	6212	6222	1	2	3	4	5	6	7	8	9
42	·6232	6243	6253	6263	6274	6284	6294	6304	6314	6325	1	2	3	4	5	6	7	8	9
43	·6335	6345	6355	6365	6375	6385	6395	6405	6415	6425	1	2	3	4	5	6	7	8	9
44	·6435	6444	6454	6464	6474	6484	6493	6503	6513	6522	1	2	3	4	5	6	7	8	9
45	·6532	6542	6551	6561	6571	6580	6590	6599	6609	6618	1	2	3	4	5	6	7	8	9
46	·6628	6637	6646	6656	6665	6675	6684	6693	6702	6712	1	2	3	4	5	6	7	7	8
47	·6721	6730	6739	6749	6758	6767	6776	6785	6794	6803	1	2	3	4	5	5	6	7	8
48	·6812	6821	6830	6839	6848	6857	6866	6875	6884	6893	1	2	3	4	4	5	6	7	8
49	·6902	6911	6920	6928	6937	6946	6955	6964	6972	6981	1	2	3	4	4	5	6	7	8
50	·6990	6998	7007	7016	7024	7033	7042	7050	7059	7067	1	2	3	3	4	5	6	7	8
51	·7076	7084	7093	7101	7110	7118	7126	7135	7143	7152	1	2	3	3	4	5	6	7	8
52	·7160	7168	7177	7185	7193	7202	7210	7218	7226	7235	1	2	2	3	4	5	6	7	7
53	·7243	7251	7259	7267	7275	7284	7292	7300	7308	7316	1	2	2	3	4	5	6	6	7
54	·7324	7332	7340	7348	7356	7364	7372	7380	7388	7396	1	2	2	3	4	5	6	6	7
	0	1	2	3	4	5	6	7	8	9	1	2	3	4	5	6	7	8	9

LOGARITHMS 3 LOG

	0	1	2	3	4	5	6	7	8	9	1	2	3	4	5	6	7	8	9
55	·7404	7412	7419	7427	7435	7443	7451	7459	7466	7474	1	2	2	3	4	5	5	6	7
56	·7482	7490	7497	7505	7513	7520	7528	7536	7543	7551	1	2	2	3	4	5	5	6	7
57	·7559	7566	7574	7582	7589	7597	7604	7612	7619	7627	1	2	2	3	4	5	5	6	7
58	·7634	7642	7649	7657	7664	7672	7679	7686	7694	7701	1	1	2	3	4	4	5	6	7
59	·7709	7716	7723	7731	7738	7745	7752	7760	7767	7774	1	1	2	3	4	4	5	6	7
60	·7782	7789	7796	7803	7810	7818	7825	7832	7839	7846	1	1	2	3	4	4	5	6	6
61	·7853	7860	7868	7875	7882	7889	7896	7903	7910	7917	1	1	2	3	4	4	5	6	6
62	·7924	7931	7938	7945	7952	7959	7966	7973	7980	7987	1	1	2	3	3	4	5	6	6
63	·7993	8000	8007	8014	8021	8028	8035	8041	8048	8055	1	1	2	3	3	4	5	5	6
64	·8062	8069	8075	8082	8089	8096	8102	8109	8116	8122	1	1	2	3	3	4	5	5	6
65	·8129	8136	8142	8149	8156	8162	8169	8176	8182	8189	1	1	2	3	3	4	5	5	6
66	·8195	8202	8209	8215	8222	8228	8235	8241	8248	8254	1	1	2	3	3	4	5	5	6
67	·8261	8267	8274	8280	8287	8293	8299	8306	8312	8319	1	1	2	3	3	4	5	5	6
68	·8325	8331	8338	8344	8351	8357	8363	8370	8376	8382	1	1	2	3	3	4	4	5	6
69	·8388	8395	8401	8407	8414	8420	8426	8432	8439	8445	1	1	2	2	3	4	4	5	6
70	·8451	8457	8463	8470	8476	8482	8488	8494	8500	8506	1	1	2	2	3	4	4	5	6
71	·8513	8519	8525	8531	8537	8543	8549	8555	8561	8567	1	1	2	2	3	4	4	5	5
72	·8573	8579	8585	8591	8597	8603	8609	8615	8621	8627	1	1	2	2	3	4	4	5	5
73	·8633	8639	8645	8651	8657	8663	8669	8675	8681	8686	1	1	2	2	3	4	4	5	5
74	·8692	8698	8704	8710	8716	8722	8727	8733	8739	8745	1	1	2	2	3	4	4	5	5
75	·8751	8756	8762	8768	8774	8779	8785	8791	8797	8802	1	1	2	2	3	3	4	5	5
76	·8808	8814	8820	8825	8831	8837	8842	8848	8854	8859	1	1	2	2	3	3	4	5	5
77	·8865	8871	8876	8882	8887	8893	8899	8904	8910	8915	1	1	2	2	3	3	4	4	5
78	·8921	8927	8932	8938	8943	8949	8954	8960	8965	8971	1	1	2	2	3	3	4	4	5
79	·8976	8982	8987	8993	8998	9004	9009	9015	9020	9025	1	1	2	2	3	3	4	4	5
80	·9031	9036	9042	9047	9053	9058	9063	9069	9074	9079	1	1	2	2	3	3	4	4	5
81	·9085	9090	9096	9101	9106	9112	9117	9122	9128	9133	1	1	2	2	3	3	4	4	5
82	·9138	9143	9149	9154	9159	9165	9170	9175	9180	9186	1	1	2	2	3	3	4	4	5
83	·9191	9196	9201	9206	9212	9217	9222	9227	9232	9238	1	1	2	2	3	3	4	4	5
84	·9243	9248	9253	9258	9263	9269	9274	9279	9284	9289	1	1	2	2	3	3	4	4	5
85	·9294	9299	9304	9309	9315	9320	9325	9330	9335	9340	1	1	2	2	3	3	4	4	5
86	·9345	9350	9355	9360	9365	9370	9375	9380	9385	9390	1	1	1	2	3	3	4	4	5
87	·9395	9400	9405	9410	9415	9420	9425	9430	9435	9440	0	1	1	2	2	3	3	4	4
88	·9445	9450	9455	9460	9465	9469	9474	9479	9484	9489	0	1	1	2	2	3	3	4	4
89	·9494	9499	9504	9509	9513	9518	9523	9528	9533	9538	0	1	1	2	2	3	3	4	4
90	·9542	9547	9552	9557	9562	9566	9571	9576	9581	9586	0	1	1	2	2	3	3	4	4
91	·9590	9595	9600	9605	9609	9614	9619	9624	9628	9633	0	1	1	2	2	3	3	4	4
92	·9638	9643	9647	9652	9657	9661	9666	9671	9675	9680	0	1	1	2	2	3	3	4	4
93	·9685	9689	9694	9699	9703	9708	9713	9717	9722	9727	0	1	1	2	2	3	3	4	4
94	·9731	9736	9741	9745	9750	9754	9759	9763	9768	9773	0	1	1	2	2	3	3	4	4
95	·9777	9782	9786	9791	9795	9800	9805	9809	9814	9818	0	1	1	2	2	3	3	4	4
96	·9823	9827	9832	9836	9841	9845	9850	9854	9859	9863	0	1	1	2	2	3	3	4	4
97	·9868	9872	9877	9881	9886	9890	9894	9899	9903	9908	0	1	1	2	2	3	3	4	4
98	·9912	9917	9921	9926	9930	9934	9939	9943	9948	9952	0	1	1	2	2	3	3	4	4
99	·9956	9961	9965	9969	9974	9978	9983	9987	9991	9996	0	1	1	2	2	3	3	3	4
	0	1	2	3	4	5	6	7	8	9	1	2	3	4	5	6	7	8	9

Citation: C. Godfrey & A. W. Siddons, *Four Figure Tables*, Cambridge, Cambridge University Press, 1913, new edition 1947, many reprints. This was a standard work for use in British schools, colleges and universities up until the early 1970s. The table has columns of differences on the right hand side to be used for interpolation.

10.3 The Gregory-Newton Formula

We saw earlier how mathematicians used linear interpolation to compute values and so draw up numerical tables. In fact, this was what Ptolemy proposed his reader to do to determine the chords of arcs in steps of minutes of arc. But linear interpolation is not always sufficiently accurate for obtaining good intermediate values. For this reason mathematicians developed a more refined process, through the use of finite differences, which we have only made reference to earlier with the examples of Indian and Arabic trigonometric tables and the logarithm tables produced by Briggs.

The initial steps in the process, using today's notation, are as follows. Suppose that a function f takes the values $y_0, y_1, \ldots y_n$ for equidistant values of a variable $x_k = x_0 + ck$, in other words $y_k = f(x_k)$ for $0 \leq k \leq n$. Then the sequence of values y_k can be associated with the sequence of first finite differences:

$$\Delta y_0 = y_1 - y_0, \quad \Delta y_1 = y_2 - y_1, \quad \Delta y_{n-1} = y_n - y_{n-1}.$$

To this new sequence Δy_k we can associate the sequence of its first finite differences, which consist of the second finite differences of the initial sequence, namely:

$$\Delta^2(y_k) = \Delta(\Delta y_k) = \Delta(y_{k+1} - y_k) = (y_{k+2} - y_{k+1}) - (y_{k+1} - y_k)$$
$$= y_{k+2} - 2y_{k+1} + y_k.$$

And so on ... The third finite differences lead to:

$$\Delta^3(y_k) = y_{k+3} - 3y_{k+2} + 3y_{k+1} - y_k.$$

And in general:

$$\Delta^r(y_k) = \sum_{0 \leq s \leq r} (-1)^{r-s} \binom{r}{s} y_{k+s}$$

We can set out these differences as a table like this:

$$
\begin{array}{cccccc}
y_0 & y_1 & y_2 & y_3 & y_4 & \cdots \\
\Delta y_0 & \Delta y_1 & \Delta y_2 & \Delta y_3 & \cdots & \\
\Delta^2 y_0 & \Delta^2 y_1 & \Delta^2 y_2 & \cdots & & \\
\Delta^3 y_0 & \Delta^3 y_1 & \cdots & & & \\
\Delta^4 y_0 & \cdots & & & &
\end{array}
$$

in which each element is found as the difference of the two neighbouring terms appearing in the row above. From the table, the following result is evident: the sum of the terms of any row is always equal to the difference between the extreme values of the preceding row.

In the short passage that appears below, which is an extract from a letter from Gregory to Collins, we find a formula that gives the interpolated value of a function f at a point $x = a$, using the finite differences of the function at a point x_0. The text may appear at first strange because the notation is unfamiliar. Here the letters $d, f, h, i, \ldots$ represent successive finite differences taken at a point x_0, and here $x_0 = 0$ and $f(0) = 0$. The expression $\alpha\gamma$ gives the value of the function at $x = a$. Transcribed into modern notation Gregory's formula would become the classic Gregory-Newton formula:

$$\alpha\gamma = f(a) - f(0) = \frac{a}{c}\Delta(0) + \frac{a(a-c)}{2c^2}\Delta^2 f(0) + \frac{a(a-c)(a-2c)}{6c^3{}^2}\Delta^3 f(0)$$
$$+ \frac{a(a-c)(a-2c)(a-3c)}{24c^4}\Delta^4 f(0)$$

Sir Thomas Harriot (c. 1610), in an unpublished text *De Numeris Triangularibus et inde De Progressionibus Arithmeticis Magisteria magna* (see [17]), had already explained the use of finite differences. He had used a table of differences which corresponds to the previous formula taken as far as fifth finite differences, and so was able to interpolate for fractional values of the variable the values of functions whose fifth differences were constant. It is probable that Briggs knew about Harriot's method, and he used a similar method to interpolate square roots from which he calculated his logarithms [2], to the extent that according to Goldstine [9], the Gregory-Newton formula ought more properly to be called the Harriot-Briggs formula. Gregory and Newton had all the same rediscovered the formula independently. The former alluded to it in 1670 in the letter to John Collins quoted below. The latter developed his ideas in a letter to John Smith dated 8 May 1675. We should also add that this formula was known to Mercator [18] and to Leibniz [16]. But it was Gregory who first gave it in the form of a polynomial.

The extract of the letter below is written in a mixture of English and Latin, the latter language being the more usual for the mathematical parts of correspondence.

James Gregory

Extract of the letter from James Gregory to John Collins, 23 November 1670.

in *The correspondence of Isaac Newton*, vol. 1 (1661–1675), p. 46, ed. H.W. Turnbull, Cambridge: Cambridge University Press, 1959.

... I remember ye did once desire of me my methode of finding the proportional parte in tables, which is this: *In figura octava mearum exercitationum, in recta *AI* imaginetur quælibet *Aα*, cui sit perpendicularis *αγ*, sitque *γ* in curva *ABH*, reliquis stantibus ut prius, sit series infinita $\frac{a}{c}, \frac{a-c}{2c}, \frac{a-2c}{3c}, \frac{a-3c}{4c}$, etc., fiatque productum ex

duobus primis seriei terminis $\frac{b}{c}$, ex tribus primis $\frac{k}{c}$, ex quatuor primis $\frac{l}{c}$, ex quinque primis

$\frac{m}{c}$, etc., in infinitum; erit recta $\alpha\gamma = \frac{ad}{c} + \frac{bf}{c} + \frac{kh}{c} + \frac{li}{c} +$ etc. in infinitum: this methode, as I

apprehend, is both more easie and universal than either Briges or Mercators, and also performed without tables.

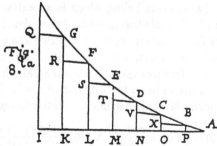

Figure 8 of *Exercitationes Geometricae*, London, 1668.

*[In figure 8 of my exercises, on the straight line *AI* is imagined any segment *Aα*, to which there is a perpendicular *αγ*, such that *γ* is on the curve *ABH*, the rest remaining as before; let there be the infinite series ... etc., and let the product of the first two terms of the series be ..., of the first three terms be ..., of the first four terms be ..., of the first five terms be ..., etc., *in infinitum*; the straight line *αγ* = ... etc. *in infinitum*: ...]

We have already remarked that these formulas can be transposed to:

$$f(a) - f(0) = \frac{a}{c}\Delta f(0) + \frac{a(a-c)}{2c^2}\Delta^2 f(0) + \frac{a(a-c)(a-2c)}{6c^3}\Delta^3 f(0)$$
$$+ \frac{a(a-c)(a-2c)(a-3c)}{24c^4}\Delta^4 f(0) + \ldots$$

If we put x_0 for 0, $f(x_0)$ for $f(0)$ and ct for a, the formula becomes:

$$f(x_0 + ct) = f(x_0) + t\Delta f(x_0) + \frac{t(t-1)}{2!}\Delta^2 f(x_0) + \frac{t(t-1)(t-2)}{3!}\Delta^3 f(x_0) + \ldots$$

When truncated at the nth step, this formula corresponds to the process of interpolation of the values of a function f, using a polynomial function of degree n having the same finite differences at x_0, up to order n, as the initial function f. This can be verified by substituting for t the successive values $0, 1, 2, \ldots, n$.

This last statement is the same as saying that the polynomial function takes the same values as f for the $n + 1$ points $x_0 + kc$, where $k = 0, 1, 2, \ldots, n$. The values $\Delta^k f(x_0)$ have in fact been found from the values $f(x_0 + ck)$ for $k = 0, 1, 2, \ldots, n$; but conversely, given the values $\Delta^k f(x_0)$ we may reconstruct the table given in the introduction and so, in particular, the values $f(x_0 + ck)$.

Today we use the notation $\binom{X}{n}$ for $\dfrac{X(X-1)\cdots(X-n+1)}{n!}$, so that the Gregory-Newton formula becomes:

$$f(x) = \sum_{k \geq 0} \Delta^k f(0) \binom{X}{n}$$

a formula that is analogous to the Taylor-Maclaurin formula:

$$f(x) = \sum_{k \geq 0} \frac{f^{(k)}(0)}{k!} X^k$$

where derivatives correspond to finite differences (see Section 13. 1).

We can add that if a polynomial of degree n is a 'good approximation' to a function that has to be interpolated, then the nth finite differences of that function will be 'almost' all equal. The nth finite differences of a polynomial of degree n are, of course, constant and equal to the nth derivative. And it is this result that acts as the guiding principle in deciding the order of differences to be considered when carrying out interpolations.

Another way of writing the Gregory-Newton formula is to put it in the nested form:

$$f(x_0 + ct) = f(x_0) + t\left\{\Delta f(x_0) + \frac{t-1}{2}\left[\Delta^2 f(x_0) + \frac{t-2}{3}\left[\Delta^3 f(x_0) + \ldots\right]\right]\right\}$$

In other words, we can start with the difference, of order n, which is assumed to be constant:

$$u_n = \Delta^n f(x_0),$$

and then we calculate in steps:

$$u_{n-1} = \Delta^{n-1}f(x_0) + u_n(t - n + 1)/n,$$

$$\cdots\cdots\cdots\cdots\cdots\cdots$$

$$u_k = \Delta^k f(x_0) + u_{k+1}(t - k)/(k + 1),$$

$$\cdots\cdots\cdots\cdots\cdots\cdots$$

$$f(x_0 + ct) = f(x_0) + u_1 t.$$

10.4 Newton's Interpolation Polynomial

It can safely be said that Newton was the mathematician who made the greatest contribution to the theory of interpolation and finite differences. He touched on this question for the first time, it appears, in a letter to John Smith on 8 May 1675 (see [25] pp. 14–21). He wished to offer his help in the construction of tables of square roots, cube roots and fourth roots of numbers from 1 to 10 000, calculated to 8 decimal figures. He proposed doing this by calculating the roots of integers, starting from 100 and in intervals of 100, to 10 decimal places, and then determining the roots of the intermediate numbers in 10s using interpolation with the help of the third finite differences.

Other references to interpolation can be found in Newton's works, for example in the *Methodus Differentialis*, which appeared in 1711 but had been written in 1676 [22], and also in the famous letter to Oldenburg of 24 October 1676 (see [24], vol. II, p. 119 and also [8]).

Here we shall consider a lemma taken from the *Principia* which presents the question in geometric context. The question is to "find a curved line of the parabolic kind which shall pass through any given number of points". A "line of a parabolic kind" does not here refer necessarily to a parabola, but to a curve of any polynomial function of degree n. Newton's solution uses a formula that is similar, except for notation, to that of Gregory. He distinguishes, however, between two situations. The first, as with Gregory, is where the sub-intervals are all equal, and he obtains the formula given in Section 10.3 above for the abscissa p. Newton labels the finite differences $b, c, d, e, \ldots$ and the polynomial coefficients, in terms of p, are written $p, q, r, s, t, \ldots$, where $q = p(p - 1)/2, r = p(p - 1)(p - 2)/6$, etc. (except for sign changes).

More interesting is the situation where the intervals are not equal, since this is often the case with results from experiments. Astronomical observations, for example, are often irregular since the weather may interfere with observations. Instead of the finite differences considered above, Newton considered what are called, since De Morgan ([6], p. 550), *divided differences*:

$$\underline{\Delta}(f(x_k)) = \frac{f(x_{k+1}) - f(x_k)}{x_{k+1} - x_k},$$

such that the integers 2, 3, 4 … corresponding to the progression along the interval, and dividing the polynomial quantities, disappear from the formula which, with present notation, becomes:

$$f(x_0 + t) = f(x_0) + t\Delta f(x_0) + t(t - 1)\Delta^2 f(x_0) + t(t - 1)(t - 2)\Delta^3 f(x_0) + \dots$$

Isaac Newton
Philosophiae naturalis principia mathematica, London, 1687.
tr. Andrew Motte (1729), revised by Florian Cajori, *Sir Isaac Newton's Mathematical Principles of Natural Philosophy and His System of the World,* Berkeley, California: University of California Press, 1947, pp. 499–500.

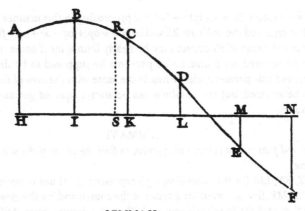

LEMMA V

To find a curved line of the parabolic kind which shall pass through any given number of points.

Let those points be A, B, C, D, E, F, &c., and from the same to any right line HN, given in position, let fall as many perpendiculars AH, BI, CK, DL, EM, FN, &c.

$$
\begin{array}{ccccc}
b & 2b & 3b & 4b & 5b \\
 & c \quad 2c & 3c & 4c & \\
 & d \quad 2d & 3d & & \\
 & & e \quad 2e & & \\
 & & f & &
\end{array}
$$

CASE 1. If HI, IK, KL, &c., the intervals of the points H, I, K, L, M, N, &c.; are equal, take b, $2b$, $3b$, $4b$, $5b$, &c., the first differences of the perpendiculars AH, BI, CK, &c.; their second differences, c, $2c$, $3c$, $4c$, &c.; their third d, $2d$, $3d$, &c., that is to say, so as AH – BI may be = b, BI – CK = $2b$, CK – DL = $3b$, DL + EM = $4b$, – EM + FN = $5b$, &c.; then $b - 2b = c$, &c., and so on to the last difference, which is here f. Then, erecting any perpendicular RS, which may be considered as an ordinate of the curve required, in order to find the length of this ordinate, suppose the intervals HI, IK, KL, LM, &c., to be units, and let AH = a, – HS = p, $\frac{1}{2}p$ into – IS = q, $\frac{1}{3}q$ into + SK = r, $\frac{1}{4}r$ into + SL = s, $\frac{1}{5}s$ into + SM = t; proceeding in this manner, to ME, the last perpendicular but one, and prefixing negative signs before the terms HS, IS, &c., which lie from S towards A; and positive signs before the terms SK, SL, &c., which lie on the other side of the point S; and, observing well the signs, RS will be = $a + bp + cq + dr + es + ft +$ &c.

CASE 2. But if HI, IK, &c., the intervals of the points H, I, K, L, &c., are unequal, take $b, 2b, 3b, 4b, 5b,$ &., the first differences of the perpendiculars AH, BI, CK, &c., divided by the intervals between those perpendiculars; $c, 2c, 3c, 4c,$ &c., their second differences, divided by the intervals between every two; $d, 2d, 3d,$ &c., their third differences, divided by the intervals between every three; $e, 2e,$ &c., their fourth differences, divided by the intervals between every four; and so forth; that is, in such manner, that b may be $= \dfrac{\text{AH} - \text{BI}}{\text{HI}}$, $2b = \dfrac{\text{BI} - \text{CK}}{\text{IK}}$,

$3b = \dfrac{\text{CK} - \text{DL}}{\text{KL}}$, &c., then $c = \dfrac{b - 2b}{\text{HK}}$, $2c = \dfrac{2b - 3b}{\text{IL}}$, $3c = \dfrac{3b - 4b}{\text{KM}}$, &c., then $d = \dfrac{c - 2c}{\text{HL}}$,

$2d = \dfrac{2c - 3c}{\text{IM}}$, &c. And those differences being found, let AH be $= a,$ $-$ HS $= p, p$ into $-$ IS $= q$, q into $+$ SK $= r, r$ into $+$ SL $= s, s$ into $+$ SM $= t$; proceeding in this manner to ME, the last perpendicular but one; and the ordinate RS will be $= a + bp + cq + dr + es + ft +$ &c.

COR. Hence the areas of all curves may be nearly found; for if some number of points of the curve to be squared are found, and a parabola be supposed to be drawn through those points, the area of this parabola will be nearly the same with the area of the curvilinear figure proposed to be squared: but the parabola can be always squared geometrically by methods generally known.

LEMMA VI

Certain observed places of a comet being given, to find the place of the same at any intermediate given time.

Let HI, IK, KL, LM (in the preceding fig.) represent the times between the observations; HA, IB, KC, LD, ME, five observed longitudes of the comet; and HS the given time between the first observation and the longitude required. Then if a regular curve ABCDE is supposed to be drawn through the points A, B, C, D, E, and the ordinate RS is found out by the preceding Lemma, RS will be the longitude required..

By the same method, from five observed latitudes, we may find the latitude at a given time.

If the differences of the observed longitudes are small, let us say 4 or 5 degrees, then three or four observations will be sufficient to find a new longitude and latitude; but if the differences are greater, as of 10 or 20 degrees, five observations ought to be used.

Thus, Lemma VI provides an immediate application of Lemma V. And we shall return to the corollary about calculating the arc under a curve, in Chap. 11 of this book when we look at approximate methods of integration.

Let us consider the construction of this interpolation polynomial. We shall use $x_0,$ $x_1, x_2, ..., x_n$ and x for the abscissas of the points H, I, K, L, M, N and S, and $y_0, y_1, y_2,$ $..., y_n$ and y for the ordinates of the points A, B, C, D, E, F, and R.

The successive divided finite differences will then be:

$$\Delta y_k = \frac{y_{k+1} - y_k}{x_{k+1} - x_k},$$

$$\Delta^2 y_k = \frac{\dfrac{y_{k+2} - y_{k+1}}{x_{k+2} - x_{k+1}} - \dfrac{y_{k+1} - y_k}{x_{k+1} - x_k}}{x_{k+2} - x_k}, \dots$$

and the formula will be written:

$$y = y_0 + \underline{\Delta} y_0 (x - x_0) + \underline{\Delta}^2 y_0 (x - x_0)(x - x_1) + \underline{\Delta}^3 y_0 (x - x_0)(x - x_1)(x - x_2) + \ldots$$

Truncated at the nth step, the right hand side is again a polynomial of degree n, which takes the same values as the function at $x_0, x_1, \ldots, x_n$. Written in this form, it is known as the Newton interpolation polynomial. Furthermore, these polynomials can be constructed one after the other; the interpolation polynomial of degree n, truncated at step $n - 1$ becomes the interpolation polynomial of degree $n - 1$.

These divided differences can themselves be found from a table and they do not depend on the order in which the points $x_0, x_1, x_2, \ldots, x_n$ are taken. Moreover:

$$\underline{\Delta}^n y_0 = \sum_{0 \leq k \leq n} \left[\frac{f(x_k)}{\prod_{i \neq k}(x_k - x_i)} \right].$$

Another way of using the formula is to write it in the nested form:

$$y = y_0 + (x - x_0)[\underline{\Delta} y_0 + (x - x_1)[\underline{\Delta}^2 y_0 + (x - x_2)[\underline{\Delta}^3 y_0 + \ldots]]].$$

In other words, we can start with the nth divided difference, taken to be constant:

$$v_n = \underline{\Delta}^n y_0,$$

and then calculate successively:

$$v_{n-1} = \underline{\Delta}^{n-1} y_0 + (x - x_{n-1}) v_n,$$

$$\ldots\ldots\ldots\ldots\ldots\ldots\ldots\ldots\ldots\ldots$$

$$v_k = \underline{\Delta}^k y_0 + (x - x_k) v_{k+1},$$

$$\ldots\ldots\ldots\ldots\ldots\ldots\ldots\ldots\ldots\ldots$$

$$y = y_0 + (x - x_0) v_1.$$

Up to now, we have always considered the interpolation points as ranged in an increasing order and the finite differences are therefore *forward differences* as compared with the *backward differences* which arise when proceeding in the opposite direction (for simplicity, we shall only consider finite differences):

$$\nabla y_k = y_k - y_{k-1}, \quad \nabla^2 y_k = y_k - 2 y_{k-1} + y_{k-2}, \ldots$$

The formula considered earlier corresponds to what is called Newton's forward difference algorithm, as compared to the backward difference algorithm, given by the formula:

$$f(x_0) + t \nabla f(x_0) + \frac{t(t+1)}{2!} \nabla^2 f(x_0) + \frac{t(t+1)(t+2)}{3!} \nabla^3 f(x_0) + \ldots$$

$$= \sum_{k \geq 0} \nabla^k f(x_0) \binom{t+k-1}{k}.$$

The forward difference algorithm is used when the value of f we want to determine by interpolation corresponds to a value of the variable x which lies close to the beginning of the interval over which the interpolation is carried out, whereas we use the backward difference algorithm for a value of x lying towards the end of the interval.

When the value of the variable lies close to the centre of the interval, the *Newton-Gauss algorithm* is more suitable:

$$f(x_0 + ct) = f(x_0) + t\Delta f(x_0) + \frac{t(t-1)}{2!}\Delta^2 f(x_0 - c) + \frac{(t+1)t(t-1)}{3!}\Delta^3 f(x_0 - c)$$
$$+ \frac{(t+1)t(t-1)(t-2)}{4!}\Delta^4 f(x_0 - 2c) + \frac{(t+2)(t+1)t(t-1)(t-2)}{5!}\Delta^5 f(x_0 - 2c) + \dots$$

The formula can be rewritten in terms of finite central differences, with a notation introduced by Sheppard in 1899 [27]:

$$\delta f(x_0) = f(x_0 + c/2) - f(x_0 - c/2)$$

where c is an interval increment, assumed constant.

In the same way as with the Gregory-Newton formula, we can also write the Newton-Gauss algorithm both in terms of forward differences and backward differences. A combination of the two leads to the *Newton-Stirling formula*:

$$f(x_0 + ct) = f(x_0) + t\left[\frac{\Delta f(x_0) + \Delta f(x_0 - c)}{2}\right] + \frac{t^2}{2!}\Delta^2 f(x_0 - c)$$

$$+ \frac{t(t^2 - 1^2)}{3!}\left[\frac{\Delta^3 f(x_0 - c) + \Delta^3 f(x_0 - 2c)}{2}\right]$$

$$+ \frac{t^2(t^2 - 1^2)}{4!}\Delta^4 f(x_0 - 2c)$$

$$+ \frac{t(t^2 - 1^2)(t^2 - 2^2)}{5!}\left[\frac{\Delta^5 f(x_0 - 2c) + \Delta^5 f(x_0 - 3c)}{2}\right] + \dots$$

This formula, which can be found in Newton's *Methodus Differentialis* ([22], Prop. iii, Case 1) was studied by Stirling in 1730 ([29], Prop. xx). It is useful when the value to be found by interpolation corresponds to a point $x = x_0 + ct$ very close to x_0. There are also a number of other variations of the Newton formulas, such as the Newton-Bessel formula (Newton [22], Prop. iii, Case ii) and the Laplace-Everett formula (Laplace [15], p. 15; Everett [7], p. 648).

10.5 The Lagrange Interpolation Polynomial

In one of his lectures at the Ecole Normale in Year III of the Revolution (1795), Lagrange referred to the question, treated previously by Newton, of using a 'parabolic curve' to interpolate a curve, that is interpolating a function by means of a polynomial function. Newton applied his result to the trajectory of a comet, whereas Lagrange was motivated by a practical problem of surveying:

From a point whose position is unknown, are observed three objects whose relative distances are known, and the three angles formed by the visual rays from the eye of the observer to these three have been determined. We wish to find the position of the observer with respect to these same objects. ([14], p. 280)

Lagrange proposed a solution by the use of trial and error; this gives rise to a curve of the errors through a finite number of points, the points corresponding to the trials; finally, to solve the problem, the curve needs to be approximated by a polynomial. This Lagrange interpolation polynomial, is none other than Newton's polynomial, but it is expressed in a different way, and this difference is interesting.

J.-L. Lagrange

From Leçons élémentaires de mathématiques donnés à l'Ecole Normale en 1795, *Journal de l'École Polytechnique*, VIIe et VIIIe cahiers, t. II (1812). *Oeuvres*, t. VII, Paris: Gauthier-Villars, 1877 (pp. 284–286).

Newton was the first to consider this Problem; this is the solution he gives for it:

Let P, Q, R, S, ... be the values of the ordinates y which correspond to the values $p, q, r, s, ...$ of the abscissas x; we shall have the following equations

$$P = a + bp + cp^2 + dp^3 + ...,$$
$$Q = a + bq + cq^2 + dq^3 + ...,$$
$$R = a + br + cr^2 + dr^3 + ...,$$
$$...,$$

the number of these equations being equal to those of the undetermined coefficients $a, b, c, ...$ Subtracting these equations one from the other, the remainders will be divisible by $q - p$, $r - q, ...,$ and we shall have, after dividing,

$$\frac{Q-P}{q-p} = b + c(q+p) + d(q^2 + qp + p^2) + ...,$$

$$\frac{R-Q}{r-q} = b + c(r+q) + d(r^2 + rq + q^2) + ...,$$

$$...$$

Let
$$\frac{Q-P}{q-p} = Q_1, \quad \frac{R-Q}{r-q} = R_1, \quad \frac{S-R}{s-r} = S_1, ...;$$

then we shall find in like manner, by subtraction and division,

$$\frac{R_1 - Q_1}{r-p} = c + d(r+q+p) + ...,$$

$$\frac{S_1 - R_1}{s-q} = c + d(s+r+q) + ...,$$

$$...$$

Similarly, let
$$\frac{R_1 - Q_1}{r-p} = R_2, \frac{S_1 - R_1}{s-q} = S_2, ...;$$

and we shall find
$$\frac{S_2 - R_2}{s-r} = d + ...,$$

$$.................................,$$

and so on.

In this way we shall find the values of the coefficients $a, b, c, \ldots$, starting with the last and substituting them in the general equation

$$y = a + bx + cx^2 + dx^3 + \ldots,$$

then, after cancellations, the following formula is derived, which it is easy to continue as far as is wished

$$y = P + Q_1(x - p) + R_2(x - p)(x - q) + S_3(x - p)(x - q)(x - r) + \ldots$$

But this solution can be reduced to much greater simplicity by the following consideration.

Since y has to take the values $P, Q, R, \ldots$, when x becomes $p, q, r, \ldots$, it is easy to see that the expression for y will be of the form

$$y = AP + BQ + CR + DS + \ldots,$$

where the quantities $A, B, C, \ldots$ must be expressible in terms of x in such a way that in putting $x = p$ we would have

$$A = 1, \quad B = 0, \quad C = 0, \ldots,$$

and likewise, on putting $x = q$, we would have

$$A = 0, \quad B = 1, \quad C = 0, \quad D = 0, \ldots,$$

and on putting $x = r$, we have similarly,

$$A = 0, \quad B = 0, \quad C = 1, \quad D = 0, \ldots, \text{ etc.},$$

from which it is easy to conclude that the values of $A, B, C, \ldots$ must be of the form

$$A = \frac{(x - q)(x - r)(x - s)\cdots}{(p - q)(p - r)(p - s)\cdots},$$

$$B = \frac{(x - p)(x - r)(x - s)\cdots}{(q - p)(q - r)(q - s)\cdots},$$

$$C = \frac{(x - p)(x - q)(x - s)\cdots}{(r - p)(r - q)(r - s)\cdots},$$

$$\ldots\ldots\ldots\ldots\ldots\ldots\ldots\ldots\ldots\ldots\ldots\ldots\ldots,$$

in taking as many factors, in the numerators and in the denominators, as there are given points of the curve, less one.

This is certainly the same polynomial since the *Lagrange polynomial* and Newton's polynomial take the same values for a number of values of the variable, equal to the degree of the polynomial plus one. Lagrange's ploynomial has the advantage of being expressed much more simply since its coefficients are given by a rule that is global and non-iterative. Newton's polynomial, on the other hand, turns out to be better suited to the case, which happens in practice, where, after having interpolated the function using n values, we wish to take account of a further value: here we simply have to add to the initial polynomial of degree $n - 1$ the term $\Delta^{n+1} y_0(x - x_0)\ldots(x - x_n)$, the coefficient $\Delta^{n+1} y_0$ being obtained by extending the table of differences by one line to the right. By contrast, with the Lagrange polynomial all the terms have to be changed.

However, if the points x_k are equidistant, $x_k = x_0 + ck$ ($k = 0, \ldots, n$), then P_n, the Lagrange polynomial of degree n, of the function f is given by the following formula where the $g_k(t)$ are independent of the point x_0 and of the increment step c and can so be tabulated:

$$P_n(x_0 + ct) = \sum_{0 \le k \le n} f(x_k) g_k(t) \text{ where } g_k(t) = \prod_{\substack{0 \le i \le n \\ i \ne k}} \frac{(t-i)}{(k-i)}$$

In 1821 Cauchy [4] proposed the use of rational functions, being more general than polynomials. Noting that there is only one rational function u, having numerator of degree p and denominator of degree q, taking the determined values u_i for $p + q + 1$ values x_i of the variable, he gives a general formula as a solution to the question. For example, for $p = q = 1$, he obtains:

$$u = \frac{u_0 u_1 \dfrac{x-x_2}{(x_0-x_2)(x_1-x_2)} + u_0 u_2 \dfrac{x-x_1}{(x_0-x_1)(x_2-x_1)} + u_1 u_2 \dfrac{x-x_0}{(x_1-x_0)(x_2-x_0)}}{u_0 \dfrac{x_0-x}{(x_0-x_1)(x_0-x_2)} + u_1 \dfrac{x_1-x}{(x_1-x_0)(x_1-x_2)} + u_2 \dfrac{x_2-x}{(x_2-x_0)(x_2-x_1)}}$$

It is interesting to note that in the text by Lagrange cited above, he wrote: "... it is clear that, for whatever proposed curve, the parabolic curve that is thus drawn will always be closer to it as the number of given points becomes greater, and their distance apart lessens." While such an assertion accords with intuition, it would be well perhaps to look at this a little more closely (see the Runge phenomenon, Section 13.3). There is also the problem of trying to determine limits to the error arising from using interpolation. Cauchy addressed this problem, and we shall look at how he dealt with it in Section 13.4.

Gaspard de Prony

It has to be said that Lagrange was himself more interested in the algebraic aspect of interpolation than in putting the theory to practical use. In contrast, one of his colleagues, Gaspard de Prony, *professeur* at the Ecole Polytechnique from 1794, held that mathematical theories ought to lead to practical calculations. To put this into practice, de Prony organised the production of tables of great accuracy by adopting the principle of division of labour in order "to manufacture logarithms as one manufactures pins". A first group of mathematicians chose the formulas (an adaptation of the calculation of finite differences); a second group of calculators prepared instructions and the lay-out of the tables (in such a·way that only additions and subtractions remained to be done); finally the calculations themselves were carried out by a third group of at least 60 to 80 persons (of which many were hairdressers reduced to unemployment by the Revolution). The calculations, completed in 1801, produced tables of logarithms from 1 to 200 000 with 14 places of decimals (as well as tables of sines). But the vast size of the tables made their publication far too expensive. In the event they were only printed partially, and even that a century later. It is possible that the 'industrial' approach adopted by Prony had an influence on Babbage and his project to carry out calculations accurately and repetitively by machine (see Grattan-Guinness [10]).

Charles Babbage and his Difference Machine

Babbage's desire to mechanise calculation arose from the exasperation he felt at the inaccuracies in printed mathematical tables. Scientists, engineers and the like relied on such tables to perform calculations requiring accuracy to more than a few figures. But the production of tables was tedious and prone to error at eachstage of preparation, from calculation to transcription to type-setting. Dionysius Lardner, a well known populiser of science, wrote in 1834 that a random selection of 40 volumes of mathematical tables incorporated 3,700 acknowledged errata, some of which themselves contained errors.

Babbage was both a connoisseur of tables and a fastidious analyst of tabular errors. He traced clusters of errors common to different editions of tables and deduced where pieces of loose type had been incorrectly replaced after falling out. On one occasion, he collaborated with John Herschel, the renowned British astronomer, to check two independently prepared sets of calculations for astronomical tables; the two men were dismayed by the numerous discrepancies. "I wish to God these calculations had been executed by steam!" Babbage exclaimed.

In 1822 Babbage built an experimental model intended to carry him towards his goal. He called his mechanical calculator a "difference engine" because it is based on the method of finite differences.

The size and complexity of the engine was monumental and roughly 25,000 parts were required. Babbage sought and obtained finance from the government to begin the project but progress was slow. Difficulties multiplied over a period of 20 years and the expenses incurred reached ten times the anticipated amount, and the project was finally abandoned. Moreover, during this time Babbage envisaged another machine, the Analytical Engine. Nonetheless, between 1847 and 1849, Babbage designed a new, and more accurate, difference machine, this one capable of providing up to 30 significant figures. This time, the government showed no interest.

In order to test if it would work, the Science Museum in London began the construction of the second Difference Machine in 1989. This was completed in 1991, in time for the bicentenary of Babbage's birth and, after some initial difficulties, the machine worked well. The machine weighed 3 tonnes, and was 3.3 m long, 2 m high and 45 cm in breadth. Babbage also envisaged a printer to go with the machine, but up to now this has not been made.

The Difference Machine uses the decimal system. Numbers are stored and operated on in eight vertical columns, each of which contains 31 engraved figure wheels. The machine contains a mechanism that prevents the wheels stopping between two integer values. Babbage claimed that his machine would either give a correct result or become stuck; it would never give a false result. It is now acknowledged that Babbage's two difference machines are the first machines to calculate automatically, while his Analytical Engine is the early precursor of our computers.

Taken from: Doron D. Swade, *Redeeming Charles Babbage's Mechanical Computer*, *Scientific American*, February 1991.

10.6 An Error Upper Bound

If we want the interpolated value to be a good approximation to the exact value, as is natural, then we need a way of determining an upper limit to the error that occurs when using interpolation. Cauchy provided formulas for calculating error upper bounds and this is the subject of the text by Cauchy that follows.

We should note that, without altering the x_i and the corresponding $f(x_i)$ values, we can change the other values of f. Doing this carries the risk that the unaltered interpolation polynomial ends up being a worse approximation to these other values unless it is known, a priori, that the function f possesses a certain regularity. We could, for example, require f to be n times differentiable in the interpolation interval, and this is precisely what Cauchy imposes on f in the text that follows.

Having shown that there exists an upper bound to the error then, as with approximations of numbers (Chapter 7), the procedure is not finite unless we predetermine the margin of error or the required degree of accuracy.

Following Ampère [1], Cauchy associates with every function its *interpolation functions*. If $a, b, c, ..., h, k$ designate the values of the variable x, he puts:

$$f(a,b) = \frac{f(a)-f(b)}{a-b}, \; f(a,b,c) = \frac{f(a,b)-f(a,c)}{b-c}, \; ...$$

If, further, we consider the divided finite differences associated with the sequence $a, b, c, ..., h, k$ we have, using the notation of Section 10.4:

$$f(a, b) = \Delta f(a), f(a, b, c) = \Delta^2 f(a), f(a, b, c, d) = \Delta^3 f(a), ...$$

This notation for the interpolation functions showing more clearly the symmetric nature of the relationship with the variables $a, b, c, ...$.
The Newton interpolation polynomial can therefore be written as:

$$P(x) = f(a) + (x - a)f(a, b) + (x - a)(x - b)f(a, b, c) + ...$$
$$+ (x - a)(x - b)...(x - h)f(a, b, c, ..., h, k).$$

On the other hand, we may also consider the quantities $f(a, x), f(a, b, x), ...$ as functions of x defined by the same formulas as above and, by definition, we shall have:

$$f(x) = f(a) + (x - a)f(a, b) + (x - a)(x - b)f(a, b, c) + ...$$
$$+ (x - a)(x - b)...(x - h)f(a, b, c, ..., h, x).$$

Furthermore, just as for a differentiable function, the mean-value theorem states the existence of a value u lying between a and x such that:

$$f(a, x) = f'(u).$$

Cauchy shows that, for a function that is n times differentiable, there exists a value u lying in every interval containing $a, b, c, ..., h$ and x such that:

$$f(a, b, c, ..., h, x) = f^{(n)}(u)/n!$$

A. L. Cauchy

Sur les fonctions interpolaires, *Comptes rendus de l'Académie des Sciences*, vol. XI (16 November 1840), p. 775. *Oeuvres*, series 1, vol. V, Paris: Gauthier-Villars, 1885, pp. 421–423.

[...]

$f(x)$ designating a given function of the variable x, and the letters

$$a, b, c, ..., h$$

representing the n particular values of this variable, the nth of the formulas (3) of the 1st Sect. gives

$$(1) \quad \begin{cases} f(x) = f(a) + (x-a)f(a,b) + (x-a)(x-b)f(a,b,c) + ... \\ \quad + (x-a)(x-b)(x-c)...(x-h)f(a,b,c,...,h,x). \end{cases}$$

If $f(x)$ is an integer function of degree n, then the interpolation function

$$f(a, b, c, ..., h, x),$$

being of degree zero with respect to x, simplifies just to a constant; and, using k for a new particular value of x, we shall have

$$f(a, b, c, ..., h, x) = f(a, b, c, ..., h, k),$$

and so

$$(2) \quad \begin{cases} f(x) = f(a) + (x-a)f(a,b) + (x-a)(x-b)f(a,b,c) + ... \\ \quad + (x-a)(x-b)(x-c)...(x-h)f(a,b,c,...,k). \end{cases}$$

Now, equation (2) will produce the expansion of $f(x)$ as a series of terms which will be proportional to the products of linear functions, and of degree, with respect to x, respectively equal to the different terms of the arithmetic progression

$$0, 1, 2, 3, ..., n.$$

To find a similar series in the case where the function $f(x)$ ceases to be an integer function, the last of the terms contained on the right hand side of the equation (2) must be neglected. Now, to determine whether this term can be neglected, it is important to know at least the limits of the error that would arise from its omission. This can be found, in a large number of cases, by the use of Theorem IX of Section 1. In effect, let us suppose that the quantities

$$a, b, c, ..., h$$

are found to be contained within the limits x_0, X, between which the function $f(x)$ remains continuous. The theorem concerned, will give for a value of x lying between these same limits,

$$f(a,b,c,...,h,x) = \frac{f^{(n)}(u)}{1.2...n};$$

and thus from equation (2) we derive

$$(3) \quad \begin{cases} f(x) = f(a) + (x-a)f(a,b) + (x-a)(x-b)f(a,b,c) + ... \\ \quad + (x-a)(x-b)(x-c)...(x-h)\dfrac{f^{(n)}(u)}{1.2...n}, \end{cases}$$

where u stands for a median quantity of the values attributed to

$$a, b, c, ..., h, x.$$

If, with the variable x and the function $f(x)$ being real, A and B stand for the least and the greatest of the values that can be attained by the derived function

$$f^{(n)}(x),$$

while x is allowed to vary between its limits x_0, X, the last term of the right hand side of the formula (3) will itself be contained within limits, equivalent to the products formed by the ratio

$$\frac{(x-a)(x-b)(x-b)...(x-h)}{1.2...n}$$

with the coefficients A and B. Hence the greatest numerical values of these two products will be the limit of the error that would be made in neglecting the term to which it applies.

Thus, when the function f to be interpolated is of class C^n in the interpolation interval I, the Lagrange polynomial P_{n-1} of degree $n-1$ corresponding to the points x_1, $x_2, ..., x_n$ satisfies

$$\left|f(x) - P_{n-1}(x)\right| \le \left|(x - x_1)(x - x_2)...(x - x_n)\right| \sup_{t \in I}\left|f^{(n)}(t)\right| / n!$$

This can be generalised in the following way. We can require of the interpolation polynomial, not only that it should take the given values at the given points, but also that its derivatives, up to a certain order, should also take given values at these points. In other words, in terms of curves, we can require that the graph of the polynomial should not only pass through given points $(x_i, f(x_i))$ of the graph of the function f to be interpolated, but also that the nature of the contacts with the graph of f at these points should be of a given order α_i. This was proposed by Hermite in 1878 [12].

The upper bound for the error made using a *Hermite polynomial* – it is also called a *Lagrange-Sylvester polynomial* – is of the same type as for the ordinary Lagrange polynomial. If the interpolation points are $x_1, x_2, ..., x_k$ and the required orders of contact are given by $\alpha_1, \alpha_2, ..., \alpha_k$, then the corresponding Hermite interpolation polynomial P_{n-1} of degree $n-1$, where $n = \Sigma(\alpha_i + 1)$, will satisfy:

$$\left|f(x) - P_{n-1}(x)\right| \le \left|(x - x_1)^{\alpha_1}(x - x_2)^{\alpha_2}...(x - x_k)^{\alpha_k}\right| \sup_{t \in I}\left|f^{(n)}(t)\right| / n!$$

10.7 Neville's Algorithm

We have already made the remark that the Lagrange interpolation polynomial is ill-suited to an iterative calculation that goes from n to $n + 1$ points. We can quote Cauchy on this point who, in a memoir of 1837 on interpolation, wrote:

and what is most annoying, is that the approximate values of different orders corresponding to different cases where we retain just one term, then two terms, then three terms ... of a series are found by calculations that are almost independent of each other, such that each new approximation, far from being easier than the ones that come before it, require on the contrary more time and more work.

E. H. Neville proposes here a procedure which remedies this inconvenience. His approach is this: being given $n + 1$ points, it is easy to construct the corresponding interpolation polynomial by using two interpolation polynomials corresponding to two sub-sets of n points having $n - 1$ points in common.

E. H. Neville

Iterative interpolation, *The Journal of the Indian Mathematical Society*, vol. 20 (1933), 87–120.

6. Iterative Computation of Lagrangian Approximations.

From two values u_a, u_b of $u(x)$ we have one first approximation L_{ab} to $u(X)$, given by

$$L_{ab} = \frac{b-X}{b-a}u_a + \frac{X-a}{b-a}u_b .$$

This is the value at X of the linear function $l_{ab}(x)$ which agrees with $u(x)$ at a and b. If we are to combine with L_{ab} a different first approximation, the latter must come from a linear function which agrees with $u(x)$ at one of the two points a, b. If this second function is $l_{bc}(x)$, the corresponding first approximation L_{bc} is given by

$$L_{bc} = \frac{c-X}{c-b}u_b + \frac{X-b}{c-b}u_c ,$$

and from L_{ab} and L_{bc} we have an approximation M_{abc} of the second order; the points at which $l_{ab}(x)$ and $l_{bc}(x)$ differ being a and c, we have to interpolate between L_{ab} and L_{bc} by regarding X as belonging to the interval ac:

$$M_{abc} = \frac{c-X}{c-a}L_{ab} + \frac{X-a}{c-a}L_{bc} .$$

Similarly a fourth tabulated value u_d gives another first approximation,

$$L_{cd} = \frac{d-X}{d-c}u_c + \frac{X-c}{d-c}u_d ,$$

another second approximation,

$$M_{bcd} = \frac{d-X}{d-b}L_{bc} + \frac{X-b}{d-b}L_{cd} ,$$

and an approximation of the third order,

$$N_{abcd} = \frac{d-X}{d-a}M_{abc} + \frac{X-a}{d-a}M_{bcd} .$$

It follows from our fundamental theorem that the number N_{abcd} is precisely the value at X of the cubic which agrees with $u(x)$ at a, b, c, d.

More generally, the value at x of the polynomial of degree n which agrees with $u(x)$ at $n + 1$ distinct points is obtained by a set of $n(n + 1)/2$ linear interpolations. We have already remarked that computation from the explicit expression for this polynomial, as given by Lagrange, is laborious if not impracticable. Linear interpolation, however, is one of the most rapid of numerical operations. What is required for computation is a process; whether or not a general formula corresponds to the process is irrelevant.

Let us consider the last paragraph, which concludes that an interpolation of degree n can be obtained by using $n(n + 1)/2$ linear interpolations. The procedure can be set out as a table similar to the one that is used for finite differences of order n for Newton's interpolation polynomial. Consider the table below where x_0, x_1, ..., x_n stand for the interpolation points and where x represents either a value of the variable, or the variable itself:

$$
\begin{aligned}
f(x_0) &= P_{0,0} \;\; P_{1,0} \;\; P_{2,0} \;\; P_{3,0} \;\; P_{4,0} \;\; P_{5,0} \\
f(x_1) &= P_{0,1} \;\; P_{1,1} \;\; P_{2,1} \;\; P_{3,1} \;\; P_{4,1} \\
f(x_2) &= P_{0,2} \;\; P_{1,2} \;\; P_{2,2} \;\; P_{3,2} \\
f(x_3) &= P_{0,3} \;\; P_{1,3} \;\; P_{2,3} \\
f(x_4) &= P_{0,4} \;\; P_{1,4} \\
f(x_5) &= P_{0,5}
\end{aligned}
$$

Each column is filled in turn, the term $P_{i+1,j}$ being found from two terms of the previous column, namely $P_{i,j}$ of the same line, and $P_{i,j+1}$ of the line below, according to the formula:

$$
P_{i+1,j} = \frac{(x_{j+i+1} - x)P_{i,j} - (x_i - x)P_{i,j+1}}{x_{j+i+1} - x_i}
$$

From the text by Neville, we can see that $P_{i,j}$ represents the interpolated value at x using the values $x_j, x_{j+1}, ..., x_{j+i}$ ($i \geq 0, j \geq 0, i + j \leq n$), or we can say that $P_{i,j}$ is the interpolation polynomial of degree i based on the $i + 1$ points beginning with row j.

The passage from one column to the next can also be defined in other ways. Thus, in Aitken's algorithm, the element $P_{i,j}$ of the table ($i \geq 0, j \geq 0, i + j \leq n$) represents the Lagrange interpolation polynomial of degree i based on the points $x_0, x_1, ..., x_{i-1}$, x_{i+j}. In this case, the passage from one column to the next is by use of the formula:

$$
P_{i+1,j} = \frac{(x_{j+i+1} - x)P_{i,0} - (x_i - x)P_{i,j+1}}{x_{j+i+1} - x_i}
$$

The CORDIC algorithm

Until recently the values of the usual functions given by calculators have been obtained by the use of approximate functions, most usually polynomials or rational functions. This type of calculation often requires a very large number of multiplications and divisions to be carried out on numbers possessing a large number of digits; all this is costly in terms of time and space.

In 1959, J. Volder [30] proposed a method for calculating trigonometric functions that used only additions, subtractions and shifts (a shift to the left or to the right corresponding to a multiplication or a division by a power of the base to which the number was expressed). This was the so called CORDIC algorithm (Coordinate Rotations on a Digital Computer) which practically revolutionised the calculations of values of functions. We give a brief explanation of how it works in the case of the tangent function.

Suppose we wish to calculate the value of $\tan \alpha$. First, we put into the memory bank the values of the angles whose tangents have simple fixed values; for calculating in base 10, we take those angles α_i such that $\tan \alpha_i = 10^{-i}$ for $1 \leq i \leq k$ where the integer k changes according to the degree of accuracy required.

Then the angle α is expressed in the form of a sum $\alpha = \sum\limits_{1 \leq i \leq k} d_i a_i$ where the d_i are natural numbers.

Finally, $\tan \alpha$ is calculated by successive approximations, noting that if $\tan \beta = \dfrac{y}{x}$, then $\tan(\beta + \alpha_i) = \dfrac{y'}{x'}$, where $x' = x - y \tan \alpha_i$ and $y' = y + x \tan \alpha_i$.

It is easy to formulate this algorithm in such a way as to carry out the two approximations of α and $\tan \alpha$ together:

$$
\begin{aligned}
&x \leftarrow 1,\ y \leftarrow 0,\ t \leftarrow 10 \text{ [Initialisation]} \\
&\text{For } i = 0 \text{ to } k \\
&\qquad t \leftarrow t/10 \\
&\qquad \text{while } \alpha > \alpha_i,\ \text{do } \alpha \leftarrow \alpha - \alpha_i \\
&\qquad\qquad x \leftarrow x - ty \\
&\qquad\qquad y \leftarrow y + tx \\
&\text{Set } \tan \alpha = y/x
\end{aligned}
$$

What is interesting about this process is that the multiplications by t only involve shifts of i digits to the right and, with the exception of the final division y/x, the only operations to be done are additions and subtractions. Thus the memory bank of some values of the function is sufficient to ensure a rapid calculation of other values to the required degree of accuracy. In 1971, J. Walther [31] extended the method to the calculations of logarithms, exponentials and square roots. In fact, all of this recalls the principle behind certain of the techniques used by Briggs for his calculations of decimal logarithms

Bibliography

[1] Ampère, A.-M., *Annales de Gergonne*, vol. 26 (1826), 329.

[2] Briggs H., *Arithmetica Logarithmica sive Logarithmorum Chiliades Triginta*, London, 1624: French tr. Adrien Vlacq, *Algorithmique logarithmétique*, Gouda, 1628.

[3] Bürgi, J. *Arithmetische und geometrische Progress Tabulen, sambt gründlichem unterricht wie solche nützlich in allerley Rechnungen zugerbrauchen und verstanden werden sol*, Prague, 1620.

[4] Cauchy, L.-A., Sur la formule de Lagrange relative à l'interpolation, *Cours d'Analyse de l'Ecole Royale Polytechnique*, note V, 1821. Oeuvres, series 2, vol. III (pp. 429–433), Paris: Gauthier-Villars, 1897.

[5] Cauchy, L.-A., Sur les fonctions interpolaires, *Comptes rendus de l'Académie des Sciences*, vol. XI (16 novembre 1840), p. 775. Oeuvres, series 1, vol. V (pp. 409–424), Paris: Gauthier-Villars, 1885.

[6] De Morgan, A., *The Differential and Integral Calculus*, London, 1842.

[7] Everett, J. D., On a central-difference interpolation formula, *British Association Reproductions* (1900), 648–650.

[8] Fraser, D. C. *Journal of the Institute of Actuaries*, vol. 51 (1918-19), 77, 211.

[9] Goldstine, H. H., *A History of Numerical Analysis, from the 16th through the 19th Century*, New York: Springer, 1977.

[10] Grattan-Guinness, I., Work for the Hairdressers: The Production of de Prony's Logarithmic and Trigonometric Tables, *Annals of the History of Computing*, vol. 12 (1990), 177–185.

[11] Gregory, J., *Exercitationes Geometricae*, London, 1668.

[12] Hermite, Ch., Extrait d'une lettre de M. Ch. Hermite à M. Borchardt sur la formule d'interpolation de Lagrange, *Journal de Crelle*, vol. 84 (1878), 70–79.

[13] Knuth, D. E., *The Art of Computer Programming*, vol. 1, Fundamental algorithms, Reading, Massachussets: Addison-Wesley, 1968; 2nd ed. 1973.

[14] Lagrange, J. L., Leçons élémentaires de mathématiques données à l'Ecole Normale en 1795, *Journal de l'Ecole Polytechnique*, VIIe et VIIIe cahiers, vol. II (1812). *Oeuvres*, vol. VII, Paris: Gauthier-Villars, 1877.

[15] Laplace, P.-S., *Théorie analytique des probabilités*, 1812.

[16] Leibniz, G.W., Lettre à Oldenburg du 3 février 1672, in *Commercium Epistolicum D. Johannis Collins*, London, 1712.

[17] Lohne, J. A., Thomas Harriot als Mathematiker, *Centaurus*, vol. 11 (1965), 19–45.

[18] Mercator, N., *Logarithmotechnia*, London, 1668.

[19] Napier, J., *Mirifici Logarithmorum Canonis descriptio*, Edinburgh, 1614.

[20] Naux, Ch., *Histoire des logarithmes de Neper à Euler*, Paris: Blanchard, 1966.

[21] Neville, E. H., Iterative interpolation, *The Journal of the Indian Mathematical Society*, vol. 20 (1933), 87–120.

[22] Newton, I., *Methodus Differentialis*, 1711: *The Mathematical Works of Isaac Newton*, ed. D.T. Whiteside, vol. II (pp. 165–173), Cambridge: Cambridge University Press, 1968.

[23] Newton, I., *Philosophiae naturalis Principia mathematica*, London, 1687. English tr. Andrew Motte (1729), revised by Florian Cajori, *Sir Isaac Newton's Mathematical Principles of Natural Philosophy and His System of the World*, Berkeley, California: University of California Press, 1947.

[24] Newton, I., *The correspondence of Isaac Newton*, ed. H.W. Turnbull, Cambridge: Cambridge University Press, 1959.

[25] Newton, I., *The mathematical papers of Isaac Newton*, ed. D.T. Whiteside, vol. IV (1674–1684), Cambridge: Cambridge University Press.

[26] Ptolemy, *Syntaxis*, i 10, ed. Heiberg i, §§ i. 31. 7–32. 9. Tr. Ivor Thomas, *Greek Mathematical Works*, 1941, 1993, Cambridge, Massachussets and London: Loeb Classical Library, Harvard University Press, pp. 423–445.

[27] Sheppard, W. F., *Proceedings of the London Mathematical Society*, vol. 31 (1899), 459.

[28] Smith, D. E., *History of Mathematics*, vol. II, 1925, reprnt. New York: Dover, 1959.

[29] Stirling, J., *Methodus Differentialis: sive Tractatus de Summatione et Interpolatione Serierum Infinitarum*, London, 1730.

[30] Volder, J., The CORDIC trigonometric computing technique, *IRE Transactions on Computers*, vol. Ec-8 (Sept. 1959), 330–334.

[31] Walther, J., A Unified Algorithm for Elementary Functions, *Spring Joint Computer Conference Proceedings*, vol. 38 (1971), 379–385.

[32] Whittaker, E. T. and G. Robinson, G., *The calculus of observations, A treatise on Numerical Mathematics*, 1924; 2nd ed. London and Glasgow: Blackie & Son, 1932.

[33] Youschkevitch, A., *Les mathématiques arabes (VIIIe-XVe siècles)*, Paris: Vrin, 1976.

11 Approximate Quadratures

The origin of the meaning of quadrature, or squaring, was to find a square having the same area as that of some given geometrical figure. The quadrature of a circle, or 'squaring the circle', is the best known example. The meaning of quadrature quickly began to take on an extended meaning of comparing the areas of two plane figures, one of which was already known – for example, a figure bounded by straight lines. Thus quadrature came to mean establishing a ratio between two plane figures, rather than measuring an area in the proper sense of the word.

To prove quadratures, the Greek geometers used a 'method of exhaustion', as it became later known, which involved the double use of 'reductio ad absurdum'. This is how Archimedes proved that the area of a circle is equal to the area of a right-angled triangle [1], and that the area of a parabolic segment (that is the area between a parabola and a chord) can also be shown to be equal to that of a triangle. The way that he divided the figure into small parts is similar to the methods that were used later to obtain approximate areas of figures. But this was not Archimedes' purpose: his intention was to establish equalities, not approximations – and that is why we shall not pursue his methods here.

This was also the case with the 9th century Arab mathematicians like Banū Mūsā, Thābit ibn Qurra, Ibrāhīm ibn Sinān, and the 10th century mathematicians Ibn al-Haytham and al-Mutaman. From a partial knowledge of Archimedes' work, and by using the method of exhaustion, they found the quadratures or cubatures of a parabolic segment, of a sphere and of a section of a parabaloid. For example, Thābit ibn-Qurra divided the parabola into trapeziums, whose bases were chosen in such a way as to yield series that he already knew how to handle.

In the 17th century, geometers again explored other quadratures but objected that the method of exhaustion did not provide a suitable means for finding these quadratures. To remedy this fault, they introduced the 'method of indivisibles'. Cavalieri, for example, developed a method for comparing the areas of two figures, by comparing their indivisible parts, that is segments of each figure found by dividing them up with lines parallel to a given straight line. In 1639, Cavalieri used a method which, in its geometrical form, corresponds to what we now know as Simpson's method [2].

The invention of the infinitesimal calculus by Newton and Leibniz, at the end of the 17th century, established the connection between finding tangents and finding areas, and provides an algorithm for an exact calculation of the area under a curve, provided that we know what is now called a 'primitive' of the function that describes the curve. But many functions do not satisfy this requirement. Also we may only know the values of the functions for some restricted number of points. What is needed, therefore, is a formula that will give us an approximation to the area, using only these limited number of known values. What had been learned from work on interpolation can be used to improve the methods: thus, if the area under a curve is divided up into strips of equal width, then the approximation is usually better if we replace the curve, not by straight line segments, but by 'parabolic' curves – that is, curves given by polynomial expressions. It was by using these 'parabolas', in the 17th century, that formulas for areas were found by Gregory, and then by Newton, Cotes and Stirling (Sections 11.1, 11.2, 11.3 and 11.4 below), although these latter did not explain their methods of calculation. We have attempted to provide these, based on the interpolation formulas that each of these mathematicians had respectively established. These formulas were given in Chapter 10, which needs to be referred to before reading what follows. However, the method has its limitations when the degree of the polynomial used for interpolation increases.

Independently of each other, and in different contexts, Euler, exploring sums of series, and Maclaurin, working on quadratures, each derived a formula which could be used to find the difference between the area under a curve over an interval, and the sum of the series of the values of the function at equidistant points of the interval. The sum of the series can be interpreted as finding an approximation to the area by the summation of rectangles or trapeziums (see Chapter 14).

To move on to the 18th century, we find Gauss having the idea of choosing the abscissas for the interpolation points in such a way that, for a given number of points, we can obtain a quadrature formula that is exact for polynomials of the highest possible degree. His explanation is not easy to follow and, for that reason, we reproduce below Jacobi's explanation of the method (Section 11.6). The Russian Chebyshev became interested in another problem: the absolute value of the coefficients in the Newton-Cotes formula, or in the Gauss formula, can sometimes be so large as to generate significant errors when we only have approximate values for the function. He determined a choice of points that gives a formula for which all the coefficients will be equal (Section 11.7).

11.1 Gregory's Formula

In 1670 James Gregory, professor of mathematics at St. Andrews and a member of the Royal Society of London, wrote to James Collins, Secretary to the Society. The letter mentions topics he was working on, and compares his work with that of his contemporaries – Newton, Barrow, Briggs, Mercator, Kepler, Huygens, etc. The letter is written in English, with the mathematical results given in Latin, which was common at the time, but the letter does not contain the proofs of the results since these were considered to be either 'lengthy' or 'tedious'. Gregory was interested in the series expansion of functions, and their applications to geometrical problems, such as the rectification of curves.

In the extract given here, Gregory makes a correction to a formula he had used for constructing tables of logarithms, trigonometric tables and for finding areas, results which he had published in 1668 in his *Geometrical Exercises* [7]. The new formula, written with the notation of Chapter 10, becomes:

$$\int_0^c f(x_0 + u)\, du = \frac{c}{2}\Delta f_0 - \frac{c}{12}\Delta^2 f_0 + \frac{c}{24}\Delta^3 f_0 - \frac{19c}{720}\Delta^4 f_0 + \frac{3c}{160}\Delta^5 f_0 - \frac{863c}{60480}\Delta^6 f_0 + \dots$$

J. Gregory

Letter from Gregory to Collins, 23 November, 1670
The Correspondence of Isaac Newton, vol. I, ed. H. W. Turnbull,
Cambridge: Cambridge University Press, 1959, pp. 45–49.

In the end of my Geometrical exercitations I fail exceedinglie, for wher I speak of trilinea quadratica, cubica, quadrato quadratica, etc., I sould say trilinea æquationibus quadraticis, cubicis, biquadraticis, inservientia; and hence in place of any thing I have described there; ponendo $AP = PO = c$, $PB = d$, primam ex differentiis primis $= f$, ex differentiis secundis $= h$, ex differentiis tertiis $= i$, ex differentiis quartis $= k$, ex differentiis quintis $= \ell$ et omnes differentias affici signo $+$; erit $ABP = \dfrac{dc}{2} - \dfrac{fc}{12} + \dfrac{hc}{24} - \dfrac{19ic}{720} + \dfrac{3kc}{164} - \dfrac{863\ell c}{60480} +$ etc. in infinitum. how ever the differencies be affected, I can easilie square the figure, and by this means al figures imaginable; ...

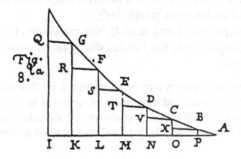

Figure 8 of Gregory's *Exercitationes Geometricae*, London, 1668

This passage immediately precedes the one given in Section 10.3 above. In the absence of any explanations by Gregory, we can suppose that the formula was derived by the term by term integration of the interpolation formula, since this was a procedure much favoured by the 'English School'. We can transcribe the interpolation formula as:

$$f(x_0 + ct) - f(x_0) = t\Delta f_0 + \frac{t(t-1)}{2!}\Delta^2 f_0 + \frac{t(t-1)(t-2)}{3!}\Delta^3 f_0 + \dots$$

Thus we obtain:

$$\text{area}(ABP) = \int_0^c f(x_0 + u)\, du = c\int_0^1 f(x_0 + ct)\, dt$$

$$= c\left[\Delta f_0 \int_0^1 t\, dt + \Delta^2 f_0 \int_0^1 \frac{t(t-1)}{2!}\, dt + \Delta^3 f_0 \int_0^1 \frac{t(t-1)(t-2)}{3!}\, dt + \Delta^4 f_0 \int_0^1 \frac{t(t-1)(t-2)(t-3)}{4!}\, dt + \dots\right]$$

$$= \frac{c}{2}\Delta f_0 - \frac{c}{12}\Delta^2 f_0 + \frac{c}{24}\Delta^3 f_0 - \frac{19c}{720}\Delta^4 f_0 + \frac{3c}{160}\Delta^5 f_0 + \dots$$

We detect here an error in the original text, where 160 should be read for 164.

This formula has a geometric interpretation. Let f be the function that defines the curve AB. Gregory supposes that the function vanishes at A. In other words, if the x value at A is x_0 and at B is $x_0 + c$, then $f(x_0) = 0$ and $f(x_0 + c) = \Delta f_0$; the area of triangle ABP is therefore $c\Delta f_0/2$ and the other terms are correction terms. We shall revisit this formula later, in Chapter 14, when we consider the Euler-Maclaurin formula.

11.2 Newton's Three-Eighths Rule

Newton described his theory of fluxions in some detail in *De Analysi per Aequationes Numero Terminorum Infinitas* (see Chapter 6). The text was published in 1711, but it had already been circulated among colleagues since 1669. Newton shows that if the area under a given curve, between two lines parallel to the ordinate axis, is $z = ax^m$, then the rate of change of z – which he later calls the 'fluxion' of z, to distinguish it from the variable z, called the 'fluent' – is $y = max^{m-1}$, which also represents the ordinate of the curve at the point x. By inverting the formula, he is able to calculate the area under a curve for which he knows the ordinate, when it can be expressed as a polynomial (he called such curves 'parabolas').

In 1687, in the *Principia* (see Section 10.4), Newton shows, in Lemma V, how "to find a curved line of the parabolic kind which shall pass through any given number of points" and he deduces:

Corollary: Hence the areas of all curves may be nearly found; for if some number of points of the curve to be squared are found, and a parabola be supposed to be drawn through those points, the area of this parabola will be nearly the same with the area of the curvilinear figure proposed to be squared: but the parabola can be always squared geometrically by methods generally known.

In the *Methodus Differentialis*, published in 1711, Newton gives the corresponding formula for the case where the 'ordinates' of a function are known at four equidistant points, which we would write as:

$$\int_{x_0}^{x_3} f(t)\, dt = \frac{3}{8}\big[f(x_0) + 3f(x_1) + 3f(x_2) + f(x_3)\big](x_3 - x_0)$$

I. Newton

Methodus Differentialis, in *Analysis Per Quantitatum Series, Fluxiones, ac Differentias: cum Enumeratione Linearum Tertii Ordinis*, London, 1711, p. 100. D. T. Whiteside, *The mathematical papers of Isaac Newton*, vol. VIII, Cambridge, 1981, p. 253.

Proposition VI

To square to a close approximation any curvilinear figure whatever, some number of whose ordinates can be ascertained.

Through the end-points of the ordinates draw a curve of Parabolic kind with the aid of the preceding problems. For this will bound a figure which can always be squared, and whose area will be equal to the area of the figure proposed with close approximation.

Scholium

[...] If, for instance, there be four ordinates positioned at equal intervals, let A be the sum of the first and the fourth, B the sum of the second and the third, and R the interval between the first and the fourth, and then ... the total area between the first and the fourth will be

$$\frac{A+3B}{8}R$$

Newton offers no justification for his formula, doubtless considering the calculation to be too elementary. It only requires a knowledge of the primitives of positive integer powers of x. What follows is a reconstruction of the proof in modern notation.

The curve is defined by $y = f(x) = f(x_0 + t\Delta x)$, and we use $y_0, y_1, y_2,$ and $y_3,$ to stand for the four ordinates corresponding to x values $x_0, x_1 = x_0 + \Delta x, x_2 = x_0 + 2\Delta x,$ and $x_3 = x_0 + 3\Delta x$. The 'parabola' of interpolation, here a cubic, is defined by the interpolation polynomial (see Section 10.4):

$$y_0 + t\Delta y_0 + \frac{1}{2}t(t-1)\Delta^2 y_0 + \frac{1}{6}t(t-1)(t-2)\Delta^3 y_0.$$

The value of the area is:

$$\int_{x_0}^{x_3} f(x)\,dx = \int_0^3 f(x_0 + t\Delta x)\,dt$$

which is approximated by

$$\Delta x \int_0^3 \left[y_0 + t\Delta y_0 + \frac{1}{2}t(t-1)\Delta^2 y_0 + \frac{1}{6}t(t-1)(t-2)\Delta^3 y_0 \right] dt$$

$$= \Delta x \left(3y_0 + \frac{9}{2}\Delta y_0 + \frac{9}{4}\Delta^2 y_0 + \frac{3}{8}\Delta^3 y_0 \right)$$

$$= \Delta x \left(3y_0 + \frac{9}{2}(y_1 - y_0) + \frac{9}{4}(y_2 - 2y_1 + y_0) + \frac{3}{8}(y_3 - 3y_2 + 3y_1 - y_0) \right)$$

$$= \frac{3\Delta x}{8}(y_0 + 3y_1 + 3y_2 + y_3) = \frac{A+3B}{8}R$$

11.3 The Newton-Cotes Formulas

Although Roger Cotes' treatise on Newton's Differential Method (*De Methodo Differentiali Newtoniana*) was only published posthumously in 1722, it was probably written in 1709 (see [6]). The author says in a Post-Script that he had composed his theorems in 1707, before reading Newton's text. He gives a little more explanation of the method, and he provides the formulas for the areas under curves for the cases where we know from 3 to 11 equidistant points, all without any supporting calculations.

R. Cotes
De Methodo Differentiali Newtoniana, in Harmonia Mensurarum sive Analysis & Synthesis Per Rationum & Angulorum Mensuras Promotae: accedunt alia Opuscula Mathematica, Cambridge, 1722. pp. 30–33.

Proposition VII. Problem. To determine the approximate areas of all curves.

We know how to pass through a certain number of points of a proposed curve, a parabolic curve, defined with the aid of the preceding problem [To find a parabolic curve to pass through any given points]; the area of this latter (which is always given by well known methods) will be very close to that of the proposed curve.

Let the given curve be *ABCD*. Erect on the line *KR* the ordinates *AK, BL, CN, DR*, etc. as is desired, and from them their respective abscissas measured from the point *K* and we determine the equation of the closest curve of the form $y = f + gx + hxx + kx^3 + $ etc. Let the abscissa *KP* be x, and the ordinate *PM* be y, the corresponding area will be $fx + \frac{1}{2}gxx + \frac{1}{3}hx^3 + \frac{1}{4}kx^4 + $ etc. [...]

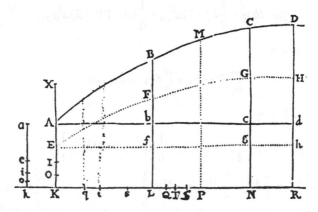

Post-script

I presented the preceding propositions, composed in 1707, to the Audience of the Academy in a Public Lecture in 1709. I had not known that a Treatise on the same subject had been composed by Newton until I received a copy, printed in 1711, from the very learned Master Publisher Jones, to whom, moved by his humanity, I immediately sent my works. But on reading the Propositions of this illustrious man, I perceived that for most of them I had followed a different way, which can be seen on comparing his second Proposition with my fourth. The rule by which, using four ordinates at equal intervals, the area between the first and the last ordinates can be calculated, is one of the most elegant and very useful. I add here the series of Rules produced by continuing in the same way. The first produces the area lying between the extremes of three equidistant ordinates, the second the area between the extremes of four equidistant ordinates, and Newton gave this, by the third we find the area between five given [ordinates] in the same way, etc. In all these rules, *A* is used for the sum of the extreme ordinates, that is to say the sum of the first and the last; *B* for the sum of the [ordinates] closest to the extremes, that is to say the sum of the second and the next to last, *C* for the sum of the next ones, that is to say the sum of the third and the antepenultimate [ordinates] and so on in the order of the letters *D, E, F*, etc. until we arrive at the middle ordinate, if there are an odd number of them, or at the sum of the two middle ones, if there are an even number of them. Finally, for the interval between the extreme ordinates, that is the base of the required area, we use the letter R.

Number of ordinates	Area
III	$\dfrac{A+4B}{6}R$
IV	$\dfrac{A+3B}{8}R$
V	$\dfrac{7A+32B+12C}{90}R$
VI	$\dfrac{19A+75B+50C}{288}R$
VII	$\dfrac{41A+216B+27C+272D}{840}R$
VIII	$\dfrac{751A+3577B+1323C+2989D}{17280}R$
IX	$\dfrac{989A+5888B-928C+10496D-4540E}{28350}R$
X	$\dfrac{2857A+15741B+1080C+19344D+5778E}{89600}R$
XI	$\dfrac{16067A+106300B-48525C+272400D-260550E+427368F}{598752}R$

Although he says in his introduction that he will present a proof of Newton's results, Cotes offers no explanation for his formulas. Doubtless he used interpolation polynomials, but probably not in the form given in Proposition VII, since the calculations would have been too arduous.

11.4 Stirling's Correction Formulas

James Stirling's *Treatise on the Summation and Interpolation of Infinite Series* was published in 1730. As an application of interpolation, he gives the Newton-Cotes formulas, without mentioning Cotes, and also, which was new, a Table of Corrections to be applied to the approximation. No justification for the results was offered, but simply a description of how to use the table, together with some examples.

J. Stirling
Tractatus de Summatione et Interpolatione Serierum Infinitarum, London, 1730, pp. 146–147.

Proposition XXXI

To find very nearly the Area of any Curve starting from being given a certain number of its equidistant Ordinates.

Draw through the extremities of the Ordinates the parabolic figure, whose Area which can be found by well known methods, is equal very nearly to that of the proposed Curve.

[...]

Table of Areas

3	$\dfrac{A+4B}{6}R$
5	$\dfrac{7A+32B+12C}{90}R$
7	$\dfrac{41A+216B+27C+272D}{840}R$
9	$\dfrac{989A+5888B-928C+10496D-4540E}{28350}R$

Table of Corrections

3	$\dfrac{P-4A+6B}{180}R$
5	$\dfrac{P-6A+15B-20C}{470}R$
7	$\dfrac{P-8A+28B-56C+70D}{930}R$
9	$\dfrac{P-10A+45B-120C+210D-252E}{1600}R$

In these Tables, A is the sum of the first and the last Ordinate, B of the second and the one before the last, C of the third and the antepenultimate, and so on until we arrive at the middle Ordinate, which is designated by the last of the letters A, B, C, etc. R is the base on which the Area rests, that is the part of the Abscissa between the first and the last Ordinate. P is the sum of the two Ordinates, one of which comes before the first, and the other after the last, at distances equal to the common interval of the other Ordinates. Moreover, the number of the ordinates, which here is odd, is indicated to the side of the Table. The expressions in the Table of Areas are the Areas contained between the Base, the Curve and the extreme Ordinates. In fact, those in the Table of Corrections are of about the same magnitude as the differences between the true Areas and the Areas given by the Table: therefore if the first digit of the Correction is found, it is then added on when the Correction is negative, and taken away when the Correction is positive; we can therefore conclude with complete certainty that the Area thus corrected is exact up to the position of the decimal of the first digit of the Correction, and no further. Therefore, thanks to the Table of Corrections, the Area that is found is correct, and at the same time we know the number of exact digits.

Example

Let $\dfrac{1}{1+x}$ be the Ordinate of the equilateral Hyperbola, and it is required to find its Area above the Abscissa equal to the unit. For x, write in order:

$$\frac{0}{8},\frac{1}{8},\frac{2}{8},\frac{3}{8},\frac{4}{8},\frac{5}{8},\frac{6}{8},\frac{7}{8},\frac{8}{8},$$

and we arrive at the nine Ordinates

$$\frac{8}{8},\frac{8}{9},\frac{8}{10},\frac{8}{11},\frac{8}{12},\frac{8}{13},\frac{8}{14},\frac{8}{15},\frac{8}{16}.$$

We therefore have

$$A = \frac{8}{8} + \frac{8}{16} = \frac{3}{2}, \quad B = \frac{8}{9} + \frac{8}{15} = \frac{64}{45}, \quad C = \frac{8}{10} + \frac{8}{14} = \frac{48}{35}, \quad D = \frac{8}{11} + \frac{8}{13} = \frac{192}{143}, \quad E = \frac{8}{12} = \frac{2}{3}:$$

and, in substituting these values in the last of the expressions for the Area, and by taking R to be one, the Area comes to .69314721. Next write in order $-\frac{1}{8}$ and $+\frac{9}{8}$ for x in the Ordinate $\frac{1}{1+x}$, and we arrive at the two Ordinates, $\frac{8}{7}$ and $\frac{8}{17}$, the former coming before the first and the latter coming after the last [ordinate]; we have therefore

$$P = \frac{8}{7} + \frac{8}{17} = \frac{192}{119}$$

which you put in the place of P, and for A, B, C, D, E, their values, and the Correction for nine Ordinates gives $+ .00000003$, which, since it is positive, will be subtracted from the Area previously found, and their remains: .69314718, exact as far as the last digit.

Why are formulas only given for an odd number of points? Stirling explains in an article published in 1719 ([14], p. 1063): "I have not given the Table for an even number, since the area for the remaining even numbers is given more exactly starting from an odd number between them." This is a very pertinent remark. In fact, if the number of points used is n, then in putting $h = R/n$, the error is in h^{n+2} for odd n, and in h^{n+1} for even n (see [16], p. 134). Just as for the formulas, we do not know how Stirling arrived at this result. On the other hand, he advises: "if eleven Ordinates do not give the area with sufficient precision, start again, and consider that the area is divided into a number of parts, for which you will have the separate values with the required accuracy" [14].

We can offer a possible interpretation for the formulas. Suppose that the function is interpolated starting from a mid-point x_0 with differences about this centre. Stirling proved this formula in the same work (Prop. XX, First case, pp. 104–105). It can also be found in Newton's *Methodus Differentialis* ([11], Prop. III, First case).

For n points, the area is obtained by using a polynomial of degree $n - 1$ that passes through these n points. The correction is what has to be added when considering a polynomial of degree $n + 1$ which passes as well through two extra points outside the range. Consider $n = 3$. Let the x values of five points be $x_{-2}, x_{-1}, x_0, x_1, x_2$, and let f_{-2}, f_{-1}, etc., be the values of the function for these x values. We have $A = f_{-1} + f_1, B = f_0, P = f_{-2} + f_2, R = 2$. The central differences are calculated by using the following table:

f_{-2}		f_{-1}		f_0		f_1		f_2
	$\Delta f_{-\frac{3}{2}}$		$\Delta f_{-\frac{1}{2}}$		$\Delta f_{\frac{1}{2}}$		$\Delta f_{\frac{3}{2}}$	
		$\Delta^2 f_{-1}$		$\Delta^2 f_0$		$\Delta^2 f_1$		
			$\Delta^3 f_{-\frac{1}{2}}$		$\Delta^3 f_{\frac{1}{2}}$			
				$\Delta^4 f_0$				

and the interpolation polynomial is written as:

$$f_0 + \frac{x\left(\Delta f_{\frac{1}{2}} + \Delta f_{\frac{1}{2}}\right) + x^2\Delta^2 f_0}{2!} + \frac{\left[2x\left(\Delta^3 f_{\frac{1}{2}} + \Delta^3 f_{\frac{1}{2}}\right) + x^2\Delta^4 f_0\right](x^2-1)}{4!}.$$

Since we integrate from -1 to 1, the odd powered terms have an integral of zero and we obtain:

$$\int_{-1}^{1} f(x)\,dx \approx 2f_0 + \frac{\Delta^2 f_0}{3} - \frac{2\Delta^4 f_0}{180}$$

$$= 2\frac{(f_1 + 4f_0 + f_{-1})}{6} - 2\frac{(f_2 - 4f_1 + 6f_0 - 4f_{-1} + f_{-2})}{180}$$

$$= \frac{(A+4B)R}{6} - \frac{(P-4A+6B)R}{180}.$$

11.5 Simpson's Rule

The formula we know as Simpson's Rule is simply the first of the Newton-Cotes formulas. Simpson's contribution was only to supply a (geometrical) proof of the result. His idea was to divide up the interval under consideration into equal parts, sufficiently small for a parabola to be a good approximation to the degree of accuracy required, and he applied the first formula to each of the sections. This is the so-called 'composite' method that is generally used today.

T. Simpson

Of the Areas of Curves, &c. by Approximation, in *Mathematical dissertations on a variety of physical and analytical subjects*, London, 1743, pp. 109–110.

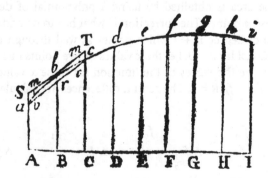

Proposition 1

Supposing *abc* to be a small Portion of any Curve *abfi*, and A*a*, B*b*, C*c*, three equidistant Ordinates; to find an Expression in Terms of those Ordinates, and the common Distance AB, that shall nearly exhibit the included Area AC*cba*A.

Let a common Parabola, having its Axis parallel to the given Ordinates, be described thro' the three Points a, b, c, of the proposed Curve, or rather, to avoid confusing the Figure, let that Curve itself represent a Portion of such a Parabola; join A and C with a Right-Line, and make SbT parallel thereto, producing Aa, and Cc to meet SbT in S and T, and drawing vm from any Point v, in the Parabola, parallel to AS.

Then vm, by the Property of the Parabola being to Sa, as bm^2 to bS^2, or, in the duplicate Ratio of bm, the Space $baSb$, included by the Parabola and the Right-Lines Sa, and Sb, will be 1/3 of the Parallelogram $braSb$, for the same Reason that a Pyramid, whose Sections made by a Plane parallel to the Base, are in a duplicate Ratio of their Distances from the Vertex, is known to be 1/3 of its circumscribing Prism.

Wherefore, seeing Bb × 2AB is equal to the Area ACTbSA and (Aa + Cc)×AB to that of AC$c$ra$A$, the former of these Quantities must exceed the Parabolic Area AC$c$ba$A$ by just half what the latter wants of it; and therefore twice the former added to once the latter, will be just three times this Area, and consequently the Area itself equal to $\dfrac{Aa+4Bb+Cc}{3}\times AB$; which Quantity, since a Parabola admits of infinite Variation of Curvature, so as to nearly coincide with any Curve for a small Distance, must be equal also to the Area sought very nearly.

Q.E.I.

Corollary

Hence may the Area of the whole Curve be also nearly found; for let the Abscissa be divided into any even Number of equal Parts, at the Points B, C, D, &c. according as a lesser or greater Degree of Accuracy is required, and let Bb, Cc, Dd, &c. be Ordinates to the Curve at those Points; then, for the same Reason that $\dfrac{Aa+4Bb+Cc}{3}\times AB$ is the Area of AC$c$ba$A$, will

$\dfrac{Cc+4Dd+Ee}{3}\times CD$ be the Area of CEedcC, and $\dfrac{Ee+4Ff+Gg}{3}\times EF$ that of EGgfeE, &c. &c.

But the Sum of all these Areas taken together, or

$\dfrac{AB}{3}\times(Aa+4Bb+2Cc+4Dd+2Ee+4Ff+2Gg$, &c.) is the Area of the whole Curve.

This idea of dividing up the interval into equal parts was also used at the same period by Maclaurin (see Section 14. 2).

11.6 The Gauss Quadrature Formulas

In a communication to the Göttingen Society of 1816 [5], Gauss begins by re-proving the Cotes formulas, which are equivalent to using interpolation polynomials that coincide, at equidistant points of the abscissa, with the function to be integrated. He then gives an evaluation of the error that is inherent in the method. For a division of the interval into $n - 1$ parts, the formulas are exact for polynomials of order less than or equal to $n - 1$ that coincide with their interpolation function. But if we are allowed to choose the n interpolation points, then we gain n 'degrees of freedom'. He then shows that, for a given number n of points of interpolation, we can find values $x_1, x_2, ..., x_n$, independent of the function to be integrated, satisfying n well chosen conditions, so that the formula we obtain turns out to be exact for all polynomials of order $\leq 2n - 1$. What this does, in a way, is to optimise the approximation.

In the case where the interval for integration is $[-1, 1]$, these $x_1, x_2, ..., x_n$ values are the roots of the Legendre polynomial of degree n. In the particular case where $n = 2$, we obtain the well known formula:

$$\int_{-1}^{1} f(x)\, dx \approx f\left(\frac{1}{\sqrt{3}}\right) + f\left(\frac{-1}{\sqrt{3}}\right),$$

a formula that is exact when f is a polynomial of degree ≤ 3. This formula is, then, a rather special case of Gauss's result.

The article by Gauss, which uses hypergeometric functions, is not easy to read. Jacobi felt the need to provide a simpler approach, and this he did in an article in 1826, extracts from which we give below. In the text, the integration is carried out over the interval $[0, 1]$, and the values of the function that are supposed to be known are given as $\alpha_1, \alpha_2, ..., \alpha_n$. We put $\varphi(x) = (x - \alpha_1)(x - \alpha_2)...(x - \alpha_n)$ and for any polynomial $f(x)$ we have $f = V\varphi + U$, where U and V are the remainder and quotient of dividing f by φ. Thus U is none other than the Lagrange interpolation polynomial of f corresponding to the values $\alpha_1, \alpha_2, ..., \alpha_n$.

C. G. Jacobi
Ueber Gauss neue Methode die Werthe der Integrale näherungsweise zu finden [On Gauss's new method for finding the value of an integral by approximation].
Journal für die reine und angewandte Mathematik, t. 1 (1826), 301–308.

[...]

Gauss showed, in the Göttingen Commentaries, that a judicious choice of the abscissas for which the ordinates are calculated will enable the degree of the approximation to be doubled; and, since this determination of the abscissas does not depend on the nature of the curve to be squared, it is possible to improve tables of values by this improved method, and he gives an example of this.

[...]

Newton's approximation method consists of substituting for $y = f(x)$ the function $U = G\varphi(x)$. The error, or the difference between the integrals of the given function and the function to be substituted, is

$$\Delta = \int y\, dx - \int U\, dx = \int \varphi(X).V\, dx.$$

The problem that now arises is how to determine the values $\alpha_1, \alpha_2, ..., \alpha_n$ to make the error as small as possible, or so that the approximation should be as accurate as possible [...] For this, $\varphi(x)$ has to be determined so that the integrals

$$\int \varphi(x)\, dx, \quad \int x\varphi(x)\, dx, \quad \int x^2\varphi(x)\, dx, \quad ..., \quad \int x^{n-1}\varphi(x)\, dx,$$

between the limits $x = 0$ and $x = 1$, between which the integral $\int y\, dx$ is taken, should be zero. Determining $\varphi(x)$ is now our task.

The integral $\int x^m \varphi(x)\, dx$ can be transformed, by a well known reduction formula, into multiple integrals of $\varphi(x)$. In general, these are:

$$\int uv\, dx = u\int v\, dx - \int du \int v\, dx$$

$$\int du \int v\, dx = du \iint v\, dx - \int d^2u \iint v\, dx$$

$$\int d^2u \iint v\, dx = d^2u \iiint v\, dx - \int d^3u \iiint v\, dx$$

$$\cdots$$

$$\int d^m u \int \cdots \int v\, dx = d^m u \iint \cdots \int v\, dx - \int d^{m+1} u \iint \cdots \int v\, dx$$

where each formula is deduced from the previous one by replacing u by du/dx and v by $\int v\, dx$. From this it immediately follows:

[...]

$$\int x^{n-1} \varphi(x)\, dx = x^{n-1} \int \varphi(x)\, dx - (n-1) x^{n-2} \iint \varphi(x)\, dx^2 + (n-1)(n-2) x^{n-3} \iiint \varphi(x)\, dx^3 +$$
$$\cdots + (-1)^{n-1} (n-1)(n-2)\ldots 1 \int \cdots \int \varphi(x)\, dx^n$$

These are formulas well known. It can be seen that if $\int \varphi(x)\, dx$, $\int x\varphi(x)\, dx$, $\int x^2 \varphi(x)\, dx$, ..., $\int x^{n-1} \varphi(x)\, dx$ vanish for given limits, then, between the same limits, $\int \varphi(x)\, dx$, $\iint \varphi(x)\, dx^2$, ..., $\int \cdots \int \varphi(x)\, dx^n$ must also vanish, and reciprocally.

Our task now comes down to determining the function $\varphi(x)$ so that its 1st, 2nd, ..., nth integrals, between the limits 0 and 1, vanish; that is, that if the successive integrals are so determined so as to vanish for $x = 0$, then they must also vanish for $x = 1$.

Let us put $\int \cdots \int \varphi(x)\, dx^n = \pi(x)$, and suppose that the successive integrals all vanish for $x = 0$, the problem can then be stated thus: to find a function $\pi(x)$, that vanishes for $x = 0$ and for $x = 1$ as well as for its 1st, 2nd, 3rd, ..., $(n-1)$th derivatives. This means the function $\pi(x)$ contains the factors x^n and $(x-1)^n$; and conversely, any function having the factor $x^n(x-1)^n$ will fulfil the desired conditions. We must therefore put $\pi(x) = x^n(x-1)^n M$. But since $\varphi(x) = (x - \alpha_1)(x - \alpha_2)\ldots(x - \alpha_n)$ is a rational integral function of degree n, it follows that $\pi(x) = \int \cdots \int \varphi(x)\, dx^n$ is a polynomial function of degree $2n$, and consequently, in our case, M is a constant. In this way, we obtain

$$\varphi(x) = \frac{M d^n [x^n (x-1)^n]}{dx^n}$$

$$= x^n - \frac{n^2}{2n} x^{n-1} + \frac{n^2(n-1)^2}{1.2.2n(2n-1)} x^{n-2} - \frac{n^2(n-1)^2(n-2)^2}{1.2.3.2n(2n-1)(2n-2)} x^{n-3} + \cdots$$

$$\cdots + (-1)^n \frac{n(n-1)(n-2)\ldots 1}{2n(2n-1)(2n-2)\ldots(n+1)}$$

where we put $M = \dfrac{1}{2n(2n-1)(2n-2)\ldots(n+1)}$.

The roots of the equation $\varphi(x) = 0$, where $\varphi(x)$ is the above found expression, thus give the values $\alpha_1, \alpha_2, \ldots, \alpha_n$ which therefore achieves the greatest possible degree of approximation.

As we know from the theory of equations, if the roots of an equation $\pi(x) = 0$ are all real, then all the roots of $d^m\pi(x)/dx^m = 0$ will also be real and will come between those of the first equation, from which, since all the roots of the equation $\pi(x) = 0$, or of the equation $x^n(1 - x)^n = 0$, are real, and in fact n of them are equal to zero, and the others $= 1$, then also the roots of $\varphi(x) = 0$, or the quantities $\alpha_1, \alpha_2, ..., \alpha_n$, are all real and lie between 0 and 1; as Gauss found in the examples he worked. [...]

To summarise, the Gauss method consists in choosing, for the x values of the n interpolation points, the roots $\alpha_1, \alpha_2, ..., \alpha_n$ of the polynomial of degree n obtained by differentiating n times, the polynomial $x^n(x - 1)^n$ in the case of the interval $[0, 1]$, or the polynomial $(x - a)^n(x - b)^n$ in the case of the interval $[a, b]$. The Gauss approximate quadrature formula can be expressed in the form:

$$\int_a^b f(x)\, dx \approx \sum_{i=1}^n c_i f(\alpha_i)$$

where the c_i are coefficients independent of f. In fact, using the Lagrange expression for the interpolation polynomial of f:

$$\sum_{i=1}^n L_i(x)f(\alpha_i) \text{ where } L_i = \prod_{j \neq i} \frac{(x - x_j)}{(x_i - x_j)}$$

then,

$$\int_a^b f(x)\, dx \approx \sum_{i=1}^n \left(f(\alpha_i) \int_a^b L_i(x)\, dx \right)$$

and we put:

$$c_i = \int_a^b L_i(x)\, dx .$$

By construction, that is through the choice of the α_i and the calculation of the c_i, this formula is exact when f is a polynomial of degree $\leq 2n - 1$. In fact, it can be shown that if f is a function of the class C^{2n}, that is it has $2n$ continuous derivatives, the error can only be terms of the order $\geq 2n$.

By way of example, let us find the three interpolation points on the interval $[-1, +1]$. Differentiating the polynomial $(x^2 - 1)^3$ three times, we obtain the polynomial $120x^3 - 72x$, whose roots are $\alpha_1 = -\frac{\sqrt{3}}{5}$, $\alpha_2 = 0$ and $\alpha_3 = \frac{\sqrt{3}}{5}$. Next, integrating the polynomial $L_2(x) = -\frac{3}{5}\left(x^2 - \frac{3}{5} \right)$, for example, over the interval $[-1, +1]$, gives $c_2 = 8/9$ ... From which:

$$\int_{-1}^1 f(x)\, dx \approx \frac{5}{9}\left[f\left(-\frac{\sqrt{3}}{5} \right) + f\left(\frac{\sqrt{3}}{5} \right) \right] + \frac{8}{9}f(0) .$$

We notice, finally, that the geometric aspect has completely disappeared from the proofs, even if Jacobi appeals to the Lagrange interpolation polynomial.

11.7 Chebyshev's Choice

In an article published in 1874 in the *Journal de Liouville*,
Chebyshev sets out to determine a quadrature formula,
where the approximation is a linear combination of the values of the function at some particular points, but each of
these values is given the same weight. More precisely, he want
to find a formula of the type

$$\int_{-1}^{+1} F(x)\varphi(x)\,dx \approx k\big[\varphi(x_1)+\varphi(x_2)+...+\varphi(x_n)\big]$$

where the $x_1, x_2, ..., x_n$ depend, not on φ but on F.

This formula appeared to him to offer more potential than
the Gauss formula, given in Section 11. 6, in the case where the values taken by the
function are only known approximately. In fact, with the Gauss formula there is a
risk of too much weight being given to a value that is seriously in error.

Granting that such a formula could exist, he determines the coefficient k and the
values $x_1, x_2, ..., x_n$. In the case where $F = 1$, he shows that $k = 2/n$ and that the $x_1, x_2,$
$..., x_n$ are solutions of the equation

(5)
$$\left[z^n e^{-\dfrac{n}{2.3z^2} - \dfrac{n}{4.5z^4} - \dfrac{n}{6.7z^6} -...} \right] = 0$$

where [ ] indicates the integer part.

The mathematical proof goes beyond the scope of this book. What we give here is
the introduction, together with the calculations that Chebyshev gives for the values
of $x_1, x_2, ..., x_n$ for the first few values of n.

P. Chebyshev
Sur les quadratures [On quadratures]
Journal de mathématiques pures et appliquées, Series 2, XIX (1874), 19–22.

1. In the very important Work that M. Hermite has just published on mathematical Analysis,
the illustrious geometer gives a new formula for approximately evaluating the integral

$$\int_{-1}^{1} \frac{\varphi(x)}{\sqrt{1-x^2}}\,dx$$

In this formula, all the values of the function $\varphi(x)$ appear with the same coefficient; this provides an essential difference between M. Hermite's formula and that of Gauss, and it makes it
very useful in numerical applications. The usefulness of approximation formulas of this type
is why I wish to present some reflections on researching these formulas.

We shall suppose that, the function $F(x)$ being given, we are looking to express, as closely
as possible, integrals of the form $\int_{-1}^{1} F(x)\varphi(x)\,dx$, whatever the function $\varphi(x)$, by the formula
$k\big[\varphi(x_1)+\varphi(x_2)+...+\varphi(x_n)\big]$, where $k, x_1, x_2, ..., x_n$ are values that do not depend on the function $\varphi(x)$.

[...]

6. In giving to n the simpler values, such that $n = 2, 3, 4, 5, 6, 7$, we find that, for these values of n, equation (5) becomes, respectively:

$$z^2 - \frac{1}{3} = 0$$

$$z^3 - \frac{1}{2}z = 0$$

$$z^4 - \frac{2}{3}z^2 + \frac{1}{45} = 0$$

$$z^5 - \frac{5}{6}z^3 + \frac{7}{72}z = 0$$

$$z^6 - z^4 + \frac{1}{5}z^2 - \frac{1}{105} = 0$$

$$z^7 - \frac{7}{6}z^5 + \frac{119}{360}z^3 - \frac{149}{6480}z = 0$$

and, on solving these equations, we obtain the following systems of values of $x_1, x_2, ..., x_n$:

$n = 2$	$n = 3$	$n = 4$
$x_1 = -0.577350$	$x_1 = -0.707107$	$x_1 = -0.794654$
$x_2 = +0.577350$	$x_2 = 0$	$x_2 = -0.187592$
	$x_3 = +0.707107$	$x_3 = +0.187592$
		$x_4 = +0.794654$

$n = 5$	$n = 6$	$n = 7$
$x_1 = -0.832497$	$x_1 = -0.866247$	$x_1 = -0.883862$
$x_2 = -0.374541$	$x_2 = -0.422519$	$x_2 = -0.529657$
$x_3 = 0$	$x_3 = -0.266635$	$x_3 = -0.323912$
$x_4 = +0.374541$	$x_4 = +0.266635$	$x_4 = 0$
$x_5 = +0.832497$	$x_5 = +0.422519$	$x_5 = +0.323912$
	$x_1 = +0.866247$	$x_6 = +0.529657$
		$x_7 = +0.883862$

With these values of $x_1, x_2, ..., x_n$, the formula

$$\frac{2}{n}\left[\varphi(x_1) + \varphi(x_2) + ... + \varphi(x_n)\right]$$

gives an approximate expression for the integral $\int_{-1}^{1} \varphi(x)\, dx$ which, in certain cases, is easier to use in applications than that of Gauss; for, in the latter formula, the values $\varphi(x_1)$, $\varphi(x_2)$, ..., $\varphi(x_n)$ have to be entered with different coefficients. Since our expression for the integral $\int_{-1}^{1} \varphi(x)\, dx$ is only exact as far as the terms in $\varphi^{(n+1)}(0)$, $\varphi^{(n+2)}(0)$, ..., more terms need to be taken, in general, than in the Gauss formula. Nevertheless, in the case where the values of $\varphi(x_1)$, $\varphi(x_2)$, ..., $\varphi(x_n)$, from which the integral $\int_{-1}^{1} \varphi(x)\, dx$ is determined, are affected by unknown errors, notably greater than that which comes from the rejected terms, the approximation formula that we have just found must be preferred to that of Gauss, even with respect to the degree of accuracy, seeing that, in this formula, the sum of the squares of the coefficients, on account of their equality, has the smallest possible value.

We note that, for $n = 2$, we retrieve the Gauss rule, but this is not the case for the following values of n. The values of $x_1, x_2, ..., x_n$, are always the roots of a polynomial, but they do not belong to the same family of polynomials used by Gauss, today called Legendre polynomials, and which are orthogonal for the scalar product

$$(P,Q) = \int_{-1}^{1} P(x)Q(x)\,\mathrm{d}x$$

(see Chapter 13). Furthermore, the roots of the polynomials used by Chebyshev are only real for $n = 1$ to $n = 7$, and for $n = 9$ (see [12]).

The example we chose corresponds to the function $F = 1$, but the method indicated was then applied by Chebyshev to a number of integrals of the form

$$\int_{-1}^{1} F(x)\varphi(x)\,\mathrm{d}x.$$

In particular, he retrieves the values given by Hermite in the article to which he refers, values which allow him to calculate an approximation for

$$\int_{-1}^{1} \frac{\varphi(x)}{\sqrt{1-x^2}}\,\mathrm{d}x.$$

The polynomials used in that case are now called Chebyshev polynomials and they form an orthogonal family for the scalar product:

$$(P,Q) = \int_{-1}^{1} \frac{1}{\sqrt{1-x^2}} P(x)Q(x)\,\mathrm{d}x.$$

11.8 Epilogue

Research into approximate quadrature formulas has been of constant interest to mathematicians and we had to leave out a good deal. Nor have we mentioned work on evaluation of the accuracy of approximations.

R. Radau published a very interesting synthesis of these formulas in 1880, in the *Journal de Liouville*. There, in particular, he defines the 'degree of precision':

We say that a quadrature formula possesses a *degree of precision $p - 1$*, to express that it is rigorously exact whenever the function $f(x)$ is an integer function of degree less than p.

An integer function is, here, a polynomial. When the function is expanded as an integer series of the type $f(x) = a_0 + a_1 x + ... + a_{p-1}x^{p-1} + a_p x^p + a_{p+1}x^{p+1} + ...$, he adds correction terms $a_p \varepsilon_p + a_{p+1}\varepsilon_{p+1} + ...$, where the ε_k were calculated once and for all. The order of magnitude of the error that is made was, however, only calculated for particular functions.

We reproduce below an extract from Radau's table, comparing the efficiency of formulas for the following numerical example:

$$\int_0^1 \frac{\mathrm{d}x}{1+x} = \ln 2 = 0.69314718056...$$

(Villarceau's method is a composite method using Cotes' formula with 5 ordinates.)

Extract from Radau's Table

Ordinates			Error in the 8th decimal	Degree of precision
Cotes	8	0.69314773.3	+ 55.3	7
"	10	0.69314720.28	+ 2.22	9
"	11	0.69314718.20	+ 0.14	11
Simpson	9	0.69315453	+ 735	3 with 4 divisions
Villarceau	9	0.69314790	+ 72	5 with 2 divisions
Gauss	3	0.69312169.3	− 2548.7	5
"	5	0.69314715.78	− 2.27	9
Chebyshev	4	0.69312796.2	− 1921.8	5
"	8	0.69314667.0	− 51.0	5 with 2 divisions

R. Radau, Valeur numérique d'une intégrale définie,
Journal de Mathématiques Pures et Appliquées, Series 3, VI (1880), 334–335.

Bibliography

[1] Archimedes, Measurement of a Circle, in *The Works of Archimedes Edited in Modern Notation* (1897) with supplement *The Method of* Archimedes (1912), tr. T. L. Heath, Cambridge: Cambridge University Press, 1912, repr., New York: Dover, n.d.

[2] Cavalieri, B., *Centuria di varii problemi*, Bologna, 1639.

[3] Chebyshev, P.L., Sur les quadratures, *Journal de Mathématiques Pures et Appliquées*, Series 2, vol. XIX (1874), 19–34. *Collected Works*, vol. III, 1859, pp. 289–323.

[4] Cotes, R., De Methodo Differentiali Newtoniana, in *Harmonia Mensurarum sive Analysis & Synthesis Per Rationum & Angulorum Mensuras Promotae: accedunt alia Opuscula Mathematica*, Cambridge, 1722.

[5] Gauss, C.F., Methodus nova integralium valores per approximationem inveniendi, in *Commentationes Societatis regiae Scientiarium Gottingensis*, vol. III (1816), 39–76. *Werke*, vol. III, 1876, pp. 165–196.

[6] Gowing, R., *Roger Cotes - Natural Philosopher*, Cambridge: Cambridge University Press, 1983.

[7] Gregory, J., *Exercitationes geometricae*, London, 1668.

[8] Gregory, J., Letter of 23 November 1670 to Collins, in *The correspondence of Isaac Newton*, vol. I, ed. H.W. Turnbull, Cambridge: Cambridge University Press, 1959.

[9] Hermite, C., *Cours d'Analyse de l'Ecole Polytechnique*, Paris, 1873.

[10] Jacobi, C.G., Ueber Gauss neue Methode die Werthe der Integrale näherungsweise zu finden, *Journal für die reine und angewandte Mathematik*, vol. I (1826), 301–308. *Werke*, vol. VI, 1891, pp. 1–11.

[11] Newton, I., Methodus Differentialis, in *Analysis per Quantitatum, Series, Fluxiones ac Differentias: cum enumeratione Linearum Tertii Ordinis*, London, 1711. D. T. Whiteside, *The mathematical papers of Isaac Newton*, vol. VII, 1697-1722, Cambridge: Cambridge University Press, 1981.

[12] Radau, R., Valeur numérique d'une intégrale définie, *Journal de Mathématiques Pures et Appliquées*, Series 3, vol. VI (1880), pp. 334–335.

[13] Simpson, T., Of the Area of Curves, &c. by approximation, in *Mathematical Dissertation on a Variety of Physical and analytical Subjects*, London, 1743.

[14] Stirling, J., Methodus Differentialis Newtoniana Illustrata, *Philosophical Transactions of the Royal Society*, vol. 30 (1719), 1050–1070.

[15] Stirling, J., *Methodus Differentialis sive Tractatus de Summatione et Interpolatione Serierum Infinitarum*, London, 1730.

[16] Stoer, J. and Bulirsch R., *Introduction to Numerical Analysis*, 2nd ed. New York: Springer Verlag, 1993.

12 Approximate Solutions of Differential Equations

> The three body problem has such importance for Astronomy, and at the same time it is so difficult, that all the efforts of geometers have for a long time been directed towards it. A complete and rigorous integration being manifestly impossible, appeal has to be made to approximation procedures.
>
> Henri Poincaré, *Les méthodes nouvelles de la Mécanique céleste*, 1892, vol. I, p. 1.

Right from the birth of the infinitesimal calculus in the 17th century, certain problems arose, in geometry and mechanics, that gave rise to first order differential equations. There followed a systematic study of differential equations in the 18th century, and particularly in the 19th century, as a consequence of applying Newton's Laws of mechanics to determine the trajectories of celestial bodies, since the study of a mechanical system of n 'material points' is equivalent to the study of $3n$ second order differential equations.

The simplest and most pleasing solutions were those that could be expressed in terms of elementary functions, or as the primitives of such functions. Now, this type of 'integration by quadrature' is only possible in practice for a relatively small number of differential equations, even when they are first order ones and of the form

$$\frac{dy}{dx} = f(x, y).$$

An alternative solution method, using Newton's technique, is to try to find solutions in the form of sums of polynomial series whose coefficients can be found by a recurrence relation derived directly from the differential equations. The basis of this technique can be found with Leibniz [11], for solving the inverse problem of tangents. "But", explains Lagrange [10], "this method has the disadvantage of giving infinite sequences, even when these sequences can be represented by finite rational expressions".

Also, Lagrange himself proposed a procedure based on continuous fractions, which he was fond of using. The determination of the successive terms appearing in the continued fraction solution was carried out by a process not unlike that of Newton's polygon (see Section 6.6). Although, in contrast to the technique of using series, "by this method, we are assured of finding directly the rational and finite value of the required quantity, when it has one", Lagrange's method did not enjoy the later developments that its author had hoped for.

This explains the interest in the approximate methods of solution that we report on here. First used by Euler in the 18th century in a relatively elementary way, to find a polygonal approximation to the solution curve (Section 12.1), approximate methods were perfected towards the end of the 19th century. The stimulus for these improvements came directly from the methods for approximate quadrature described in the previous chapter, since, to find a solution φ of the differential equation

$$\frac{dy}{dx} = f(x, y)$$

which satisfies $\varphi(x_0) = y_0$, is to calculate, for each value of x, the quantity

$$y = y_0 + \int_{x_0}^{x} f(t, \varphi(t)) \, dt .$$

The method of finite differences which leads to Gregory's formula (Section 11.1) has its parallel in work by Adams (Section 12.6). The trapezium method, Simpson's method (Section 11.5) and the Gauss method (Section 11.6) were the inspiration for successive improvements by Runge, Heun and Kutta (see Sections 12.3, 12.4 and 12.5). These last methods are now called 'one-step' whereas the method proposed by Adams is a 'multi-step' method.

The use of these methods, or those derived from them, is today markedly expanded, partly because more and more disciplines of mathematical modelling require the use of differential equations, and partly because those numerical techniques which are based on a discretisation of differential equations are particularly well adapted to being handled by computer. This is the reason why these methods, which for the most part date from a century ago, have only been developed from the 1950s onwards. In particular, approximate numerical integration of differential systems, carried out by computer, can be used to study the stability of the solar system over long time periods.

In the interval between, more theoretical studies, like those by Cauchy in particular, have led to the general proof of the existence of solutions. The first, by Cauchy in 1820, is based on the polynomial approximation of Euler, infinitely refined; it was rediscovered by Lipschitz in 1868 with hypotheses that were rather more general. However, we give here the lucid proof by Picard, a proof by successive approximations (Section 12.2).

12.1 Euler's Method

In a chapter entitled "On the integration of differential equations by approximation" in his work *Institutionum Calculi Integralis* of 1768, Euler notes the unsatisfactory aspect of using series, and he describes there the following method, known now as 'Euler's method'. Considering a first order differential equation of the form

$$\frac{dy}{dx} = V(x, y)$$

and with 'initial conditions':

$$x = a, \ y = b,$$

Euler determines the corresponding solution values by a closer and closer approximation method. To find an approximate value y of the solution, for the value x of the variable, he progresses from a towards x through intermediary points a', a'', ..., 'sufficiently close' for the function V to be considered as a constant over the subintervals he defines and therefore, using b', b'', ... for the corresponding solution values, he is able to put:

$$b' - b = (a' - a) \ V(a, b)$$
$$b'' - b' = (a'' - a') \ V(a', b')$$
$$\dotfill,$$

in other words, this comes down to assimilating the curve solution to a straight sided polygon, the solution curve being the same as its tangent in the neighbourhood of each intermediary point.

L. Euler

Institutionum Calculi Integralis, St. Petersburg, 1768, vol. 1, Chap. 7, (§§ 650–653).
Opera Omnia, 1st series, vol. 11, Teubner, 1913, pp. 424–425.

CHAPTER VII

On the integration of differential equations by approximation

Problem 85

Being given a differential equation, to find a complete integral very nearly.

Solution

Let x and y be a pair of variables connected by a differential equation, that equation being of the form $\dfrac{dy}{dx} = V$, where V stands for any function of the variables x and y. And now, we look for a complete integral, which can be interpreted thus: whenever a determined value is assigned to x, say $x = a$, the second variable y has to take a given value, say $y = b$. Let us be led then in our study in such a way as to examine the value of y when a value a little different from a is given to x, let us put $x = a + \omega$ to find y. Now, since ω is a very small quantity, the value of y itself will differ very little from b; consequently, as x varies just from a to $a + \omega$, it is allowed to consider the quantity V as being constant in this interval. From this fact, if we put $x = a$, and $y = b$, then $V = A$ and by reason of this minuscule variation we shall have $\dfrac{dy}{dx} = A$ and from this, by integrating, $y = b + A(x - a)$, a constant being added, as is clear, so that for $x = a$, $y = b$. So, for $x = a + \omega$, $y = b + A\omega$.

Hence, in the same way that from the given values at the beginning $x = a$, $y = b$, we found next closest following values of $x = a + \omega$ and $y = b + A\omega$, it is therefore possible to progress by very small intervals until one arrives at values as distant as one wishes from the initial values. In order to be able to look at these operations more clearly let us establish, step by step, what ensues.

Variable	Successive values
x	$a,\ a',\ a'',\ a''',\ a^{iv},\ ...,\ 'x, x$
y	$b,\ b',\ b'',\ b''',\ b^{iv},\ ...,\ 'y, y$
V	$A,\ A',\ A'',\ A''',\ A^{iv},\ ...,\ 'V, V$

It is clear that starting from the initial given values $x = a$ and $y = b$ we shall have $V = A$, then for the second values we shall have $b' = b + A(a' - a)$, the difference $a' - a$ being supposed to be as small as is wished. From this, in putting $x = a'$ and $y = b'$, we obtain $V = A'$ and, starting from that, for the third values $b'' = b' + A'(a'' - a')$, and, at that point, if we put $x = a''$ and $y = b''$, we find $V = A''$. Now for the fourth values, we shall have $b''' = b'' + A''(a''' - a'')$ and, from this, in putting $x = a'''$ and $y = b'''$, we shall find $V = A'''$, and so we can advance as far as values that are as distant from the initial values as we wish. But the first sequence, giving the successive values of the variable x, can be taken as one wishes, provided that it ascends, or equally well descends, by very small intervals.

Corollary 1

So, for very small intervals taken one at a time, the calculations can be done in the same way, and we thus obtain values on which the succeeding ones depend. In this way, for the values of the variable x chosen one at a time, we can therefore indicate the corresponding values of the variable y.

Corollary 2

The smaller the intervals over which the values of x are supposed to progress, the more accurate will be the values obtained one by one. This being the case, however, the errors made for each value, even though they are very small indeed, will accumulate on account of their multiplicity.

Corollary 3

Furthermore, in this calculation the errors arise from the fact that, in the intervals taken one at a time, we consider the two quantities x and y, and so the function V, as constants. Therefore, the more the value of the variable V changes from one interval to the next, the more we should fear large errors.

Let us consider the differential equation $y' = f(x, y)$ with the initial conditions $(x_0, y_0) = (a, b)$. We let φ stand for the solution, whose existence and uniqueness we shall consider later. In Euler's method, the solution curve is approximated by a polygonal line as shown in the figure below.

To simplify the notation and procedure, suppose that we progress from a to x by a constant step h. We use $x_i = a + ih$ for the points of the subdivision, with $x_0 = a, x_n = x$, and $h = (x - a)/n$. Thus, in each interval $[x_i, x_{i+1}]$, the solution curve is the same as its tangent at the point x_i. To start with, at the point $(x_0, y_0) = (a, b)$, the tangent has the equation $(y - y_0)/(x - x_0) = f(x_0, y_0)$, from which, for the approximate value $\varphi(x_1)$ we have:

$$y_1 = y_0 + h\, f(x_0, y_0).$$

In fact, from the next step, the point (x_1, y_1) is not exactly on the graph of φ and $f(x_i, y_i)$ is no longer the gradient of the tangent to the graph of φ at $(x_1, \varphi(x_1))$. However, we go on to approximate $\varphi(x_2)$ by

$$y_2 = y_1 + h\, f(x_1, y_1)$$

and so on ...

The algorithm can be summarised thus:

for $i = 0$ to $n - 1$:
$x_{i+1} = x_i + h$
$y_{i+1} = y_i + h\, f(x_i, y_i)$.

As Euler remarks, the error at each stage can be made very small by choosing a sufficiently small step; on the other hand, there is an accumulation of errors as we go from a to x. The method is, however, a reasonable one.

In 1820 Cauchy provided a rigorous proof of the method in his *Leçons données à l'Ecole Royale Polytechnique* [2]. This appears in the ninth *leçon*, "Limit of the Errors that can be made in using the Method described in the seventh *leçon*", where Cauchy shows that, assuming f, f_x' and f_y' to be continuous, the value Y, found by using what we call

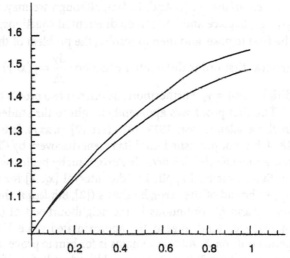

Example of a solution by Euler's method of the differential equation $\dfrac{dy}{dx} = \dfrac{y-x}{y+x}$ with initial conditions $x = 0$, $y = 1$, in the interval $[0, 1]$ and a step $h = 0.1$. For $x = 1$, the exact value is 1.498278 (see texts by Runge and Kutta below).

Euler's method, as an approximation at the value X of the solution φ of the differential equation $\dfrac{dy}{dx} = f(x, y)$ which satisfies $\varphi(x_0) = y_0$, is defective by an error $|Y - \varphi(X)|$ which is at most equal to

$$(B + AC)|X - x_0|e^{C|X - x_0|}h$$

where A, B and C represent respectively the upper bounds of $|f|$, $|f_x'|$ and $|f_y'|$, and h is the incremental step in going from x_0 to X. We can write, in particular:

$$|Y - \varphi(X)| \le Kh$$

where K is a constant dependent only on X (and of course the given f, x_0, y_0).

Thus, the inequality just stated justifies Euler's method in showing that, with the value X being fixed, it is theoretically possible to approach $\varphi(X)$ as close as is wished, on condition that the incremental step h is chosen to be sufficiently small. But this inequality does not in itself prove the existence of a solution. However, Cauchy also made progress here at a theoretical level.

12.2 The Existence of a Solution

"It is A. -L. Cauchy who has set the general theory of differential equations on an indestructible base" affirms Paul Painlevé in an article in the *Encyclopédie des Sciences mathématiques* [15]. And, in fact, although we may recall the earlier works of Lagrange, Laplace and Poisson on differential equations, it was indeed Cauchy who was the first to pose, and then to resolve, the problem of the existence of a solution φ of a general first order differential equation $\dfrac{dy}{dx} = f(x, y)$ satisfying a given initial condition $\varphi(x_0) = y_0$. Furthermore, he offered two proofs of the result.

The first proof was apparently taught to the students of the Ecole Polytechnique in the academic year 1819–1820 (see [2], p. *xxv*). It is described in a text printed in 1824, but not published until its recent discovery by Christian Gilain in the library of the *Académie des Sciences de Paris*. Cauchy begins with the approximation Y given by Euler's method applied to the interval $[x_0, X]$ for a given subdivision. Using the upper bound of the error, he shows ([2], 8th *leçon*) that, with quite general hypotheses – f and f_y' continuous in the neighbourhood of (x_0, y_0) – and letting the incremental step tend to zero, the approximated value Y tends towards a limit that depends only on X. All that remains is for him to prove that the function defined in this way, in making X vary in the neighbourhood of x_0, is in fact the solution of the proposed differential equation. Cauchy thus proves the existence of a solution in the neighbourhood of the initial conditions by a sort of limiting process of Euler's method.

This unpublished proof was referred to by Cauchy in 1835 [3] in a memoir where he describes the 'method of upper bounds', a second proof valid only for analytical equations, leading therefore to a less general statement. The first proof was given again by Coriolis in 1837 [4] and given again by the *abbé* Moigno in a treatise in 1844 (see [13], vol. 2, Part 1, *leçons* 26, 27). Despite this, Cauchy's first existence theorem remained little known, doubtless because of the greater importance given at that time to functions of complex variables.

Thus, Lipschitz seems not know about it when, on coming back to functions of real variables, in an article in 1868 [12], he provides an improvement to Cauchy's existence theorem. What he does is to assume that the function f remains continuous, but he replaces the continuity of f_y' by the 'Lipschitz condition' on f with respect to the variable y in the neighbourhood of (x_0, y_0), that is he supposes the existence of a constant k, independent of x, such that:

$$|f(x, y_1) - f(x, y_2)| \le k|y_1 - y_2|.$$

It was, however, Picard's work which provided the completion of the theory of the existence of local solutions. In a memoir of 1890, Picard gives a third existence proof: here he constructs a function, which is a local solution, by a method of successive global approximations. In volume 2 of his *Traité d'Analyse*, Picard describes these three types of proof: the 'Cauchy-Lipschitz', the 'method of upper bounds' and 'successive approximations'.

C. E. Picard

Mémoire sur la théorie des équations aux dérivées partielles et la méthode des approximations successives, *Journal de Mathématiques pures et appliquées*, 4th series, vol. 6 (1890), 145–210, 231 (pp. 197–200).

Chapter V

Some Remarks on Ordinary Differential Equations

1. The approximation methods that we have just been using can clearly be used in the case of ordinary differential equations; it is a point on which it appears to me to be not without value to pause, although we have only to apply the methods studied above.

Let us take, first of all, a single first order equation

$$\frac{dy}{dx} = f(y,x);$$

we can thus establish the fundamental theorem about the existence of the integral of this equation, taking for $x = x_0$ the value $y = y_0$.

For this purpose, we consider the equations

$$\frac{dy_1}{dx} = f(y_0,x),$$

$$\frac{dy_2}{dx} = f(y_1,x),$$

$$\cdots\cdots\cdots\cdots\cdots ,$$

$$\frac{dy_n}{dx} = f(y_{n-1},x),$$

carrying out each of the quadratures, such that for $x = x_0$ we would have $y_n = y_0$. What is required is to ensure that y_n tends, for infinite n, towards a limit y which represents the required integral, provided that x remains in the neighbourhood of x_0. We impose on the function $f(y, x)$ the hypothesis that it is defined and continuous for the values of x and y taken, respectively, between $x_0 - a$ and $x_0 + a$ in the one case, and between $y_0 - b$ and $y_0 + b$ in the other case; further, we can determine a positive constant k, such that

$$|f(y_2, x) - f(y_1, x)| < k|y_2 - y_1|$$

and we suppose the function and the variables to be real.

Let M be the maximum modulus of $f(y, x)$ when x and y remain within the limits indicated. We have

$$y_1 = \int_{x_0}^{x} f(y_0,x)\, dx + y_0 .$$

Let ρ be a quantity at most equal to a; y_1 will remain within the required limits if

$$M\rho < b,$$

and it is obvious that this will be the same for $y_2, ..., y_n$. Using δ to stand for a quantity at least equal to ρ, we shall suppose that x remains between $x_0 - \delta$ and $x_0 + \delta$.

We now have, in putting $y_n - y_{n-1} = z_n$,

$$\frac{dz_1}{dx} = f(y_0, x),$$

$$\frac{dz_2}{dx} = f(y_1, x) - f(y_0, x),$$

$$\cdots\cdots\cdots\cdots\cdots\cdots\cdots\cdots,$$

$$\frac{dz_n}{dx} = f(y_{n-1}, x) - f(y_{n-2}, x),$$

and all the z will disappear for $x = x_0$. We have

$$|z_1| < M\delta,$$
$$|z_2| < kM\delta^2,$$

then

$$|z_3| < k^2 M\delta^3,$$

and, in general,

$$|z_n| < M\delta(k\delta)^{n-1}.$$

We see, therefore, in writing

$$y_n = y_0 + z_1 + z_2 + ... + z_n,$$

that y_n will tend towards a limit if $k\delta < 1$. The series

$$y = y_0 + z_1 + ... + z_n + ...$$

will be convergent, like a decreasing geometric progression. Thus y_n converges towards a limit y, when x remains between $x_0 - \delta$ and $x_0 + \delta$, with δ *being the least of the quantities*

$$a, \frac{b}{M}, \frac{1}{k}.$$

In this interval, y manifestly represents a continuous function of x. We have, furthermore,

$$y_n = \int_{x_0}^{x} f(y_{n-1}, x)\, dx + y_0,$$

and, since y_n and y_{n-1} tend to y, it follows that

$$y = \int_{x_0}^{x} f(y, x)\, dx + y_0$$

and, therefore,

$$\frac{dy}{dx} = f(y, x),$$

that is to say the limit y satisfies the differential equation. The existence of the integral has thus been established. We could clearly use the same method of proof if f was an analytical function of complex variables x and y.

From the way that Picard introduces into his formulas the approximations $y_1, ..., y_n$ we are reminded of Euler's method of primitives. But whereas, in Euler, the $y_1, ..., y_n$ are numbers, here the $y_1, y_2, ..., y_n$ are functions. Thus, the Euler formula

$$y_{i+1} = y_i + (x_{i+1} - x_i)f(x_i, y_i)$$

is replaced in Picard's proof by

$$y_{i+1}(x) = y_i(x) + \int_{x_0}^{x} f(x, y_i(x)) \, dx .$$

Here we can see the difference between the two problems, and their resolution. With Euler, we look for successive approximations in order to calculate the value of a function at a point x; with Picard we look for successive approximations to prove the existence of a function throughout an interval $[x_0, x]$. In other words, the variable that is the object of successive approximations is, in the first case a number, and in the second case a function.

Even though the purpose of the second method is to provide a proof of the existence of a solution, we can also interpret it as a calculation algorithm for that solution, to the extent that we are able to carry out the quadratures that successively appear, that is that we know how to calculate the integrals

$$\int_{x_0}^{x} f(x, y_i(x)) \, dx .$$

From this we have an algorithm for solving differential equations by successive quadratures, which allows us to approximate to the exact solution over the whole of the local existence interval.

12.3 Runge's Methods

The development of the science of celestial mechanics, with Lagrange, Laplace, Gauss and Le Verrier in particular, called for more powerful procedures for the solution of differential equations. The theory of ballistics also required similar tools.

Now, we have seen, in the justification of Euler's method, that the error involved in the approximation is bounded above by Kh where K is a constant and h is the incremental step; the method can be said to be of order 1. Therefore, to achieve an accuracy as good as is required, it is sufficient, in principle, to choose h sufficiently small, but reducing the value of h also means increasing the number of operations and, consequently, increasing the rounding errors. After a certain point, the accuracy does not improve and ends up by getting worse; there is a need therefore to improve the method and, in particular, to raise its order.

Instead of assimilating a curve to its tangent at certain points, we could improve the accuracy by using the Taylor expansion (see Section 13.1) of the solution φ to an order greater than 1:

$$\varphi(x) = \varphi(x_0) + (x - x_0)\varphi'(x_0) + (x - x_0)^2\varphi''(x_0)/2! + \dots$$
$$+ (x - x_0)^n\varphi^{(n)}(x_0)/n! + \dots$$

In this way, we can use as an approximation for $\varphi(x_1)$ the quantity

$$y_1 = \varphi(x_0) + (x_1 - x_0)\varphi'(x_0) + (x_1 - x_0)^2\varphi''(x_0)/2! + \dots$$
$$+ (x_1 - x_0)^n\varphi^{(n)}(x_0)/n! + \dots$$

since the quantities $\varphi^{(j)}(x_0)$ can be calculated by successive differentiation from the identity:

$$\varphi'(x) = f(x, \varphi(x)).$$

Thus,

$$\varphi''(x) = f_x'(x, \varphi(x)) + f_y'(x, \varphi(x))\,\varphi'(x) = f_x'(x, \varphi(x)) + f(x, \varphi(x))\,f_y'(x, \varphi(x)).$$

Now, the calculations of these derivatives quickly become very tedious, and Runge proposed other procedures, inspired by calculating integrals. Evaluating an integral, and integrating a first order differential equation, are two processes that are closely linked, since if φ is the solution of the differential equation $y' = f(x, y)$, then:

$$\varphi(x_{i+1}) - \varphi(x_i) = \int_{x_i}^{x_{i+1}} \varphi'(x)\,dx = \int_{x_i}^{x_{i+1}} f(x, \varphi(x))\,dx$$

In Euler's method, the last integral is evaluated assuming f is constant over the interval $[x_i, x_{i+1}]$. This is associated with the rectangle method for evaluating an in-

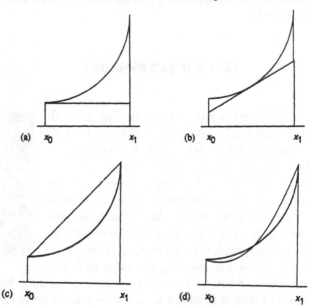

Approximation to the area under a curve by the methods of (a) rectangles, (b) tangent trapeziums, (c) chord trapeziums, (d) Simpson.

tegral. The method can, of course, be improved with f being linear, as in the trapezium methods, using tangents or chords (see figures below), or f can be a quadratic function, as in Simpson's Rule (see Section 11.5). These improvements suggested alternative methods to Runge for the approximation of solutions to differential equations, which were more powerful, since they were of order 2 or 3, as he explains in an article of 1895.

It should be added that the method inspired by chord trapeziums, described below by Runge, had already been partially formulated by Coriolis in 1837 [4].

C. Runge
Ueber die numerische Auflösung von Differentialgleichungen,
Mathematische Annalen, vol. 46 (1895), 167–178.

On the numerical solution of differential equations

The calculation of any sort of numerical solution of a given differential equation whose analytical solution is not known appears to have excited little interest among mathematicians up till now, if we exclude singularities, where the particular circumstances allow the calculation to be dealt with by quadratures. It appears that it is not known that the methods of numerical calculation of integrals can be generalised, so that they can be applied, with the same success, to any differential equation. I shall describe a generalisation of the well known Simpson's Rule, whose application appears to me to be particularly useful, which is not to say that other methods of mechanical integration might not also themselves provide usable generalisations.

Euler makes the remark in his Introduction, that a solution of the differential equation $dy/dx = f(x, y)$ can be calculated by approximations, starting from a pair of values (x_0, y_0), and with an increment in y of $\Delta y = f(x_0, y_0)\Delta x$ corresponding to a small increment Δx. From the newly found point $x_1 = x_0 + \Delta x$, $y_1 = y_0 + \Delta y$ we calculate anew the increment of y corresponding to a small increment Δx of x, which we put equal to $f(x_1, y_1)\Delta x$. Continuing in this way, we obtain a polygonal line which approaches, as closely as we wish, a solution of the differential equation, provided that each Δx and Δy are sufficiently small. Later Cauchy provided a rigorous proof of this, by the use of certain hypotheses which I shall not recall here. Euler's method is not very accurate, as can be seen immediately on applying the calculations to an integral of the form $\int_{x_0}^{x} f(t)\, dt$ which can be considered as the solution of the differential equation $\dfrac{dy}{dx} = f(x)$.

In fact, Euler's procedure would give the approximate value

$$f(x_0)\,(x_1 - x_0) + f(x_1)\,(x_2 - x_1) + \ldots$$

whose deviation from the true value, as can immediately be seen geometrically, is in general of the same order as the intervals $x_1 - x_0, x_2 - x_1, \ldots$. It is known that the values

$$f(\tfrac{x_0 + x_1}{2})(x_1 - x_0) + f(\tfrac{x_1 + x_2}{2})(x_2 - x_1) + \ldots$$

and
$$\frac{f(x_0) + f(x_1)}{2}(x_1 - x_0) + \frac{f(x_1) + f(x_2)}{2}(x_2 - x_1) + \ldots$$

are much better, and whose deviations from the true values are of the second order with respect to the intervals. The first of these two represents, geometrically, the sum of the 'tangent' trapeziums, that is, the trapeziums that are bordered above by the tangents at the mid-point of the intervals. The second aproximate value is the sum of chord trapeziums, that is, the sum of the trapeziums bordered above by the chords joining two neighbouring points. Using N_1 for the first approximate value and N_2 for the second, then even better approximation can be obtained in using

$$N_1 + \tfrac{1}{3}(N_2 - N_1)$$

the geometrical significance of which is the sum of 'parabola' strips, that is strips bordered above by arcs of parabolas which each pass through three points of the curve, corresponding to the ends and the mid-point of each interval. This is the approximation that is given by 'Simpson's Rule', and the error is of the fourth order with respect to the intervals.

An analogous consideration leads to a substantial improvement in Euler's method for differential equations. I shall restrict myself firstly to first order equations.

Instead of putting

(1) $\Delta y = f(x_0, y_0)\Delta x$, etc. ...

it is already preferable to put

(2) $\Delta y = f(x_0 + \tfrac{1}{2}\Delta x, y_0 + \tfrac{1}{2} f(x_0, y_0)\Delta x).\Delta x$, etc. ...

This calculation corresponds to that of tangent trapeziums for the integral and coincides exactly with it when $f(x, y)$ is supposed independent of y.

We could also use, as with 'chord' trapeziums,

(3) $\Delta y = \dfrac{f(x_0, y_0) + f(x_0 + \Delta x, y_0 + f(x_0, y_0)\Delta x)}{2} \Delta x$, etc. ...

If the true value of Δy, which we can imagine as expanded as a series of increasing powers of Δx, is compared with the values given by the approximation process, these also expanded as a series in Δx, it can be see that the primitive method of Euler gives errors of a second order and the other two errors of a third order.

In fact, the true expression for Δy is

$$\Delta y = f\,\Delta x + (f_1 + f_2 f)\frac{\Delta x^2}{2!} + (f_{11} + 2f_{12}f + f_{22}f^2 + f_2(f_1 + f_2 f))\frac{\Delta x^3}{3!} + \dots$$

where f_1 and f_2 are the partial derivatives of the 1st order of $f(x, y)$ and f_{11}, f_{12}, f_{22} those of the second order, with the usual notation.

Compared with this, the approximations for Δy are respectively

(1) $f\Delta x$

(2) $f\,\Delta x + (f_1 + f_2 f)\dfrac{\Delta x^2}{2} + (f_{11} + 2f_{12}f + f_{22}f^2)\dfrac{\Delta x^3}{8} + \dots$

(3) $f\,\Delta x + (f_1 + f_2 f)\dfrac{\Delta x^2}{2} + (f_{11} + 2f_{12}f + f_{22}f^2)\dfrac{\Delta x^3}{4} + \dots$

If, by analogy with Simpson's Rule, we add to (2) one third of the difference between (2) and (3), we obtain a new approximate value

$$f\,\Delta x + (f_1 + f_2 f)\frac{\Delta x^2}{2} + (f_{11} + 2f_{12}f + f_{22}f^2)\frac{\Delta x^3}{6} + \dots$$

which coincides with the true value as far as the third order, in the case where $f(x, y)$ is not dependent on y, and so $f_2 = 0$, but not in general.

The analogy with Simpson's Rule cannot therefore be maintained in this form. But it can be given another form. The expression (3) has to be replaced by another expression which corresponds to the chord trapezium form where $f(x, y)$ does not depend on y.

In fact, let us take for Δy

$$\tfrac{1}{2}(\Delta'y + \Delta'''y)$$

where $\Delta'y = f(x_0, y_0)\,\Delta x$, and $\Delta'''y$ is related to $\Delta'y$ by the formulas

$$\Delta''y = f(x_0 + \Delta x, y_0 + \Delta'y)\,\Delta x$$
$$\Delta'''y = f(x_0 + \Delta x, y_0 + \Delta''y)\,\Delta x$$

This approximation, expanded as a series of powers of Δx, gives

(3a) $$f\,\Delta x + (f_1 + f_2 f)\frac{\Delta x^2}{2} + (f_{11} + 2f_{12}f + f_{22}f^2 + 2f_2(f_1 + f_2 f))\frac{\Delta x^3}{4} + \dots$$

The difference between (3a) and (2) is therefore:

$$\left[\tfrac{1}{8}(f_{11} + 2f_{12}f + f_{22}f^2) + \tfrac{1}{2}f_2(f_1 + f_2 f)\right]\Delta x^3 + \dots\,.$$

And when a third of this difference is added to (2), we obtain a new approximation

(4) $$f\,\Delta x + (f_1 + f_2 f)\frac{\Delta x^2}{2} + (f_{11} + 2f_{12}f + f_{22}f^2 + f_2(f_1 + f_2 f))\frac{\Delta x^3}{6} + \dots,$$

an expression that coincides with the true value up to order 3.

What is also interesting to observe in practical applications is better shown through an example

Consider the differential equation $\quad \dfrac{dy}{dx} = \dfrac{y - x}{y + x}$

We wish to find a solution that gives $y = 1$ for $x = 0$.

This can be expressed analytically. Using polar coordinates, the differential equation takes the form

$$d\varphi = -\frac{dr}{r}$$

from which, when φ has to be equal to $\dfrac{\pi}{2}$ for $r = 1$,

$$r = e^{\frac{\pi}{2} - \varphi}.$$

This formula allows us to check our procedure.

$$f(x, y) = \frac{y - x}{y + x}$$

Tangent trapeziums		Chord trapeziums	
x	y	x	y
0	1	0	1
$\Delta x = \underline{0.1}$	$f(0,1)\Delta x = \underline{0.1}$	$\Delta x = \underline{0.2}$	$f(0,1)\,\Delta x = \underline{0.2}$
0.1	1.1	0.2	1.2
$\Delta x = \underline{0.2}$	$f(0.1, 1.1)\Delta x = \underline{0.167}$		$f(0.2, 1.2)\Delta x = \underline{0.143}$
0.3	1.167		1.143
			$f(0.2, 1.143)\Delta x = 0.140$

$$\frac{f(0,1)+f(0.2,1.113)}{2}\Delta x = \frac{0.2+0.140}{2}=0.170$$

Therefore, we obtain by the tangent trapezium method $y = 1.167$, and by the chord trapezium method $y = 1.170$. We must add a third of the difference to the first value and so we put $y = 1.168$.

When the sequence of operations has been well understood, it is no longer necessary to write them out in all the detail that we have done here. I will do the calculations for two further stages, which can be understood without commentary.

x	y		x	y	
0.2	1.168		0.2	1.168	
0.15	0.106		0.3	0.212	←
0.35	1.274		0.5	1.380	
0.3	0.170			0.140	
0.5	1.338	Tangent trapeziums		1.308	
	1.341			0.134	←
	0.003	Chord trapeziums		0.346	
				0.173	
				1.341	

x	y		x	y	
0.5	1.339		0.5	2.339	
0.25	0.114		0.5	0.228	←
0.75	1.453		1	1.567	
0.5	0.160			0.110	
1	1.499	Tangent trapeziums		1.449	
	1.499	Chord trapeziums		0.092	←
				0.320	
				0.160	
				1.499	

The sizes of the last two steps have been taken greater than the first one. The closeness of the two values corresponding respectively to tangent trapeziums and chord trapeziums guarantees that the error will not be significantly greater than of the order of 1 unit in the third decimal place.

In fact, we find that, for $x = 1$, $r = e^{\frac{\pi}{2}-\varphi}$ gives $1.498 < y < 1.499$

We can now summarise the three methods proposed by Runge. The calculation of the value $y = \varphi(x)$ where φ satisfies the differential equation $\frac{dy}{dx} = f(x,y)$, with the initial conditions (x_0, y_0), can be carried out, with a constant incremental step of $h = (x - x_0)/n$, in the following way:

for $i = 0$ to $n - 1$:

$x_{i+1} = x_i + h$

method inspired by 'tangent trapeziums':

$y_{i+1} = y_i + hf(x_i + \frac{h}{2}, y_i + \frac{h}{2}f(x_i, y_i))$

method inspired by 'chord trapeziums':

$y_{i+1} = y_i + \frac{h}{2}[f(x_i, y_i) + f(x_i + h, y_i + hf(x_i, y_i))]$

method inspired by Simpson's Rule:

$y'_{i+1} = y_i + hf(x_i, y_i)$ (Euler)

$y''_{i+1} = y_i + hf(x_i + \frac{h}{2}, y_i + \frac{h}{2}f(x_i, y_i))$ (tangent trapeziums)

$\bar{y}'''_{i+1} = y_i + hf(x_i + h, y_i + hf(x_i, y_i)) = y_i + hf(x_{i+1}, y'_{i+1})$

$y'''_{i+1} = y_i + hf(x_{i+1}, \bar{y}'''_{i+1})$

$\dfrac{y'_{i+1} + \bar{y}'''_{i+1}}{2}$ (chord trapeziums), $\dfrac{y'_{i+1} + y'''_{i+1}}{2}$ (modified chord trapeziums)

$y_{i+1} = \frac{2}{3}y''_{i+1} + \frac{1}{3}[\frac{1}{2}(y'_{i+1} + y'''_{i+1})] = \frac{1}{6}[y'_{i+1} + 4y''_{i+1} + y'''_{i+1}]$

We should point out that the first two algorithms do not strictly translate one or other of the trapezium methods since an application of these would lead, respectively, to the equations:

$$y_{i+1} = y_i + hf(x_i + \frac{h}{2}, \varphi(x_i + \frac{h}{2}))$$

and $y_{i+1} = y_i + \frac{h}{2}[f(x_i, y_i) + f(x_{i+1}, y_{i+1})]$

The first requires knowing $\varphi(x_i + \frac{h}{2})$ and the second is an equation in y_{i+1} that is *a priori* difficult to solve. This is why Runge approximates these unknown values by the values produced by the Euler method itself. This was also the reason why Runge calculates y'''_{i+1} by using intermediary steps in the application of the modified Simpson method.

Cauchy, in his 12th *leçon* at the Ecole Polytechnique [2], repeated again by Moigno in his 28th *leçon* [13], proposes a calculation which recalls this method of the chord trapeziums, but where he indicates, in effect, a solution, by successive approximations, of the equation:

$$y'''_{i+1} = y_i + hf(x_{i+1}, y'''_{i+1}),$$

before evaluating the arithmetic mean:

$$y_{i+1} = \frac{y'_{i+1} + y'''_{i+1}}{2}, \text{where } y'_{i+1} = y_i + h f(x_i, y_i).$$

If the function f is 'sufficiently regular', then the first two algorithms correspond to methods of order 2 and the last to a method of order 3, that is, roughly, that when the incremental step h is divided by 2, the error will be divided respectively by $4 = 2^2$ and by $8 = 2^3$.

In this respect, we should note that the order of approximation of a method is different according as it is a quadrature or a solution of a differential equation. Thus, Simpson's Rule is of order 4 for quadratures but only of order 3 for the solution of differential equations.

In his article, Runge goes on to consider the case of a second order differential equation, which, as we know, reduces to the solution of a pair of first order differential equations. This is how he handles an example of a differential equation that describes a raindrop, which is known not to have an analytical solution, taking the example from Adams who had explained his method of dealing with the same equation (see 12.6 below).

12.4 Heun's Methods

We have seen, in the article by Runge, that the order of approximation solution methods can be improved by considering combinations of approximations. For example, Simpson's Rule can be improved to an order 3 method when we consider

$$\Delta y = \tfrac{1}{2}(\Delta' y + \Delta''' y)$$

where $\Delta''' y$ is found using $\Delta' y$ and then $\Delta'' y$. In 1900, Karl Heun systematised this possibility.

But more than that, he drew inspiration from the methods of quadrature presented by Gauss (see Section 11.6). For, evaluating $\int_x^{x+\Delta x} f(x)\,dx$ is the same as calculating the value $y + \Delta y$ at $x + \Delta x$ of the function φ which satisfies the particular differential equation

$$\frac{dy}{dx} = f(x)$$

where the function f does not depend on y.

Thus, Heun proposed obtaining Δy in the form of a linear combination of the various increments $\Delta_v y$ of y, with coefficients α_v, obtained by considering the values of $f(x, y)$ calculated at the same time for different values $x + \varepsilon_v \Delta x$ of the variable x in the interval $[x, x + \Delta x]$ (the method of Gauss) and for the corresponding values of y calculated by successive approximations (combinations of approximations).

Next, as with Gauss, the coefficients are required to satisfy conditions which ensure that, both the Taylor's series expansion of Δy, and the expansion of $\int_x^{x+\Delta x} f(x, y)\,dx$ according to the formulas proposed by Heun, should coincide to the highest possible order.

The extract below shows, in the case of a single approximation to y, the different formulas for obtaining order 2 approximations, and we find here, of course, as the first particular cases, the Runge 'trapezium method' formulas. But we also see here, how increasing the number of well chosen parameters may increase the order of approximation.

We need to be aware, in reading Heun, that the successive approximations of y are found from the increment $\Delta^{(m)}y$, defined by the last relation, which itself determines $\Delta^{(m-1)}y$, and so on back to $\Delta''y$, and then $\Delta'y$. In the text, the symbols Df and D^2f stand for:

$$Df = f'_x + ff'_y \quad \text{and} \quad D^2f = f''_{x^2} + 2ff''_{xy} + f^2 f''_{y^2}$$

So that, if φ is a solution of the differential equation, its first derivatives can be expressed in the form:

$$\varphi'(x) = Df \quad \text{and} \quad \varphi''(x) = D^2f + f'_y Df \,.$$

K. Heun

Neue Methode zur approximativen Integration der Differentialgleichungen einer unabhängigen Veränderlichen (New method for approximate integration of differential equations of an independent variable), *Zeitschrift für Mathematik und Physik*, vol. 45 (1900), 23–38.

2. The problem setting. The following interpretation can be given to the Gauss quadrature method. If the function y is given by the equation

$$\frac{dy}{dx} = f(x),$$

then the numbers α and ε have to be determined such that

$$\Delta y = \sum_{v=1}^{v=n} \alpha_v f(x + \varepsilon_v \Delta x).\Delta x$$

provides the most exact representation possible of y within the interval Δx. Corresponding to this requirement, there are a certain number of equations for determining the sought values α and ε. The method of integration for differential equations that we are going to develop here, considered first in their simplest form, is based on the following extension of the fundamental idea of Gauss.

We are going to try to determine a system of magnitudes

$$\alpha_1, \alpha_2, ..., \alpha_n\,;\, \varepsilon_1, \varepsilon_2, ..., \varepsilon_n\,;\, \varepsilon_1', \varepsilon_2', ..., \varepsilon_n'\,;\, \varepsilon_1'', \varepsilon_2'', ..., \varepsilon_n''\,; \text{etc.}$$

in such a way that the expression

$$\Delta y = \sum_{v=1}^{v=n} \{\alpha_v f(x + \varepsilon_v \Delta x, y + \Delta'_v y)\}.\Delta x \,,$$

where

$$\Delta'_v y = \{\varepsilon_v f(x + \varepsilon'_v \Delta x, y + \Delta''_v y)\}.\Delta x$$

$$\Delta''_v y = \{\varepsilon'_v f(x + \varepsilon''_v \Delta x, y + \Delta'''_v y)\}.\Delta x$$

$$\cdots\cdots\cdots\cdots\cdots\cdots\cdots\cdots\cdots\cdots$$

$$\Delta^{(m)}_v y = \varepsilon^{(m-1)}_v f(x, y).\Delta x \,,$$

represents, throughout the interval Δx, the function $y = F(x)$ given by the differential equation

$$\frac{dy}{dx} = f(x, y),$$

with the best approximation that can be achieved in this way. To this end, we can make the initial restrictive hypothesis that, by Taylor's theorem, $F(x)$ can be represented in the chosen interval by a convergent series of the argument. In establishing the final formulas, we shall only go as far as the fourth order of approximation, since this will perfectly suffice for practical purposes.

3. Second order approximations. For $m = 1$, we have the representation

$$\Delta y = \sum_{v=1}^{v=n} \{\alpha_v \, f(x + \varepsilon_v \, \Delta x, y + \Delta'_v y)\}. \Delta x$$

where $\Delta'_v \, y = \varepsilon_v \, f(x, y) \Delta x$

or, omitting the index v and substituting $\Delta' y$,

$$\Delta y = \sum \{\alpha \, f(x + \varepsilon \Delta x, y + \varepsilon f \Delta x)\}. \Delta x$$

The expansion of Δy by Taylor's theorem gives

$$\Delta y = \Sigma \alpha \, f . \Delta x + \Sigma \alpha \varepsilon \, Df . \Delta x^2 + \frac{1}{2!} \Sigma \alpha \varepsilon^2 \, D^2 f . \Delta x^3 + \dots$$

But, on the other hand,

$$\Delta y = y' \, \Delta x + \frac{1}{2!} y'' \, \Delta x^2 + \frac{1}{3!} y''' \, \Delta x^3 + \dots$$

If the two values of y must agree up to the second order (inclusive), then α and ε will have to satisfy the equalities

$$\Sigma \alpha = 1, \quad \Sigma \alpha \varepsilon = \tfrac{1}{2}, \quad \dots \tag{a}$$

A complete agreement up to the third order is not accessible by this means. The equalities (a) contain $2n$ unknowns. Hence, $n \leq 1$. For $n > 1$, arbitrary conditions have to be attached. In the easiest case, $n = 1$, we obtain $\alpha = 1$, $\varepsilon = \tfrac{1}{2}$ and the corresponding approximation formula

(I) $\Delta y = f(x + \tfrac{1}{2} \Delta x, y + \tfrac{1}{2} f \Delta x). \Delta x$.

If we put $n = 2$, then the equalities

$$\alpha_1 + \alpha_2 = 1, \quad \alpha_1 \varepsilon_1 + \alpha_2 \varepsilon_2 = \tfrac{1}{2}$$

have to be satisfied.

The arbitrary hypothesis $\alpha_1 = \alpha_2$ leads to the formula:

(II) $\Delta y = \tfrac{1}{2} \{f(x, y) + f(x + \Delta x, y + f \Delta x)\}. \Delta x$.

For $\varepsilon_1 = \tfrac{1}{3}, \varepsilon_2 = \tfrac{2}{3}$, and so for equal divisions of the interval of the argument, the result is:

(III) $\Delta y = \tfrac{1}{2} \{f(x + \tfrac{1}{3} \Delta x, y + \tfrac{1}{3} f \Delta x) + f(x + \tfrac{2}{3} \Delta x, y + \tfrac{2}{3} f \Delta x)\}. \Delta x$.

When using a value of the function that comes before, it is convenient to take $\varepsilon_1 = -1, \varepsilon_2 = +1$. We then obtain

(IV) $\Delta y = \tfrac{1}{4} \{f(x - \Delta x, y - f \, \Delta x) + 3f(x + \Delta x, y + f \, \Delta x)\}. \Delta x$.

Thus the value that comes after is given here three times the weight of the one that comes before. If we wish to achieve for $n = 2$ the closest possible agreement with the analogous quadrature formula of Gauss, we would then need to add the following to the equalities (a):

$$\Sigma \alpha \varepsilon^2 = \tfrac{1}{3}, \quad \Sigma \alpha \varepsilon^3 = \tfrac{1}{4}.$$

From this we then have $\quad \alpha_1 = \alpha_2 = \tfrac{1}{2},$

$$\varepsilon_1 = \tfrac{1}{2}(1 - \tfrac{1}{3}\sqrt{3}), \quad \varepsilon_2 = \tfrac{1}{2}(1 + \tfrac{1}{3}\sqrt{3})$$

and we obtain the symmetrical formula

(V) $$\Delta y = \frac{1}{2} \{ f(x + \varepsilon_1 \Delta x, y + \varepsilon_1 f \Delta x) + f(x + \varepsilon_2 \Delta x, y + \varepsilon_2 f \Delta x) \} . \Delta x$$

where we use the values $\varepsilon_1 = 0.2113...$, and $\varepsilon_2 = 0.7887...$

The values of the numbers ε are the same as those which Gauss (Werke 3, 193) gave in the corresponding quadrature formula. The error to the third order (correctio proxima) is determined by the expression

$$C_3 = \frac{1}{3!} f_{01} \, Df . \Delta x^3 .$$

If $f(x, y)$ does not depend on y, then we have $f_{01} = f_{02} = 0$ and equation (V) is transformed directly into Gauss's formula, but then at the same time we leave the domain of differential equations.

In applications, it is easiest to use equation (1), if we do not require the use of a representation that is more strictly convergent, or, for this weak degree of accuracy, we could just as well use graphical integration which is almost equivalent.

Heun then goes on to study the case where $m = 2$, where two successive approximations to Δy have to be made. He proceeds by the same method of identifying expansions, notes that it is not possible to make the Δx^4 terms disappear and shows that the approximation of order 3 leads to four conditions:

$$\Sigma \alpha = 1, \quad \Sigma \alpha \varepsilon = \tfrac{1}{2}, \quad \Sigma \alpha \varepsilon^2 = \tfrac{1}{3}, \quad \Sigma \alpha \varepsilon \varepsilon' = \tfrac{1}{6}$$

Thus, for $n = 2$ and in putting for example $\varepsilon_1 = 0$, we obtain the formulas

$$\Delta y = \tfrac{1}{4} \{ f(x, y) + 3f(x + \tfrac{2}{3}\Delta x, y + \Delta'y) \} . \Delta x$$

$$\Delta'y = \tfrac{2}{3} f(x + \tfrac{1}{3}\Delta x, y + \tfrac{1}{3}\Delta x) . \Delta x .$$

In the case of approximations of order 4, evaluating the formulas quickly becomes more laborious, but the refinement proposed later by Kutta can simplify the calculations.

Heun later saw the possibility of handling the case of systems of equations. Moreover it is Heun's method which was used for the first integration of a system of differential equations by the analogue calculator ENIAC ([7], p. 287).

12.5 Kutta's Methods

Following on from the articles by Runge and Heun, Kutta proposed, in 1901, a way of perfecting Heun's formulas by a complication which, in the end, simplifies the calculations and leads to a better approximation. The extract from the article by Kutta given below begins just after he had reviewed Heun's technique, where he remarks that in the different approximations of Δy the calculation of the gradient of $\Delta^{(k+1)}y/\Delta x$, is carried out from a point situated on the straight line of gradient $\Delta^{(k)}y/\Delta x$ passing through the point (x, y), since

$$\Delta^{(k+1)}y = f(x + \varepsilon_k \Delta x, y + \varepsilon_k \Delta^{(k)}y).\Delta x$$

(with a number of changes of notation).

M. W. Kutta

Beitrag zur näherungweisen Integration totaler Differentialgleichungen (Contribution to the approximate integration of ordinary differential equations), *Zeitschrift für Mathematik und Physik*, vol. 46 (1901), 435–453.

The difference between the method presented here and the one which I have just described is such that, after a certain number of directions have been calculated, the point where the next direction is to be calculated is no longer supposed to be situated on one of the earlier directions, issuing from the starting point, but will itself be reached by a polygonal line whose directions have already been calculated. The analytical expression is the following: let Δ', Δ'', ..., $\Delta^{(v-1)}$ be already calculated, but we calculate the next value as:

$$\Delta^{(v)} = f(x + \kappa \Delta x, y + \kappa_1 \Delta' + \kappa_2 \Delta'' + ... + \kappa_{v-1} \Delta^{(v-1)}) \Delta x,$$

where the κ are arbitrary numerical coefficients, which are only related to each other by the requirement that our point should be, up to the second order, situated on the tangent of the integral curve at the starting point, that is by the relation:

$$\kappa = \kappa_1 + \kappa_2 + ... + \kappa_{v-1}.$$

As concerns the simple numerical values, as we shall find later, the complication of the calculation due to the necessary summation of the Δ', ..., $\Delta^{(v-1)}$ is quite small. On the other hand, we obtain a greater choice of numerical coefficients, which allows us to establish approximations of a given order, by calculating the least possible values of the function, and to take rational coefficients. The formulation is therefore as follows: We set down:

$$\begin{aligned}
\Delta' &= f(x, y)\, \Delta x, \\
\Delta'' &= f(x + \kappa \Delta x, y + \kappa \Delta')\, \Delta x, \\
\Delta''' &= f(x + \lambda \Delta x, y + \rho \Delta'' + (\lambda - \rho)\Delta')\, \Delta x, \\
\Delta'''' &= f(x + \mu \Delta x, y + \sigma \Delta''' + \tau \Delta'' + (\mu - \sigma - \tau)\Delta')\, \Delta x, \\
\Delta^{v} &= f(x + v \Delta x, y + \varphi \Delta'''' + \chi \Delta''' + \psi \Delta'' + (v - \varphi - \chi - \psi)\Delta')\, \Delta x,
\end{aligned}$$

and for the required approximation, we put:

$$\Delta y = a\Delta' + b\Delta'' + c\Delta''' + d\Delta'''' + e\Delta^{v} + ...$$

Here the quantities κ, λ, μ, v, ... ; ρ, σ, τ, φ, χ, ψ, ... ; a, b, c, d, e, ... are numerical coefficients that can be chosen arbitrarily. These must be determined in such a way that, in expanding Δy

by Taylor's formula, we obtain an agreement with the true expansion, which comes from the integral equation, as far as the required order.

For an approximation of the first order, it is sufficient to calculate one value of the function f, and from $a = 1$, we have the trivial approximation

$$\Delta y = f(x, y) \, \Delta x.$$

For the second order two values of the function have to be calculated, and we obtain the already known condition equations (see Herr Heun, p. 28).

$$a + b = 1, \quad b\kappa = \tfrac{1}{2},$$

which we do not need to discuss.

For the third order, three values of the function have to be calculated, and Taylor's formula provides, for the required approximation

$$\Delta y = a\Delta' + b\Delta'' + c\Delta''', \text{ where}$$
$$\Delta' = f(x, y) \, \Delta x,$$
$$\Delta'' = f(x + \kappa \, \Delta x, y + \kappa \, \Delta') \, \Delta x,$$
$$\Delta''' = f(x + \lambda \, \Delta x, y + \rho \, \Delta'' + (\lambda - \rho)\Delta') \, \Delta x,$$

the four following conditions:

$$a + b + c = 1, \quad b\kappa + c\lambda = \tfrac{1}{2}, \quad b\kappa^2 + c\lambda^2 = \tfrac{1}{3}, \quad c\rho\kappa = \tfrac{1}{6},$$

from which we derive the system of solutions:

$$\rho = \frac{\lambda(\lambda - \kappa)}{\kappa(2 - 3\kappa)}; \quad a = \frac{6\kappa\lambda - 3(\kappa + \lambda) + 2}{6\kappa\lambda}$$

$$b = \frac{2 - 3\lambda}{6\kappa(\kappa - \lambda)}; \quad c = \frac{2 - 3\kappa}{6\lambda(\lambda - \kappa)}$$

This system gives a double infinity of solutions, of which there are an infinity of solutions for $\rho = \lambda$.

(I) $\qquad\qquad \lambda = 3\kappa(1 - \kappa); \quad a = \dfrac{2 - 12\kappa + 27\kappa^2 - 18\kappa^3}{18\kappa^2(1 - \kappa)}$

$$b = \frac{3\kappa - 1}{6\kappa^2}; \quad c = \frac{1}{18\kappa^2(1 - \kappa)}$$

from which we can obtain the relations given by Herr Heun.

As particular cases which can be obtained either through a passage to the limit from the general formulas given above, or directly from the system of equations, we can mention:

(II) $\qquad \kappa = \tfrac{2}{3}; \quad \lambda = 0; \quad b = \tfrac{3}{4}; \quad c = \tfrac{1}{4\rho}; \quad a = \tfrac{1}{4} - \tfrac{1}{4\rho}.$

(III) $\qquad \kappa = \tfrac{2}{3}; \quad \lambda = \tfrac{2}{3}; \quad a = \tfrac{1}{4}; \quad c = \tfrac{1}{4\rho}; \quad b = \tfrac{3}{4} - \tfrac{1}{4\rho}.$

And then the cases:

(IV) $\qquad \lambda = \tfrac{2}{3}; \quad a = \tfrac{1}{4}; \quad b = 0; \quad c = \tfrac{3}{4}; \quad \rho = \tfrac{2}{9\kappa}.$

(V) $\qquad \lambda = \dfrac{3\kappa - 2}{3(2\kappa - 1)}; \quad a = 0.$

The last two cases only require two values of the function to be entered in the final result, which reduces the accuracy with which the third needs to be calculated by an order, as mentioned above. Clearly the quantity c cannot become zero, nor can a and b simultaneously.

Even when calculating more than three values of the function, the number of parameters that come into the final formula cannot be made to be less than two.

Each of the cases (I) to (V) again yield an infinite number of solutions. As an example, let us choose:

$$\kappa = \tfrac{2}{3}; \quad \lambda = \tfrac{2}{3}; \quad \rho = \tfrac{2}{3},$$

which fit both cases (I) and (III), and we then find:

$$a = \tfrac{1}{4}; \quad b = \tfrac{3}{8}; \quad c = \tfrac{3}{8},$$

and the corresponding approximation:

$$\Delta y = \frac{2\Delta' + 3\Delta'' + 3\Delta'''}{8},$$
$$\Delta' = f(x, y)\,\Delta x,$$
$$\Delta'' = f(x + \tfrac{2}{3}\Delta x, y + \tfrac{2}{3}\Delta')\Delta x,$$
$$\Delta''' = f(x + \tfrac{2}{3}\Delta x, y + \tfrac{2}{3}\Delta'')\Delta x.$$

With the system $\kappa = \tfrac{1}{3}$; $\lambda = \tfrac{2}{3}$; $\rho = \tfrac{2}{3}$, we can obtain with the formulas (I) or (IV) the results already obtained by Herr Heun:

$$\Delta y = \frac{\Delta' + 3\Delta'''}{4},$$
$$\Delta' = f(x, y)\,\Delta x,$$
$$\Delta'' = f(x + \tfrac{1}{3}\Delta x, y + \tfrac{1}{3}\Delta')\Delta x,$$
$$\Delta''' = f(x + \tfrac{2}{3}\Delta x, y + \tfrac{2}{3}\Delta'')\Delta x.$$

The system $\kappa = \tfrac{2}{3}$; $\lambda = \tfrac{2}{3}$; $\rho = \tfrac{1}{3}$, gives:

$$\Delta y = \frac{\Delta' + 3\Delta'''}{4},$$
$$\Delta' = f(x, y)\,\Delta x,$$
$$\Delta'' = f(x + \tfrac{2}{3}\Delta x, y + \tfrac{2}{3}\Delta')\Delta x,$$
$$\Delta''' = f(x + \tfrac{2}{3}\Delta x, y + \tfrac{\Delta' + \Delta''}{3})\Delta x.$$

The system $\kappa = \tfrac{1}{2}$, $\lambda = 1$ gives:

$$\Delta y = \frac{\Delta' + 4\Delta'' + \Delta'''}{6},$$
$$\Delta' = f(x, y)\,\Delta x,$$
$$\Delta'' = f(x + \tfrac{1}{2}\Delta x, y + \tfrac{1}{2}\Delta')\Delta x,$$
$$\Delta''' = f(x + \Delta x, y + 2\Delta'' - \Delta')\Delta x.$$

The last of these formulas has a certain analogy to Simpson's Rule, but it is of a less precise order.

Kutta went on to consider approximations of order 4 and order 5 in Δx. He obtained, in particular, for order 4, the now established formulas:

$$\Delta y = \frac{\Delta' + 3\Delta'' + 3\Delta''' + \Delta''''}{8},$$

$$\Delta' = f(x, y)\,\Delta x,$$

$$\Delta'' = f(x + \tfrac{1}{3}\Delta x, y + \tfrac{1}{3}\Delta')\Delta x,$$

$$\Delta''' = f(x + \tfrac{2}{3}\Delta x, y + \tfrac{1}{3}(3\Delta'' - \Delta'))\Delta x,$$

$$\Delta'''' = f(x + \Delta x, y + \Delta''' - \Delta'' + \Delta')\Delta x.$$

This formula reminds us, of course, of Newton's three-eighths rule for approximate quadratures (Section 11.2).

At the end of his article, Kutta compares the different methods applied to an example, actually the one used by Runge, namely the differential equation:

$$\frac{dy}{dx} = \frac{y - x}{y + x},$$

with the initial conditions $x = 0, y = 1$, which is known to have the solution given by:

$$\ln(x^2 + y^2) = 2\arctan\frac{x}{y}.$$

Kutta compares the different methods to the exact solution and, so as to make the comparison fair, for each method, he takes account of 4 values of the function for each interval Δx. In order, he illustrates:

- Taylor's Formula to order 4,
- Euler's method, by dividing the interval Δx into 4, that is using a polygonal line with four segments,
- Runge's method in the order 3 version, otherwise called the modified Simpson's method,
- Heun's method in the order 3 version, which we referred to in our commentary following Heun's text; this requires the use of 3 values of the function,
- Kutta's method in the order 4 version, which we have just looked at, and which uses 4 values of the function.

This last turns out to be the most effective, as can be seen from the tables given below, which are taken from Kutta's article.

The first table gives the values of Δy produced by the different methods, for $x = 0.2$ (I), $x = 0.5$ (II) and $x = 1$ (III), calculated from respective initial values of 0, 0.2

Δy	I	II	III
Taylor	0.1666667	0.3368533	0.4936913
Euler	0.1754353	0.3573505	0.5367900
Runge	0.1678487	0.3393690	0.4991167
Heun	0.1680250	0.3395806	0.4990390
Kutta	0.1678449	0.3392158	0.4982940
true value	0.1678417	0.3392094	0.4982784

and 0.5 for x and the previously calculated values for y. The second table gives the deviations from the exact values.

errors	I	II	III
Taylor	– 11750	– 23561	– 45871
Euler	+ 75936	+ 181411	+ 385116
Runge	+ 70	+ 1596	+ 8383
Heun	+ 1833	+ 3712	+ 7606
Kutta	+ 32	+ 64	+ 156

In the book by Runge and König published in 1924, we see that the method that Runge had developed is referred to as the Runge-Kutta method ([19], p. 286). It is not easy to make a clear distinction between the methods that Heun established and those of Kutta and Runge, as is often done in some numerical analysis books. It is rather that there was a progression over time, with successive refinements of methods, each author referring back to the work of a predecessor. Taking Euler's method as a starting point Runge proposes new formulas to improve it, by analogy with trapezium methods and Simpson's Rule for integration. Heun uses Gauss's idea in order to improve the accuracy of quadratures without having to increase the number of points being taken. Kutta, finally, introduces supplementary parameters which, by a judicious choice, improves accuracy without introducing too much extra complication. Each proposed new extension is compared each time to the Taylor series expansion. This is not done, however, to determine an upper bound for the remainder, but simply making the assumption that agreement with Taylor's expansion up to order k will give an indication of the size of the error. Moreover, it will be exact if the function f is 'sufficiently regular'.

Reasoning by analogy with approximate quadratures could also lead one to think of using Newton's interpolation polynomials, and therefore finite differences as used by Gregory (see Section 11.1). And it so happened, even before the methods of Runge, Heun and Kutta had been developed, that John Adams in 1883 proposed using finite differences, as we shall see in the next section.

12.6 John Adams and the Use of Finite Differences

The quadrature approximation formulas that Runge transposed for calculating the solutions of differential equations corresponded to approximating the functions to be integrated by polynomials of degree 1 or 2. We can continue in this vein by considering Newton interpolation polynomials of degree n and the corresponding interpolation formulas. From this we can deduce numerical integration formulas for differential equations where the differences $y_{i+1} - y_i$ will be expressed in the form of linear combinations of finite differences deduced from the successive values of (x_i, y_i), in the same way that Gregory obtained integration formulas through using finite differences (Section 11.1).

$$\int_0^c f(x_0 + u)\, du = \frac{c}{2}\Delta f_0 - \frac{c}{12}\Delta^2 f_0 + \frac{c}{24}\Delta^3 f_0 - \frac{19c}{720}\Delta^4 f_0 + \frac{3c}{160}\Delta^5 f_0 - \frac{863c}{60480}\Delta^6 f_0 + \ldots$$

John Adams who had, with Le Verrier, discovered the planet Neptune by calculation in 1846, developed this method of integration in 1883. He described it in connection with determining the shape of a water droplet as part of a study on capillary action. The meridian of the droplet satisfies a differential equation which he did not know how to solve, and so he calculated the shape by approximation using finite differences. The method was used twice: first to determine, by any method, the approximate values of the solutions at consecutive points, then to use the method proper on these calculated values.

We should beware that, in the extract given below, what Adams refers to as Δq_0 does not correspond with the forward difference $q_1 - q_0$, but to the backward difference $q_0 - q_{-1}$, usually noted by ∇q_0 (see Section 10.4).

J. C. Adams
An explanation of the method of integration employed, in Francis Bashforth, *An attempt to test the theories of capillary action,* Cambridge: Cambridge University Press, 1883, pp. 17–21.

[...]

After a few points of the curve, in the neighbourhood of the starting point, have been determined by the foregoing or some equivalent method, it will usually be found more convenient to determine other points of the curve in succession by making use of a series of successive values of the differential coefficient which is given immediately by the differential equation, rather than by employing the values of the successive differential coefficients of higher orders which are found by means of the several derived equations.

To fix the ideas we will suppose, with especial reference to our present problem, that the given differential equation is one of the first order, say

$$\frac{dy}{dt} = q = f(y, t)$$

Let $\ldots, t_{-4}, t_{-3}, t_{-2}, t_{-1}, t_0, t_1,$ &c. be a series of values of the independent variable t, forming an arithmetical progression with the common difference ω.

Let $\ldots, y_{-4}, y_{-3}, y_{-2}, y_{-1}, y_0, y_1,$ &c.

denote the corresponding values of y, and let

$$\ldots, q_{-4}, q_{-3}, q_{-2}, q_{-1}, q_0, q_1, \text{&c.}$$

be the corresponding values of q, or of $\dfrac{dy}{dt}$,

and suppose ω to be so small that the successive differences of these values of q soon become small enough to be neglected.

Let $t = t_0 + n\omega,$

and suppose that we have already found the values of

$$\ldots, y_{-4}, y_{-3}, y_{-2}, y_{-1}, \text{up to } y_0$$

and therefore also those of $\ldots, q_{-4}, q_{-3}, q_{-2}, q_{-1}, \text{up to } q_0,$

and that the successive differences of these quantities are taken according to the following scheme:

n	q				
...	...		...		
			...	...	
-4	q_{-4}		...	...	
		Δq_{-3}		...	
-3	q_{-3}		$\Delta^2 q_{-2}$	...	&c.
		Δq_{-2}		$\Delta^3 q_{-1}$	
-2	q_{-2}		$\Delta^2 q_{-1}$		$\Delta^4 q_0$ &c.
		Δq_{-1}		$\Delta^3 q_0$	
-1	q_{-1}		$\Delta^2 q_0$		
		Δq_0			
0	q_0				

Then the general value of q found by the ordinary formula of interpolation, for any value of n, will be

$$q = q_0 + \Delta q_0 \frac{n}{1} + \Delta^2 q_0 \frac{n(n+1)}{1.2} + \Delta^3 q_0 \frac{n(n+1)(n+2)}{1.2.3} + \Delta^4 q_0 \frac{n(n+1)(n+2)(n+3)}{1.2.3.4} + \&c.$$

provided that n be taken between limits for which this series remains convergent.

Hence the general value of y will be

$$y = \int q \, dt = \omega \int q \, dn,$$

or, substituting the above value of q, and adding a constant to the integral so as to make $y = y_0$ when $n = 0$,

$$y = y_0 + \omega \left\{ q_0 n + \Delta q_0 \frac{n^2}{2} + \Delta^2 q_0 \int \frac{n(n+1)}{1.2} \, dn + \Delta^3 q_0 \int \frac{n(n+1)(n+2)}{1.2.3} \, dn + \&c. \right\}$$

where all the integrals are supposed to vanish when $n = 0$.

If, in particular, we put $n = -1$, and substitute the several values of the definite integrals

$$\int_0^{-1} \frac{n(n+1)}{1.2} \, dn, \quad \int_0^{-1} \frac{n(n+1)(n+2)}{1.2.3} \, dn, \quad \&c.$$

we shall have, by changing the signs throughout,

$$y_0 - y_{-1} = \omega \left\{ q_0 - \frac{1}{2} \Delta q_0 - \frac{1}{12} \Delta^2 q_0 - \frac{1}{24} \Delta^3 q_0 - \frac{19}{720} \Delta^4 q_0 - \frac{3}{160} \Delta^5 q_0 - \frac{863}{60480} \Delta^6 q_0 \right.$$
$$\left. - \frac{275}{24192} \Delta^7 q_0 - \frac{33953}{3628800} \Delta^8 q_0 - \frac{8183}{1036800} \Delta^9 q_0 - \&c. \right\}$$

Similarly, putting $n = 1$ and substituting the values of the definite integrals

$$\int_0^1 \frac{n(n+1)}{1.2} \, dn, \quad \int_0^1 \frac{n(n+1)(n+2)}{1.2.3} \, dn, \quad \&c.$$

we shall have

$$y_1 - y_0 = \omega\left\{q_0 + \frac{1}{2}\Delta q_0 + \frac{5}{12}\Delta^2 q_0 + \frac{3}{8}\Delta^3 q_0 + \frac{251}{720}\Delta^4 q_0 + \frac{95}{288}\Delta^5 q_0 + \frac{19087}{60480}\Delta^6 q_0\right.$$

$$\left. + \frac{5257}{17280}\Delta^7 q_0 + \frac{1070017}{3628800}\Delta^8 q_0 + \frac{2082753}{7257600}\Delta^9 q_0 + \&c.\right\}$$

It will usually be found expedient to choose ω so small as to render it unnecessary to proceed beyond the fourth order of differences.

The series last found gives the value of y_1 in terms of quantities which are all supposed to be already known, that is, the value of the variable y which was previously known for values of the independent variable extending as far as $t = t_0$, now becomes known for the value $t = t_0 + \omega$, or at the end of an additional interval ω.

It will be remarked, however, that the coefficients of the series above found for $y_0 - y_{-1}$, after the first two terms, are much smaller and diminish much more rapidly than the corresponding coefficients of the series for $y_1 - y_0$. Hence by taking into account the same number of terms of the series in the two cases, the value of $y_0 - y_{-1}$ will be determined with much greater accuracy than that of $y_1 - y_0$.

In what has gone before, the successive values of y up to y_0 are supposed to be already known, and therefore the equation which gives the value of $y_0 - y_{-1}$ may be regarded as merely supplying a verification of former work. If, however, we suppose that the value of y_0 is only approximately known, while the successive values as far as y_{-1} have been found with the degree of accuracy desired, we may use the equation for $y_0 - y_{-1}$ to give the corrected value of y_0, in the following manner.

Suppose that (y_0) is an approximate value of y_0, and let $y_0 = (y_0) + \eta$, where η is so small that its square may be neglected.

Also let (q_0) be the corresponding approximate value of q_0 found from the equation

$$q = f(y, t)$$

by putting $y = (y_0)$ and $t = t_0$.

Then we may put
$$q_0 = (q_0) + k\eta,$$

where k denotes the value of the partial differential coefficient $\dfrac{dq}{dy}$ or $\dfrac{df(y,t)}{dy}$ found by substituting (y_0) for y and t_0 for t after the differentiation.

Let $\Delta(q_0)$, $\Delta^2(q_0)$, $\Delta^3(q_0)$, $\Delta^4(q_0)$, &c. denote the values of the successive differences formed with the approximate value (q_0) and the known values q_{-1}, q_{-2}, &c. which immediately precede it, then we have

$$\Delta q_0 = \Delta(q_0) + k\eta,$$
$$\Delta^2 q_0 = \Delta^2(q_0) + k\eta,$$
$$\Delta^3 q_0 = \Delta^3(q_0) + k\eta,$$
$$\&c. = \&c.$$

But, by the equation before obtained,

$$y_0 - y_{-1} = \omega\left\{q_0 - \frac{1}{2}\Delta q_0 - \frac{1}{12}\Delta^2 q_0 - \frac{1}{24}\Delta^3 q_0 - \frac{19}{720}\Delta^4 q_0 - \frac{3}{160}\Delta^5 q_0 - \frac{863}{60480}\Delta^6 q_0\right.$$

$$\left. - \frac{275}{24192}\Delta^7 q_0 - \frac{33953}{3628800}\Delta^8 q_0 - \frac{8183}{1036800}\Delta^9 q_0 - \&c.\right\}$$

Or, substituting for $y_0, q_0, \Delta q_0, \Delta^2 q_0$, &c. their values in terms of η and known quantities,

$$(y_0) - y_{-1} + \eta = \omega\left\{(q_0) - \tfrac{1}{2}\Delta(q_0) - \tfrac{1}{12}\Delta^2(q_0) - \tfrac{1}{24}\Delta^3(q_0) - \tfrac{19}{720}\Delta^4(q_0) - \&c.\right\}$$
$$+ \omega k\eta\left\{1 - \tfrac{1}{2} - \tfrac{1}{12} - \tfrac{1}{24} - \tfrac{19}{720} - \&c.\right\}.$$

Hence if ε denote the excess of the quantity

$$\omega\left\{(q_0) - \tfrac{1}{2}\Delta(q_0) - \tfrac{1}{12}\Delta^2(q_0) - \tfrac{1}{24}\Delta^3(q_0) - \tfrac{19}{720}\Delta^4(q_0) - \&c.\right\}$$

over the quantity $(y_0) - y_{-1}$, we shall have

$$\eta = \varepsilon + \omega k\eta\left\{1 - \tfrac{1}{2} - \tfrac{1}{12} - \tfrac{1}{24} - \tfrac{19}{720} - \&c.\right\}$$

or

$$\eta = \frac{\varepsilon}{1 - \omega k\left[1 - \tfrac{1}{2} - \tfrac{1}{12} - \tfrac{1}{24} - \tfrac{19}{720} - \&c.\right]},$$

which determines η, and therefore $y_0 = (y_0) + \eta$, and $q_0 = (q_0) + k\eta$ both become known.
If in finding ε we stop at the term involving $\Delta^4(q_0)$, we shall have

$$\eta = \frac{\varepsilon}{1 - \tfrac{251}{720}\omega k},$$

and

$$k\eta = \frac{k\varepsilon}{1 - \tfrac{251}{720}\omega k}.$$

[...]

The numerical operations will be greatly facilitated by the use of Tables which exhibit the values of

$$\tfrac{19}{720}\Delta^4 q, \quad \tfrac{3}{160}\Delta^5 q, \quad \tfrac{863}{60480}\Delta^6 q, \quad \&c.$$

for given values of $\Delta^4 q, \Delta^5 q, \Delta^6 q, \&c.$

Such tables have been formed by Mr. Bashforth for this purpose, and are given at the end of this Chapter.

Having made these preliminary observations on the general method of finding successive small portions of a curve by means of its differential equation, we will now proceed to apply the method to the problem under consideration, viz. to the tracing of the curve formed by a meridional section of a drop of fluid ...

John Adams thus proposes two formulas. The second allows him to calculate the value of the solution at a point t_1 from the already calculated values at $t_{-r}, t_{-r+1}, ...,$ t_{-1}, t_0, and is sometimes called a predictive formula or an extrapolation formula. The first, ties the value of the solution taken at t_0 to those taken at $t_{-r}, t_{-r+1}, ..., t_{-1}, t_0$, and can therefore be considered to be an implicit formula, to be solved finally by successive approximations or, to borrow the term from Adams, as a corrective formula, which allows us to improve the approximate value used at t_0. Moreover, these two formulas can be used successively in order to obtain sufficiently accurate values.

Nowadays, the second formula, the explicit formula, corresponds to what is often referred to in numerical analysis courses as the Adams-Bashforth method. In actual fact, the text given above appeared in a work by Bashforth [1], but the method itself would appear to be due to Adams alone. The first formula, the implicit formula, corresponds to what is called the Adams-Moulton method, from the name of an author of a number of works using this formula and its variants [14, Chap. 12].

Whereas the methods of Runge, Heun and Kutta were inspired by the works of Euler and Gauss, located within the German tradition, the multi-step methods presented by Adams derive from the Gregory-Newton formulas, and were developed uniquely within a British tradition. It is significant to note, at the end of the 19th century, the use of such methods by Sheppard to improve numerical tables and by the astronomer Darwin for calculating orbits of planets (see [20]).

The methods of finite differences are, of course, susceptible to many variations, and so methods of approximate solution of differential equations by the use of finite differences have been the subject of much attention. We can, for example, mention the names of Nyström and of Milne – and their formulas. These methods, based on finite differences, are particularly effective when dealing with regular functions.

12.7 Epilogue

We cannot end this chapter on approximate solutions of differential equations without mentioning, even if only briefly, the difficult question of the stability of the solar system. In fact, the system of differential equations governing evolution of the solar system has not been solved, since Poincaré showed that the same problem restricted to three bodies cannot be resolved. All calculations carried out by astronomers have thus been, and always will be, approximate ones. Poincaré explains:

It should not be believed that, in order to obtain an ephemeris [astronomical table] with great accuracy over a large number of years, it is sufficient to calculate a large number of terms in the expansions [...].

[...] The greater part of these expansions are not convergent in the sense that geometers give to that word. Doubtless, this is of little importance for the time being, since we are assured that the calculation of the first few terms gives a very satisfactory approximation; but it is no less true that these series are not able to provide an indefinite approximation. There will come a time, then, when they will become insufficient. Moreover, certain theoretical consequences that one might be tempted to draw from the form of these series are not legitimate, on account of their divergence. It is for this reason that they cannot be used to resolve the question of the stability of the solar system.   ([17], pp. 2–3)

Now, following the demonstration of the existence of the Lorenz attractor in 1963, physicists became aware of the possibility of there being very simple differential systems – three equations of the first order – which could produce chaotic behaviour, namely an infinite variation of initial positions could lead to extremely rapidly divergent corresponding trajectories.

Moreover, according to the astronomer Jacques Laskar, "in 1988, the Americans G. Sussman and J. Wisdom at MIT showed that the movement of Pluto is chaotic, by means of a numerical integration by computer of the movement of the exterior planets (Jupiter, Saturn, Uranus, Neptune and Pluto) over a duration of 875 million years." He continued: "A little later, using a very different method, I obtained similar results for the inferior planets (Mercury, Venus, Earth and Mars) which have not yet been taken into account in long numerical integrations of the solar system." ([5], p. 201)

Bibliography

[1] Adams, J.C., An explanation of the method of integration employed, in Francis Bashforth, *An attempt to test the theories of capillary action*, Cambridge: Cambridge University Press, 1883.

[2] Cauchy, A.-L., *Résumé des Leçons données à l'Ecole Royale Polytechnique, Suite du Calcul Infinitésimal*, in *Equations différentielles ordinaires*, intr. Christian Gilain, Paris: Etudes Vivantes, 1981.

[3] Cauchy, A.-L., Mémoire sur l'intégration des équations différentielles, Prague, 1835, in *Exercices d'Analyse et de Physique mathématique*, vol. I, Paris, 1840. *Oeuvres complètes*, 2nd series, vol. XI, Paris: Gauthier-Villars, pp. 399 et seq.

[4] Coriolis, G., Mémoire sur le degré d'approximation qu'on obtient pour les valeurs numériques d'une variable qui satisfait à une équation différentielle, en employant pour calculer ces valeurs diverses équations aux différences plus ou moins approchées, *Journal de Mathématiques pures et appliquées*, vol. 2 (1837), 229–244.

[5] Dahan Dalmedico, A. et al., *Chaos et déterminisme*, Paris: Points Sciences, Seuil, 1992.

[6] Euler, L., *Institutionum Calculi integralis*, vol. I, St Petersbourg, 1768. *Opera omnia*, 1st series, vol. 11, Teubner, 1913.

[7] Goldstine, H. H., *A history of numerical analysis from the 16th through the 19th Century*, New York: Springer Verlag, 1977.

[8] Heun K., Neue Methode zur approximativen Integration der Differentialgleichungen einer unabhängigen Veränderlichen, *Zeitschrift für Mathematik und Physik*, vol. 45 (1900), 23–38.

[9] Kutta, W., Beitrag zur näherungweisen Integration totaler Differentialgleichungen, *Zeitschrift für Mathematik und Physik*, vol. 46 (1901), 435–453.

[10] Lagrange, J.-L., Sur l'usage des fractions continues dans le calcul intégral, *Nouveaux Mémoires de l'Académie royale des Sciences et Belles-Lettres de Berlin*, 1776. *Oeuvres*, vol. IV, Paris: Gauthier-Villars, 1869, pp. 301–332.

[11] Leibniz, G. W., Supplementum geometriae practicae sese ad problemata transcendentia extendens, ope novae methodi generalissimae per series infinitas, *Acta Eruditorum*, April 1693. *Mathematische Schriften*, vol. V, 285–288. French tr. Marc Parmentier, Extension de la géométrie pratique aux problèmes transcendants grâce à une méthode nouvelle absolument générale par des séries infinies, in *Leibniz, Naissance du Calcul différentiel*, Paris: Vrin, 1989, pp. 240–246.

[12] Lipschitz, R., Disamina della possibilità d'integrare completamente un dato sistema di equazioni differenziali ordinarie, *Annali di Matematica pura ed applicata*, 2nd series, vol. 2 (1868/69), pp. 288–302. French tr., Sur la possibilité d'intégrer complètement un système donné d'équations différentielles, *Bulletin des Sciences Mathématiques et Astronomiques*, vol. X (1876), 149–159.

[13] Moigno (l'Abbé), *Leçons de calcul différentiel et de calcul intégral rédigées principalement d'après les méthodes de M. A.-L. Cauchy et étendues aux travaux les plus récents des géomètres*, vol. 2, Calcul intégral, part 1, Paris: Bachelier, 1844.

[14] Moulton, F.R., *Differential Equations*, New York, 1930. Repub. Dover, 1958.

[15] Painlevé, P., Existence de l'intégrale générale, in *Encyclopédie des Sciences mathématiques*, t. II, vol 3, Paris, 1910, fasc. 1, pp. 1–57.

[16] Picard, E., Mémoire sur la théorie des équations aux dérivées partielles et la méthode des approximations successives, *Journal de Mathématiques pures et appliquées*, 4th series, vol. 6 (1890), pp. 145–210 and 231.

[17] Poincaré, H., *Les Méthodes nouvelles de la Mécanique céleste*, vol. 1, Paris: Gauthier-Villars, 1892.

[18] Runge, C., Ueber die numerische Auflösung von Differentialgleichungen, *Mathematische Annalen*, vol. 46 (1895), 167–178.

[19] Runge, C. & König, H., *Vorlesungen über numerisches Rechnen*, Berlin: Springer, 1924.

[20] Tournès, D., L'intégration approchée des équations différentielles ordinaires (1671–1914), Thesis, University Paris 7, June 1996.

13 Approximation of Functions

> ... to show the flaws in the results which they
> (physicists) accept without question, we do not
> need to root out functions as 'monstrous' as
> continuous functions which have no derivatives;
> the 'Runge phenomenon' shows that the classic
> polynomial interpolation procedure can cer-
> tainly be divergent for analytical functions
> which are as 'excellent' as is wished.
>
> Jean Dieudonné, *Calcul infinitésimal*, 1968

The idea of approximating a function has already been raised when we considered interpolation. We noted there the difficulty in certain cases of making a clear distinction between two points of view. In principle, with interpolation, we construct a function which has to take specific values for a finite number of values of the variable, whereas, with approximating, we look for a function which approximates 'as well as possible' the values of the other function, and this for all values of the variable over a certain interval. We can say that approximation is global in nature, since the approximating function has a relation to all the values of the variable over an interval, which is not the case with an interpolation function, only required to be equivalent to the interpolated function for a finite number of values (which must necessarily be the case when the values of the function are obtained experimentally).

As a general rule, in both cases, we require the approximate function and the interpolation function to belong to a certain predetermined class of simple, or sufficiently regular, functions. Moreover, this is the purpose of finding a function that approximates to a function that is otherwise perfectly determined, since the approximate function is chosen so as to be more easily manageable. The polynomial functions play a special part here, as they do with interpolation, since both differentiation and integration of these functions can be easily performed. Other classes of functions can also be considered, rational functions for example or, especially for periodic functions, we can consider trigonometric functions.

Nevertheless, with regard to approximations, a specific choice has to be exercised, that of the measure of the extent of the deviation of the function f being approximated from the approximating function φ, and this deviation must be minimised with respect to a certain number of constraints. Depending on the choice of the measure (a norm over a functional space) the optimal approximation will vary. Where we have a finite number of values, we have already indicated the importance of this choice, in the method of least squares (see Section 9.2).

The simplest choice to make to start with, is to measure the approximation error between f and φ over the interval $[a, b]$ by:

$$\|f - \varphi\| = \sup_{a \leq x \leq b} |f(x) - \varphi(x)|$$

and this measure corresponds to 'uniform approximation' over the interval $[a, b]$. For interpolation, the quantity $|f(x) - \varphi(x)|$ is zero for a finite number of points, but can be large elsewhere. In the case of uniform approximation, Taylor's formula (Section 13.1) gives, for a function $n + 1$ times differentiable, a polynomial of degree n, where the degree of uniform approximation is measured by the 'Lagrange remainder' (see Section 13.2). The Newton interpolation polynomial with the Cauchy error upper bound can also be considered as an approximating function, and this can become optimal, thanks to Chebyshev, who showed that the interpolation points can be changed so as to be optimal with respect to the remainder (Section 13.3). On the other hand, if it is desired to keep the interpolation points regularly spaced, and also to be sure that interpolation polynomials uniformly approach the function being considered, we can use the spline functions introduced by Schoenberg (Section 13.4).

Uniform approximation presents the particularity – which can sometimes turn out to be a disadvantage – of placing too great an emphasis on 'local accidents'. This does not happen when we use the 'mean approximation' associated with the measure:

$$\|f - \varphi\| = \int_a^b |f(x) - \varphi(x)|\, dx$$

but this perhaps disregards extreme values too much. The 'mean quadratic approximation', associated with the measure:

$$\|f - \varphi\| = \sqrt{\int_a^b |f(x) - \varphi(x)|^2\, dx}$$

provides a suitable middle way. It also has the advantage that it leads in general to relatively simple calculations (being the usual transposition of the Euclidean norm). This approximation, when we have continuity, is analogous to the method of least squares, since the latter corresponds, in a discrete context, to the minimisation of the quantity $\sum_{i=1}^{n} [f(x_i) - \varphi(x_i)]^2$ (see Chapter 9 on linear systems). Thus De la Vallée-Poussin shows that a trigonometric function which approximates a periodic function as much as possible, in the least squares sense, corresponds to the terms of its Fourier series (Section 13.5). Finally, we shall see how the algorithm for the discrete Fourier transformation was considerably improved with the discovery of the algorithm known as the Fast Fourier Transform (Section 13.6).

Uniform Approximation

The development of the idea of approximation of functions took place, of course, at the same time as the gradual evolution of the idea of a function, such as we understand it today. The sort of regular functions considered by Euler or Lagrange are far removed from the 'pathological' functions that were described towards the end of the 19th century, which could be continuous but not necessarily 'well-behaved'.

However, in a memoir of 1885, Weierstrass stated the theorem of uniform approximation that applies to all continuous functions of whatever type: every continuous function over an interval [a, b] can be uniformly approximated by polynomials [25]. Powerful though this result is, it says nothing about how to construct a polynomial that is an approximation to a given function to a given degree of uniform approximation.

In the same memoir, Weierstrass also shows that every continuous periodic function – of period 2π – can be uniformly approximated by finite trigonometric expressions, that is linear combinations of cos kx and sin kx, where k is any integer. In fact these two theorems are interchangeable, since the approximation of a function $f(x)$ by polynomials is equivalent to approximating $f(\cos \varphi)$ by trigonometric expressions in cos φ.

The texts that follow, of which the first two came before Weierstrass's theorems, all relate to polynomials that constitute approximations, over a certain interval, of functions that are continuous, or more usually, sufficiently differentiable, with an indication of the upper bound of the approximation error between the function being approximated and the approximating function.

13.1 Taylor's Formula

The various variants of 'Taylor's Formula' produce polynomials as approximations for functions, provided they are sufficiently differentiable. Such a formula was certainly known to James Gregory: in his letter to Collins of 15 February 1671 [20], for example, he gives the series expansion for the function tan x, from which we can deduce that he saw the relation to the successive derivatives of the function (see Section 5.4). The formula itself, however, was published for the first time in 1715, by Brook Taylor in his *Methodus incrementorum directa et inversa* [24]. In his work, Taylor develops a calculus of finite differences (see Section 10.3) and he establishes 'Taylor's formula' (Corollary II below) directly from the 'Newton-Gregory formula' (Proposition VII below).

In his text, Taylor uses a somewhat complicated notation with accents (acute and grave) and dots used as superscripts and subscripts, which we have reproduced here: z stands for the finite difference Δz, $\underset{\cdot}{z}$ for $\Delta^2 z$, etc. As z increases uniformly, Δz is constant and $\Delta^2 z$ is zero. In the corollary, $\dot{x}$, $\ddot{x}$ and $\dddot{x}$ correspond to successive derivatives of x.

B. Taylor
Methodus incrementorum directa et inversa, London, 1715, pp. 21–23. English tr. D. J. Struik, *A Source Book in Mathematics 1200-1800*, Cambridge, Massachusetts: Harvard University Press, 1969, pp. 329–332.

Proposition VII, *Theorem* III. Let z and x be two variable quantities, of which z increases uniformly with given increments $\underset{\cdot}{z}$ Let $n \underset{\cdot}{z} = v, v - \underset{\cdot}{z} = \grave{v}, \grave{v} - \underset{\cdot}{z} = \grave{\grave{v}}$, etc. Then I say that when z grows into $z + v$, then x grows into

$$x + x\frac{v}{1\underset{\cdot}{z}} + x\frac{v\grave{v}}{1.2\underset{\cdot}{z}^2} + x\frac{v\grave{v}\grave{\grave{v}}}{1.2.3\underset{\cdot}{z}^3} + \text{etc.}$$

[...]

Corollary II. If we substitute for evanescent increments the fluxions proportional to them, then all $\grave{\grave{v}}$, $\grave{v}$, v, $\underset{\prime}{v}$, $\underset{\prime\prime}{v}$ become equal. When z flows uniformly into $z + v$, x becomes

$$x + \dot{x}\frac{v}{1\dot{z}} + \ddot{x}\frac{v^2}{1.2\dot{z}^2} + \dddot{x}\frac{v^3}{1.2.3\dot{z}^3} + \text{etc.}$$

or with v changing its sign, when z decreases to $z - v$, x becomes

$$x - \dot{x}\frac{v}{1\dot{z}} + \ddot{x}\frac{v^2}{1.2\dot{z}^2} - \dddot{x}\frac{v^3}{1.2.3\dot{z}^3} + \text{etc.}$$

We can recognise the Gregory-Newton formula in Proposition VII: z corresponds to the increment of the variable, x, $\underset{\cdot}{x}$, $\underset{\cdot\cdot}{x}$, etc., correspond to Δx, $\Delta^2 x$, $\Delta^3 x$, etc., $v/z = n$, $\grave{v}/z = n - 1$, $\grave{\grave{v}}/z = n - 2$, and so we have:

$$x + \frac{n}{1}\Delta x + \frac{n(n-1)}{1.2}\Delta^2 x + \frac{n(n-1)(n-2)}{1.2.3}\Delta^3 x + \ldots$$

And we recognise Taylor's formula in Corollary II on replacing $\dot{x}/\dot{z}$ by dx/dz, $\ddot{x}/\dot{z}^2$ by $d^2 x/dz^2$, etc.... and we then have:

$$x + v\frac{dx}{dz} + \frac{v^2}{1.2}\frac{d^2 x}{dz^2} + \frac{v^3}{1.2.3}\frac{d^3 x}{dz^3} + \ldots$$

The use of finite differences at this period served as a means for going forward from a secure base. The change from one formula to another corresponds to a limit-

ing process where $\Delta z \to 0$ and $n \to \infty$ in such a way that $v = n\Delta z$ as well as $\underset{\cdot}{v} = v - \Delta z$
$= (n - 1)\Delta z$ and $\overset{\cdot\cdot}{v} = \overset{\cdot}{v} - \Delta z = (n - 2)\Delta z, \dots$

Taylor offers no further justifications for his results, nor any indication of the number of terms that need to be considered – for a polynomial of degree n the formula stops after n additional terms – nor any indication of an upper bound for the terms "etc." in his formulas. The possible size of the remainder term was considered later by Lagrange.

13.2 The Lagrange Remainder

Joseph-Louis Lagrange published a major didactic work in 1797, *La Théorie des Fonctions analytiques* [17], in which he proposed to set the foundations of the differential calculus beyond "all considerations of infinitely small or evanescent quantities", and to use instead "the algebraic analysis of finite quantities". With this intent, he 'proves' Taylor's theorem: every function f, that is every analytical expression obtainable as a linear combination of elementary functions, can have an expansion as a polynomial series of the form

$$f(x + i) = f(x) + ip + i^2q + i^3r + \dots$$

Lagrange proposed using this formula to define the derivatives of f by putting:

$$f'(x) = p, \ f''(x) = 2q, \ f'''(x) = 6r, \dots$$

and so prove the usual algorithms about the derivatives of functions.

Moving on from this formal treatment to numerical considerations, Lagrange envisages a bounded interval for the 'remainder' in Taylor's expansion. He writes: "The perfection of the methods of approximation in which series are used depends not only on the convergence of the series, but also on being able to estimate the error that arises from the terms that are neglected, and in this respect it can be said that almost all the methods of approximation that are used in the solution of geometrical and mechanical problems are still very imperfect."

Lagrange returns to this again in his *Leçons sur le Calcul des Fonctions* [16], and in *Lesson Nine*, partly reproduced below, he gives the following bounds for the remainder:

$$\frac{i^n}{n!}f^{(n)}(p) \leq f(x+i) - \sum_{k=0}^{n-1}\frac{i^k}{n!}f^{(k)}(x) \leq \frac{i^n}{n!}f^{(n)}(q)$$

where p and q lie in the interval $[x, x + i]$. He first obtains the result starting for $n = 1$ and then iterates the argument.

J.-L. Lagrange

Leçons sur le calcul des fonctions, presented to the Ecole Polytechnique, year VII (1799). *Séances des Ecoles Normales*, vol. X, year IX (1801), published in one volume Paris: Courcier, 1806.

LESSON NINE

Every function $f(x + i)$ can be expanded, as has been seen, as a series

$$f(x) + i f'(x) + \frac{i^2}{2} f''(x) + \frac{i^3}{2.3} f'''(x) + \ldots,$$

which extends naturally to infinity, unless the derivative functions of $f(x)$ become zero, which happens when $f(x)$ is a polynomial function of x.

When this expansion is only used to generate derived functions, it does not matter whether the series goes to infinity or not; this is also the case when the series is only considered as a simple analytic transformation of the function; but if it is wished to use it to obtain the value of the function in particular cases, as in giving an expression of a simpler form because of the quantity i which is removed from the function, then, only taking account of a certain, large or small, number of terms, it is important to have a means of evaluating the remainder of the series which is neglected, or at least to find limits to the error that is made in neglecting that remainder.

[...]

First, let p and q be the values of $x + i$ which make the derived function $f'(x + i)$ the least and the greatest, taking x as given, and letting i vary from zero to some given value of i. Then $f'(p)$ will be the least value of $f'(x + i)$, and $f'(q)$ will be the greatest value; consequently, $f'(x + i) - f'(p)$ and $f'(q) - f'(x + i)$ will always be positive quantities.

Regarding these two quantities as derived functions with respect to the variable i, their primitive functions, chosen so as to be zero when $i = 0$, will be, since x, p, and q are supposed constant

$$f(x + i) - f(x) - i f'(p) \quad \text{and} \quad i f'(q) - f(x + i) + f(x).$$

Thus, provided that $f'(x + i)$ never becomes infinite from $i = 0$ up to the given value of the variable i, which will be the case if $f'(p)$ and $f'(q)$ are not infinite quantities, we shall have by the preceding principle, if i is positive,

$$f(x + i) - f(x) - i f'(p) > 0 \quad \text{and} \quad f(x) - f(x + i) + i f'(q) > 0,$$

from which we have

$$f(x + i) > f(x) + i f'(p) \quad \text{and} \quad f(x + i) < f(x) + i f'(q).$$

Suppose next that p and q are values of $x + i$ which make the second derivative of the function, $f''(x + i)$, the least and the greatest, as we let i vary from zero to a given value; we shall have $f''(p)$ and $f''(q)$ as the least and greatest value of $f''(x + i)$; consequently, $f''(x + i) - f''(p)$ and $f''(q) - f''(x + i)$ will always be positive quantities.

Regarding these functions as derived functions with respect to the variable i, their primitive functions, chosen so that they are zero when $i = 0$, will be

$$f'(x + i) - f'(x) - i f''(p),$$
$$i f''(q) - f'(x + i) + f'(x).$$

Thus, provided that $f''(x + i)$ never becomes infinite for the whole range of i, which will be the case if $f''(p)$ and $f''(q)$ are not infinite, these two quantities will be, by the same principle, always positive and finite, i being supposed positive; and in regarding them as derived functions with respect to i, their primitive functions, chosen so as to be zero when $i = 0$, will be, since x, p, and q are supposed constant,

$$f(x+i) - f(x) - i\,f'(x) - \frac{i^2}{2} f''(p),$$

$$\frac{i^2}{2} f''(q) - f(x+i) + f(x) + i f'(x).$$

These new quantities will also be, by the same principle, always positive; we shall thus have

$$f(x+i) - f(x) - i f'(x) - \frac{i^2}{2} f''(p) > 0,$$

$$f(x) - f(x+i) + i f'(x) + \frac{i^2}{2} f''(q) > 0,$$

from which we have

$$f(x+i) > f(x) + i f'(x) + \frac{i^2}{2} f''(p),$$

$$f(x+i) < f(x) + i\,f'(x) + \frac{i^2}{2} f''(q).$$

[...]

Thus, in general, the quantity $f(x+i)$, whether i be positive or negative, will always lie between these two:

$$f(x) + i f'(x) + \frac{i^2}{2} f''(x) + \frac{i^3}{2.3} f'''(x) + ... + ... \frac{i^\mu}{2.3...\mu} f^\mu(p),$$

$$f(x) + i f'(x) + \frac{i^2}{2} f''(x) + \frac{i^3}{2.3} f'''(x) + ... + ... \frac{i^\mu}{2.3...\mu} f^\mu(q),$$

in taking p and q as the values of $x + i$ which correspond to the least and the greatest of the values of $f^\mu(x + i)$, over the whole range of i, from $i = 0$, provided that the two quantities $f^\mu(p)$ and $f^\mu(q)$ are not infinite.

In his *Théorie des Fonctions analytiques*, Lagrange obtains the remainder in its integral form. He writes:

"... $f(x) = f(x - xz) + xzf'(x - xz) + \frac{x^2z^2}{2} f''(x - xz) + x^3 R$, R being a function of z,

which vanishes when $z = 0$. We find, on taking the primitive functions with respect to z, and removing terms which mutually cancel,

$$R' = \frac{z^2}{2} f'''(x - xz),$$

the function represented by f''' being the third function of $f(x)$ with respect to x, transformed by the substitution of $x - xz$ in the place of x." ([17], art. 17)

The standard formula follows immediately, on substituting $x - xz = x_0$:

$$f(x) = f(x_0) + (x - x_0)f'(x_0) + \frac{(x - x_0)^2}{2}f''(x_0) + \int_{x_0}^{x} \frac{(x - t)^2}{2}f'''(t)\,dt.$$

In his 35th *Leçon* given at the Ecole Polytechnique in 1823 [3], Cauchy stated and proved the Taylor-Lagrange formula with the remainder in the integral form which we use today. In the same course, he warned that the Taylor's series of a function may not converge and also that, even if it does converge, the sum of the series may differ from the function. It was in the 38th *Leçon* that he provided the now classic example of the function $f(x) = e^{-1/x^2}$.

For suitable cases however, Taylor's formula with the Lagrange remainder can be used to define an approximation polynomial, on an interval I, for a function f of class C^{n+1} (that is $n + 1$ times continuously differentiable) by a polynomial of degree n:

$$\left| f(x) - \sum_{k=0}^{n} \frac{(x - a)^k}{k!}f^{(k)}(a) \right| \leq \frac{|x - a|^{n+1}}{(n+1)!}M_{n+1}$$

where
$$M_{n+1} = \sup_{t \in I}\left| f^{(n+1)}(t) \right|.$$

A little later, in 1874, Chebyshev wrote:

"Thus, the approximate representation of a function $f(x)$ in the neighbourhood of $x = a$ can be found in the form of a polynomial of degree n by stopping the Taylor series expansion

$$f(x) = f(a) + \frac{x - a}{1}f'(a) + \frac{(x - a)^2}{1.2}f''(a) + ...$$

at the term $\dfrac{(x - a)^n}{1.2...n}f^{(n)}(a)$. [...] If we are only concerned with values of x that lie in the neighbourhood of a, these expressions of $f(x)$ give a representation of it with the greatest precision that can be provided by expressions this form. But, this is not the case, if the variable x is only subject to the requirement to lie within, more or less wide, limits. In this case, a search for approximate values of $f(x)$ require methods that are essentially different from those we have just been discussing ..." ([7], §1).

13.3 Chebyshev's Polynomial of Best Approximation

When a function f is interpolated by the Lagrange polynomial L_f of degree $n - 1$ by using n values corresponding to n points $a_1, a_2, ..., a_n$, the error made, or rather the difference $f(x) - L_f(x)$ is bounded above, over the interpolation interval I, according to Cauchy (see Section 10.6), by

$$|(x - a_1)(x - a_2)...(x - a_n)|\frac{\sup_{t \in I}|f^{(n)}(t)|}{n!}.$$

This interpolation function L_f can be interpreted as an approximation to f over I. However, the size of this upper bound depends not only on the number n of the points considered, but also on the approximated function f with the involvement of the upper bound of the nth derivative. In fact, if one is tempted to think, along with Lagrange (see Section 10.5), that the more the difference between the points diminishes, the better is the approximation, one should be on one's guard.

Méray raised the problem in 1884 [19] in identifying the difficulty posed by the existence of complex poles in the proximity of the interpolation set; he came back to this again in 1896 [18]. But it was Runge, in an article in 1901 entitled "On empirical functions and interpolation between equidistant ordinates", who clearly demonstrated the existence of the following phenomenon: there exist functions which are indefinitely differentiable, for which the sequence of interpolation polynomials corresponding to equidistant points of increasing number do not tend to the function. In the *Encyclopédie des Sciences Mathématiques* ([11], vol. II, p. 215), Maurice Fréchet summarises the astonishing nature of the example provided by Runge:

"C. Runge showed that if we evaluate a polynomial $P_n(x)$ of degree n coinciding with the continuous function $\dfrac{1}{1+x^2}$, at the successive end-points of the n equal divisions of the interval $[-5, +5]$, then as n increases indefinitely, not only does this polynomial not necessarily tend towards $\dfrac{1}{1+x^2}$ for the whole of the interval $[-5, +5]$, but even more, it will diverge outside the interval $[-3.63..., +3.63...]$."

So, not only do we not necessarily have convergence, but we may even have divergence to infinity. At each point, the deviation between the value of the function being 'approximated' and that of the interpolation polynomials may tend to infinity as the number of interpolation points increases (see figure 13.1).

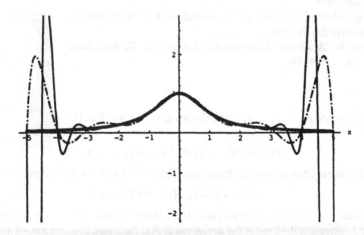

Fig. 13.1. The Runge Phenomenon: interpolation of the function $1/(1+x^2)$ (in bold) for the interval $[-5, +5]$ using for interpolation 11 points (dotted line) and 21 points (continuous line); we can observe the divergence at the ends of the interval.

However, if we note that the upper bound of the error obtained by Cauchy depends also on the choice of the points $a_1, a_2, ..., a_n$, and if we shift the point of view from interpolation to approximation by choosing the points $a_1, a_2, ..., a_n$ in such a way as to minimise the error, we shall see with Chebyshev that, independently of the function f being approximated, there is an optional choice for the points $a_1, a_2, ..., a_n$ in the sense that this choice minimises the quantity:

$$\sup_{x \in I} |(x - a_1)(x - a_2)...(x - a_n)|.$$

Finally, we wish to reduce the error to 0 for a polynomial of degree n. This question, first tackled by Chebyshev in 1854 [5], was dealt with in a more general way by him in 1859 [7]. We give some extracts from this later text below. Chebyshev poses the following problem: "being given the approximate value of $f(x)$, deduced by ordinary methods, either in the form of a polynomial, or in the form of a fraction, to find the changes to be made to the coefficients, when it is wished to minimise the limit of its errors between $x = a - h$ and $x = a + h$, h being a sufficiently small value", and he replied to this in a general way, with the help of the Theorem 2 given below.

This theorem shows, in essence, that if a polynomial of degree $n - 1$ constitutes the best uniform approximation of a function Y in the interval $[a - h, a + h]$, then the maximum error L between the polynomial and the function must necessarily be attained $n + 1$ times. Applying this to the case of a polynomial of degree n deviating the least possible from 0, this theorem leads to what are now called *Chebyshev polynomials*. To do this, Chebyshev had to consider the expansion of the function $\sqrt{x^2 - h^2}$ as a continued fraction, the required approximation polynomial being one of the numerators of the fractions so obtained; but we shall not go into the detail of this here.

P. L. Chebyshev
Sur les questions de minima qui se rattachent à la représentation approximative des fonctions,
Mémoires de l'Académie Impériale des sciences de St. Petersburg,
vol. VII (1859), 199–291.

Theorem 2

The quantities $p_1, p_2, ..., p_{n-1}, p_n$ being chosen in such a way that the function

$$F(x) = p_1 x^{n-1} + p_2 x^{n-2} + ... + p_{n-1} x + p_n - Y$$

should deviate as little as possible from zero, from $x = -h$ to $x = +h$, the equations

$$F^2(x) - L^2 = 0, \quad (x^2 - h^2) F'(x) = 0$$

have at least $n + 1$ distinct common solutions, which lie between $x = -h$ and $x = +h$. The quantity L represents the limit of the deviations of $F(x)$ from zero between $x = -h$ and $x = +h$.

[...]

On the function which, among those of the form $x^n + p_1 x^{n-1} + p_2 x^{n-2} + ... + p_{n-1} x + p_n$, deviates the least possible from zero between the limits $x = -h$ and $x = +h$.

Since the function

$$x^n + p_1 x^{n-1} + p_2 x^{n-2} + \ldots + p_{n-1} x + p_n,$$

is simply the value of

$$p_1 x^{n-1} + p_2 x^{n-2} + \ldots + p_{n-1} x + p_n - Y$$

in the case where $Y = -x^n$, we conclude, by virtue of theorem 2, that the coefficients $p_1, p_2, \ldots, p_{n-1}, p_n$ being chosen in such a way that the expression

$$F(x) = x^n + p_1 x^{n-1} + p_2 x^{n-2} + \ldots + p_{n-1} x + p_n$$

deviates the least possible from zero between $x = -h$ and $x = +h$, the equations

(*) $$F^2(x) - L^2 = 0, \quad (x^2 - h^2) F'(x) = 0$$

have at least $n + 1$ distinct common solutions.

Suppose now that $x = x_0$ is one of these solutions. It is not difficult to see that the expression $(x^2 - h^2)(F^2(x) - L^2)$ will therefore be divisible by $(x - x_0)^2$. In fact, from the first of the preceding equations, the expression

$$(x^2 - h^2)(F^2(x) - L^2)$$

will vanish for $x = x_0$. Further, since its first derivative is

$$2(x^2 - h^2)F(x) F'(x) + 2x(F^2(x) - L^2),$$

it also reduces to zero for $x = x_0$ by virtue of the same equations, which proves to us that the expression $(x^2 - h^2)(F^2(x) - L^2)$ is divisible by $(x - x_0)^2$.

The same thing holds for the other common solutions of the equations (*), and since the number of these distinct solutions cannot be less than $n + 1$, it follows that the expression $(x^2 - h^2)(F^2(x) - L^2)$ is divisible by $n + 1$ distinct factors $(x - x_0)^2$, $(x - x_1)^2$, $(x - x_2)^2$, $\ldots$, $(x - x_n)^2$, and therefore by their product $(x - x_0)^2(x - x_1)^2(x - x_2)^2 \ldots (x - x_n)^2$. But the expression $(x^2 - h^2)(F^2(x) - L^2)$ where

$$F(x) = x^n + p_1 x^{n-1} + p_2 x^{n-2} + \ldots + p_{n-1} x + p_n,$$

being only of degree $2n + 2$, the quotient of the division of this expression by the product $(x - x_0)^2(x - x_1)^2(x - x_2)^2 \ldots (x - x_n)^2$ can only be a constant. Therefore,

$$(x^2 - h^2)(F^2(x) - L^2) = C(x - x_0)^2(x - x_1)^2(x - x_2)^2 \ldots (x - x_n)^2.$$

This equation can only hold, evidently, if $x + h$ and $x - h$ are among the factors $x - x_0, x - x_1, x - x_2, \ldots, x - x_n$. Now, if we suppose $x - x_0 = x + h, x - x_1 = x - h$, this equation, divided by $(x + h)(x - h) = x^2 - h^2$, becomes

$$F^2(x) - L^2 = C(x^2 - h^2)(x - x_2)^2 \ldots (x - x_n)^2,$$

or

(**) $$F^2(x) - L^2 = (x^2 - h^2)\Phi^2(x)$$

[...]

From which we have

$$\frac{F(x)}{\Phi(x)} = \sqrt{x^2 - h^2} + \frac{L^2}{\Phi(x)\left[F(x) + \Phi(x)\sqrt{x^2 - h^2}\right]}$$

which proves that the fraction $\dfrac{F(x)}{\Phi(x)}$ is the value $\sqrt{x^2-h^2}$ exactly as far as terms of the or-

der of $\dfrac{1}{\Phi^2(x)}$ inclusive. But this can only be so if $\dfrac{F(x)}{\Phi(x)}$ is one of the convergents of

$\sqrt{x^2-h^2}$ which can be found from its continued fraction expansion.

[...]

... we find, definitely

$$F(x) = \frac{(x+\sqrt{x^2-h^2})^n + (x-\sqrt{x^2-h^2})^n}{2^n}.$$

The function $F(x)$ thus found is certainly a polynomial of degree n over the interval $[-h, +h]$. In fact, if we expand the expression given for $F(x)$, we can see easily that the square roots disappear. Another way of writing the polynomial is to put $x = h \cos\varphi$; we then have:

$$F(x) = \frac{h^n}{2^n}[(\cos\varphi + i\sin\varphi)^n + (\cos\varphi - i\sin\varphi)^n] = \frac{h^n}{2^n}(e^{in\varphi} + e^{-in\varphi})$$

or $$F(x) = \frac{h^n}{2^{n-1}}\cos n\varphi = \frac{h^n}{2^{n-1}}\cos(n\arccos\frac{x}{h}).$$

We then obtain what is called the *Chebyshev polynomial* T_n of degree n by substituting $F(x) = T_n(\frac{x}{h})$. This is the polynomial with leading coefficient unity and of degree n that deviates the least from 0 over the interval $[-1, +1]$. Since

$$\cos nu = 2\cos u\cos(n-1)u - \cos(n-2)u,$$

the polynomial $$T_n(x) = \frac{1}{2^{n-1}}\cos(n\arccos x)$$

satisfies the recurrence: $$T_n(x) = xT_{n-1}(x) - \tfrac{1}{4}T_{n-2}(x)$$

with $$T_0(x) = 2 \quad\text{and}\quad T_1(x) = x$$

(see Commentary on Orthogonal Polynomials below).

If we go back now to the initial question of interpolating a function by means of a Lagrange polynomial of degree $n-1$, the text above shows that there is an optimal choice for the interpolation points. If $[a-h, a+h]$ is the interval under consideration then we should choose, for the x values of the interpolation points, the roots of the polynomial $F(x)$, plus a. Now the roots of $F(x)$ are:

$$h\cos\frac{2k+1}{2n}\pi \quad\text{where}\quad 0 \le k < n.$$

Furthermore, this is how Chebyshev, at the end of his memoir of 1854 [5], proposes obtaining these values:

From the centre of the interval $x = a - h, x = a + h$, taken on the abscissa axis, and with a radius equal to half that interval, we draw a circle; we inscribe in this circle a regular polygon of $2n$ sides, arranged so that two of its sides are perpendicular to the x axis; the vertices of this polygon, up to quantities of the order h inclusive, determine the abscissas of the points ....

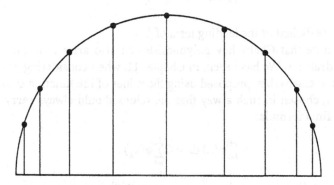

Optimal distribution for 13 interpolation points. The division of the x axis with
$x_k = a + h \cos [(2k + 1)/2n]\pi$ is not regular; there is a bunching of points towards the
extremities of the interval.

The polynomial with leading coefficient unity and of degree n which approximates closest to 0 over the interval $[a - h, a + h]$ – or, rather, which deviates from 0 the least – has a maximum value of $\dfrac{h^n}{2^{n-1}}$, and this it attains $n + 1$ times. Thus, with the choice of interpolation points corresponding to the roots of this polynomial, the upper bound of the Lagrange interpolation error, according to Cauchy, becomes:

$$|f(x) - L_{n-1}(x)| \leq \frac{h^n}{2^{n-1}} \frac{\sup_t |f^{(n)}(t)|}{n!},$$

whereas the Taylor-Lagrange formula has an upper bound:

$$\left| f(x) - \sum_{k=0}^{n-1} \frac{(x - a)^k}{k!} f^{(k)}(a) \right| \leq h^n \frac{\sup_t |f^{(n)}(t)|}{n!}.$$

If we adopt the Chebyshev choice of interpolation points, we are assured that in going from n to $n + 1$ points, the deviation from the function being approximated will not increase, and the divergence shown by Runge will not be produced. However, convergence itself is not guaranteed. In fact, in 1914, Faber showed that there exist continuous functions for which Chebyshev interpolation does not converge to the function.

In what has been said so far, we were concerned with optimising the approximation of all continuous n times differentiable functions, functions belonging to the class C^n, by a polynomial of degree $n - 1$, and this in a way that is independent of the function being approximated. A much more difficult problem is that of determining, for any given function, the polynomial of degree $n - 1$ which is the best polynomial to give uniform approximation. In the particular case where f is a polynomial of degree n, we can however show that the best uniform approximation polynomial of degree $n - 1$ over the interval $[a - h, a + h]$ is given by:

$$f(x) - ch^n T_n\left(\frac{x-a}{h}\right)$$

where c is the coefficient of the leading term of f.

Finally, we note that Chebyshev polynomials can also arise in a quite different way, from quadratures. We have seen, in Chapter 11, when considering approximate quadratures, that Chebyshev proposed using the values of the function φ to integrate at the points x_k, chosen in such a way that the values should always carry the same weights in the final formula:

$$\int_{-1}^{+1} \varphi(x)\,dx = C\sum_{k=1}^{n} \varphi(x_k)\,.$$

If, following Hermite, we wish to investigate $\int_{-1}^{+1}\dfrac{\varphi(x)}{\sqrt{1-x^2}}\,dx$, then the x_k are simply the values of the roots of the polynomials $T_n(x)$. (see [6], § 4)

13.4 Spline-Fitting

Investigations into the best uniform approximation of functions, initiated by Chebyshev, were pursued again at the beginning of the 20th century, in particular by Charles De la Vallée-Poussin [9] and by Sergei Bernstein [1]. Bernstein showed that Chebyshev type interpolation always converges towards the function being interpolated as the number of interpolation points is increased provided that f is sufficiently regular, a Lipschitz function for example. In a more general, and simpler, way Bernstein links every continuous function f over the interval [0, 1] with the sequence of polynomials defined by

$$Q_n(x) = \sum_{p=0}^{n} f\left(\frac{p}{n}\right)\binom{n}{p} x^p (1-x)^{n-p}$$

which are now called *Bernstein polynomials* of f, and this sequence converges uniformly towards f over [0, 1]. The proof requires the Weierstrass Approximation Theorem.

In the approximation of functions by polynomials, Chebyshev was interested in the choice of interpolation points. An alternative approach is to consider the length of the intervals between the interpolation points. More precisely, by interpolating over increasingly smaller intervals, we can limit the degree of the polynomials to be used, and thereby considerably reduce the calculations required. This is what happens when we use *spline functions*.

The spline functions are simple generalisations of polynomials being piece-wise defined polynomials required to satisfy certain conditions of continuity and differentiability. Although they had been used before in particular special cases, these functions were defined and studied systematically for the first time by Schoenberg in the 1940s. While Schoenberg is considered to be the father of spline functions, after the appearance of his seminal paper in 1946 [23], there were both grand fathers and great grand fathers of these functions [2]. Schoenberg himself cites Hermite, Peano and Euler [22], but we may also recognise L. Maurer (1896), A. Sommerfeld (1904), M. Lerch (1908), G. Pólya (1913), V. Brun (1932) and K. Fränz (1941) [2].

Notwithstanding these many historical references, the subject is a relatively recent one, and we shall limit ourselves to quoting a brief introductory paragraph from Schoenberg's work.

S. N. Schoenberg
Contributions to the problem of approximation of equidistant data by analytic functions. *Quarterly of Applied Mathematics*, vol. IV (1946), 67–68.

Polynomial spline curves of order k. A spline is a simple mechanical device for drawing smooth curves. It is a slender flexible bar made of wood or some other elastic material. The spline is placed on the sheet of graph paper and held in place at various points by means of certain heavy objects (called "dogs" or "rats") such as to take the shape of the curve we wish to draw. Let us assume that the spline is so placed and supported as to take the shape of a curve which is nearly parallel to the x-axis. If we denote by $y = y(x)$ the equation of this curve then we may neglect its small slope y', whereby its curvature becomes

$$1/R = y''/(1 + y'^2)^{3/2} \approx y''.$$

The elementary theory of the beam will then show that the curve $y = y(x)$ is a polygonal line composed of cubic arcs which join continuously, with a continuous first and second derivative. These junction points are precisely the points where the heavy supporting objects are placed.

Description of spline curves of order k. Our last remark suggests the following definition.

DEFINITION 4. A real function $F(x)$ defined for all real x is called a spline curve of order k and denoted by $\Pi_k(x)$ if it enjoys the following properties:

1) It is compressed of polynomial arcs of degree at most $k - 1$.
2) It is of class C^{k-2}, i. e., $F(x)$ has $k - 2$ continuous derivatives.
3) The only possible function points of the various polynomial arcs are the integer points $x = n$ if k is even, or else the points $x = n + 1/2$ if k is odd.

Thus a $\Pi_1(x)$ is a step function with possible discontinuities at the points $x = n + 1/2$. A $\Pi_2(x)$ has an ordinary polygonal graph with vertices only at the integer points $x = n$. A $\Pi_4(x)$ corresponds to the elementary mathematical description of an ordinary (infinite) spline with the "dogs" placed at all or only some of the points with $x = n$.

It should be noticed that if a $\Pi_k(x)$ is of class C^{k-1}, then $\Pi^{(k-1)}(x)$ must necessarily be constant for all x. Thus such a $\Pi_k(x)$ reduces to a polynomial of degree $k - 1$. It is just this relaxation of the requirement of the continuity of the $(k - 1)$-order derivative of $\Pi_k(x)$ which turns the spline curve into a flexible and versatile instrument of approximation. Likewise, only the "dogs" (or "rats") enable the ordinary spline to trace curves differing from the graph of a cubic polynomial.

The special importance of spline curves will be due to the fact that by the addition of several spline curves of successive orders we may get any desired polygonal line of given degree m and class C^μ.

A *cubic spline function* over an interval $[a, b]$ corresponds to a function $\Pi_4(x)$, using the notation just given. It is a function twice continuously differentiable over $[a, b]$ for which there is a sub-division $a = x_0 < x_1 < ... < x_n = b$ such that, for each of the sub-intervals $[x_k, x_{k+1}]$ the function is a polynomial of at most degree 3.

If f is a continuous function over $[a, b]$ we can interpolate by means of a cubic spline function s such that:

$$\text{for } k = 0, ..., n, \quad s(x_k) = f(x_k).$$

However, the function s is not unique, and has two further degrees of indetermination. If f is of class C^1, then s is entirely determined by the supplementary conditions:

$$s'(a) = f'(a) \quad \text{and} \quad s'(b) = f'(b).$$

It can be shown that, by increasing the number of interpolation points, which may be equally spaced for example, the approximation of a function f, supposed to be of class C^4, by cubic spline functions, will always converge towards f, which is contrary to the situation with using a Lagrange interpolation polynomial. The convergence is, moreover, of order δ^4, where δ is the sub-division interval.

We can also use two-dimensional spline functions to approximate flexible thin surfaces ...

Mean Quadratic Approximation

When the functions being considered belong to a normed space, that is a vector space endowed with a norm, defined in quadratic form, we can attempt to find mean quadratic approximations.

In the discrete case, this reduces to the method of least squares (see Chapter 10). In effect, if we wish to represent a function f by a polynomial φ of degree m, and we know the values of f at $n + 1$ points $x_0, x_1, ..., x_n$ with $n > m$, we need to be able to minimise $\sum_{k=0}^{n} (f(x_k) - \varphi(x_k))^2$. Both Chebyshev in 1855 [4] and Hermite in 1859 [15] addressed this question.

Furthermore, mean quadratic approximations provide a natural way of thinking about families of orthogonal polynomials, such as those of Legendre and Chebyshev, which we have been considering. The polynomial of best approximation for the norm in question can be expressed in a simple way using the family of corresponding orthogonal polynomials (see below).

While the first part of this chapter has looked at approximation by polynomials, in this second part we are particularly interested in periodic functions, and their approximation by trigonometric functions. In actual fact, interpolation of a function f over the interval $[0, 2\pi]$ by the trigonometric summation

$$\sum_{0 \leq k \leq N-1} (a_k \cos kx + b_k \sin kx)$$

can also be expressed as interpolation by the summation:

$$\sum_{0 \leq k \leq N-1} c_k e^{ikx}.$$

Now, if we put $\omega = e^{ix}$, this reduces to interpolating the function f by a polynomial in ω, namely:

$$P(\omega) = c_0 + c_1\omega + c_2\omega^2 + \dots + c_{N-1}\omega^{N-1},$$

the interpolation points, which provide the values x_k of the variable x, becoming the points $\omega^k = e^{ix_k}$.

Orthogonal polynomials

When the scalar product of two continuous functions f and g is given by the integral:

$$\int_a^b w(x) f(x) g(x)\, dx$$

where $w(x)$ stands for a certain weight function ≥ 0, there is associated with it a unique family of *orthogonal polynomials* P_n which satisfy the following conditions: P_n is of degree n, with leading coefficient 1, such that for all polynomials Q of degree $\leq n - 1$, we have

$$\int_a^b w(x) P_n(x) Q(x)\, dx = 0.$$

For example, the different scalar products

$$\int_{-1}^{+1} f(x) g(x)\, dx \qquad \int_{-1}^{+1} \frac{1}{\sqrt{1-x^2}} f(x) g(x)\, dx$$

$$\int_0^{+\infty} e^{-x} f(x) g(x)\, dx \qquad \int_{-\infty}^{+\infty} e^{-x^2} f(x) g(x)\, dx$$

lead respectively to the classic polynomials of Legendre (Section 11.6), Chebyshev (Sections 11.7 and 13.3), Laguerre and Hermite.

The polynomial φ of best approximation of degree n of a function f for one of these norms can thus be written simply, relative to the corresponding family of orthogonal polynomials P_n, as:

$$\varphi = \sum_{0 \leq k \leq n} \lambda_k P_k$$

where the coefficients λ_k are given by:

$$\lambda_k = \frac{\int_a^b w(x) P_k(x) f(x)\, dx}{\int_a^b w(x) P_k^2(x)\, dx}.$$

In the case of periodic functions and the Hermitian scalar product $\int_{-\pi}^{+\pi} f(x)\overline{g(x)}\,dx$, an analogous role is played by the trigonometric functions e^{inx}, as we shall see below.

13.5 Fourier Series

Euler investigated trigonometric series from the 1730s and he produced expansions of periodic functions as trigonometric series in his Memoir of 1749 on the irregularities of the orbits of Saturn and Jupiter. Soon afterwards, Clairaut produced expressions for the 'Fourier coefficients' of a function.

At the same period, d'Alembert was investigating the problem of vibrating strings. A string, fixed at its extremities, vibrates according to the equation

$$\frac{\partial^2 u}{\partial t^2} = c^2 \frac{\partial^2 u}{\partial x^2}$$

where $u(x, t)$ is the transversal displacement, at time t, of the point at position x with respect to its point of equilibrium. Solutions to this equation are sought, and d'Alembert provides some particular solutions. He was of the view that the form of the string would be bound to yield analytically simple solutions. It was through trying to determine all the solutions of this equation, that Euler came to enlarge the concept of what was meant by a function. In 1753, Daniel Bernoulli set out to find the solutions in the form of trigonometric series. In this way, he thought to resolve the problem in all its generality, which presupposes that every function can be written as the sum of a trigonometric series.

The theory was not put on a sound basis until Fourier's work. The essence of his ideas is contained in a memoir submitted to the Academy of Sciences in Paris in 1807 entitled "Theory of the propagation of heat in solids" [14], and Fourier took the idea up again in his celebrated work 1822, *Théorie analytique de la Chaleur* [10]. Here again, the problem was to solve a partial differential equation and Fourier obtains the expressions for the coefficients, in the form that we know them today, in section VI of chapter III: "the expansion of an arbitrary function as a trigonometric series". Fourier series were to close the debate between Euler, d'Alembert and Bernoulli about the generality of Bernoulli's solution to d'Alembert's problem. These series were also to become indispensable to the theory of partial differential equations with boundary conditions.

Interested, like Fourier, in the application of mathematics to physics, both Poisson and Cauchy contributed to the study of Fourier series. Dirichlet also embarked on a rigorous study of the convergence Fourier series to the function, from 1822 to 1826.

But we should further note here that these coefficients can also be interpreted in terms of the best mean quadratic approximation. Thus, for functions of period 2π, we consider the quadratic form:

$$f \rightarrow \int_0^{2\pi} f^2(x)\, dx\,.$$

This case is explained very clearly in Charles de la Vallée-Poussin's *Leçons sur l'approximation des fonctions d'une variable réelle* [9]. The presentation requires no further commentary.

C. De la Vallée-Poussin
Leçons sur l'approximation des fonctions d'une variable réelle, Paris: Gauthier-Villars, 1919, pp. 10–11.

Chapter I.
Approximation by Fourier Series

8. Fourier series and constants. Minimum property which defines them. – Let $f(x)$ be a bounded and integrable function of period 2π. Let us consider the finite trigonometric series of order n

$$S_n = \tfrac{1}{2}a_0 + \sum_{k=1}^{n}(a_k \cos kx + b_k \sin kx).$$

This sequence is called the *Fourier series of order n* relative to $f(x)$, if the coefficients a_k and b_k are determined by the condition to minimise the integral

$$\int_0^{2\pi}[f(x) - S_n]^2\, dx\,.$$

This integral is a quadratic expression, positive in a and b, which must necessarily have a minimum. To obtain it, the partial derivatives in a and b must be made to disappear, that is we must put

$$\int_0^{2\pi}[f(x) - S_n]\cos kx\, dx = 0, \qquad \int_0^{2\pi}[f(x) - S_n]\sin kx\, dx = 0$$

$$(k = 0, 1, 2, ..., n),$$

from which, without difficulty we have,

(1)
$$a_k = \frac{1}{\pi}\int_0^{2\pi} f(x)\cos kx\, dx, \qquad b_k = \frac{1}{\pi}\int_0^{2\pi} f(x)\sin kx\, dx$$

$$(k = 0, 1, 2, ..., n),$$

It is important to make the remark that, because of its periodicity, the interval for integration can be replaced by any other of the same amplitude 2π, without any change to the value of these integrals.

The constants a_k and b_k determined by the formulas (1) are the *Fourier constants* of $f(x)$. The infinite series

$$\tfrac{1}{2}a_0 + \sum_{1}^{\infty}(a_k \cos kx + b_k \sin kx),$$

considered from a purely formal point of view, whether convergent or not, is the *Fourier series of $f(x)$*. The sum S_n is that of the first $n + 1$ terms of this series.

We note simply, as a matter of transition from one view point to the other, that interpolation of a function $g(x)$ by a polynomial, when the interpolation points are taken as the roots

$$x_j = \cos \frac{2j-1}{2N} \pi$$

of the Chebyshev polynomial T_N, corresponds to interpolation of the periodic function $f(t) = g(\cos t)$ by the trigonometric polynomial

$$\sum_{0 \leq k \leq n-1} a_j \cos jt$$

with equally distributed points

$$t_j = \frac{2j-1}{2N} \pi .$$

13.6 The Fast Fourier Transform

The discrete Fourier transformation plays an important role in the theory of handling signals. Now, the evaluation of the N corresponding coefficients requires, a priori, N multiplications for each one of them, or N^2 multiplications in total, a number that increases, of course, rapidly with N. However, an important paper by Cooley and Tukey [8], written in 1965, describes an improved evaluation method that only requires the number of multiplications to be of order $N \log N$, provided particular values of N are chosen. This method, and its variants, is known as the *Fast Fourier Transform* or the FFT algorithm.

We know that Gauss was much occupied with interpolation and approximation, and particularly with trigonometric interpolation [12]. According to Goldstine [13], there is evidence in Gauss's manuscripts of, not only investigation into finite Fourier series, but also of what we now call the FFT algorithm ([12], p. 303).

Consider using trigonometric interpolation for a periodic function f with period 2π, that is representing the function by expressions of the type:

$$p(x) = \sum_{0 \leq k \leq N-1} A(k) e^{ikx} .$$

Suppose the n interpolation points x_k are uniformly distributed over the interval $[0, 2\pi]$, that is:

$$x_k = 2k\pi/N \quad \text{for} \quad k = 0, 1, 2, ..., N-1$$

and let:

$$f(x_k) = f_k .$$

For the coincident points:

$$p(x_k) = f_k \quad \text{for} \quad k = 0, 1, ..., N-1$$

we have the formulas:

$$A(k) = \frac{1}{N} \sum_{j=0}^{N-1} f_j \, e^{-2i\pi jk/N} \quad \text{for } k = 0, 1, \ldots, N-1.$$

The expressions for $A(k)$ can be read as discrete approximations of the Fourier transform of f:

$$f(t) = \int_{-\infty}^{+\infty} f(x) \, e^{-2i\pi t x} \, dx \, .$$

To simplify, put $\omega = e^{2i\pi/N}$, and so $e^{ix_k} = e^{2ik\pi/N} = \omega^k$

and

$$A(k) = \frac{1}{N} \sum_{j=0}^{N-1} f_j \, \omega^{-jk} \quad \text{for } k = 0, 1, \ldots, N-1.$$

Formulas of this type arise in signal theory, and the principle of the method proposed by Cooley and Tukey, and its variants, rests on the following remark:

The N^2 elements in the array of ω^{jk} for $0 \leq j \leq N-1$ and $0 \leq k \leq N-1$ only take N distinct values, namely the Nth roots of unity. By regrouping the values to take account of equivalence, we can make considerable economies on the number of operations.

In the extract given below, from the fundamental article by Cooley and Tukey, it is the inverse transformation that is being considered, that is the problem of calculating the values

$$p(x_j) = \sum_{0 \leq k \leq N-1} A(k) \omega^{jk}$$

from the values of the coefficients $A(k)$. The calculations are, however analogous.

J. W. Cooley and J. W. Tukey
An Algorithm for the Machine Calculation of Complex Fourier Series
Mathematics of Computation, vol. XIX (1965), 297–298.

Consider the problem of calculating the complex Fourier series

(1)
$$X(j) = \sum_{k=0}^{N-1} A(k) . \omega^{jk} \, , \quad j = 0, 1, \ldots, N-1,$$

where the given Fourier coefficients $A(k)$ are complex and ω is the principal Nth root of unity,

(2)
$$\omega = e^{2\pi i/N}.$$

A straightforward calculation using (1) would require N^2 operations where "operation" means, as it will throughout this note, a complex multiplication followed by a complex addition.

The algorithm described here iterates on the array of given complex Fourier amplitudes and yields the result in less than $2N \log_2 N$ operations without requiring more data storage than is required for the given array A. To derive the algorithm, suppose N is a composite, i.e., $N = r_1 . r_2$. Then let the indices in (1) be expressed

$$j = j_1 r_1 + j_0, \qquad j_0 = 0, 1, \ldots, r_1 - 1, \qquad j_1 = 0, 1, \ldots, r_2 - 1,$$

(3)

$$k = k_1 r_2 + k_0, \qquad k_0 = 0, 1, \ldots, r_2 - 1, \qquad k_1 = 0, 1, \ldots, r_1 - 1.$$

Then, one can write

(4)
$$X(j_1, j_0) = \sum_{k_0} \sum_{k_1} A(k_1, k_0) . \omega^{j k_1 r_2} \omega^{j k_0} .$$

Since

(5)
$$\omega^{j k_1 r_2} = \omega^{j_0 k_1 r_2} ,$$

the inner sum, over k_1, depends only on j_0 and k_0 and can be defined as a new array,

(6)
$$A_1(j_0, k_0) = \sum_{k_1} A(k_1, k_0) . \omega^{j_0 k_1 r_2} .$$

The result can then be written

(7)
$$X(j_1, j_0) = \sum_{k_0} A_1(j_0, k_0) . \omega^{(j_1 r_1 + j_0) k_0} .$$

There are N elements in the array A_1, each requiring r_1 operations, giving a total of $N r_1$ operations to obtain A_1. Similarly, it takes $N r_2$ operations to calculate X from A_1. Therefore, this two-step algorithm, given by (6) and (7), requires a total of

(8)
$$T = N(r_1 + r_2)$$

operations.

It is easy to see how successive applications of the above procedure, starting with its application to (6), give an m-step algorithm requiring

(9)
$$T = N(r_1 + r_2 + \ldots + r_m)$$

operations, where

(10)
$$N = r_1 . r_2 \ldots r_m .$$

If $r_j = s_j t_j$ with $s_j, t_j > 1$, then $s_j + t_j < r_j$ unless $s_j = t_j = 2$, when $s_j + t_j = r_j$. In general, then, using as many factors as possible provides a minimum to (9), but factors of 2 can be combined in pairs without loss. If we are able to choose N to be highly composite, we may make very real gains. If all r_j are equal to r, then, from (10) we have

(11)
$$m = \log_r N$$

and the total number of operations is

(12)
$$T(r) = r N \log_r N.$$

As Cooley and Tukey show in the rest of their article, taking numbers of the form 2^m offers considerable advantages with the use of computers with binary arithmetic. Finally, another advantage of the FFT algorithm is that it reduces the size of accumulated errors.

Bibliography

[1] Bernstein, S., Sur l'ordre de la meilleure approximation des fonctions continues par des polynômes de degré donné, *Mémoires de l'Académie Royale de Belgique*, 2nd series, vol. IV (1912).

[2] Butzer, P.L., Schmidt, M. and Stark, E.L., Observations on the History of Central B-Splines, *Archives for the History of Exact Sciences*, vol 39 (1988), 137–156.

[3] Cauchy, L.-A., *Résumé des Leçons données à l'Ecole Polytechnique sur le Calcul infinitésimal*, 1823. *Oeuvres*, Series 2, vol. IV.

[4] Chebyshev, P.L. Sur les fractions continues, *Journal de Mathématiques pures et appliquées*, 2nd series, vol. III (1858), 289–323. *Oeuvres*, vol. I, pp. 201–230, New York: Chelsea Publishing Company, French tr. of Complete Works (in Russian) by A. A. Markov & N. Y. Simon (ed.), 2 vols., St. Petersburg, 1899–1907.

[5] Chebyshev, P.L. Théorie des mécanismes connus sous le nom de parallélogrammes, *Mémoires présentés à l'Académie Impériale des sciences de St-Pétersbourg par divers savants*, vol. VII (1854), pp. 539–568. *Oeuvres*, vol. I, pp. 109–143, New York: Chelsea Publishing Company.

[6] Chebyshev, P.L., Sur les quadratures, *Journal de mathématiques pures et appliquées*, 2nd series, vol. XIX (1874), 19–34. *Oeuvres*, vol. II, pp. 165–180, New York: Chelsea Publishing Company.

[7] Chebyshev, P.L., Sur les questions de minima qui se rattachent à la représentation approximative des fonctions, *Mémoires de l'Académie Impériale des sciences de St. Pétersbourg*, 6th series, Sciences mathématiques et physiques, vol. VII (1859), 199–291. *Oeuvres*, vol. I, New York: Chelsea Publishing Company, pp. 271–378.

[8] Cooley, J.W. and Tukey, J. W., An algorithm for the Machine Calculation of Complex Fourier Series, *Mathematics of Computation*, vol. XIX (1965), 297–301.

[9] De la Vallée-Poussin, Ch., *Leçons sur l'approximation des fonctions d'une variable réelle*, Paris: Gauthier-Villars, 1919.

[10] Fourier, J.-B., *Théorie analytique de la Chaleur*, Paris, 1822. *Oeuvres*, vol. I, Paris: Gauthier-Villars, 1888. English tr. by A. Freeman, Cambridge, 1878, rprnt. New York, 1955.

[11] Fréchet M., *Encyclopédie des Sciences Mathématiques*, volume II.

[12] Gauss, C.F., Theoria interpolationis methodo novo tractata, *Nachlass, Werke*, vol. III, pp. 265–327, Göttingen, 1866.

[13] Goldstine, H.H., *A history of numerical analysis from the 16th through the 19th century*, New York: Springer, 1977.

[14] Grattan-Guiness, I. & Ravetz, J. R., *Joseph Fourier 1768-1830. A Survey of the Life and Work, Based on a Critical Edition of his Monograph on the Propogation of Heat, Presented to the Institute de France in 1807*, Cambridge, Massachusetts: MIT Press, 1972.

[15] Hermite, Ch., Sur l'interpolation, *Comptes Rendus de l'Académie des Sciences*, vol. 48 (1859), p. 62. *Oeuvres*, vol. II, pp. 87–92, Paris: Gauthier-Villars, 1908.

[16] Lagrange, J.-L., *Leçons sur le calcul des fonctions*, professed at the Ecole Polytechnique in year VII (1799); in *Séances des Ecoles Normales*, vol. X, year IX (1801),Paris: Courcier, 1806. *Oeuvres*, vol. X, Paris: Gauthier-Villars, 1884.

[17] Lagrange, J.-L., *Théorie des fonctions analytiques*, 1797; 2nd edn. 1813. *Oeuvres*, vol. IX, Paris: Gauthier-Villars, 1881.

[18] Méray, Ch., Nouveaux exemples d'interpolations illusoires, *Bulletin des Sciences mathématiques*, 2nd series, vol. 20 (1896), 266–270.

[19] Méray, Ch., Observations sur la légitimité de l'interpolation, *Annales Scientifiques de l'Ecole Normale Supérieure*, 3rd series, vol. 1 (1884), 165–176.

[20] Newton, I., *The Correspondence of*, ed. H. W. Turnbull, Cambridge: Cambridge University Press, 1959.

[21] Runge, C., Über empirische Funktionen und die Interpolation zwischen äquidistanten Ordinaten, *Zeitschrift für Mathematik und Physik*, vol. 46 (1901), 224–243.

[22] Schoenberg, I.J., Cardinal Spline Interpolation, CBMS 12, Philadelphia: SIAM, 1973.

[23] Schoenberg, I.J., Contributions to the problem of approximation of equidistant data by analytic functions, *Quarterly of Applied Mathematics*, vol. IV (1946), 45–99 and 112–141.

[24] Taylor, B., *Methodus incrementorum directa et inversa*, London, 1715, Cambridge, Massachusetts: Harvard University Press, 1969, pp. 329–332. reprnt. Princeton, New Jersey: Princeton University Press, 1986.

[25] Weierstrass, K., Über die analytische Darstellbarkeit sogenannter willkürlicher Functionen reeller Argumente, *Sitzungsberichte der Königl. Preuss. Akad. Wissenschaften Berlin* (1885), 633–639. *Mathematische Werke*, vol. III, Berlin: Mayer & Müller, 1903, pp. 1–37.

14 Acceleration of Convergence

Although the first use of the term convergence can be attributed to Gregory in 1668, the concept of convergence as we use it today was not defined explicitly until the 19th century. Ever since the invention of the infinitesimal calculus, the use of divergent series had been surrounded by controversy. They were effectively banned at the beginning of the 19th century, but were rehabilitated at the end of the century by introducing the idea of summation. Poincaré played a fundamental role in the theory of summation. In *Les méthodes nouvelles de la Mécanique céleste*, he explains the different approaches to the meaning of convergence:

There is a sort of misunderstanding between geometers and astronomers concerning the meaning of the word convergence. Geometers, preoccupied with perfect rigour, and often quite indifferent to the extent of the inextricable calculations which they can conceive of, without thinking about how they could be carried out in practice, say that a series is convergent when the sum of the terms tends towards a determined limit, even when the first terms diminish very slowly. Astronomers, on the other hand, are accustomed to say that a series converges when the first twenty terms, for example, diminish very rapidly, even when the subsequent terms may increase indefinitely.

The history of these ideas goes beyond the scope of this book and the reader is referred to, for example, Kline ([11], Chapter 47). We wish only to underline the fact that it is very important for mathematicians and users of mathematics to consider carefully the behaviour of the terms of a series. The fact that a summation is convergent is neither a necessary nor a sufficient condition for the series to be of interest. It is more important that the summation process should be rapid and that it can be easily done.

While the methods described here are written in the style of the time, and use the mathematical tools that were then available, all use a series expansion which is asymptotic, whether this is the expansion of a general term, or the remainder of order n of the series, or of its deviation from a given quantity. When only a finite number of terms is considered, it is possible to make some evaluation of the neglected quantities. Alternatively, the order of magnitude of the neglected quantities can be reduced by means of suitable, and astute, manipulation.

Different ways of treating the problem will be found in this chapter. Stirling (see Section 14.1) and Kummer evaluated the remainders of the series of order n. Euler and Maclaurin, at the same time but independently, established a formula giving the deviation between an integral and a finite sum of terms in the form of an infinite series. Euler used this to determine the approximate value of various sums of numerical series and the limits of sequences, like the famous Euler constant (see Section 14.3), while Maclaurin used his formula to determine approximate quadratures (see Section 14.2). Furthermore, by separating out the principal part of the deviation by linear combinations, he invented a procedure that was later worked on in the 20th century by Richardson (see Section 14.5) and Romberg (see Section 14.6). We also include Aitken's method, which uses sequences that are equivalent to geometric sequences (see Section 14.4).

14.1 Stirling's Method for Series

An extract from Stirling's *Treatise on the Summation and Interpolation of Infinite Series* appears above in Section 11.4. In the Preface to that work, he clearly announced his objectives:

Since it often happens that series converge so slowly towards the truth, that they go on to the proposed end no more than if they were in fact divergent, I have supplied some theorems in the first part of this treatise, by which one quickly attains the values of just those series which approach [their value] the slowest.

In the introduction to the first part of the work, *On the Summation of Series*, Stirling explains:

In this first part I have attempted to shorten the calculations for the Quadrature of Curves, including for the most difficult problems, and thereby I obtain the values of Infinite Series more quickly than by the simple addition of terms, as is generally done.

The principle behind Stirling's method is the decomposition of the general term of a series into a sum of terms whose summation is easy to find, that is one where the zth term is of the form

$$z(z-1)(z-2) \dots (z-p),$$

or

$$\frac{p}{z(z+1)(z+2)\dots(z+k)}.$$

In proposition I below, the partial sums of series of the first type are discussed, and in proposition II, the sums of series of the second type. In both cases, the use of 'etc.' refers to an unspecified, but finite, number of terms. We see, in an example, that when the series cannot be written in this way, that is as a finite sum of series, Stirling neglects terms of the expansion after a certain point, according to the degree of accuracy that he wants. This is where we see acceleration of convergence in action.

In Stirling's text, z stands for an integer and T is the zth general term of the series. In proposition I, S is the sum of the first z terms (i.e. the *partial sum* to z terms), while in proposition II, S stands for the sum of the terms from the zth term to infinity (i.e. the $(z-1)$th *remainder term*).

J. Stirling
Tractatus de Summatione et Interpolatione Serierum Infinitarum, London, 1730, pp. 20–29.

PROPOSITION I

If the terms of any series are formed by writing the numbers 1, 2, 3, 4, 5, etc. in place of z in the quantity $A + Bz + Cz(z-1) + Dz(z-1)(z-2) + Ez(z-1)(z-2)(z-3) +$ etc. then the Sum of the Terms from the first to the one whose number is z will be

$$Az + (z+1) \text{ into } \left(\tfrac{1}{2} Bz + \tfrac{1}{3} Cz(z-1) + \tfrac{1}{4} Dz(z-1)(z-2) + \text{etc.} \right).$$

The proposition is proved thus.

Form the sum $S = Az + (z + 1)$ into $\left(\tfrac{1}{2} Bz + \tfrac{1}{3} Cz(z-1) + \tfrac{1}{4} Dz(z-1)(z-2) + \text{etc.} \right)$, namely

$$S = Az + \tfrac{1}{2} B(z+1)z + \tfrac{1}{3} C(z+1)z(z-1) + \tfrac{1}{4} D(z+1)z(z-1)(z-2) + \text{etc.}$$

Next substitute the following values of the variables for those presently there; that is $S - T$ in place of S, and $z - 1$ in place of z. You will thus obtain

$$S - T = A(z-1) + \tfrac{1}{2} Bz(z-1) + \tfrac{1}{3} Cz(z-1)(z-2) + \tfrac{1}{4} Dz(z-1)(z-2)(z-3) + \text{etc.}$$

Subtracting this equation from the earlier one gives

$$T = A + Bz + Cz(z - 1) + Dz(z - 1)(z - 2) + \text{etc.}$$

Whence, if conversely the values are given to the terms as is the case in the Proposition, the sum will be as has been given. Further, this sum is zero when z is zero. Q. E. D.

$$[\ldots]$$

EXAMPLE V

If we have the cubes $1, 8, 27, 64, 125, 216$, etc., which is given by z^3, and if we decompose z^3 to the required form $z + 3z(z - 1) + z(z -1)(z -2)$, we shall have: $A = 0, B = 1, C = 3, D = 1$, and all the others zero. On account of this, the sum is $(z + 1)$ times $\tfrac{1}{2} z + z(z-1) + \tfrac{1}{4} z(z-1)(z-2)$

which, collected together, gives $\dfrac{z^2}{4}(z+1)^2$.

$$[\ldots]$$

PROPOSITION II

If the terms of any series are formed by writing any numbers which differ by unity

$$\frac{A}{z(z+1)} + \frac{B}{z(z+1)(z+2)} + \frac{C}{z(z+1)(z+2)(z+3)} + \frac{D}{z(z+1)(z+2)(z+3)(z+4)} + \text{etc.}$$

the sum of all the terms to infinity, starting from any given term of (rank z) will be

$$\frac{A}{z} + \frac{B}{2z(z+1)} + \frac{C}{3z(z+1)(z+2)} + \frac{D}{4z(z+1)(z+2)(z+3)} + \text{etc.}$$

Let the sum be

$$S = \frac{A}{z} + \frac{B}{2z(z+1)} + \frac{C}{3z(z+1)(z+2)} + \frac{D}{4z(z+1)(z+2)(z+3)} + \text{etc.}$$

Then, for the values taken by S and z, put the following in their place, namely $S - T$ for S and $z + 1$ for z, since what we have here is an infinite number of terms: we have

$$S - T = \frac{A}{z+1} + \frac{B}{2(z+1)(z+2)} + \frac{C}{3(z+1)(z+2)(z+3)} + \frac{D}{4(z+1)(z+2)(z+3)(z+4)} + \text{etc.}$$

After subtracting this equation from the earlier one, we have

$$T = \frac{A}{z(z+1)} + \frac{B}{z(z+1)(z+2)} + \frac{C}{z(z+1)(z+2)(z+3)} + \frac{D}{z(z+1)(z+2)(z+3)(z+4)} + \text{etc.}$$

Whence, conversely, if this is the general Term, the sum will be as stated in the Proposition.

COROLLARY

If the Term is $\dfrac{p}{z(z+1)(z+2)(z+3)(z+4) \text{ etc.}}$, reject the last factor, then divide what remains

by the number of remaining factors, and you will have the Sum of the Terms. If the Term is
A/z, reject the factor z, and since nothing remains, divide A by zero, and you will have for the
sum a Quantity that is infinitely large, as is known. And the first that I know to have studied
this is Dr Taylor.

[...]

EXAMPLE VI

To sum the Series $1 + \frac{1}{4} + \frac{1}{9} + \frac{1}{16} + \frac{1}{25} + \frac{1}{36} +$ etc. where the Denominators are the Squares of
the numbers 1, 2, 3, 4, etc. and the general Term is $1/zz$. This, reduced to a summable form, be-
comes:

$$\frac{1}{z(z+1)} + \frac{1}{z(z+1)(z+2)} + \frac{1.2}{z(z+1)(z+2)(z+3)} + \frac{1.2.3}{z(z+1)(z+2)(z+3)(z+4)} + \text{etc.}$$

from which we have the Sum

$$\frac{1}{z} + \frac{1}{2z(z+1)} + \frac{1.2}{3z(z+1)(z+2)} + \frac{1.2.3}{4z(z+1)(z+2)(z+3)} + \text{etc.}$$

By letting

$$A = \frac{1}{z}, B = \frac{A}{z+1}, C = \frac{2B}{z+2}, D = \frac{3C}{z+3}, E = \frac{4D}{z+4}, \text{etc.}$$

we get

$$A + \frac{1}{2}B + \frac{1}{3}C + \frac{1}{4}D + \text{etc.}$$

Now, if we substitute 13 for z, we shall have the Sum of all the Terms which come after the
twelfth; and in this case we shall have $A = \frac{1}{13}$, $B = \frac{1}{14}A$, $C = \frac{2}{15}B$, $D = \frac{3}{16}C$, $E = \frac{4}{17}F$, etc. And
$A + \frac{1}{2}B + \frac{1}{3}C + \frac{1}{4}D +$ &c. will be the sum of the terms $\frac{1}{169} + \frac{1}{196} + \frac{1}{225} + \frac{1}{256} +$ &c.
And thus we have the comptation:

A	=	.076923077	.076923077
B	=	5494505	2747252
C	=	732601	244200
D	=	137363	34341
E	=	32321	6464
F	=	8978	1496
G	=	2835	405
H	=	992	124
I	=	378	42
K	=	155	16
L	=	67	6
M	=	31	3
N	=	15	1
			.079957427

From which we obtain .079957427 for the sum of the terms $\frac{1}{169}+\frac{1}{196}+\frac{1}{225}+$ &c. which, added to the twelve first terms, that is 1.564976638, makes 1.644934065 for the value of the whole series $1+\frac{1}{4}+\frac{1}{9}+\frac{1}{16}+$ &c.

In Proposition I, if we let $S(z)=\sum_{k=1}^{z}T(k)$, then clearly $S(z)-S(z-1)=T(z)$ for all integer $z\geq 1$, and $S(0)=T(0)$. Conversely, these equalities define the sum to the zth term of the series S.

Stirling is content therefore simply to verify these equalities, using the identity:

$$[(z+1)z(z-1)\ldots(z-k+2)]-[z(z-1)\ldots(z-k+1)]=kz(z-1)\ldots(z-k+2).$$

He can therefore easily deduce the sums of the powers of integers.

In Proposition II, let $S(z)=\sum_{k=z}^{\infty}T(k)$, then we have again $S(z)-S(z+1)=T(z)$. Stirling is again content simply to verify this relation. If the series is taken to be convergent, then the behaviour of S is effectively determined, since $S(z)$ tends to zero as z tends to infinity. The identity he uses is

$$\frac{1}{z(z+1)\ldots(z+k-1)}-\frac{1}{(z+1)\ldots(z+k)}=\frac{k}{z(z+1)\ldots(z+k)}.$$

The principle of the proof, where the general term of the series appears as the first finite difference of the sum function, connects it with the works of Newton and the 'English' School, to which he always makes reference in his prefaces (see Chapter 10). We should also point out that Stirling published an alternative procedure of accelerated summation of series in 1719 [21].

In the second example, Stirling's example VI, the coefficients of the expansion of $\frac{1}{z^2}$ are obtained by recurrence using the identities:

$$\frac{1}{z\cdot z}=\frac{1}{z(z+1)}+\frac{1}{z(z+1)z},\ldots$$

$$\frac{1}{z(z+1)\ldots(z+k)z}=\frac{1}{z(z+1)\ldots(z+k+1)}+\frac{k+1}{z(z+1)\ldots(z+k+1)z}.$$

We may recall that the sum of the series is equal to $\pi^2/6\approx 1.644934067$ and the nth remainder is of order $1/n$. The efficiency of Stirling's method can be gauged from the fact that calculating this result by summing successive terms would require at least 500 000 000 terms.

14.2 The Euler-Maclaurin Summation Formula

In 1734, Bishop George Berkeley published a tract that criticised the absence of rigorous foundations to Newton's works. This work, *The Analyst*, carried the rather long, but explanatory subtitle:

Or A Discourse Addressed to an Infidel Mathematician. Wherein It Is Examined Whether the Object, Principles, and Inferences of the Modern Analysis are More Distinctly Conceived, or More Evidently Deduced, than Religious Mysteries and Points of Faith. "First Cast the Beam Out of Thine Own Eye: and Then Shall Thou See Clearly to Cast Out the Mote Out of Thy Brother's Eye".

It was to answer these criticisms that Maclaurin wrote his *Treatise on Fluxions* in 1737 which appeared in 1742.

The theorems which interest us here are stated in Book I (Articles 352 and 353) and proved in Book II. The proofs reflect the methods and language of the time. Like Newton, Maclaurin considers the problem in a geometric context, and sets out to determine the area contained between the curve, the abscissa axis, and two lines parallel to the ordinate axis. The term 'fluxion', corresponds to the derivative of the function we start with, which is therefore called the 'fluent'.

The traditional practice of the 'English' school, since Wallis, was the frequent use of infinite series. Thus, Maclaurin, in Article 745, advises: "When a fluent cannot be represented exactly in algebraic terms, it should then be expressed by a convergent series". But, in Article 827, he adds:

When an area or a fluent is reduced to a series by the methods described in Art. 745, these series, in some cases, converge at so slow a rate as to be of little use for finding the area ...

Maclaurin then goes on to give the example of the equilateral hyperbola $y = \dfrac{1}{1+x}$, and adds:

But this series converges so slowly that the sum of the first 1000 terms of it is found deficient from the true value of the area in the fifth decimal; and other examples similar to this might be brought, wherein the area may be more easily computed from the inscribed polygons than from the series. Some further artifice is therefore necessary in order to compute the area in such cases [...]. The following theorems, derived from the method of fluxions, may be of use for this purpose; and serve for the resolution of many problems that are usually referred to what is called Sir Isaac Newton's *differential method*.

As we would express it today, what these theorems do, is to compare the area under the curve, measured by the definite integral $\int_a^b y(t)\,dt$, with the area obtained by the 'trapezium method', measured by the finite summation $h\left(\dfrac{y(a)+y(b)}{2} + \sum_{k=1}^{n-1} y(a+kh) \right)$.

Today we would express the theorems by:

$$\int_a^b y(t)\,dt = h\left(\frac{y(a)+y(b)}{2} + \sum_{k=1}^{n-1} y(a+kh) \right) - \sum_{j=1}^{etc.} h^{2j} \frac{B_{2j}}{2j!} [y^{(2j-1)}(b) - y^{(2j-1)}(a)].$$

The difference is the sum of a series, not necessarily convergent, in which the B_{2j} are Bernoulli numbers (identified as such later by Euler). The index notation and sign used for Bernoulli numbers differ according to authors.

After numerous examples where he applies his theorems to different summations, Maclaurin then obtains the Newton-Cotes quadrature formulas by a procedure that we call Romberg's method (see Section 14.6). He was thus able to evaluate the error involved in using the Newton-Cotes formulas.

A little earlier, in 1732, on the Continent, and in a totally independent way, Euler obtained these same formulas (published in 1738). We shall come back to this when we look at the Euler constant (Section 14.3).

C. Maclaurin
A Treatise on Fluxions, Edinburgh, 1742. 2nd edn. London, 1801, vol. II, pp. 260–276.

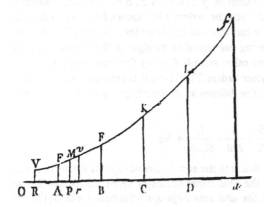

828. Suppose the base AP $= z$ (fig. 314), the ordinate PM $= y$ and, the base being supposed to flow uniformly, let $\dot{z} = 1$. Let the first ordinate AF be represented by a, AB $= 1$, and the area ABEF $= A$. As A is the area generated by the ordinate y, so let B, C, D, E, etc. represent the areas upon the same base AB, generated by the respective ordinates $\dot{y}, \ddot{y}, \dddot{y}, \ddddot{y}$, &c. Then

$$AF = a = A - \frac{B}{2} + \frac{C}{12} - \frac{E}{720} + \frac{G}{30240} - \&c.$$

For, by art. 752,
$$A = a + \frac{\dot{a}}{2} + \frac{\ddot{a}}{6} + \frac{\dddot{a}}{24} + \frac{\ddddot{a}}{120} + \&c.$$

whence we have the equation (Q):
$$a = A - \frac{\dot{a}}{2} - \frac{\ddot{a}}{6} - \frac{\dddot{a}}{24} - \frac{\ddddot{a}}{120} - \&c.$$

In like manner
$$\dot{a} = B - \frac{\ddot{a}}{2} - \frac{\dddot{a}}{6} - \frac{\ddddot{a}}{24} - \&c.$$

$$\ddot{a} = C - \frac{\dddot{a}}{2} - \frac{\ddddot{a}}{6} - \&c.$$

$$\ddot{a} = D - \frac{\dddot{a}}{2} - \&c.$$

$$\dddot{a} = E - \&c.$$

by which latter equations, if we exterminate $\dot{a}, \ddot{a}, \dddot{a}, \ddddot{a},$ &c. from the value of a in the equa-

tion Q, we shall find that $\qquad a = A - \frac{B}{2} + \frac{C}{12} - \frac{E}{720} + e\&.$

The coefficients are continued thus: let $k, l, m, n,$ &c. denote the respective coefficients of $\dot{a}, \ddot{a}, \dddot{a},$ &c. in equation Q; that is, let $k = \frac{1}{2}, l = \frac{1}{6}, m = \frac{1}{24}, n = \frac{1}{120},$ &c.; suppose $K = k = \frac{1}{2},$ $L = kK - l = \frac{1}{12}, M = kL - lK + m = 0, N = kM - lL + mK - n = -\frac{1}{720},$ and so on; then

$$a = A - KB + LC - MD + NE - \&c.$$

where the coefficients of the alternate areas $D, F, H,$ &c. vanish.

829. As A is the fluent of $y \dot{z}$, so B is the fluent of $\dot{y} \dot{z}, C$ of $\ddot{y} \dot{z}, E$ of $\dddot{y} \dot{z},$ &c. Therefore, since $\dot{z} = 1$, and these areas are generated while the ordinate PM moves from AF to BE, the area B will be expressed as the excess of the last ordinate BE above the first AF; C, by the difference of the fluxions of the ordinates (having due regard to the signs of these fluxions), E by the difference of their third fluxions, and the other areas $G, I,$ &c. by the respective differences of their fluxions of the corresponding higher orders. Therefore, if α represent BE − AF, the difference of the ordinates, and $\beta, \delta, \zeta,$ &c. the differences of their fluxions of the first, third, fifth orders, &c. then

$$a = A - \frac{\alpha}{2} + \frac{\beta}{12} - \frac{\delta}{720} + \frac{\zeta}{30240} + \&c.$$

830. Supposing now the base Aa to be divided into the equal successive parts AB, BC, CD, &c. and each part equal to unit, let the sum of the equidistant ordinates AF, BE, CK, &c., exclusive of the last ordinate af, be represented by S, the total area AFfa upon the base Aa by A, the excess of af above AF by α, and the respective excesses of their first, third, fifth fluxions, &c. by $\beta, \delta, \zeta,$ &c. the fluxion of the base being supposed equal to 1, then it follows, from the last article, that

$$S = A - \frac{\alpha}{2} + \frac{\beta}{12} - \frac{\delta}{720} + \frac{\zeta}{30240} + \&c. \text{ and } A = S + \frac{\alpha}{2} - \frac{\beta}{12} + \frac{\delta}{720} - \frac{\zeta}{30240} + \&c.$$

which give two of the theorems mentioned in art. 352 and 353, where we had hyperbolic figures chiefly in view. This proposition more generally expressed, without supposing z or $\dot{z}$ equal to unit, is that

$$S = \frac{A}{z} - \frac{\alpha}{2} + \frac{\beta z}{12\dot{z}} - \frac{\delta z^3}{720\dot{z}^3} + \frac{\zeta z^5}{30240\dot{z}^5} - \frac{\theta z^7}{1209600\dot{z}^7} + \&c.$$

[...]

848. The base Aa being divided into any number of equal parts represented by n, let the area AFfa = Q, the sum of the extreme ordinates AF + af = A, the sum of all the intermediate ordinates BE + CK + &c. = B, the base Aa = R, and the same quantities be represented by $\beta, \delta, \zeta,$ etc as formerly; then the area

$$\mathrm{AF}fa = Q = \frac{A}{2n+2} + \frac{nB}{nn-1} \times R - \frac{R^4\delta}{720nn} + \frac{R^6\zeta}{30240} \times \frac{nn+1}{n^4} - \&c.$$

For, supposing, in art. 830, $e = \dfrac{R}{n}$, $S + \dfrac{af - \mathrm{AF}}{2} = B + \dfrac{A}{2} = \dfrac{nQ}{R} + \dfrac{R\beta}{12n} - \dfrac{R^3\delta}{720n^3} + \dfrac{R^5\zeta}{30240n^5} - \&c.$

and supposing, $e = R$ in the same theorem, $\dfrac{\mathrm{AF} + af}{2} = \dfrac{A}{2} = \dfrac{Q}{R} + \dfrac{R\beta}{12} - \dfrac{R^3\delta}{720} + \dfrac{R^5\zeta}{30240} - \&c.$ then,

if we exterminate β by these two equations, the proposition will appear. If we neglect δ, ζ, θ, &c.

$$\mathrm{AF}fa = \frac{AR}{2n+2} + \frac{nBR}{nn-1}.$$

Suppose $n = 2$, or that there are three ordinates only (in which case B denotes the middle ordinate), then the area

$$\mathrm{AF}fa = \frac{A+4B}{6} \times R - \frac{R^4\delta}{4\times720} + \frac{5R^6\zeta}{16\times30240} - \&c.$$

If we suppose $n = 3$, or that there are four ordinates only, B will represent the sum of the second and third, and the area

$$\mathrm{AF}fa = \frac{A+3B}{8} \times R - \frac{\delta R^4}{9\times720} + \frac{\zeta R^6}{81\times3024} - \&c.$$

By neglecting δ, ζ, θ, &c. we shall have two of the theorems given by Sir *Isaac Newton* and others for computing the area from four equidistant ordinates, the latter of which (viz. $\mathrm{AF}fa = \dfrac{A+3B}{8} \times R$) is much recommended by Mr. *Cotes*.

849. By exterminating δ, ζ, θ, &c., successively, other theorems will be found by which the area will be more and more accurately determined from the ordinates. Let there be five ordinates, A the sum of the first and last, B the sum of the second and fourth, and C the middle ordinate; then the area

$$\mathrm{AF}fa = \frac{7A+32B+12C}{90} \times R - \frac{31R^6\zeta}{6\times16\times16\times30240} + \&c.\ ;$$

for by the rule for three ordinates,

$$\frac{Q}{R} = \frac{A+4C}{6} - \frac{R^3\delta}{4\times720} + \frac{5R^5\zeta}{16\times30240}.$$

By dividing the base into two equal parts, and computing from the same rule the area that stands upon each part, and adding these areas together,

$$\frac{2Q}{R} = \frac{A+4B+2C}{6} - \frac{R^3\delta}{32\times720} + \frac{5R^5\zeta}{2\times16\times16\times30240} - \&c.$$

then, by exterminating δ by these two equations, the proposition appears. These Theorems may be continued in like manner, and some judgement formed of the accuracy of the several rules, by comparing the quantities that are neglected in them.

The start of the proof is based precisely on what we call Maclaurin's theorem, a particular case of Taylor's theorem (see Section 13.1), and which had been proved in article 752.

If we suppose that for $t = 0$, the moving point is at F, that is $z(0) = 0$, the condition $\dot{z} = 1$ gives $z = t$, and the abscissa B to E gives $z(1) = 1$.

The first equations are obtained by applying Maclaurin's theorem successively to the functions

$$x \mapsto \int_0^x y(t)\,dt\,,\ x \mapsto \int_0^x y'(t)\,dt\,,\ x \mapsto \int_0^x y''(t)\,dt\,,\text{etc.}$$

between 0 and 1, and with the notation $a = y(0)\,,\ \dot{a} = y'(0)\,,\ \ddot{a} = y''(0)\,$, etc.

$$A = \int_0^1 y(t)\,dt\,,\ B = \int_0^1 y'(t)\,dt\,,\ C = \int_0^1 y''(t)\,dt\,,\text{etc.}$$

The second step is to 'exterminate' $a, \dot{a}, \ddot{a}, \dddot{a}$, etc. to obtain recurrence relations between (the unknown) K, L, M, N, etc. and (the known) k, l, m, n, etc., since Maclaurin wants to obtain a in the form $a = A - KB + LC - MD + NE \ldots$. This entails purely symbol manipulation, which is set out in an infinite table:

$$
\begin{array}{lll}
a = A - k\dot{a} - l\ddot{a} - m\dddot{a} - n\ddddot{a} - \text{ etc.} & \text{to be multiplied by 1} \\
\dot{a} = B - k\ddot{a} - l\dddot{a} - m\ddddot{a} - \text{ etc.} & \times & (-K) \\
\ddot{a} = C - k\dddot{a} - l\ddddot{a} - \text{ etc.} & \times & (L) \\
\dddot{a} = D - k\ddddot{a} - \text{ etc.} & \times & (-M) \\
\ddddot{a} = E - \text{ etc.} & \times & (N)
\end{array}
$$

For example, to 'exterminate' $\ddot{a}$, we write the coefficient of $\ddot{a}$, which is that of D (here $-M$) in the fourth equation, as the sum of the coefficients of $\ddot{a}$ appearing in the three previous equations, taking account of the multipliers. That is: $-kL + lK - m$. These recurrence relations can be used to obtain K, L, M, etc. one after the other.

Paragraph 829 can be translated simply as

$$A = \int_0^1 y(t)\,dt\,,\ B = \int_0^1 y'(t)\,dt = y(1) - y(0) = \alpha\,,\text{ etc.}$$

In paragraph 830, we simply sum the preceding equalities for each of the intervals obtained by subdivision. The values taken for $y, y', y'', \ldots$ at the intermediary points cancel each other in pairs so that only the extreme values remain.

If S is interpreted as the area obtained by the rectangle method, then $z(S + \alpha/2)$ corresponds to the area obtained by the trapezium method. Using current notation, we write the formula as:

$$\int_a^b y(t)\,dt = h\left(\frac{y(a) + y(b)}{2} + \sum_{k=1}^{n-1} y(a + kh)\right) - \sum_{j=1}^{\text{etc.}} h^{2j}\frac{B_{2j}}{2j!}[y^{(2j-1)}(b) - y^{(2j-1)}(a)]$$

Paragraph 848 deduces more effective numerical integration formulas from this formula. Using the same approximate numerical integration formula for two different ways of dividing up the interval, it is possible to eliminate the first correction term (which is considered to be of a higher order than the following one) by a linear combination of the two results. When the interval is divided in half, we have what is called Romberg's method (Section 14.6). We see here how we can progress from the trapezium method to Simpson's method, and the first three of Cotes's formulas are found by Maclaurin (see chapter 10).

14.3 The Euler Constant

The extract by Euler that is given here was published in the *Commentaries of the Academy of Sciences of St. Petersburg* for the year 1736. Leonhard Euler had come to St. Petersburg in 1727, where he rejoined Daniel Bernoulli. Despite severe problems with his health (he lost the use of his right eye in 1735), he generated an incredible amount of mathematics, the results of which are to be found in *Introductio in Analysin Infinitorum* and *Institutiones Calculis Differentialis*.

The article, of which we give an extract below, is concerned with the proof and application of the 'Euler-Maclaurin' formula. While Maclaurin uses series to evaluate quadratures, Euler uses quadratures to evaluate sums of series (possibly even divergent ones). Euler had already published the formula, without proof, in 1732 [6], and he was to offer further justifications and applications later ([3], chapter X and [4], chapter VII).

Euler uses X and S respectively for the general term and the partial sum of the series to the xth term. Thus X is a function of x which is sometimes an integer index (in the series), and sometimes a variable (of the function X). In paragraphs 1 to 20 he establishes the formula

$$S = \int X\,dx + \frac{X}{1.2} + \frac{dX}{1.2\,3.2\,dx} - \frac{d^3X}{1.2.3.4.5.6\,dx^3} + \frac{d^5X}{1...7.6\,dx^5} - \frac{3\,d^7X}{1...9.10\,dx^7}$$
$$+ \frac{5\,d^9X}{1...11.6\,dx^9} - \frac{69\,d^{11}X}{1...13.210\,dx^{11}} + \frac{35\,d^{13}X}{1...15.2\,dx^{13}} - \frac{3617\,d^{15}X}{1...17.30\,dx^{15}} + \text{etc.}$$

specifying that in $\int X\,dx$ "a constant must be added so that, on putting $x = 0$, we shall have $S = 0$".

The extract here shows the formula being applied to calculating the 'Euler constant', that is the calculation of the limit, as p tends to infinity, of the quantity

$$\sum_{n=1}^{p} \frac{1}{n} - \ln p.$$

The calculation produces 16 decimals, of which 15 are exact. (Note that Euler uses just ℓ for logarithm.)

L. Euler
Inventio summae cuiusque seriei ex dato termino generali, *Commentarii academiae scientiarum Petropolitanae*, 47 (1736), 1741. *Leonhardi Euleri Opera Omnia*, Vol. XIV, pp. 117–119.

24. [...] Meanwhile, I observed that, thanks to this method, series of this nature can easily be summed in a way that approximates well, which is of remarkable use for series which converge slowly and are otherwise difficult to sum. How this is to be done, I shall teach through examples.

25. I shall therefore consider first the harmonic series and, before all the others, this one:

$$1 + \tfrac{1}{2} + \tfrac{1}{3} + \tfrac{1}{4} + \text{ etc.}$$

whose general term is $\dfrac{1}{x}$. We wish to find the sum term S.

We therefore have $X = \dfrac{1}{x}$ and $\int X \, dx = \text{Const.} + \ell x$, and, it follows:

$$\frac{dX}{dx} = \frac{-1}{x^2}; \quad \frac{d^3X}{dx^3} = \frac{-1.2.3}{x^4}; \quad \frac{d^5X}{dx^5} = \frac{-1.2.3.4.5}{x^6}; \text{ etc.}$$

On substituting, we have:

$$S = \text{Const.} + \ell x + \frac{1}{2x} - \frac{1}{12x^2} + \frac{1}{120x^4} - \frac{1}{252x^6} + \frac{1}{240x^8} - \frac{1}{132x^{10}} + \frac{691}{32760x^{12}} - \frac{1}{12x^{14}} + \text{ etc.},$$

where the constant to be added must be chosen so that, on putting $x = 0$, we shall have $S = 0$. But, using this, because of all the infinitely large terms, the constant cannot be determined.

26. In fact, in order to determine the constant, another case must be considered, for which the sum of the series is known; the case that is obtained when a fixed number of terms is collected to form a single sum. Let us therefore add up the first 10 terms $1 + \tfrac{1}{2} + \tfrac{1}{3} + \ldots + \tfrac{1}{10}$ and we find the sum = 2.92896 82539 68253 9, to which has to be added the sum of the terms given by the formula, namely:

$$\text{Const.} + \ell 10 + \frac{1}{20} - \frac{1}{1\,200} + \frac{1}{1\,200\,000} - \frac{1}{252\,000\,000}$$

$$+ \frac{1}{24\,000\,000\,000} - \frac{1}{1\,320\,000\,000\,000} + \text{ etc.}$$

That being done, since $\ell 10 = 2.30258\,50929\,94045\,684$, we find that the constant to be added is = 0.57721 56649 01532 9. Having determined this, we can now find the sum of the series, no matter how many terms are taken.

To obtain the same degree of accuracy by simply adding the first p terms of $\sum\limits_{n=1}^{p} \dfrac{1}{n} - \ln p$ would require the calculation of $5 . 10^4$ terms.

The way that Euler expresses his formula makes it appear, at first sight, to be quite different from Maclaurin's. But if in Maclaurin's formula we substitute $b = x, f = X$, in the terms containing the values taken at a by f and its successive derivatives, we shall find the constant that Euler determined.

The nature of the remainder term, being a series which is not always convergent, was considered by a number of later mathematicians, such as Poisson [15], Jacobi [10] and Hermite [8]. Poisson's memoir, read to the French Academy of Sciences in 1826, also shows how Gregory's formula (Section 11.1 above) can be deduced ([7], pp. 586–587). The argument is as follows:

Beginning with

$$\int_0^c f(x)\,dx = \omega P_n - \frac{\omega^2}{12}[f'(0) - f'(c)] + \frac{\omega^4}{720}[f'''(0) - f'''(c)]$$
$$- \frac{\omega^6}{30240}[f^{(5)}(0) - f^{(5)}(c)] + \ldots$$

and

$$P_n = \tfrac{1}{2}f(0) + f(\omega) + f(2\omega) + \ldots + f(n\omega - \omega)\,,\text{ where } \omega = \frac{c}{n}\,,$$

he defines $F(z) = f(z) - f(c - z)$, which has the interpolation formula:

$$F(z) = F(0) + \frac{z}{\omega}\Delta_1 + \frac{z(z-\omega)}{2\omega^2}\Delta_2 + \frac{z(z-\omega)(z-2\omega)}{2.3\omega^3}\Delta_3 + \ldots$$

We also have $\qquad f^{(2m-1)}(0) - f^{(2m-1)}(c) = F^{(2m-1)}(0).$

It follows that: $\qquad \omega[f'(0) - f'(c)] = \Delta_1 - \tfrac{1}{2}\Delta_2 + \tfrac{1}{3}\Delta_3 - \tfrac{1}{4}\Delta_4 + \text{etc.}$

$$\omega^3[f'''(0) - f'''(c)] = \Delta_3 - \tfrac{3}{2}\Delta_4 + \text{etc.}$$

etc.

and on substituting these values in the original equation, we have

$$\int_0^c f(x)\,dx = \omega P_n + \frac{\omega}{12}\Delta_1 - \frac{\omega}{24}\Delta_2 + \frac{19\omega}{720}\Delta_3 - \frac{3\omega}{160}\Delta_4 + \text{etc.}$$

If we now put $\omega = c$ we have Gregory's formula.

The Euler-Maclaurin formula also has the advantage that it can be used to give an asymptotic expansion of the remainder term for certain series. As an example, consider the series whose general term is $a_k = \frac{1}{k^m}$ (with $m > 1$). Using the Euler-Maclaurin formula in the form:

$$\int_0^1 f(x)\,dx = \frac{f(1) + f(0)}{2} - \sum_{j=1}^p \frac{B_{2j}}{(2j)!}[f^{(2j-1)}(1) - f^{(2j-1)}(0)] + R_{2p+1}$$

with the function $f_k : x \mapsto \dfrac{1}{(k+m)^m}$ and $m > 1$, we obtain:

$$f_k^{(p)}(x) = (-1)^p \frac{(m+p-1)!}{(m-1)!} \frac{1}{(k+x)^{m+p}}$$

and so $\displaystyle \int_0^1 f_k(x)\,dx = \left[\frac{-1}{(m-1)(k+x)^{m-1}} \right]_{x=0}^{x=1} = \frac{-1}{(m-1)} \left[\frac{1}{(k+1)^{m-1}} - \frac{1}{k^{m-1}} \right]$

Substituting in the Euler-Maclaurin formula we therefore have, for all natural number k:

$$\frac{1}{(m-1)} \left[\frac{1}{k^{m-1}} - \frac{1}{(k+1)^{m-1}} \right] = \frac{1}{2} \left[\frac{1}{(k+1)^m} + \frac{1}{k^m} \right]$$

$$+ \sum_{j=1}^{p} \frac{B_{2j}}{(2j)!} \frac{(m+2j-2)!}{(m-1)!} \left[\frac{1}{(k+1)^{m+2j-1}} - \frac{1}{k^{m+2j-1}} \right] + R_{2p+1} .$$

Summing this for $k = K$ to ∞, we find:

$$\frac{1}{(m-1)} \frac{1}{K^{m-1}} = \frac{1}{2K^m} + \sum_{k=K}^{\infty} \frac{1}{(k+1)^m} - \sum_{j=1}^{p} \frac{B_{2j}}{(2j)!} \frac{(m+2j-2)!}{(m-1)!} \frac{1}{K^{m+2j-1}} + \sum_{k=K}^{\infty} R_{2p+1}$$

which then gives:

$$a_{k+1} + a_{k+2} + a_{k+3} + \ldots =$$

$$\frac{1}{(m-1)} \frac{1}{K^{m-1}} - \frac{1}{2K^m} + \sum_{j=1}^{p} \frac{B_{2j}}{(2j)!} \frac{(m+2j-2)!}{(m-1)!} \frac{1}{K^{m+2j-1}} - \sum_{k=K}^{\infty} R_{2p+1} .$$

In a memoir of 1835, Kummer [12] proposed a more general method valid for the case where the ratio $\dfrac{a_k}{a_{k+1}}$ can be expanded in the form

$$\frac{a_k}{a_{k+1}} = 1 + \frac{v_1}{k} + \frac{v_2}{k^2} + \frac{v_3}{k^3} + \ldots$$

In particular, he obtained the formula:

$$a_{k+1} + a_{k+2} + a_{k+3} + \ldots = a_k \left(ck + c_1 + \frac{c_2}{k} + \frac{c_3}{k^2} + \ldots \right)$$

where the c_k are defined from the v_k by simple recurrence relations. He concludes:

This is a new summation formula, quite general and, among all the others, one that really depends on the most elementary methods. It also has other advantages over the usual formula which, with the present object in mind, can be presented thus:

$$a_{k+1} + a_{k+2} + a_{k+3} + \ldots = \int_k^{\infty} a_k\,dk - \frac{a_k}{2} - \frac{B_1\,da_k}{1.2\,dk} + \frac{B_2\,d^2a_k}{1.2.3.4\,dk^2} + \ldots$$

where $B_1 = 1/6$, $B_2 = 1/30$, etc. are the Bernoulli numbers. This formula can often be used legitimately for finding the numerical sum of series which converge very slowly. But when a_k is not a function which can be expanded, [...] its integral and derivatives are meaningless. It is only when $a_k = 1/k^m$ that the two summation formulas are identical.

14.4 Aitken's Method

In an article published in 1926, Aitken proposed an extension to Bernoulli's method of recurrent series (see Section 7.7). This provided for not only an approximation to all the roots of a given equation but also a means of "deriving from a sequence of approximations to a root successive sequences of ever-increasing rapidity of convergence". Although the method can be generalised to all convergent sequences, the idea seems to have come from considering this particular case. We may recall the way in which the Bernoulli sequence is generated by considering an example used by Aitken in his article (§ 3.).

In order to solve an algebraic equation, for example

$$z^4 - 10z^3 - 92z^2 + 234z + 315 = 0$$

we introduce a sequence whose general term is $f(t)$ (t being an integer), defined by the linear recurrence relation associated with the equation:

$$f(t + 4) - 10f(t + 3) - 92f(t + 2) + 234f(t + 1) + 315f(t) = 0,$$

and with the first terms chosen arbitrarily, for example:

$$f(0) = f(1) = f(2) = 0 \text{ and } f(3) = 1.$$

When it turns out, as in this easy case, that the moduli of the roots are distinct from each other, and can therefore be ranked in decreasing order:

$$|z_1| > |z_2| > |z_3| > |z_4|$$

the ratio $Z(t) = \dfrac{f(t+1)}{f(t)}$ tends to z_1 as t tends to infinity. In this example, where $z_1 = 15$ is the exact value of the root, we obtain the sequence:

$Z(4) = 19.2 \quad Z(5) = 13.57 \quad Z(6) = 15.76 \quad Z(7) = 14.67$
$Z(8) = 15.16 \quad Z(9) = 14.93 \quad Z(10) = 15.03 \quad Z(11) = 14.98 \ldots$

The Aitken extension is to prove that the ratio

$$(*) \quad \frac{\begin{vmatrix} f(t+1) & f(t+2) & \dots & f(t+m) \\ f(t) & f(t+1) & \dots & f(t+m-1) \\ \dots & \dots & \dots & \dots \\ f(t-m+2) & f(t-m+3) & \dots & f(t+1) \end{vmatrix}}{\begin{vmatrix} f(t) & f(t+1) & \dots & f(t+m-1) \\ f(t-1) & f(t) & \dots & f(t+m-2) \\ \dots & \dots & \dots & \dots \\ f(t-m+1) & f(t-m+2) & \dots & f(t) \end{vmatrix}} \xrightarrow[t\to\infty]{} z_1 z_2 z_3 \dots z_m$$

The sequence in the example above, which is the Bernoulli sequence, is found when we take $m = 1$. In the extract from Aitken's article given here, he uses the result $(*)$ in the case $m = 2$ again applied to this simple example, to generate sequences that converge to the root more and more rapidly.

A. C. Aitken

On Bernoulli's Numerical Solution of Algebraic Equations, *Proceedings of the Royal Society of Edinburgh*, vol. 46 (1926), 289–305 (§8, 300–302).

§8. Derived Sequences of more Rapid Convergence

So far we have not obtained a very great degree of accuracy in the numerical examples. We shall now proceed to derive from the primary sequences successive sequences of increasing approximative power.

Consider first the sequence of approximations z_1, namely

$$Z_1(t) = \frac{f(t+1)}{f(t)}$$

and construct their finite differences of first and second order. It is an immediate consequence of $(*)$ that $\Delta Z_1(t)$ tends to become a geometric sequence (provided z_2 is not one of a pair of conjugate complex roots, or of a pair equal in value but opposite in sign) of common ratio $\frac{z_2}{z_1}$. Hence the *deviations* of $Z_1(t)$ from z_1 will also tend to become a geometric sequence with the same common ratio. Thus a further approximate relation is suggested, viz.

$$\frac{z_1 - Z_1(t+2)}{z_1 - Z_1(t+1)} = \frac{\Delta Z_1(t+1)}{\Delta Z_1(t)}$$

and solving for z_1 we are led to investigate the *derived* sequence

$$Z_1^{(1)}(t) = \frac{\begin{vmatrix} Z_1(t+1) & Z_1(t+2) \\ Z_1(t) & Z_1(t+1) \end{vmatrix}}{\Delta^2 Z(t)}.$$

Retention of terms of higher order shows that $Z_1^{(1)}(t)$ tends to z_1, but its first differences $\Delta Z_1^{(1)}(t)$ and the deviations $z_1 - Z_1^{(1)}(t)$ tend to form geometric sequences each of ultimate common ratio $\left(\dfrac{z_2}{z_1}\right)^2$ or $\dfrac{z_3}{z_1}$, whichever is the greater. Thus the convergency of the derived sequence $Z_1^{(1)}(t)$ is more rapid than that of the primary sequence $z_1(t)$, and it also commences at a more advanced stage of approximation. In the same way we can form a second derived sequence

$$Z_1^{(2)}(t) = \frac{\begin{vmatrix} Z_1^{(1)}(t+1) & Z_1^{(1)}(t+2) \\ Z_1^{(1)}(t) & Z_1^{(1)}(t+1) \end{vmatrix}}{\Delta^2 Z_1^{(1)}(t)}.$$

Retention of terms of still higher order shows that the deviations and differences of the second derived sequence possess similar properties to those of the earlier sequences, and indeed it may be shown generally that when the equation possesses only real roots we can proceed indefinitely forming derived sequences in this way, each converging to z_1 with greater rapidity than the preceding.

[...]

Example. To obtain a more accurate value for the greatest root z_1 of the example of §3.

The table below shows the primary and derived sequences obtained from the last five values of Z_1, to eleven significant digits. Recalling that z_1 was 15, we can compare the sequences and observe the notable improvement in the approximation.

Z_1	$Z_1^{(1)}$	$Z_1^{(2)}$
14.667121186	15.001418373	14.999999987
15.159777145	15.000304169	
14.926402783	15.000065221	
15.034550817		
14.983920543		

A similar improvement can be made in the sequences for the remaining roots.

The starting point for Aitken seems to have been prompted by considering particular sequences. In the particular case of $m = 2$, the formula (*) gives:

$$Z_2(t) = \frac{\begin{vmatrix} f(t+1) & f(t+2) \\ f(t) & f(t+1) \end{vmatrix}}{\begin{vmatrix} f(t) & f(t+1) \\ f(t-1) & f(t) \end{vmatrix}} \xrightarrow{t \to \infty} z_1 z_2$$

and therefore:

$$\frac{\Delta Z(t+1)}{\Delta Z(t)} = \frac{\dfrac{f(t+3)}{f(t+2)} - \dfrac{f(t+2)}{f(t+1)}}{\dfrac{f(t+2)}{f(t+1)} - \dfrac{f(t+1)}{f(t)}} = \frac{Z_2(t+1)}{Z(t)\,Z(t+1)} \xrightarrow{t \to \infty} \frac{z_1 z_2}{z_1^2} = \frac{z_2}{z_1}.$$

This technique, called now *Aitken's Δ^2 method* can, in fact, be applied to more general situations. Consider a sequence with general term u_n which tends to a limit u such that

$$\frac{u_{n+1}-u}{u_n-u} \xrightarrow[n\to\infty]{} |a| < 1.$$

Using a reasoning analogous to Aitken's, the 'other approximation relation' can then be written

$$\frac{u-u_{n+2}}{u-u_{n+1}} \approx \frac{\Delta u_{n+1}}{\Delta u_n}$$

and solving with respect to u we obtain:

$$u \approx \frac{\begin{vmatrix} u_n & u_{n+1} \\ u_{n+1} & u_{n+2} \end{vmatrix}}{\Delta u_{n+1} - \Delta u_n} = \frac{u_n u_{n+2} - u_{n+1}^2}{\Delta^2 u_n} = v_n.$$

Note a change of sign compared with Aitken. We can verify that:

$$\frac{v_n-u}{u_n-u} \xrightarrow[n\to\infty]{} 0.$$

The sequence v_n defined this way therefore converges to u more rapidly than the sequence u_n.

In order to avoid the propagation of rounding errors, it is better to use the sequence v_n in the form:

$$v_n = u_n - \frac{(u_{n+1}-u_n)^2}{u_{n+2} - 2u_{n+1} + u_n} = u_n - \frac{\Delta u_n^2}{\Delta^2 u_n}.$$

It is now possible to go on to iterate the process, replacing the sequence v_n by the sequence w_n defined by:

$$w_n = v_n - \frac{\Delta v_n^2}{\Delta^2 v_n}, \dots$$

When the primitive sequence u_n is found by successive approximations, that is when we try to solve $F(x) = x$ by putting:

$$u_{n+1} = F(u_n).$$

There is an improvement to the process known as Steffensen's method. Steffensen suggested first calculating v_0 by Aitken's method from the three approximations u_0, $u_1 = F(u_0)$, $u_2 = F(u_1)$, that is:

$$v_0 = u_0 - \frac{(u_1-u_0)^2}{u_2 - 2u_1 + u_0},$$

and then using this value of v_0 to obtain the iteration $v_1 = F(v_0)$, $v_2 = F(v_1)$ and then finding w_0:

$$w_0 = v_0 - \frac{(v_1 - v_0)^2}{v_2 - 2v_1 + v_0},$$

and so on.

The combination of successive iterations and the Aitken Δ^2 method generates the sequence $x_0 = u_0$, $x_1 = v_0$, $x_2 = w_0$, ... This sequence x_n can also be defined directly by the recurrence relation:

$$x_{n+1} = x_n - \frac{(F(x_n) - x_n)^2}{F(F(x_n)) - 2F(x_n) + x_n}$$

This can also be interpreted as a method of successive approximations associated with the function

$$G(x) = x - \frac{(F(x) - x)^2}{F(F(x)) - 2F(x) + x}$$

To end with, we note that the function G can, in its turn, be thought of as resulting from the application of Newton's method to the solution of the equation $f(x) = 0$ where $f(x) = x - F(x)$. This is done by considering the sequence defined by the recurrence relation:

$$x_{n+1} = x_n - \frac{f(x_n)}{f'(x_n)}$$

but, since we do not know $f'(x_n)$, we approximate it by:

$$\frac{f(x_n) - f(x_{n-1})}{x_n - x_{n-1}}.$$

The convergence of Steffensen's method is quadratic, like Newton's method, but it has the additional advantage that derivatives do not need to be calculated.

14.5 Richardson's Extrapolation Method

The principle of the method is simple. The idea, which has already been met with Maclaurin, is to use a linear combination of formulas corresponding to two different steps in order to refine the approximation.

More specifically, suppose that the required quantity, a, can be approximated by a function that has an asymptotic series expansion of the form

$$T(h) = a + bh^i + ch^j + \dots + eh^m + o(h^m), \text{ where } i < j < \dots < m.$$

Since $T(h)$ tends to a as h tends to 0, it is clear that if $h_2 < h_1$, then $T(h_2)$ is a better approximation that $T(h_1)$. But an even better approximation can be found by eliminating the h^i term by using a linear combination of $T(h_1)$ and $T(h_2)$:

$$T(h_1, h_2) = \frac{h_2^i T(h_1) - h_1^i T(h_2)}{h_2^i - h_1^i}$$

Tradition credits Huygens with the idea of using this improvement, applied to Archimedes' method for finding the circumference of the circle (see Section 5.1). In 1654 [9], he wrote "the whole circumference of the circle is greater than the perimeter of the polygon of equal sides inscribed within it, plus a third of the quantity by which this same perimeter exceeds the perimeter of another inscribed polygon with half the number of sides". Huygens also provided a similar upper bound, for which he needed to use sines. To use the notation introduced in Section 5.1, where p_n denotes the perimeter of the inscribed polygon with 3.2^{n-1} sides, the value p_n' given by Huygens as

$$p_n' = p_{n+1} + (p_{n+1} - p_n)/3 = (4p_{n+1} - p_n)/3$$

is an improved value. Here we have, $i = 2$, and putting h_1 for p_n and h_2 for p_{n+1}, where $h_2 = 2h_1$ we obtain the formula written above.

A generalised version of this method turns up in an 1859 book on arithmetic by Saigey [19]:

From a theorem which is our own, if we consider the polygons of the preceding number as the *first approximations* of the circumference, we shall obtain *second approximations* by adding to each of the first ones the third of its excess over the preceding one; *third approximations*, by adding to each of the second approximations the 15th of its excess over the preceding one; *fourth approximations* by adding to each of the thirds the 63rd part of its excess over the preceding one; and so on indefinitely, the fractions 1/3, 1/15, 1/63, ... having as denominators, respectively, $(4 - 1), (4^2 - 1), (4^3 - 1), ...$

The procedure is iterated here, as with Maclaurin.

But the method was established formally in the 20th century by Richardson. In a paper of 1911 [16], he considered establishing approximate methods for the solution of differential equations:

Both for engineering and for many of the less exact sciences, such as biology, there is a demand for rapid methods, easy to be understood, and applicable to unusual equations and irregular bodies. If they can be accurate, so much the better; but 1 per cent. would suffice for many purposes.

The theory given in the 1911 paper was presented again by Richardson in a paper of 1927, this time taking great care to explain it, including the use of 'history stories' to illustrate his arguments.

L. F. Richardson
The Deferred Approach to the Limit, *Philosophical Transactions of the Royal Society of London*, series A, vol. 226 (1927), 300–301, 305–306.

Various problems concerning infinitely many, infinitely small, parts had been solved before the infinitesimal calculus was invented; for example, Archimedes on the circumference of the circle. The essence of the invention of the calculus appears to be that the passage to the limit was thereby taken at the earliest possible stage, where diverse problems had operations like *d*/*dx* in common. Although the infinitesimal calculus has been a splendid success, yet there remain problems in which it is cumbrous or unworkable. When such difficulties are encountered it may be well to return to the manner in which they did things before the calculus was invented, postponing the passage to the limit until after the problem had been solved for a moderate number of moderately small differences.

For obtaining the solution of the difference-problem a variety of arithmetical processes are available. This memoir deals with central differences arranged in the simplest possible way [...].

Confining attention to problems involving a single independent variable *x*, let *h* be the 'step', that is to say, the difference of *x* which is used in the arithmetic, and let $\phi(x, h)$ be the solution of the problem in differences. Let *f*(*x*) be the solution of the analogous problem in the infinitesimal calculus. It is *f*(*x*) which we want to know, and $\phi(x, h)$ which is known for several values of *h*. A theory, published in 1910,* but too brief and vague, has suggested that, if the differences are 'centred' then

$$\phi(x, h) = f(x) + h^2 f_2(x) + h^4 f_4(x) + h^6 f_6(x) \ldots \text{ to infinity} \tag{1}$$

odd powers of *h* being absent. The functions $f_2(x), f_4(x), f_6(x)$ are usually unknown. Numerous arithmetical examples have confirmed the absence of odd powers, and have shown that it is often easy to perform the arithmetic with several values of *h* so small that $f(x) + h^2 f_2(x)$ is a good approximation to the sum to infinity of the series in (1). If generally true, this would be very useful, for it would mean that if we have found two solutions for unequal steps h_1, h_2, then by eliminating $f_2(x)$ we would obtain the desired *f*(*x*) in the form

$$f(x) = \frac{h_2^2 \phi(x, h_1) - h_1^2 \phi(x, h_2)}{h_2^2 - h_1^2} \tag{2}$$

This process represented by the formula (2) will be named the 'h^2 - extrapolation' DEFN.

[...]

An ancient problem retouched. To find the circumference of a circle of unit radius.

Imagine that we are back in the time of Archimedes. As a first, and obviously very crude, approximation, take the perimeter of an inscribed square $= 4\sqrt{2} = 5.6568$. As a second approximation, take the perimeter of an inscribed hexagon = 6 exactly. The errors of these two estimates should be to one another as $1/4^2 : 1/6^2$, that is 9 : 4, if the error is proportional to the square of the co-ordinate difference. Thus the extrapolated value is

$$6 + \tfrac{4}{5}(6 - 5.6568) = 6.2746 .$$

The error in the extrapolated value is thus only 1/33 of the error in the better of the two values from which it was derived; so that extrapolation seems a useful process. To get as good a result from a single inscribed regular polygon it would need to have 35 sides, and in the absence of any tables of sines, the calculation would take longer.

Napier's Exponential Base.

Next suppose that we were living at a time before logarithms or Napier's base had been calculated, and that it was required to find

$$\underset{n \to \infty}{\text{Limit}} \left(\frac{2n+1}{2n-1} \right)^n = \underset{n \to \infty}{\text{Limit}} \ \phi_n \text{ say} .$$

If we put $-n$ in place of n the function $\left(\dfrac{2n+1}{2n-1}\right)^n$ is unchanged, and so, *if* an expansion like the following exists, valid for both signs of n

$$\left(\frac{2n+1}{2n-1}\right)^n - \operatorname*{Limit}_{n\to\infty}\left(\frac{2n+1}{2n-1}\right)^n = A_0 + \frac{A_1}{n} + \frac{A_2}{n^2} + \ldots \frac{A_r}{n^r}\ldots,$$

then the odd coefficients $A_1, A_3, \ldots$ are necessarily zero. Also if the limit exists A_0 must vanish. And by making n sufficiently large the term A_2/n^2 will predominate.

So
$$\phi_n - \operatorname*{Limit}_{n\to\infty}\phi_n = \frac{A_2}{n^2} + \ldots$$

On performing the multiplications it is found easily that the function runs as follows:

$n =$	1	2	3	4	5
$\left(\dfrac{2n+1}{2n-1}\right)^n =$	3.00000	2.77777	2.74400	2.73261	2.72741

Now if the errors are proportional to n^{-2} the error of ϕ_4 is to the error of ϕ_5 as 25 is to 16. Therefore the extrapolated value is $\phi_5 + \frac{16}{9}(\phi_5 - \phi_4)$. This works out to 2.71817.

Napier showed that the correct result is 2.71828, so that the error of our extrapolation was -0.00011, whereas the error of ϕ_5 was $+0.00913$, that is 82 times greater. To get as accurate a result without extrapolation as we did with it we should have to calculate $\phi_{5\sqrt{82}}$ that is ϕ_{45}, a tedious process seeing that logarithms had not yet been tabulated. So extrapolation is a great economiser of toil.

The following table is intended to show how $n^2\{\phi_n - \operatorname{Lt}\phi_n\}$ approaches its limit which we have called A_2

n	1	2	3	4	5	∞
$n^2\{\phi_n - \operatorname{Lt}\phi_n\}$	0.2817	0.2380	0.2315	0.2295	0.2283	0.2265

To find the limit A_2 we have $1/n \log \phi = \log(1 + 1/2n) - \log(1 - 1/2n)$. On expanding the logarithms, rearranging, and taking antilogs it is found that

$$\phi_n - e = e/12n^2 + \text{terms in } n^{-4} \text{ and higher even powers.}$$

So $A_2 = e/12 = 0.2265$, which is entered in the table under ∞.

*Phil. Trans. Royal. Soc. London, A. vol. 210, 310–311, § 1.2

The remainder of the article is mainly concerned with conditions for validity and the value of applying the process to solving differential equations. The principle is clearly adaptable to cases where the asymptotic expansion of $\phi(x, h)$ does not have only even powers.

14.6 Romberg's Integration Method

Romberg's method, commonly found in modern books on numerical analysis, is simply Richardson's formula applied iteratively to the trapezium approximation function $T(h)$. In fact, using the Euler-Maclaurin formula, what we have is:

$$T(h) = \int_\alpha^\beta f(x)\,dx + \sum_{j=1}^n a_j h^{2j} + o(h^{2n}) \text{ with } h = \frac{\beta - \alpha}{k}.$$

W. Romberg
Vereinfachte numerische Integration, *Norske Videnskabers Selskabs forhandlinger*, t. 28 (1955).

Numerical integration simplified

Suppose we need to evaluate an integral

$$I = \int_a^b F(x)\,dx$$

approximately from the values taken by the function at equidistant points of division of the interval. The Trapezium Formula and Simpson's Formula are the most frequently used. Gauss also used higher degree polynomial functions passing through the points F_k and calculated the corresponding sequences of coefficients. We shall only repeat here those for 4 and 8 intervals:

Trapeziums	T	: 1, 1
Simpson	S	: 1, 4, 1
Gauss	G^I	: 7, 32, 12, 32, 7
Gauss	G^{II}	: 989, 5888, − 929, 10496, − 4540, 10496, − 929, 5888, 989

The Gauss formulas are seldom used on account of their inconvenient weightings. But, by a small modification of the calculation, we are able to calculate an approximate value of I, even better than that obtained by S and G^I, without using the different coefficients given above.

To do this, we divide the interval $a \leq x \leq b$ into $8N$ parts, with $b - a = 8Nh$. We designate $F(a + kh)$ by F_k for $k = 1, 2, \ldots, 8N - 1$, and use F_0 for the mean of $F(a)$ and $F(b)$, and we calculate first the most crude approximation, by trapeziums

$$T_1 = 8h \sum_{n=0}^{N-1} F_{8n} = (b-a) \overline{\sum_{n=0}^{N-1} F_{8n}}$$

(the bar over the top indicates that we take the arithmetic mean; the interval length is $8h$).

We now divide each interval in half and we form the mean of the F of all the new division points, namely

$$U_1 = 8h \sum_{n=0}^{N-1} F_{8n+4} = (b-a) \overline{\sum_{n=0}^{N-1} F_{8n+4}}.$$

We divide in half again:

$$U_2 = (b-a) \overline{\sum F_{8n+2} + F_{8n+6}},$$

and finally

$$U_4 = (b-a)\sum \overline{F_{8n+1} + F_{8n+3} + F_{8n+5} + F_{8n+7}} \ .$$

Successively refined approximations by trapeziums are therefore

$$T_2 = \overline{T_1 + U_1}, \ T_4 = \overline{T_2 + U_2}, \ T_8 = \overline{T_4 + U_4}$$

for sub-intervals of length $4h, 2h$ and h, respectively.

The U and the T are crude approximations of I since the remainder is proportional to h^2. Now, it is known* that, from two successive divisions, a more accurate value can be found, namely:

$$S_2 = T_2 + \frac{T_2 - T_1}{2^2 - 1}; \ S_4 = T_4 + \frac{T_4 - T_2}{2^2 - 1}; \ S_8 = T_8 + \frac{T_8 - T_4}{2^2 - 1};$$

and

$$V_2 = U_2 + \frac{U_2 - U_1}{2^2 - 1}; \ V_4 = U_4 + \frac{U_4 - U_2}{2^2 - 1}.$$

It seems that it has not been noticed that the S are exactly those which are found by Simpson's Formula.

Since the S and the V have a remainder in h^4, we obtain, on following (*),

$$R_4 = S_4 + \frac{S_4 - S_2}{2^4 - 1}; \ R_8 = S_8 + \frac{S_8 - S_4}{2^4 - 1}; \text{ and } W_4 = V_4 + \frac{V_4 - V_2}{2^4 - 1}.$$

R_4 and R_8 are precisely the approximations of I that would be obtained on using the ponderous factors of G^I. R and W have a remainder of order h^6. Using this we can continue to form

$$Q_8 = R_8 + \frac{R_8 - R_4}{2^6 - 1}, \ \text{ with a remainder in } h^8, \text{ etc.}$$

The advantage of this method is that each new division only requires the mean of the new values of the function, multiplied by $(b-a)$; the new U allows each of the sequences T, S, R, Q and U, V, W to be extended by one term. With the remainders being proportional to h^2, h^4, h^6, h^8, approximations of ever increasing order are easily obtained.

The approximation G^{II} given above will not be found in the approximations Q, since the F in U_4 are given equal weights.

[...]

Example. To find an approximation to the value of

$$I = \int_0^1 \frac{\pi}{2} \cos\frac{\pi}{2} z \, dz = 1$$

dividing the interval into 8 equal parts, $N = 1, 8h = 1$.

$F(a)$	= 1.57079 6327,	$F(b) = 0$
F_1	= 1.54061 3916	
F_2	= 1.45122 6576	
F_3	= 1.30606 9413	
F_4	= 1.11072 0735	
F_5	= 0.87268 7681	From which:
F_6	= 0.60111 7730	$U_1 = F_4, \quad T_1 = F_0,$
F_7	= 0.30644 7161	$U_2 = \overline{F_2 + F_6} = 1.02617\ 2153$
F_0	= 0.78539 8163	$U_4 = \overline{F_1 + F_3 + F_5 + F_7} = 1.00645\ 4543$

[...]

Table of approximated values by increasing order

Interval length	Remainder of order: h^2	h^4	h^6	h^8
8h	$T_1 = 0.78539\ 8163$ $U_1 = 1.11072\ 0735$			
4h	$T_2 = 0.94085\ 9449$ $U_2 = 1.02617\ 2153$	$S_2 = 1.00227\ 9878$ $V_2 = 0.99798\ 9293$		
2h	$T_4 = 0.98711\ 5801$ $U_4 = 1.00645\ 4543$	$S_4 = 1.00013\ 4584$ $V_4 = 0.99988\ 2006$	$R_4 = 0.99999\ 1566$ $W_4 = 1.00000\ 8187$	
h	$T_8 = 0.99678\ 5172$	$S_8 = 1.00000\ 8296$	$R_8 = 0.99999\ 9876$	$Q_8 = 1.00000\ 0008$

* For example, L. Collatz, *Numerische Behandlung vor Differentielgleichungen*, Springer, 1951, p. 6, equation (1.7)

Romberg's explanation is sufficiently clear not to require further comment, but further background information can be found in Dutka [2].

Bibliography

[1] Aitken, A.C., On Bernoulli's numerical solution of algebraic equations, *Proceedings of the Royal Society of Edinburgh*, vol. 46 (1926), pp. 289–305.

[2] Dutka, J., Richardson Extrapolation and Romberg Integration, *Historia Mathematica*, 11 (1984), 3–21.

[3] Euler, L., *Institutiones calculi differentialis cum ejus usu in analysi finitorum ac doctrina serierum*, St Petersburg, 1755. *Leonhardi Euleri Opera omnia*, Vol. X, 1913.

[4] Euler, L., *Institutiones calculi integralis*, St Petersburg, 1768–1769.

[5] Euler, L., Inventio summae cuiusque seriei ex dato termino generali, *Commentarii academiae scientiarum Petropolitanae*, 47 (1736), 1741. *Leonhardi Euleri Opera Omnia*, Vol. XIV, pp. 117–119.

[6] Euler, L., Methodus generalis summandi progressiones, *Commentarii academiac scientarium Petropolitanae*, 25 (1732-3), 1738, pp. 68–97.

[7] Gregory, J., Letter of 23 November, 1670 to Collins, in *Correspondence of Isaac Newton*, Vol. 1, ed. Turnbull, Cambridge: Cambridge University Press, 1959.

[8] Hermite, Ch., Sur la formule de Maclaurin, extrait d'une lettre à Borchardt, 1877, *Journal für die reine und angewandte Mathematik*, t. 84 (1878), 425–431.

[9] Huygens, C., De circuli magnitudine inventa, *Oeuvres*, t. XII, 1654.

[10] Jacobi, C.G., De usu legitimo formulae summatoriae Maclaurinianae, *Journal für die reine und angewandte Mathematik*, t. XII (1834), 263–272.

[11] Kline, M., *Mathematical thought from Ancient to Modern Times*, Oxford: Oxford University Press, 1972.

[12] Kummer, E., Eine neue Methode, die numerischen Summen langsam convergirender Reihen zu berechnen, *Journal für die reine und angewandte Mathematik*, t. XVI, fasc. 3 (1835), 206–214.

[13] Maclaurin, C., *A Treatise on Fluxions*, Edinburgh, 1742, 2nd edn., London, 1801.

[14] Poincare, H., *Les méthodes nouvelles de la mécanique céleste*, Paris, 1893.

[15] Poisson, S. -D., Mémoire sur le calcul numérique des Intégrales définies, 1826, *Mémoires de l'Académie Royale des Sciences de l'Institut de France*, t. VI (1827).

[16] Richardson, L.F., The Approximate Arithmetical Solution by Finite Differences of Physical Problems involving Differential Equations, with an Application to the Stresses in a Masonry Dam, *Philosophical Transactions of the Royal Society*, vol. 210 (1911), pp. 307–357.

[17] Richardson, L.F., The Deferred Approach to the Limit, *Philosophical Transactions of the Royal Society*, vol. 226 (1927), 299–361.

[18] Romberg, W., Vereinfachte numerische Integration, *Det Kong. Norske Videnskabernes Selskabs Forhandlinger*, Trondheim, vol. 28, no7 (1955), pp. 30–36.

[19] Saigey, M., *Problèmes d'arithmétique et exercices de calcul du second degré avec les solutions raisonnées*, Paris: Hachette, 1859.

[20] Steffensen, J.F., *Interpolation*, New York, 1927; 2nd ed., New York: Chelsea, 1950.

[21] Stirling, J., Methodus Differentialis Newtoniana Illustrata, *Philosophical Transactions of the Royal Society*, vol. 30 (1719), pp. 1050–1070.

[22] Stirling, J., *Methodus Differentialis sive Tractatus de Summatione et Interpolatione Serierum Infinitarum*, London, 1730.

15 Towards the Concept of Algorithm

So far we have looked at algorithms designed to solve particular problems. For the most part, algorithms were in use long before it was felt necessary to give a clear definition of what is meant by an algorithm. Today, an algorithm is defined as a finite and organised set of instructions, intended to provide the solution to a problem, and which must satisfy certain conditions. An example would be [10]:

1. The algorithm must be capable of being written in a certain language: a language is a set of words written using a defined alphabet.
2. The question that is posed is determined by some given data, called *enter*, for which the algorithm will be executed.
3. The algorithm is a procedure which is carried out step by step.
4. The action at each step is strictly determined by the algorithm, the entry data and the results obtained at previous steps.
5. The answer, called *exit*, is clearly specified.
6. Whatever the entry data, the execution of the algorithm will terminate after a finite number of steps.

This way of defining an algorithm is by and large sufficient for us to be able to agree on whether or not a certain procedure constitutes an algorithm. However, in the 1930s, problems concerning the foundations of mathematical logic caused mathematicians to explore what precisely constitutes an algorithm and so to propose a definition of the concept of algorithm. How and why was there a need to define the concept of algorithm, and in what way would it be different from the notion of an algorithm? The answer to this is that the meaning of the working notion of an algorithm, as described above, is implicit in its use in solving certain types of problems. The significance of a concept, however, lies in the relation it has to other concepts of a theory, in this case to the rest of mathematics. The reason why it became necessary to establish a clear definition of the concept of an algorithm was in order to show that there are certain problems for which an algorithm cannot be used.

The link between the idea of an algorithm and the approach used by mathematical logicians goes back a long way. In the 17th century, Leibniz envisaged a "universal character language" which would enable all mathematical reasoning to be reduced to simple calculations. He tried to set up a calculus of propositions and made allusion to "machines which could imitate reason" and, as a young man, he had set out plans for a machine to solve equations. Similarly, Charles Babbage, in his 1821 treatise *On the influence of symbols in Mathematical Reasoning* explained how mathematical reasoning could be set up so as to use algebraic symbols quite mechanically. Three years later he presented a small calculating machine to Cambridge Philosophical Society, built to calculate a table with constant second differences (see Section 10. 6).

Boole went further, when he proposed an 'algebrisation' of logic in his treatise *An investigation on the Laws of Thought* (1854). Here he introduced a formal notation for deciding whether propositions were true or false. His work was taken up by S. Jevons and C. S. Peirce. Jevons used Boole's algorithms to describe an automatic procedure for producing logical functions, and he used this to build a mechanical machine, presented to the Royal society in 1870, and called "piano logic". Peirce extended algebraic logic to the theory of de Morgan relations and introduced the use of Truth Tables in 1885, as an automatic decision making process [5].

A first formalised logic is given by Frege in his *Begriffschrift* (1879). His system is constructed from primitive symbols, negation, implication, the universal quantifier and the identity, and inferences made in a purely formal manner [6]. Thus, simply by examining the text it is possible to verify whether or not one formula has been correctly inferred, that is deduced, from another. Peano introduced a formalised language for arithmetic in his *Arithmeticae principia* (1889) so that every proposition could be written using symbols, and so be deduced by procedures similar to solving algebraic equations. The works of Frege and Peano inspired those of Russell, who considered all mathematics to be symbolic logic [14].

The starting point for the work that led to the definition of the concept of an algorithm can be traced to the mathematical programme of David Hilbert. In the 1890s, Hilbert developed what is called the formalist conception of mathematics. For him, the mathematician works on objects, the nature of which is not specified. Proofs are pure deductions from axioms, themselves conceived as simple rules of the game. A statement can be judged as true if it is not contradictory to the system of axioms.

In 1899, in a lecture on the concept of number, Hilbert proposed an axiomatic method for introducing numbers: numbers are conceived as simple objects between which there exist relations described by a system of axioms. This presupposes, of course, that we can be assured of the non-contradiction of the system. At the Heidelberg 3rd International Congress of Mathematicians in 1904, Hilbert explained that this would reduce all arithmetic to logic. Thus, axioms and theorems would be represented by a finite sequence of symbols, and each proof would be carried out according to well defined "mechanical rules". He made this clear in 1922 in posing the *Entscheidungsproblem*, or 'decision problem', which was to find a procedure by which, in a finite number of steps, it could be tested whether a formal expression could be deduced from a given system of axioms. These ideas led Hilbert and his followers to study formal systems, and these were presented in a work by Hilbert and Ackermann in 1928 [7]. These systems were inspired by the arithmetic systems of Peano and the logical systems of Russell and Whitehead found in *Principia Mathematica*.

At the 1928 International Congress, Hilbert proposed a programme of research in the form of problems. This programme asks, in particular, three fundamental questions of mathematics:

Is mathematics *complete*, that is, can every mathematical statement be confirmed as valid or invalid?

Is mathematics *consistent*, that is, is it impossible to prove both a mathematical statement and its contradiction?

Is mathematics *decidable*, that is, does there exist a procedure by which it can be decided, without necessarily carrying out the proof, if a mathematical statement is true (non-contradictory)?

Formalist thinking leads to the idea that the answers to all these questions are affirmative since mathematical activity is a simple manipulation of symbols, and mathematical truth reduces to a simple non-contradiction in a system of propositions. The purpose lying behind the last of the three questions, the *Entscheidungsproblem*, is to be able, in principle, to reduce all mathematical questions to a mechanical form. However, less than ten years later, it was shown that it was not possible to give a affirmative response to Hilbert's questions. In order to tackle the problem, it was necessary to be more precise about what we mean when we talk of a computable function, a computable number and a decision procedure? In other words, in essence, how are we to define a concept of algorithm?

In 1931, a young Austrian mathematician, Kurt Gödel, proved the incompleteness of arithmetic, that is that there exist some propositions in arithmetic for which it is impossible to prove whether they are true or false. Gödel's theorem marks, in a way, the limit to formalist thought and the theorem is regarded by many mathematicians as one of the most profound of the century. To prove the theorem, Gödel introduced the notion of a recursive function, which has the property of being computable. In 1934, Gödel gave a course of lectures at Princeton: he took up again the ideas of his 1931 paper and this time defined the concept of the general recursive function. In an article in 1936, the logician Kleene clarified Gödel's definitions and provided a definition of what he called a primitive recursive function and also the definition of recursive function used today.

The works of Gödel inspired the research of Alonzo Church, Stephen Kleene, Alan Turing and Emil Post. These mathematicians attacked Hilbert's *Entscheidungsproblem* and showed that there were, indeed, undecidable problems, that is mathematical statements for which *no procedure exists* by which it can be decided if the statement is true or false. To do this, each of them defined a concept of computability, that is a concept of algorithm.

It should be noted that whereas the concept of a recursive function appears incidentally in the 1931 Gödel article, it is central to the argument put forward by Church. Furthermore, the concepts of a machine, and of a procedure, are key ideas in the articles by Turing and Church. Kleene's 1936 article, entitled "General Recursive Functions of Natural Numbers" is, as the title suggests, principally concerned with an examination of the concept of a recursive function and of the equivalence between different definitions.

The question of the equivalence between all the concepts of computability stated between 1931 and 1936 is essential. First, it guarantees the coherence between the different mathematical results obtained. Second it comes into the *meta*mathematical question: does the concept of algorithm correspond to the notion of algorithm and to the intuitive idea we have of it?

It appears that this last question was raised by Gödel in a conversation he had with Church while the latter was working on a concept differing from that of Gödel. In an article of 1935/6, Church gives a positive answer. This reply, today called 'Church's Thesis', that a function is recursive if and only if it is effectively computable, lies in the realm of metamathematics. It could just as well be called 'Turing's Thesis' or 'Post's Thesis', since the question of a correspondence between the concept and the intuitive notion of algorithm plays an essential role in the works of these two mathematicians. To the point that both start from the activity of a person who computes and want their conceptions of algorithm to be shown to have 'psychological fidelity'. A rare occurrence indeed in a work of mathematics! As Gödel indicates, the intuitive idea of an algorithm played an important heuristic role in all those works which ruthlessly took their revenge on formalist Hilbertian thought.

Recursive Functions and Computable Functions

15.1 The 1931 Definition

In his famous article of 1931 entitled "On the undecidable propositions of the *Principia Mathematica*", Gödel introduced a formal system P constructed from the logical system of Russell's *Principia* and Peano's axioms of the natural numbers. He showed that, in this system, there exist undecidable propositions, that is assertions which can neither be proved nor disproved. More precisely, Gödel showed that formalised arithmetic is necessarily incomplete or inconsistent, and that consistency of arithmetic cannot be proved from within the axiomatics of the system.

The key idea used by Gödel is to codify all formulas of the system P, as well as their proofs, by the use of natural numbers: "a formula is a sequence of natural numbers and a proof is a finite sequence of finite sequences of natural numbers". It is ironic to note that Hilbert's programme was intended to replace all arithmetic by logic, that is transforming numbers into formulas, and here we have Gödel undermining the programme by codifying the formulas by numbers.

Gödel's approach is as follows. He shows that formulas can be ordered using a free variable of the system: the nth formula is labelled $R(n)$. If a is a formula in one free variable, he uses $[a; n]$ to denote the formula obtained by substituting the integer n for the free variable. Then he considers the class K of integers such that:

(1) $n \in K$ if and only if $[R(n); n]$ cannot be proved.

There exists a formula S such that the formula $[S; n]$ intuitively signifies that $n \in K$. For, let q be such that $S = R(q)$, then $[R(q); q]$ is undecidable. We have, in fact:

- if $[R(q); q]$ can be proved, then $q \in K$ by the definition of $S = R(q)$, and so $[R(q); q]$ is not provable by (1), which is a contradiction;
- if $[R(q); q]$ cannot be proved, then *not* $q \in K$ by the definition of $S = R(q)$, and so $[R(q); q]$ is provable by (1), which is a contradiction.

As Gödel observes: "the analogy of this result with the antinomy of Richard is immediately evident; there is also a close relation to the *Liar Paradox*". We may recall that the Russell paradox, exposed by Frege and Russell in 1903, caused a crisis among mathematicians and also a real fascination for a wider public, producing a whole flurry of new paradoxes [5].

To prove his theorem, Gödel defines the concepts of recursive function and recursive relation and proves that "every recursive function is arithmetic". The idea of defining a function by recurrence was not new: it can be found, for example, in the *Manuel d'Arithmétique à l'usage des établissements supérieurs* of 1861, where Grassmann defines addition and multiplication of integers by the passage from one number to the next, and also in the important book by Dedekind of 1888 entitled *Was sind und was sollen die Zahlen?* [5]. For a modern treatment of Gödel's theorem, see [1].

K. Gödel

Über formal unentscheidbare Sätze der Principia Mathematica und verwandter System I,
Monatshefte für Mathematik und Physik, t. 38 (1931), 173–198. Extract from pp. 179–180.

We shall introduce now a digression which for the moment has nothing to do with the formal system P, and immediately give the following definition: an arithmetic function[1] $\varphi(x_1, x_2, ..., x_n)$ is said to be *defined recursively* by the arithmetic functions $\psi(x_1, x_2, ..., x_{n-1})$ and $\mu(x_1, x_2, ..., x_{n+1})$ if, for all $x_2, ..., x_n, k$[2], we have:

$$\varphi(0, x_2, ..., x_n) = \psi(x_2, ..., x_n)$$
$$\varphi(k + 1, x_2, ..., x_n) = \mu(k, \varphi(x_2, ..., x_n), x_2, ..., x_n).$$

An arithmetic function φ is said to be *recursive* is there exists a finite sequence of arithmetic functions $\varphi_1, \varphi_2, ..., \varphi_m$ which terminates with φ and has the property that each of the functions φ_k of the sequence is recursively defined from the two preceding functions, either by substitution[3] from any of the preceding ones, or is a constant function or the successor function $x + 1$. The shortest length of the sequence of φ_i required to represent a recursive function is said to be its rank.

---

[1] That is, its domain of definition is the set of non negative integer numbers (or n-tuples of such numbers) and that the values of the function are non negative integers.

[2] Small Roman letters (sometimes with indices) are, in what follows, always variables of non negative integers (except where expressly otherwise indicated).

[3] More precisely: by the substitution of certain of the preceding functions in the free places of one of its preceding functions, for example $\varphi_k(x_1, x_2) = \varphi_p[\varphi_q(x_1, x_2), \varphi_r(x_2)]$ $(p, q, r < k)$. All the variables of the left hand side do not necessarily have to appear on the right hand side (similarly for the recursive schema).

The functions studied here are functions defined on n-tuples of natural numbers and take their values from the set of natural numbers. We can show that addition and multiplication are recursive functions.

Using Polish Notation, introduced by Lukasiewicz in 1920 [5], we write: $+ (x, y)$ for $x + y$.

We therefore have:
$$+ (0, y) = y$$

and
$$+ (x + 1, y) = + (x, y) + 1.$$

If we use S for the successor function, we have:

$$+ (x + 1, y) = S(+ (x, y)),$$

and so addition is defined recursively by the successor function, in the Gödel sense.

In the same way, for multiplication, we write $\times (x, y)$ for $x \times y$. We therefore have:

$$\times (0, y) = 0$$
and $$\times (x + 1, y) = + (\times (x, y), y).$$

If we use C to denote the constant zero function then we have:

$$\times (0, y) = C(y)$$
and $$\times (x + 1, y) = + (\times (x, y), y).$$

In this way, multiplication has been defined recursively, in the Gödel sense, by means of the constant function C and the addition function, a function that is itself recursive. Hence multiplication is recursive of rank 2.

15.2 General Gödel Recursive Functions

In the Spring of 1934, Gödel gave a course of lectures at Princeton where he returned to the ideas he had presented in 1931. The lectures were edited by Kleene and Rosser and appeared under the title of "On Undecidable Propositions of Formal Mathematical Systems". Here Gödel clarifies the concept of recursive function and, importantly, extends the concept to that of general recursive function. In a note, he indicates that this latter concept was suggested to him by the mathematician Jacques Herbrand. The extract below gives Gödel's initial definitions of these concepts.

The concept of the general recursive function is based on a double recursion. The introduction of a function defined by double recursion appears in the works of Ackermann in 1920. This function was presented by Hilbert in 1925 in a lecture on infinity, so as to prove the continuum hypothesis [5] and Ackermann studies it in a paper of 1928 [6].

Kurt Gödel

On Undecidable Propositions of Formal Mathematical Systems, Lectures given at the Institute of Advanced Studies, Princeton, 1934, in Martin Davis (ed.), *The Undecidable*, Raven Press, 1965, pp. 39–71.

2. Recursive functions and relations

Now we turn to some considerations which for the present have nothing to do with a formal system.

[...]

The function $\varphi(x_1, ..., x_n)$ shall be *compound* with respect to $\psi(x_1, ..., x_m)$ and $\chi_i(x_1, ..., x_n)$ ($i = 1, ..., m$) if, for all natural numbers $x_1, ..., x_n$,

(1) $$\varphi(x_1, ..., x_n) = \psi(\chi_1(x_1, ..., x_n), ..., \chi_m(x_1, ..., x_n)).$$

$\varphi(x_1, ..., x_n)$ shall be said to be *recursive* with respect to $\psi(x_1, ..., x_{n-1})$ and $\chi(x_1, ..., x_{n+1})$ if, for all natural numbers $k, x_2, ..., x_n$,

(2)
$$\varphi(0, x_2, ..., x_n) = \psi(x_2, ..., x_n)$$
$$\varphi(k + 1, x_2, ..., x_n) = \chi(k, \varphi(k, x_2, ..., x_n), x_2, ..., x_n).$$

In both (1) and (2), we allow the omission of each of the variables in any (or all) of its occurrences on the right side (e.g. $\varphi(x, y) = \psi(\chi_1(x), \chi_2(x, y))$ is permitted under (1))[1]. We define the class of *recursive* functions to be the totality of functions which can be generated by substitution, according to the scheme (1), and recursion, according to the scheme (2), from the successor function $x + 1$, constant functions $f(x_1, ..., x_n) = c$, and identity functions $U_j^n(x_1, ..., x_n) = x_j$ $(1 \leq j \leq n)$. In other words, a function φ shall be recursive if there exists a finite sequence of functions $\varphi_1, ..., \varphi_n$ which terminates with φ such that each function of the sequence is either the successor function $x + 1$ or a constant function $f(x_1, ..., x_n) = c$, or an identity function $U_j^n(x_1, ..., x_n) = x_j$, or its compound with respect to preceding functions, or is recursive with respect to preceding functions. A relation R shall be *recursive* if the representing function is recursive.

Recursive functions have the important property that, for each given set of values of the arguments, the value of the function can be computed by a finite procedure[2].

[...]

9. General recursive functions

If $\psi(y)$ and $\chi(x)$ are given recursive functions, then the function $\varphi(x, y)$, defined inductively by the relations

$$\varphi(0, y) = \psi(y), \quad \varphi(x + 1, 0) = \chi(x), \quad \varphi(x + 1, y + 1) = \varphi(x, \varphi(x + 1, y)),$$

is not in general recursive in the limited sense of §2. This is an example of a definition by induction with respect to two variables simultaneously.

[...] The consideration of various sorts of functions defined by inductions leads to the question what one would mean by "every recursive function".

One may attempt to define this notation as follows: If φ denotes an unknown function, and $\psi_1, ..., \psi_k$ are known functions, and if the ψ's and the φ are substituted in one another in the most general fashions and certain pairs of the resulting expressions are equated, then if the resulting set of functional equations has one and only one solution for φ, φ is a recursive function[3].

Thus we might have

$$\begin{aligned}
\varphi(x, 0) &= \psi_1(x), \\
\varphi(0, y + 1) &= \psi_2(y), \\
\varphi(1, y + 1) &= \psi_3(y), \\
\varphi(x + 2, y + 1) &= \psi_4(\varphi(x, y + 2), \varphi(x, \varphi(x, y + 2))).
\end{aligned}$$

---

[1] This sentence could have been omitted, since the removal of any of the occurrences of variables on the right may be effected by means of the function U_j^n.

[2] The converse seems to be true, if, besides recursions according to the scheme (2), recursions of other forms (e.g., with respect to two variables simultaneously) are admitted. This cannot be proved, since the notion of finite computation is not defined, but it serves as a heuristic principle.

[3] This was suggested by Herbrand in a private communication.

Gödel explains what exactly are the three types of basic functions which can be used to define the concept of recursive function: the successor function S, the zero constant function C, and the identity functions U_j^n. These are defined, respectively, by:

$$S(x) = x + 1$$
$$C(x) = 0$$
$$U_j^n(x_1, \ldots, x_n) = x_j.$$

With this notation, we have, for addition:

$$+ (0, y) \quad = U_1^1(y)$$
$$+ (x + 1, y) = S(U_2^3(x, +(x, y), y))$$

and for multiplication:

$$\times (0, y) \quad = C(y)$$
$$\times (x + 1, y) = +(U_2^3(x, \times(x, y), y), U_3^3(x, \times(x, y), y))$$

A recursive function is a function that can be defined from the basic functions using a finite number of operations of composition and recursion.

Gödel indicates that the concept of general recursive function, which he defines precisely later in the article, is in fact much broader than the earlier idea. The Ackermann function is an interesting example of a general recursive function. This can be defined by the following equations:

$$A(0, y) \qquad = 1 \qquad \qquad (1)$$
$$A(1, 0) \qquad = 2 \qquad \qquad (2)$$
$$A(x + 2, 0) \qquad = S(S(x)) \qquad (3)$$
$$A(x + 1, y + 1) = A(A(x, y + 1), y) \quad (4)$$

The fourth equation defines a double recursion, since the recursion depends on two variables. This function is general recursive without being recursive. In fact, it can be shown that the function $h(x) = A(x, x)$ increases faster than any recursive function defined on one variable [4].

15.3 Alonzo Church and Effective Calculability

The two papers by Gödel, as well as the research undertaken by Kleene, were cited by Church in his paper entitled: "An Unsolvable Problem of Elementary Number Theory", presented to the American Mathematical Society in April 1935. In this paper, Church defines another concept of recursive function, whose equivalence to the general recursive function of Gödel was not immediately apparent, but resulted from further work by Kleene.

A. Church

An Unsolvable Problem of Elementary Number Theory, in *The American Journal of Mathematics*, vol. 58 (1936), 345–363.

1. Introduction

There is a class of problems of elementary number theory which can be stated in the form that it is required to find an effectively calculable function f of n positive integers, such that $f(x_1, x_2, ..., x_n) = 2$ [1] is a necessary and sufficient condition for the truth of a certain proposition of elementary number theory involving $x_1, x_2, ..., x_n$ as free variables.

[...]

The purpose of the present paper is to propose a definition of effective calculability which is thought to correspond satisfactorily to the somewhat vague intuitive notion in terms of which problems of this class are often stated, and to show, by means of an example, that not every problem of this class is solvable.

[...]

4. Recursive functions

We define a class of expressions, which we shall call *elementary expressions*, and which involve, besides parentheses and commas, the symbols 1, S, an infinite set of numerical variables $x, y, z, ...$, and, for each positive integer n, an infinite set $f_n, g_n, h_n, ...$ of functional variables with subscript n. This definition is by induction as follows. The symbol 1 or any numerical variable, standing alone, is an elementary expression. If A is an elementary expression, then $S(A)$ is an elementary expression. If $A_1, A_2, ..., A_n$ are elementary expressions and f_n is any functional variable with subscript n, then $f_n(A_1, A_2, ..., A_n)$ is an elementary expression.

The particular elementary expressions 1, $S(1)$, $S(S(1))$, ... are called *numerals*. And the positive integers 1, 2, 3, ... are said to correspond to the numerals 1, $S(1)$, $S(S(1))$, ...

An expression of the form $A = B$, where A and B are elementary expressions, is called an *elementary equation*.

The *derived equations* of a set E of elementary equations are defined by induction as follows. The equations of E themselves are derived equations. If $A = B$ is a derived equation containing a numerical variable x, then the result of substituting a particular numeral for all the occurrences of x in $A = B$ is a derived equation. If $A = B$ is a derived equation containing an elementary expression C (as part of either A or B), and if either $C = D$ or $D = C$ is a derived equation, then the result of substituting D for a particular occurrence of C in $A = B$ is a derived equation.

Suppose that no derived equation of a certain finite set E of elementary equations has the form $k = l$ where k and l are different numerals, that the functional variables which occur in E are $f_{n_1}^{1}, f_{n_2}^{2}, ..., f_{n_r}^{r}$ with subscripts $n_1, n_2, ..., n_r$, respectively, and that, for every value of i from 1 to r inclusive, and for every set of numerals $k_1^{i}, k_2^{i}, ..., k_{n_i}^{i}$, there exists a unique numeral k^i such that $f_{n_i}^{i}(k_1^{i}, k_2^{i}, ..., k_{n_i}^{i}) = k^i$ is a derived equation of E. And let $F^1, F^2, ..., F^r$ be the functions of positive integers defined by the condition that, in all cases, $F^i(m_1^{i}, m_2^{i}, ..., m_{n_i}^{i})$ shall be equal to m^i, where $m_1^{i}, m_2^{i}, ..., m_{n_i}^{i}$ and m^i are the positive integers which correspond to the numerals $k_1^{i}, k_2^{i}, ..., k_{n_i}^{i}$ and k^i respectively. Then the set of equations E is said to *define*, or to be a set of *recursion equations* for, any one of the functions F^i, and the functional variable $f_{n_i}^{i}$ is said to *denote* the function F^i.

A function of positive integers for which a set of recursion equations can be given is said to be *recursive*[2].

It is clear that for any recursive function of positive integers there exists an algorithm using which any required particular value of the function can be effectively calculated. For the derived equations of the set of recursion equations E are effectively enumerable, and the algorithm for the calculation of particular values of a function F^i, denoted by a functional variable $f_{n_i}{}^i$, consists in carrying out the enumeration of the derived equations of E until the required particular equation of the form $f_{n_i}{}^i(k_1{}^i, k_2{}^i, \ldots, k_{n_i}{}^i) = k^i$ is found[3].

[...]

7. The notion of effective calculability

We now define the notion, already discussed, of an *effectively calculable* function of positive integers by identifying it with the notion of a recursive function of positive integers[4] [...]. This definition is thought to be justified by the considerations which follow, so far as positive justification can ever be obtained for the selection of a formal definition to correspond to an intuitive notion.

It has already been pointed out that, for every function of positive integers which is effectively calculable in the sense just defined, there exists an algorithm for the calculation of its values.

Conversely it is true, under the same definition of effective calculability, that every function, an algorithm for the calculation of the values of which exists, is effectively calculable. For example, in the case of a function F of one positive integer, an algorithm consists in a method by which, given any positive integer n, a sequence of expressions (in some notation) $E_{n1}, E_{n2}, \ldots, E_{nr_n}$, can be obtained; where E_{n1} is effectively calculable when n is given; where E_{ni} is effectively calculable when n and the expressions $E_{nj}, j < i$, are given; and where, when n and all the expressions E_{ni} up to and including E_{nr_n} are given, the fact that the algorithm has terminated becomes effectively known and the value of $F(n)$ is effectively calculable. [...]

---

[1] The selection of the particular positive integer 2 instead of some other is, of course, accidental and non-essential.

[2] This definition is closely related to, and was suggested by, a definition of recursive functions which was proposed by Kurt Gödel, in lectures at Princeton, N. J., 1934, and credited by him in part to an unpublished suggestion of Jacques Herbrand. The principal features in which the present definition of recursiveness differs from Gödel's are due to S. C. Kleene.

In a forthcoming paper by Kleene to be entitled, "General recursive functions of natural numbers," (abstract in *Bulletin of the American Mathematical Society*, vol. 41), several definitions of recursiveness will be discussed and equivalences among them obtained. In particular, it follows readily from Kleene's results in that paper that every function recursive in the present sense is also recursive in the sense of Gödel (1934) and conversely.

[3] The reader may object that this algorithm cannot be held to provide an effective calculation of the required particular value of F^i unless the proof is constructive that the required equation $f_{n_i}{}^i(k_1{}^i, k_2{}^i, \ldots, k_{n_i}{}^i) = k^i$ will ultimately be found. But if so this merely means that he should take the existential quantifier which appears in our definition of a set of recursion equations in a constructive sense. What the criterion of constructiveness shall be is left to the reader. [...]

[4] The question of the relationship between effective calculability and recursiveness (which it is here proposed to answer by identifying the two notions) was raised by Gödel in conversation with the author. [...]

The concept of recursive function defined by Church corresponds to the intuitive idea that an algorithm that provides for calculating a function can be given by a set of equations of a certain type. Consider, for example, the Fibonacci function. The sequence of Fibonacci numbers $1, 2, 3, 5, 8, 13, 21, \ldots$ is such that each term is the sum of the two preceding ones. The Fibonacci function which gives the xth term of this sequence is defined by the following set of elementary equations:

$$FIB(0) = 1$$
$$FIB(1) = 2$$
$$FIB(S(S(x))) = + (FIB(x), FIB(S(x))).$$

This function can also be defined as a recursive function in the Gödel sense, but this requires advanced techniques.

Church shows how we can effectively calculate the values of a given function from a set of elementary equations. All that is needed is to compute the derived equations obtained by substituting a particular value for a variable (substitution) or by substituting one term of an equation by another term (replacement). Thus, his concept certainly corresponds to the idea of a function whose values are calculable by an effective procedure, that is by an algorithm.

Consider, for example, the set of equations defining the Ackermann function:

$$A(0, y) = 1 \tag{1}$$
$$A(1, 0) = 2 \tag{2}$$
$$A(x + 2, 0) = S(S(x)) \tag{3}$$
$$A(x + 1, y + 1) = A(A(x, y + 1), y) \tag{4}$$

From the derived equations we are able to effectively calculate the values of A. We have:

$$A(2, 0) = 4, \text{ etc., } A(x, 0) = 2x \qquad \text{from (3) by substitution}$$
$$A(0, 1) = 1 \qquad \text{from (1) by substitution}$$
$$A(1, 1) = A(A(0, 1), 0) = A(1, 0) = 2 \qquad \text{from (4) by replacement}$$
$$A(2, 1) = A(A(1, 1), 0) = A(2, 0) = 4 = 2^2 \qquad \text{from (4) by replacement}$$
$$A(3, 1) = A(A(2, 1), 0) = A(4, 0) = 8 = 2^3 \qquad \text{from (4) by replacement}$$
$$A(x, 1) = A(A(x - 1, 1), 0) = A(2^{x-1}, 0) = 2 \times 2^{2x-1} = 2^{2x} \qquad \text{from (4)}$$
$$A(0, 2) = 1 \qquad \text{from (1) by substitution}$$
$$A(1, 2) = A(A(0, 2), 1) = A(1, 1) = 2 \qquad \text{from (4) by replacement}$$
$$A(2, 2) = A(A(1, 2), 1) = A(2, 1) = 2^2 \qquad \text{from (4) by replacement}$$
$$A(3, 2) = A(A(2, 2), 1) = A(2^2, 1) = 2^{2^2} \qquad \text{from (4) by replacement}$$
$$A(x, 2) = A(A(x - 1, 2), 1) = 2^{2^{\cdot^{\cdot^{\cdot^2}}}} \qquad \text{from (4) by replacement}$$
etc.

In his paper, Church explicitly poses the question as to whether the concept of recursiveness corresponds to the notion of being effectively calculable. He indicates, in a note, that this question was raised in a conversation that he had had with Gödel. The latter had already pointed out, in his 1934 paper, that every recursive function, in his sense, is computable by a finite procedure. We also find, in a footnote to the paper, doubtless added later, that "the converse appears to be true", provided that the concept of general recursive function is understood in the sense that it is used in the paper.

15.4 Recursive Functions in the Kleene Sense

In his article "General Recursive Functions of Natural Numbers", presented to the American Mathematical Society in September 1935, Stephen Kleene further developed the ideas of Gödel's 1934 lectures, of which he had been one of the editors. He starts by defining the concept of a primitive recursive function, which Gödel called a recursive function in his 1931 article and in his 1934 lecture course. He then gives several equivalent definitions for the concept of a recursive function, which corresponds to the general recursive function of Gödel's 1934 course. The table below shows how these various terms are related:

Gödel	recursive function	general recursive function
Church		recursive function
Kleene	primitive recursive function	recursive function

In his new proof of the existence of unsolvable problems, Kleene gives an example of a function that is recursive but not primitive recursive.

Thus, over the course of just four years, we have passed from a relatively imprecise definition of recursive function, given in the course of a proof, to a systematic study of the concept of recursive function itself. This shows how rapidly mathematicians came to appreciate the importance of the concept.

Today, the study of recursive functions is a fundamental part of all branches of logic, as well as arithmetic. An example is Matijasevič's proof that Hilbert's 10th problem is unsolvable [12]. The problem was to find a decision procedure to determine whether a Diophantine equation, a polynomial equation with integer coefficients requiring integer solutions, has a solution. The proof depends on showing that the set of solutions of a Diophantine equation is coincident with recursively enumerable sets, that is the empty set and the set of values of recursive functions [5].

S.C. Kleene
General Recursive Functions of Natural Numbers, *Mathematische Annalen*, vol. 112, 5 (1936), 727–729.

The substitution

(1) $\varphi(x_1, \ldots, x_n) = \theta(\chi_1(x_1, \ldots, x_n), \ldots, \chi_m(x_1, \ldots, x_n)),$

and the ordinary recursion with respect to one variable

(2) $\varphi(0, x_2, \ldots, x_n) = \psi(x_2, \ldots, x_n)$

$\varphi(y + 1, x_2, \ldots, x_n) = \chi(y, \varphi(y, x_2, \ldots, x_n), x_2, \ldots, x_n),$

where $\theta, \chi_1, \ldots, \chi_m, \psi, \chi$ are given functions of natural numbers, are examples of the definition of a function φ by equations which provide a step by step process for computing the value $\varphi(k_1, \ldots, k_n)$ for any given set $k_1, \ldots, k_n$ of natural numbers. It is known that there are other definitions of this sort, e. g. certain recursions with respect to two or more variables simultaneously, which cannot be reduced to a succession of substitutions and ordinary recursions[1]. Hence a characterization of the notion of recursive definition in general, which would include all these cases, is desirable. A definition of general recursive function of natural numbers was suggested by Herbrand to Gödel, and was used by Gödel with an important modification in a series of lectures at Princeton in 1934. In this paper we offer several observations on general recursive functions, using essentially Gödel's form of the definition.

The definition will be stated in §1. It consists in specifying the form of the equations and the nature of the steps admissible in the computation of the values, and in requiring that for each given set of arguments the computation yield a unique number as value. The operations on symbols which occur in the computation have a similarity to ordinary recursive operations on numbers. This similarity will be utilized, by the Gödel method of representing formulas by numbers, to prove that every (general) recursive function is expressible in the form $\psi(\varepsilon y[\rho(x_1, \ldots, x_n, y) = 0])$ where ψ and ρ are ordinary or "primitive" recursive functions and $(x_1, \ldots, x_n)(Ey)[\rho(x_1, \ldots, x_n, y) = 0]$ [2]. Also, it is seen directly that, for any recursive function $\rho(x_1, \ldots, x_n, y)$, $\varepsilon y[\rho(x_1, \ldots, x_n, y) = 0]$ is a recursive function, provided $(x_1, \ldots, x_n)(Ey)$ $[\rho(x_1, \ldots, x_n, y) = 0]$.

[...]

§1. The relation between primitive and general recursive functions

A recursive function (relation) in the sense of Gödel will now be called a *primitive recursive function* (relation). By using

$$
\begin{array}{lll}
& S(x) = x + 1 & \text{the successor function} \\
(3) & C(x) = 0 & \text{the constant function 0} \\
& U_i{}^n(x_1, \ldots, x_n) = x_i & \text{identity functions}
\end{array}
$$

as initial functions, the definition of primitive recursive function can be phrased thus:
Definition 1. A function is *primitive recursive* if it can be defined from the functions (3) by (zero or more) successive applications of schemas (1) and (2) ($m, n = 1, 2, \ldots; i = 1, \ldots, n$) [3].

---

[1] W. Ackermann, Zum Hilbertschen Aufbau der reellen Zahlen, *Math. Annalen* 99 (1928), 118–133; Rózsa Péter, Konstruktion nichtrekursiver Funktionen, *Math. Annalen*, 111 (1935), 42–60.

[2] In the "functions" which we consider, the arguments are understood to range over the natural numbers (i. e. non-negative integers) and the values to be natural numbers. Also, for abbreviation, we use propositional functions of natural numbers, calling them "relations" (alternatively "classes", when there is only one variable) and employing the following notations: $(x)A(x)$ [for all natural numbers, $A(x)$], $(Ex)A(x)$ [there is a natural number x such that $A(x)$], $\varepsilon x[A(x)]$ [the least natural number x such that $A(x)$, or 0 if there is no such number], - [not], $\vee$ [or], & [and], $\rightarrow$ [implies], $\equiv$ [is equivalent to].

[3] This form of the definition was introduced by Gödel to avoid the necessity of providing for omissions of arguments on the right in schemas (1) and (2). The operations in the construction of primitive recursive functions can be further restricted. See Rózsa Péter, Über den Zusammenhang der verschiedenen Begriffe der rekursiven Funktionen, *Math. Annalen*, 110 (1934), 612–632.

The functions being studied are always functions defined on n-tuples of natural numbers and which take their values from the set of integers. To define the concept of primitive recursive function, Kleene considers two operations on the set of these functions: substitution and ordinary recursion. Then, as with Gödel, he defines three basic functions: the successor function S, the constant function C and the identity functions $U_j^{\;n}$.

A primitive recursive function is a function which can be defined from these base functions with a finite number of operations of substitution and ordinary recursion.

Kleene then goes on to give two equivalent definitions for the concept of recursive function. All that is needed is to add a third operation: minimisation. A function f is defined from g by minimisation if for all $x_1, x_2, ..., x_n$ we have

$$f(x_1, x_2, ..., x_n) = \min \{y \text{ natural number} \mid g(x_1, x_2, ..., x_{n-1}, y) = x_n\}.$$

A recursive function is a function which can be defined from the base functions by a finite number of operations of substitution, ordinary recursion and minimisation.

Thus, subtraction can be defined using addition and minimisation. Using $d(x, y)$ for $x - y$, with $x \geq y$, we have:

(3) $d(x, y) = \min \{z \mid +(y, z) = x\}.$

The three formulations provided by Kleene supersede the different conceptions given by other mathematicians. In particular, the definitions given by Church are covered, as is indicated in a note in the above article.

Machines

15.5 The Turing Machine

In the spring of 1935 the young British mathematician Alan Turing followed a course on the foundations of mathematics given in Cambridge by Von Neumann. Von Neumann, who concluded his course of lectures with Gödel's theorem, had been present at the 1928 International Congress where Hilbert had presented his programme. Hilbert's third question was left as an unresolved question: the *Entscheidungsproblem*, was mathematics decidable? Did there exist a "mechanical process" by which one could determine whether or not a proposition could be proved? While still a child, Turing had dreamt of inventing a writing machine, and it was his mechanical point of view that was to be the basis for his machine [8].

In his famous 1936 paper, "On computable Numbers, with an Application to the *Entscheidungsproblem*", Turing proved that Hilbert's *Entscheidungsproblem* can have no solution. To do this, he had to clarify the expression "there exists a general process for determining ...". He makes the remark that every problem concerning a 'general process' can be expressed as a problem concerning a general process for determining if a given number n has a property $G(n)$ (n is the Gödel representation of a provable formula), that is calculating a number whose nth digit is 1 if $G(n)$ is true and 0 if not. The question then comes down to this: "What are the possible processes which may be carried out for calculating a number?". It is therefore a matter of saying precisely what is meant by an effective calculating process, or an algorithm for calculating. This question led Turing to define what we call today a Turing machine.

We have seen, that a few months earlier, Alonzo Church, also engaged with Hilbert's problem, had given an example of a non-decidable problem in elementary number theory. To do this, he defined the idea of 'effective calculability'. Turing, in an appendix to his paper, provided a proof for the equivalence of Church's effective calculability and 'computing by a Turing machine'.

At the start of his paper, Turing defines the mechanism of the machine in an arbitrary way. But later on, in the extract given below, he attempts to justify this mechanism by appealing to the intuitive idea of computing a number. He does this by reference to an individual, the "computer", who is engaged in the process of carrying out calculations on paper. All references by Turing to a computer are, of course, references to a person, not a machine.

A. M. Turing
On Computable Numbers with an Application to the Entscheidungsproblem,
Proceedings of the London Mathematical Society, series 2, vol. 42 (1936-7), 230–265.

[...]

Computing is normally done by writing certain symbols on paper. We may suppose this paper is divided into squares like a child's arithmetic book. In elementary arithmetic the two-dimensional character of the paper is sometimes used. But such a use is always avoidable, and I think that it will be agreed that the two-dimensional character of paper is no essential of computation. I assume then that the computation is carried out on one-dimensional paper, i.e. on a tape divided into squares. I shall also suppose that the number of symbols which may be printed is finite. If we were to allow an infinity of symbols, then there would be symbols differing to an arbitrarily small extent[1]. The effect of this restriction of the number of symbols is not very serious. It is always possible to use sequences of symbols in the place of single symbols. Thus an Arabic numeral such as 17 999999999999999 is normally treated as a single symbol. Similarly in any European language words are treated as single symbols (Chinese, however, attempts to have an enumerable infinity of symbols). The differences from our point of view between the single and compound symbols are that the compound symbols, if they are too lengthy, cannot be observed at one glance. This is in accordance with experience. We cannot tell at a glance whether 9999999999999999 and 999999999999999 are the same.

The behaviour of the computer at any moment is determined by the symbols which he is observing, and his "state of mind" at that moment. We may suppose that there is a bound B to the number of symbols or squares which the computer can observe at one moment. If he wishes to observe more, he must use successive operations. We will also suppose that the number of states of mind which need be taken into account is finite. The reasons for this are of the same character as those which restrict the number of symbols. If we admitted an infinity of states of mind, some of them will be "arbitrarily close" and will be confused. Again, the restriction is not one which seriously affects computation, since the use of more complicated states of mind can be avoided by writing more symbols on the tape.

Let us imagine the operations performed by the computer to be split up into "simple operations" which are so elementary that it is not easy to imagine them further divided, Every such operation consists of some change of the physical system consisting of the computer and his tape. We know the state of the system if we know the sequence of symbols on the tape, which of these are observed by the computer (possibly with a special order), and the state of mind of the computer. We may suppose that in a simple operation not more than one symbol is altered. Any other changes can be split up into simple changes of this kind. The situation in regard to the squares whose symbols may be altered in this way is the same as in regard to the observed squares. We may, therefore, without loss of generality, assume that the squares whose symbols are changed are always "observed" squares.

Besides these changes of symbols, the simple operations must include changes of distribution of observed squares. The new observed squares must be immediately recognisable by the computer. I think it is reasonable to suppose that they can only be squares whose distance from the closest of the immediately previously observed squares does not exceed a certain fixed amount. Let us say that each of the new observed squares is within L squares of an immediately previously observed square.

In connection with "immediate recognisability", it may be thought that there are other kinds of square which are immediately recognisable. In particular, squares marked by special symbols might be taken as immediately recognisable. Now, if these squares are marked only by single symbols there can be only a finite number of them, and we should not upset our theory by adjoining these marked squares to the observed squares. If, on the other hand, they are marked by a sequence of symbols, we cannot regard the process of recognition as a simple process. This is a fundamental point and should be illustrated. In most mathematical papers the equations and theorems are numbered. Normally the numbers do not go beyond (say) 1000. It is, therefore, possible to recognise a theorem at a glance by its number. But if the paper was very long, we might reach Theorem 157767733443477; then, further on in the paper, we might find "… hence (applying Theorem 157767733443477) we have …". In order to make sure which was the relevant theorem we should have to compare the two numbers figure by figure, possibly ticking the figures off in pencil to make sure of their not being counted twice. If in spite of this it is still thought that there are other "immediately recognisable" squares, it does not upset my contention so long as these squares can be found by some process of which my type of machine is capable. This idea is developed in III below.

The simple operations must therefore include:

(a) Changes of the symbol on one of the observed squares.
(b) Changes of one of the squares observed to another square within L squares of one of the previously observed squares.

It may be that some of these changes necessarily involve a change of state of mind. The most general single operation must therefore be taken to be one of the following:

(A) A possible change (a) of symbol together with a possible change of state of mind.
(B) A possible change (b) of observed squares, together with a possible change of state of mind.

The operation actually performed is determined, as has been suggested [above], by the state of mind of the computer and the observed symbols. In particular they determine the state of mind of the computer after the operation is carried out.

We may now construct a machine to do the work of this computer. To each state of mind of the computer corresponds an "m-configuration" of the machine. The machine scans B

squares corresponding to the B squares observed by the computer. In any move the machine can change a symbol on a scanned square or can change any one of the scanned squares to another square distant not more than L squares from one of the other scanned squares. The move which is done, and the succeeding configuration, are determined by the scanned symbol and the m-configuration.

---

[1] If we regard a symbol as literally printed on a square we may suppose that the square is $0 \le x \le 1, 0 \le y \le 1$. The symbol is defined as a set of points in this square, viz. the set occupied by printer's ink. If these sets are restricted to be measurable, we can define the "distance" between two symbols as the cost of transforming one symbol into the other if the cost of moving unit area of printer's ink unit distance is unity, and there is an infinite supply of ink at $x = 2, y = 0$. With this topology the symbols form a conditionally compact space.

The Turing machine therefore is a formalisation of the intuitive idea of what we do when performing a calculation algorithm. Its mechanism is very simple, and the way it is explained today is even more simple, without detracting anything from the performance of the machine [10]. We suppose that the computer examines a single square (that is $B = 1$) and that he can displace this to the left or the right (that is $L = 1$). This simplified machine is capable of carrying out the same computations described by Turing in his 1936 paper: the examination or displacement of a number of squares can be effected by a sequence of simple operations.

In its present-day presentation, we say that a Turing machine is composed of:

- a *tape* composed of cells containing symbols drawn from a finite alphabet,
- a *central unit* which can assume a number of finite states,
- a *read-write head* which permits communication between the central unit and the tape.

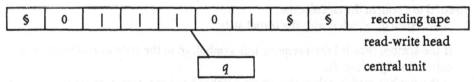

To refer back to Turing's description, the central unit is the brain of the computer (the person doing the calculations) which can assume a finite number of states of mind, and the tape is the strip of paper on which he carries out his calculations.

Each operation is determined by the state of the central unit and by the contents of the cell being examined. At each operation:

- the central unit can have a change of state,
- the read-write head can change the symbol of the cell being examined,
- the read-write head can move a cell to the right or to the left.

Let A be the finite alphabet of symbols used for the tape, with the symbol $ for an empty cell, Q the finite set of states of mind of the central unit, with i for the initial state and T the set of terminal states of the central unit. At the beginning, the central unit is in the state i, and it stops when it reaches a state t of T. The mechanism of the machine is therefore given by an application π of Q×A in Q×A×$\{r, l\}$, where r, l represent displacements to the right or left of the read-write head. Thus, for A = $\{0, |, \$\}$ and for states of Q, p and q, $\pi(q, 0) = (p, |, r)$ would have the following effect:

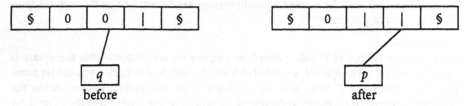

before   after

To see how a Turing machine carries out a calculation algorithm with a simple example, we shall consider subtraction. At the start, the recording tape contains, say, two whole numbers n and m, with $m > n$. At the finish, the tape has to have $m - n$ written on it. To represent the numbers m and n, just one symbol is sufficient: a vertical bar | for example. Thus, at the start, the tape contains n bars separated from m bars by some character c, say (Fig. 1):

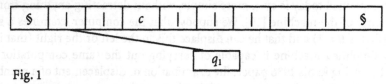

Fig. 1

In order to carry out the subtraction, the first bar of the number n has to be removed, then the last bar of the number m, then the process is repeated after having returned to the first cell containing a bar. The machine stops when there is no bar before the symbol c, and then the number of bars remaining on the tape following the symbol c is precisely $m - n$.

Consequently, the alphabet for the tape is A =$\{|, \$, c\}$ and five states (of mind) are needed to perform the calculation.

In the state q_1, which is also the initial state:

- if the symbol read is | then remove this symbol, go to the state q_2 and move to the cell to the right (Fig. 2),
- if the symbol read is c then the whole number has been read, remove the symbol c, go to state q_5 and move to the cell to the right (Fig. 6).

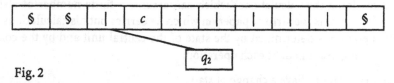

Fig. 2

In state q_2 we need to look for the last bar of the tape, so:

- if the symbol read is | or c remain in state q_2, move one cell to the right without changing the contents of the cell,
- if the symbol read is $\$$ go to state q_3, move one cell to the left without changing the symbol (Fig. 3).

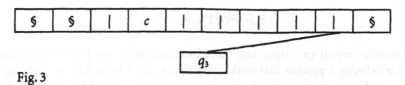

Fig. 3

In state q_3 the symbol read is | (if not, this would imply $m > n$ is false and the machine cannot proceed), this symbol is removed, pass to state q_4 and move one cell to the left (Fig. 4).

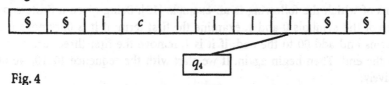

Fig. 4

In state q_4 we need go from right to left to look for the first bar on the tape, and so:

- if the symbol read is | or c remain in state q_4, move to the left without changing the contents of the cell.
- if the symbol read is $ go to state q_1, move to the cell to the right without changing the symbol (Fig. 5).

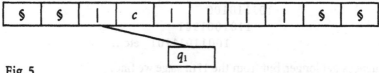

Fig. 5

When state q_5 is reached, it means that all the bars before the symbol c have been removed: state q_5 is therefore the terminal state and $m - n$ bars remain on the tape (Fig. 6).

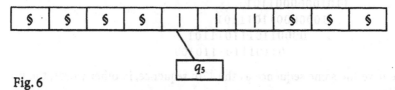

Fig. 6

The transition function of the Turing machine is a function π of Q×A in Q×A×{r, l} which can be represented in the form of the following table:

π	|	c	$
q_1	$(q_2, \$, r)$	$(q_5, \$, r)$	
q_2	$(q_2, \|, r)$	(q_2, c, r)	$(q_3, \$, l)$
q_3	$(q_4, \$, l)$		
q_4	$(q_4, \|, l)$	(q_4, c, l)	$(q_1, \$, r)$
q_5			

15.6 Post's Machine

A few months later than Turing, and quite independently, Emil Post proposed a machine for defining a process that could solve a general problem. Just like Turing, Post had been galvanised into action by the recent results due to Gödel and Church. But Post had also, for a long time, been interested in a problem which appeared to be a suitable candidate for undecidability.

In 1921, and while still a student at Princeton, Post became interested in a class of problems of which the following is an example [13]. Being given a finite sequence made up of the symbols 0 and 1, examine the first term: if it is 0, remove the first three terms and add 00 to the end; if it is 1, remove the first three terms and add 1101 to the end. Then begin again. If we start with the sequence 10010, we obtain, successively:

```
10010
  101101
    1011101
      11011101
        111011101
          0111011101
            101110100
              1101001101
                10011011101  etc...
```

The sequences get longer, but from the 17th stage we find:

```
011011101110100
  01110111010000
    1011101000000
      11010000001101
        100000011011101
          0000110111011101
            011011101110100
```

and we have the same sequence as the 17th sequence, in other words, the process is periodic.

Is this the same for all sequences of 0's and 1's? Post mentions this problem in a paper in 1943 writing: "The little progress that has been made in the solution of such problems makes them candidates for undecidability". Post's problem has been generalised to what is called the *tag system*, which can be represented by a kind of small Turing machine [13].

In 1947, Post proved the undecidability of the problem, called today *Post's Correspondence Problem*, which can be stated as follows. Let U and V be two lists of k non-empty words from an alphabet A, $U = \{u_1, u_2, ..., u_k\}$ and $V = \{v_1, v_2, ..., v_k\}$, does there exist a sequence of integers $i_1, i_2, ..., i_n$ such that the words $u_{i_1} u_{i_2} ... u_{i_n}$ and $v_{i_1} v_{i_2} ... v_{i_n}$ are the same? For example, consider:

$A = \{a, b\}$
$U = \{u_1 = b, u_2 = babbb, u_3 = ba\}$
$V = \{v_1 = bbb, v_2 = ba, v_3 = a\}$

then there is a solution $u_2u_1u_1u_3 = v_2v_1v_1v_3 = babbbbbba$ and so the answer here is positive. However, the Post correspondence problem turns out to be undecidable, that is there is no algorithm, given any arbitrary U, V, for deciding the existence of such a solution sequence [9]. The undecidability of this problem is used to prove the undecidability of many other problems.

At the beginning of his 1936 paper, Post defines a 1-process, that is a sequence of instructions for directing a worker carrying out actions in a symbolic space. At the end of the paper, Post considers a number of fundamental questions. Is his idea equivalent to Church's effective calculability? Post does not prove this equality, but he remarks that Church's conception does itself need to be justified: does it correspond to the intuitive idea of calculability, that is an algorithm? Post's conclusion is that, for the moment, the equivalence of his conception and all other formulations should be accepted as a "working hypothesis".

E. L. Post
Finite Combinatory Processes, Formulation 1,
The Journal of Symbolic Logic, vol. 1 (1936), 103–105.

The present formulation should prove significant in the development of symbolic logic along the lines of Gödel's theorem on the incompleteness of symbolic logics[1] and Church's results concerning absolutely unsolvable problems.[2]

We have in mind a *general problem* consisting of a class of *specific problems*. A solution of the general problem will then be one which furnishes an answer to each specific problem.

In the following formulation of such a solution two concepts are involved: that of a *symbol space* in which the work leading from problem to answer is to be carried out[3], and a fixed unalterable *set of directions* which will both direct operations in the symbol space and determine the order in which those directions are to be applied.

In the present formulation the symbol space is to consist of a two way infinite sequence of spaces or boxes, i. e., ordinally similar to the series of integers $\ldots, -3, -2, -1, 0, 1, 2, 3, \ldots$. The problem solver or worker is to move and work in this symbol space, being capable of being in, and operating in but one box at a time. And apart form the presence of the worker, a box is to admit of but two possible conditions, i. e., being empty or unmarked, and having a single mark in it, say a vertical stroke.

One box is to be singled out and called the starting point. We now further assume that a specific problem is to be given in symbolic form by a finite number of boxes being marked with a stroke. Likewise the answer is to be given in symbolic form by such a configuration of marked boxes. To be specific, the answer is to be the configuration of marked boxes left at the conclusion of the solving process.

The worker is assumed to be capable of performing the following primitive acts:[4]

(a) *Marking the box he is in (assumed empty)*,
(b) *Erasing the mark in the box he is in (assumed marked)*,
(c) *Moving to the box on his right*,
(d) *Moving to the box on his left*,
(e) *Determining whether the box he is in, is or is not marked*.

The set of directions which, be it noted, is the same for all specific problems and thus corresponds to the general problem, is to be of the following form. It is to be headed:

Start at the starting point and follow direction 1.

It is then to consist of a finite number of directions to be numbered 1, 2, 3, ..., n. The ith direction is then to have one of the following forms:

(A) *Perform operation O_i [O_i = (a), (b), (c), or (d)] and then follow direction j_i,*
(B) *Perform operation (e) and according as the answer is yes or no correspondingly follow direction j_i' or j_i'',*
(C) *Stop.*

Clearly but one direction need be of type C. Note also that the state of the symbol space directly affects the process only through directions of type B.

A set of directions will be said to be *applicable* to a given general problem if in its application to each specific problem it never orders operation (a) when the box the worker is in is marked, or (b) when it is unmarked.[5] A set of directions applicable to a general problem sets up a deterministic process when applied to each specific problem. This process will terminate when and only when it comes to the direction of type (C). The set of directions will then be said to set up a *finite 1-process* in connection with the general problem if it is applicable to the problem and *if the process it determines terminates for each specific problem.* A finite 1-process associated with a general problem will be said to be a *1-solution* of the problem if the answer it thus yields for each specific problem is always correct.

We do not concern ourselves here with how the configuration of marked boxes corresponding to a specific problem, and that corresponding to its answer, symbolize the meaningful problem and answer. In fact the above assumes the specific problem to be given in symbolized form by an outside agency and, presumably, the symbolic answer likewise to be received.

[...]

The writer expects the present formulation to turn out to be logically equivalent to recursiveness in the sense of the Gödel-Church development.[6] Its purpose, however, is not only to present a system of a certain logical potency but also, in its restricted field, of psychological fidelity. In the latter sense wider and wider formulations are contemplated. On the other hand, our aim will be to show that all such are logically reducible to formulation 1. We offer this conclusion at the present moment as a *working hypothesis.* And to our mind such is Church's identification of effective calculability with recursiveness.[7] Out of this hypothesis, and because of its apparent contradiction to all mathematical development starting with Cantor's proof of the non-enumerability of the points of a line, independently flows a Gödel-Church development. The success of the above program would, for us, change this hypothesis not so much to a definition or to an axiom but to a *natural law.* Only so, it seems to the writer, can Gödel's theorem concerning the incompleteness of symbolic logics of a certain general type and Church's results on the recursive unsolvability of certain problems be transformed into conclusions concerning all symbolic logics and all methods of solvability.

---

[1] Kurt Gödel, Über formal unentscheidbare Sätze der Principia Mathematica und verwandter Systeme I, *Monatshefte für Mathematik und Physik*, vol. 38 (1931), 173–198.
[2] Alonzo Church, An unsolvable problem of elementary number theory, *American Journal of Mathematics*, vol. 58 (1936), 345–363.
[3] Symbol space, and time.
[4] As well as otherwise following the directions described below.

[5] While our formulation of the set of directions could easily have been so framed that applicability would immediately be assured it seems undesirable to do so for a variety of reasons.

[6] The comparison can perhaps most easily be made by defining a 1-function and proving the definition equivalent to that of recursive function. (See Church, loc. cit., p. 350) A 1-function $f(n)$ in the field of positive integers would be one for which a finite 1-process can be set up which for each positive integer n as problem would yield $f(n)$ as answer, n and $f(n)$ symbolized as above.

[7] Cf. Church, loc. cit., pp. 346, 356–358. Actually the work already done by Church and others carries this identification considerably beyond the working hypothesis stage. But to mask this identification under a definition hides the fact that a fundamental discovery in the limitations of the mathematicizing power of Homo Sapiens has been made and blinds us to the need of its continual verification.

Post does not speak of a machine although his formulation, perhaps to a greater extent than Turing's resembles our concept of a computer. What Post calls a box can be compared to a memory which can be either empty or contain a value. Furthermore, his set of instructions can be compared to a sort of program with branches. But it is a minimal formulation of what we would call a program machine together with an operating program. In fact, just one symbol can be entered in the memory, and this can be moved only to a contiguous memory. It is certainly appropriate to describe this minimal formulation as "Formulation 1".

Using a vertical bar for the symbol, a configuration of boxes, where the box being visited is labelled 0, would look like this:

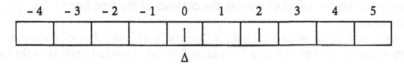

An instruction is preceded by a label, which is simply its number, and is terminated by a jump to another instruction indicated by its label. We can consider four types of instruction.

- First type: marking (a) or erasing (b). Marking would be written:
  1: <u>mark</u> ; <u>goto</u> 3 ;
  which means that the first instruction is to mark the square with | and then jump to instruction 3. Erasing would be an instruction like:
  2 : <u>erase</u> ; <u>goto</u> 1;
  meaning that the second instruction is to erase the mark in the box being visited and then jump to instruction 1.

- Second type: movement to the right (c) or left (d) in the symbolic space. These instructions would be, for example:
  3 : <u>right</u> ; <u>goto</u> 5 ;
  4 : <u>left</u> ; <u>goto</u> 2 ;.

- Third type: conditional jump. A conditional jump would be written:

  5 : <u>if</u> marked <u>then</u> 3 <u>if not</u> 2 ;

  meaning the fifth instruction is to examine the contents of the box. If the box is marked, jump to instruction 3, if it is not marked, jump to instruction 2.

- Fourth type: end instruction, which can be written:

  6 : <u>stop</u>.

  Consider, now the following set of instructions:

  1 : <u>right</u> ; <u>goto</u> 2 ;

  2 : <u>if</u> marked <u>then</u> 3 <u>if not</u> 1 ;

  3 : <u>erase</u> ; <u>goto</u> 4 ;

  4 : <u>stop</u> .

  Starting with the configuration given above, we would have the sequence:

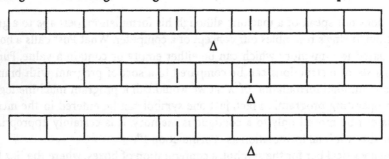

The sequence of the labels of the instructions that are carried out is: 1, 2, 1, 2, 3, 4. This program has the effect of erasing the contents of the first box to the right that contains an entry.

Let us now look at what Post calls a "general problem", that of adding strictly positive integers, and try to find a "1-solution". What is needed, then is to find a "1-process", that is a set of instructions which terminates for each "specific problem", that is to say for each two given integers n and m, it gives as solution the integer $n + m$. The initial state could be represented by a configuration of boxes of which there are n marked boxes, followed by an empty box, and then m marked boxes. We suppose that the box being visited is the first marked box at the left. For $n = 4$ and $m = 3$, the configuration would look like this:

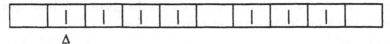

The final expected configuration is composed of $n + m$ marked boxes, with the first marked box on the left as the box being visited:

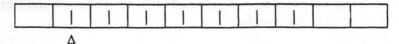

To get from the initial configuration to the final configuration, each of the bars on the right has to be slid one box to the left, and the marker has to return to the first of the marked boxes. This can be done by the following set of instructions:

1: <u>if</u> marked <u>then</u> 2 <u>if not</u> 3 ;
2: <u>right</u> ; <u>goto</u> 1 ;
3: <u>mark</u> ; <u>goto</u> 4 ;
4: <u>if</u> marked <u>then</u> 5 <u>if not</u> 6 ;
5: <u>right</u> ; <u>goto</u> 4 ;
6: <u>left</u> ; <u>goto</u> 7 ;
7: <u>erase</u> ; <u>goto</u> 9 ;
8: <u>if</u> marked <u>then</u> 9 <u>if not</u> 10 ;
9: <u>left</u> ; <u>goto</u> 8 ;
10: <u>stop</u>.

For the "specific problem" considered, the sequence of labels of the instructions is: 1, 2, 1, 2, 1, 2, 1, 2, 1, 3, 4, 5, 4, 5, 4, 5, 4, 6, 7, 9, 8, 9, 8, 9, 8, 9, 8, 9, 8, 9, 8, 9, 8, 10.

Post concludes his paper by touching on the question of the equivalence of his system with the definitions of recursive functions given by Gödel and Church. He considers more generally all the definitions which could be made to translate the idea of process or calculation. His "working hypothesis" is that they are all equivalent to his formulation, which has the merit of psychological fidelity.

15.7 Conclusion

With the introduction of the concept of an algorithm, the history of algorithms changes into the history of a new field of science: the field of algorithms. Here, it is not a question of seeking an algorithm for solving a particular problem but the solution of problems posed by the general study of algorithms. This area of research has been developed, in particular, alongside the construction of computers and the invention of programming languages.

As we have seen, the origin of the idea of an algorithm lies in the question of the existence of an algorithm for the solution of a problem. But the field of algorithmic study is also concerned with the complexity of algorithms, that is the intuitive idea of the cost of an algorithm in the time taken for it to carry out a task (temporal complexity) and in the amount of memory required (spatial complexity). The temporal and spatial complexities of an algorithm may be defined, respectively, by the number of movements and the number of cells used by the corresponding Turing machine, as a function of the length of the input data.

The concept of programming languages, and writing algorithms in these languages, poses a number of questions related to syntax: the description of the syntax of a language, the non ambiguous nature of the syntax, verification that a program conforms to the syntax of the language, the writing of a compiler for interpreting and executing a program from the syntax of the language. All these problems were responsible for the construction of a number of concepts and theories: grammars, Chomsky classification, automata theory and theories of languages, etc.

The field of algorithmic study is also concerned with the validity of algorithms, that is proving whether an algorithm terminates satisfactorily, having answered the proposed question. This problem leads to one of conceptualising the semantics of programming languages and to searching for algorithms for verifying algorithms [11]!

The ancient dream of Leibniz progressively becomes reality.

Bibliography

[1] Andrews, P., *An Introduction to Mathematical Logic and Type Theory: to Truth through Proof*, Orlando: Academic Press, 1986.

[2] Church, A., *Introduction to Mathematical Logic*, Princeton: Princeton University Press, 1956.

[3] Davis, M. (ed.), *The Undecidable*, Hewlett, New York: Raven Press, 1965. [This contains the texts quoted in this chapter.]

[4] Davis, M. & Weyuker, E., *Computability, Complexity and Languages*, Orlando: Academic Press, 1983.

[5] Guillaume, M., Axiomatique et logique, in J.Dieudonné (ed.) *Abrégé d'histoire des mathématiques 1700-1900*, t. II, Paris: Hermann, 1978.

[6] Heijenoort, J. Van, *From Frege to Gödel*, Cambridge, Massachusetts: Harvard University Press, 1967. [This contains English tr. of articles on logic by Frege and Ackermann.]

[7] Hilbert D. & Ackermann W., *Principles of Mathematical Logic*, English tr., New York: Chelsea Publishing Co., 1950.

[8] Hodges, A., *Alan Turing: the Enigma of Intelligence*, London: Burnett Books, 1983.

[9] Hopcroft, J. & Ullman, J., *Formal Languages and their Relation to Automata*, Reading Massachusetts: Addison-Wesley Publishing Company,1969.

[10] McNaughton R., *Elementary Computability, Formal Languages, and Automata*, Englewood Cliffs, New Jersey: Prentice Hall, 1982.

[11] Manna, Z., *Mathematical Theory of Computation*, New York: McGraw-Hill Inc., 1974.

[12] Matijasevič, Y.V., Dipohantine Representation of recursively enumerable Predicates, *Proceedings of the International Congress of Mathematicians in 1970*, Paris: Gauthier-Villars, 1971.

[13] Minsky, M., Computation: *Finite and Infinite Machines*, Englewood Cliffs, New Jersey: Prentice Hall, 1967.

[14] Russell, B., *The Principles of Mathematics*, Cambridge: Cambridge University Press, 1903.

[15] Uspensky, V.A., *Post's Machine*, Moscow: MIR, 1983.

Biographies

The following notes provide a brief summary of the lives of the principal characters appearing in this history of algorithms, together with some indication of the contributions they have made to mathematics. For more detailed biographical information the reader may care to consult, among others, the following works which we have used to prepare these notes:

Concise Dictionary of Scientific Biography, New York: Scribner's sons, 1981, which summarises the monumental:

Gillespie, *Dictionary of Scientific Biography*, 16 vols., New York: Scribner's sons, 1970–1980.

Hoefer, *Nouvelle biographie générale*, 46 vols., Paris: Firmin-Didot, 1855–1866.

Michaud, *Biographie universelle ancienne et moderne*, 46 vols., Paris: Desplaces, 1811–1862.

Poggendorff, *Biographisch-literarisches Handwörterbuch zur Geschichte der exakten Wissenschaften*, Leipzig, 1863 on.

Taton, *Histoire générale des sciences*, 4 vols., Paris: Presses Universitaires de France, 1957–1964.

Abū Kāmil, Shujāᶜ ibn Aslam ibn Muḥammad ibn Shujāᶜ (*c.* 850 – *c.* 930)
Nothing is known of the life of Abū Kāmil. He lived under the reign of a number of Abbasid Caliphs at a time when the Arab-Islamic world experienced considerable economic prosperity but also political and ideological ferment. He is considered as one of the foremost Arabic algebraists. Among the six mathematical works attributed to Abū Kāmil, three have come down to us: *The complete [book] of algebra, The book of original [things] in calculation* and *The book [of calculation] of inheritance with the aid of roots.*

Abu'l-Wafā' al-Buzjānī (Buzjān 940 – Baghdad 997)
Apparently of Persian descent Abu'l-Wafā' went to Baghdad in 959 where he became the most important figure in the school of mathematics and astronomy which had been established from the beginning of the 9th century. Abu'l-Wafā' made many astronomical observations and carried on the tradition of his predecessors in combining original works with commentaries on the Greek mathematical classics. In mathematics only two of his works survive: *The Book on Arithmetic necessary to Scribes and Merchants* and *The Book of Geometric Constructions.*

Adams, John (Laneast, Cornwall 1819 – Cambridge 1892)
Adams became interested in mathematics and astronomy from a very early age. In 1839 his parents sent him to Cambridge. At the age of 24, contemporaneously with Le Verrier, Adams explained perturbations in the orbit of Uranus by the presence of another, as yet unknown, planet. In 1851 he was elected President of the Royal Astronomical Society; soon after, he started his work on lunar parallax. In 1858 he took up the chair of mathematics at St. Andrews University. Like Euler and Gauss, Adams calculated the values of many mathematical constants.

Aitken, Alexander Craig (Dunedin 1895 – Edinburgh 1967)
Born in New Zealand, Aitken astounded his grand parents when at the age of nine he explained his discovery of the mating behaviour of the albatross. Not believing him, they punished him for lying. After two years at the University of Otago, Aitken volunteered for military service and was seriously injured at the Battle of the Somme. He completed his studies at Otago, graduating in 1919, after which he taught at Otago Boy's High School until 1923 when Bell persuaded him to come to Edinburgh to further his studies. He remained at Edinburgh for the rest of his life. Aitken worked mostly on numerical analysis, statistics and linear algebra. Photograph by kind permission of the Mathematisches Forschungsinstitut, Oberwolfach.

Al-Baghdādī, Abū Manṣūr ʿAbd al-Qāhir ibn Ṭāhir ibn Muḥammad ibn ʿAbdallah al-Tamīmī al-Shāfiʿī (Baghdad ? – c. 1037)
Al-Baghdādī is known as a theologian and an arithmetician. His two books on arithmetic concern measure and calculation procedures. The second of these has sections on: Indian integer arithmetic, Indian fraction arithmetic, Indian treatment of sexagesimals, the decimal system and irrational arithmetic and the practice of commercial arithmetic.

Al-Battānī, al-Ḥarrānī (Harran c. 858 – c. 929)
Known in Europe as Albategnius, al-Batani, like Thābit ibn Qurra, belonged to the Sabian sect. He practised astronomy at Raqqa, a city on the left bank of the Euphrates. The astronomy of al-Battānī, a revision of Ptolemy's *Almagest*, is justly considered to be one of the most important works of astronomy from the Middle Ages. In it, al-Battānī uses the sine in place of the chord, following the example of earlier Indian and Arab writers, and he also uses the cosine, as the sine of the complementary angle, and the versed sine for the cosine of an obtuse angle.

Al-Bīrūnī, Abū ar-Rayḥān Muḥammad ibn Ahmad (Kâth, Khwarezm 973 – Ghazni, Afghanistan, 1050)
Al-Bīrūnī was brought up by the royal family of Banu ʿIrāq who governed Khwārezm, a kingdom bordering the Aral Sea. He travelled in Iran and India before settling in Ghāzī in Afghanistan. He was accomplished in many areas of knowledge: astrology, astronomy, geography, geology, mineralogy, pharmacology, history, mathematics and philosophy. He corresponded with many distinguished thinkers, such as the philosopher Avicenna and the mathematician and astronomer Abu'l-Wafa. He made a significant contribution to the history of mathematics, in particular through the information he was able to provide on Indian mathematics.

Al-Karajī, Abū Bakr ibn Muḥammad ibn al-Ḥusayn (*fl.* 1020)
Almost nothing is known of the life of al-Karajī (also known as al-Karkhī). It is not known if he comes from the Baghdad area or from the Persian town Karaj. It appears that he lived in Baghdad for a considerable time, and it is there that he wrote his principal mathematics works. These dealt with arithmetic, in particular the *Book of Sufficiency*, and especially with algebra, like the *Original Book*, and the *Fakhri*. His contribution to algebra was to extend arithmetic operations to irrational numbers and polynomials, studying the latter independently of equations, and solving linear systems of equations.

Al-Kāshī, Ghiyāth ad-Dīn Jamshīd Masʿūd (*b.* Kāshān – Samarkand 1429)
Originating from Kāshān in Iran, al-Kāshī was an astronomer and mathematician. He published his *Compendium on Science and Astronomy*, in Persian, in 1410 and his astronomical tables in 1413. In 1417 he accepted an invitation from the Sultan Ulug Beg to teach and carry out research at the newly established Scientific College at Samarkand. In 1424 he published his *Treatise on the Circumference*, his *Epistle on the Chord* a year later, and two years prior to his death, *The Key to Calculation*.

Al-Khayyām, Abu l-Fatḥ ʿUmar ibn Ibrāhīm (Nissapur 1048 - 1131)
We have no definitive information about the life of ʿUmar al-Khayyām and contradictory legends depict him either as a profound mystic believer or as a hedonistic agnostic and libertine. He is most popularly known for his poetry, from the English translation of the *Rubáiyát* by Fitzgerald, but among his surviving mathematical works are a treatise on algebra, an *Epistle on the division of the quadrant of the circle*, both dealing with cubic equations, and *Commentaries to Difficulties in the Introduction to Euclid's Book*.

Al-Khwārizmī, Abū JaʿFar Muḥammad ibn Mūsā (*c.* 780 - *c.* 850)
Al-Khwārizmī was an astronomer and mathematician who came from Khwārezm, a region bordering the Aral Sea. From 813 to 833 he was active in the House of Wisdom, an academy founded in Baghdad by the Abbasid Caliph al-Maʿmūn. The most important scientific works by al-Khwārizmī are *The Compendium on Calculation by algebra and the muqabala*, the [*Astronomical*] *Tables of Sindhind*, a manual of arithmetic, where the Indian decimal system is presented for the first time to the Arab world, a book on mathematical geography entitled the *Book of the Form of the Earth* and a Jewish Calendar.

Al-Māhānī, Abū ʿAbdallah Muḥammad ibn ʿĪsā (*c.* 853 - Baghdad 866)
Al-Māhānī was a ninth century Persian mathematician who mostly lived in Baghdad. He correctly predicted three lunar eclipses to within an accuracy of half an hour. His main contributions were to mathematics. Al-Māhānī was the first to use a cubic equation to tackle Archimedes' problem of dividing a sphere by a plane into two parts in a given ratio (*On the Sphere and Cylinder*, Book II, Prop. 4). He also wrote an, as yet undiscovered, treatise on the quadrature of the parabola, as well as an *Epistle on the Theory of Magnitudes*.

Al-Ṭūsī, Sharaf al-Dīn al-Muẓaffar ibn Muḥammad ibn al-Muẓaffar (*fl.* 13th century)
Nothing is known of the early life of Sharaf al-Dīn al-Ṭūsī. There are reports from between 1260 and 1280 of students having received instruction from him in various places: Ṭūs in Iran, then Hamadan, Mosul, Aleppo and Damascus. His contribution to astronomy was his *Treatise on the Linear Astrolabe* and in mathematics the most important work was his *Treatise on Equations*. He also wrote an *Opuscule on the Asymptote* to a branch of the rectangular hyperbola and an *Epistle on a Problem of Geometry* treated algebraically.

Al-Uqlīdisī, Abū al-Ḥasan Aḥmad ibn Ibrāhīm (*fl.* Damascus 952–953)
Al-Uqlīdisī is known from a single copy of one of his works: *Sections on Indian Calculation*. This manuscript of 1157 is a copy of a work that was written in 952 in Damascus. This text is the most important of the hundred or so Arab arithmetics that have survived. First, it is the oldest and it uses decimal fractions. Second, there is a suggestion in what al-Uqlīdisī writes that we can see here the passage of Indian calculations from the dust table to calculations on paper.

Archimedes (Syracuse 287 - 212 BC)
Little is known of the life of Archimedes. Son of the astronomer Phidias, he studied at Alexandria with Euclid's descendants. He composed the greater part of his work at Syracuse where he was killed during a Roman invasion of the town. His mechanical inventions and his works in physics are legendary. In mathematics, he used a Euclidean approach to calculations of areas and volumes, but did not hesitate to introduce ideas from statics (in *Quadrature of the Parabola*, for example). In *The Sand-Reckoner* he developed an original system, based on powers of 10, for recording large numbers.

Aristarchus of Samos (*c.* 310 – *c.* 230 BC)

A pupil of Strato, the third principal of the Lyceum founded by Aristotle, Aristarchus of Samos is renowned for having proposed a heliocentric model for the solar system some 18 centuries before Copernicus. In his work *On the sizes and distances of the Sun and the Moon*, he made valid mathematical deductions from particular observations. Anticipating future trigonometry, he used arithmetic and geometric methods to obtain approximations to the sines of small angles.

Arnauld, Antoine (Paris 1612 – Brussels 1694)

Arnauld, an ordained priest, obtained his doctorate of theology in 1641. He was admitted to the Sorbonne in 1643, but excluded in 1656, on account of his Jansenist opinions. He spent some time in hiding, then withdrew to the Abbey of Port-Royal, before going into exile, first in Flanders and then to the Netherlands. Arnauld was a theologian and philosopher, interested in logic and in the philos phy and language of mathematics. His *Nouveaux éléments de géométrie*, published in 1667, mark an awakened interest in Euclidean geometry following on from Pascal.

Artin, Emile (Vienna 1898 – Hamburg 1962)

After studies at Vienna, Leipzig and Göttingen, Artin took up an appointment at Hamburg. From 1937 to 1958 he worked in the United States, at the University of Notre-Dame, the University of Indiana and at Princeton. He then returned to Hamburg. His studies related to the theory of numbers and he made an important contribution to the class field theory. With Otto Schreier he founded the theory of real fields, a branch of abstract algebra.

Āryabhata (*c.* 476 – ?)

Āryabhata lived in Kusumpara – now Patna, on the Ganges – the imperial capital of the Gupta dynasty during the 4th and 5th centuries. In his own work written in 510, known as the *Āryabhatiya*, Āryabhata describes science as having an honoured place in Kusumpara. Composed in Sanskrit verse, the work is in three parts: the first on mathematics, the second on planetary motion, the third on the sphere and ellipses. For many centuries it was an essential work of reference for Indian astronomers. Its 8th century Arabic translation was still being used in the 11th century, notably by al-Bīrūnī.

As-Samaw'al, ibn Yaḥyā ibn ʿAbbās al Magrhribī (Baghdad ? – ? 1175)

As-Samaw'al's father was a rabbinical scholar from Fez who had moved to Baghdad. As-Samaw'al was both a doctor and teacher of mathematics, working in the Fertile Crescent and in central Asia. In medicine, he published a treatise on sexology. Among his mathematical works were *The dazzling book of algebra*, in which he completes and improves al-Karajī's project of extending arithmetic operations to algebra, and *The book of Indian calculations* in which he explains decimal fractions.

Az-Zinjānī ʿAbdalwahhāb ibn Aḥmad ibn Ibrāhīm (*d.* Baghdad 1257)

Az-Zinjānī was a 13th century scholar known more as a philologist than as a mathematician. His only known surviving mathematical work, *The calculator's aid*, is a compilation of calculation methods and problems in number theory.

Babbage, Charles (Teignmouth 1792 – London 1871)

Entering the University of Cambridge in 1810, he was elected to a chair in 1827, which he held for a period of twelve years without ever lecturing. In 1816 he was made a Fellow of the Royal Society, of which he became an outspoken critic. Babbage was an activist who founded a number of societies. In an attempt to bring new life to the teaching of mathematics, he collaborated with Peacock and Herschel, to produce a translation of the *Calcul différentiel et in-*

tégral by Lacroix. Author of *On the Economy of Machines and Manufactures...* (1832), Babbage also tackled mathematical logic. He envisaged two calculating machines, one using differences, the other analytical.

Bachet de Méziriac, Claude Gaspard (Bourg-en-Bresse 1581 – 1638)
Born into a noble family, Bachet de Méziriac was brought up by the Jesuits. He pursued his studies at Padua and taught at Como and Milan. He was elected to the Académie française in 1635. A writer, poet and mythologist, he is known today for his richly annotated Greek and Latin edition of the *Arithmetica* of Diophantus. Bachet's *Problèmes plaisants et délectables* remains a classic of works of recreational mathematics.

Banū Mūsā (*fl.* 870)
The father of the three Banu Musa brothers was a reputed astrologer. On his death, the Caliph al-Maʿmūn acknowledged the scientific skills of the brothers and they entered the House of Wisdom. Between them they studied mathematics, astronomy, musical instruments and mechanics. They built their own observatory and set up a school of translators of Greek works to which Thābit ibn Qurra belonged. The individual contributions of each of the brothers is not known. Of the many texts that are attributed to them, the most important is the *Book on the Measurement of Plane and Spherical Figures* in which the Euclidean method of exhaustion makes its first appearance in Islamic literature.

Barrême, François (Lyon *c.* 1640 – Paris 1705)
Barrême was an arithmetician who established himself in Paris in the middle of the 17th century and whose lectures attracted crowds of admirers. Colbert acknowledged the value of Barrême's work and honoured him with his protection. Following his *Livre nécessaire pour tous les comptables contenant les calculs des intérêts* (1694), Barrême was appointed auditor to the Paris *Chambre des comptes* (Treasury). His two books *Les comptes faits du grand commerce* (1670) and *L'arithmétique ou le livre facile pour apprendre l'arithmétique soi-même et sans maître* (1677) passed through numerous editions.

Barrow, Isaac (London 1630 – 1677)
Barrow entered Trinity College, Cambridge at the age of 14 and took his degree when only 18. After taking holy orders, he travelled for some time. In 1662 he was elected to the post of professor of geometry at Gresham College, London and soon after he was elected the first Lucasian professor of mathematics at Cambridge, a post which he later gave up in favour of Newton. He became chaplain to Charles II in 1670, master of Trinity College in 1673, and vice-chancellor of the university in 1675. Barrow was, from 1662, a member of the Royal Society. Apart from his translations of Euclid, Archimedes, and other Greek mathematicians, Barrow's public lectures on geometry and optics were celebrated.

Berkeley, George (County Kilkenny, Ireland 1685 – Oxford 1753)
An Anglo-Irish Anglican bishop, philosopher and scientist. After taking holy orders, Berkeley taught Greek, Hebrew and Theology at Dublin. He travelled in France, Italy and Sicily. Berkeley was enthused with propagating Christianity and travelled to America but, due to lack of means, returned to England in 1732. He was consecrated Bishop of Cloyne, Dublin in 1734. He sought to show that God was the creative source of all objects and ideas: hence his *Analyst* is as much an *apologia* for theology as it is a critique of the theory of fluxions.

Bernoulli, Daniel (Groningen 1700 – Basel 1782)
Son of Jean Bernoulli, professor of Greek in Groningen, then of mathematics in Basel, Daniel Bernoulli received his early education from his father and from his brother Nicolas. He studied philosophy, logic and then medicine. In 1724, Daniel Bernoulli published his *Exercitatio-*

nes mathematicae which gave him entry to the St Petersburg Academy. This was the most prolific period of his life; he then returned to Basel where he was made professor. Daniel Bernoulli wrote on medicine, hydrodynamics, mechanics and probability. Portrait by kind permission of Birkhäuser-Verlag.

Bernstein, Sergei Natanovich (Odessa 1880 – Moscow 1968)
Son of a professor of anatomy at Odessa University, Bernstein went to Paris in 1898 to study at the Sorbonne and the École Supérieure d'Electronique. He went to Göttingen in 1902 but submitted his thesis to the Sorbonne in 1904, supervised by Picard. He then returned to Russia where he taught at the universities of Kharkov (1907–1933), Leningrad (1933–1941) and, finally, Moscow. His work (on differential equations, approximation of functions and probability) united the traditions of the St. Petersburg School with those of Göttingen and Paris.

Bertrand, Joseph (Paris 1822 – 1900)
Bertrand grew up in a scientific family. On his father's death, he was taken up by Duhamel who allowed him to follow courses at the École Polytechnique, to which he was officially admitted in 1841. After teaching elementary mathematics at the Collège Saint-Louis, Bertrand became a tutor in analysis at the École Polytechnique in 1844. During the 1848 revolution, he was a captain in the Garde Royale. His Traité d'arithmétique and Traité élémentaire d'algèbre, works which enjoyed considerable popularity, come from this period. Bertrand's career was a glittering one: École Polytechnique, Collège de France, Académie des Sciences and Académie Française.

Bessel, Friedrich (Minden 1784 – Königsberg 1846)
Until the age of 15, Bessel did not show any particular talent. However, as an assistant to a merchant in Bremen, he excelled in commercial accounts, and spent his evenings studying geography, Spanish and English. With a seafaring career in mind, Bessel read books on navigation and became interested in astronomy. In 1806 he became assistant in a private observatory and in 1810 was appointed director of the observatory in Königsberg. Bessel taught astronomy and worked on geodesic measurements. His contributions to mathematics, in particular the study of Bessel functions themselves, were all connected with astronomy.

Bézout, Étienne (Nemours 1739 – Basses-Loges/Avon 1783)
Member of the Académie des Sciences from 1758, Bézout taught mathematics to future Guards officers. His *Cours de mathématiques*, which went through many editions, was translated into English and had an important influence on the teaching of mathematics, not only in France but also abroad, particularly in America. Bézout's research was principally in the field of algebra, as is shown by his *Théorie générale des équations* of 1779.

Bhāskara (*c.* 1115 – *c.* 1178)
Bhāskara belonged to the priestly caste of Brahmins, the first of the four major Hindu castes, which alone could carry out religious tasks, study and recite scriptures. A popular work on astrology is attributed to his father. In 1150, Bhāskara completed a monumental work on astronomy, the *Siddhāntasiromani*. The *Lilāvatī* and the *Bijaganita* which deal with arithmetic and algebra are sometimes thought to be part of the larger work on astronomy. Bhāskara was the first to use the decimal system in written work, he invented the + and – conventions, and used letters to represent unknown quantities.

Bombelli, Raffaele (Bologna 1526 – 1572)
Engineer in the service of Alexander Rufini, the bishop of Melfi and favourite of Pope Paul III, Bombelli was engaged in the work of draining marshland areas. On a visit to Rome, he came across the work of Diophantus, which he helped to popularise. During the later years of his life, he wrote a work called *Algebra*, thus raising the status of algebra to that of a separate discipline. He had a strong influence on Simon Stevin and, later, on Leibniz.

Boole, George (Lincoln 1815 – Cork 1864)
Boole was a self-taught mathematician who also learnt Greek, French and German. He had intended taking holy orders but instead, at the age of 15, founded his own school. In 1834 he devoted himself to mathematics and discovered Newton's *Principia* and Lagrange's *Mécanique analytique*. In 1849 he was appointed to the post of professor at Queen's College, Cork. The works of Boole indicate his main interests: *Treatise on Differential Equations* (1859), *Treatise on the Calculus of Finite Differences* (1860), *The Mathematical Analysis of Logic* (1847) and *An Investigation of the Laws of Thought* (1854).

Bouquet, Jean-Claude (Morteau 1819 – Paris 1885)
A student of the École Normale Supérieure, Bouquet received his doctorate in 1852. He taught at a number of institutions including the École Normale Supérieure, the École Polytechnique and, finally, the Sorbonne. He was elected a member of the Académie des Sciences in 1875. Collaborating with Briot, he clarified Cauchy's works on analytical functions and applied them to elliptic functions.

Brahmagupta (*c.* 598 – *c.* 665)
Brahmagupta lived at Ujjain, the centre of astronomy in central India. At the age of thirty, he wrote the *Brāhmasphutasiddhānta*, a work on astronomy, several chapters of which were devoted to mathematics. Chapter twelve, *Ganita*, is about arithmetic calculations while chapter eighteen, *Kuttaka*, deals with what we would today call algebra. In another work on astronomy, the *Khandakhādyaha*, Brahmagupta gives a particular method for determining intermediate values of sines from a table of values.

Briggs, Henry (Warley Wood, Yorkshire 1561 – Oxford 1630)
Briggs studied at St. John's College, Cambridge. In 1596, he was elected to the newly created post of professor of geometry at Gresham College, London. He later taught at Merton College, Oxford. After coming into contact with Napier, he realised that an improvement to the latter's invention of logarithms could be made by changing to a decimal base, the effect of which was to rapidly expand the practical use of logarithms for calculations. The principal works of Briggs were: *Logarithmorum chiliasprima* (1617), *Arithmetica logarithmica* (1624) and *Trigonometria Britannica* (1633).

Briot, Charles (St.-Hippolyte 1817 – Bourg d'Ault 1882)
Following his education at the École Normale Supérieure, Briot taught at a number of institutions: lycée d'Orléans, universities of Lyon and Paris, lycées Bonaparte and Saint-Louis, Sorbonne and École Normale Supérieure. He was both a physicist and mathematician. He wrote a *Traité des fonctions abéliennes* and, with Bouquet, the *Théorie des fonctions doublement périodiques*, both fundamental works, going through many editions, which contributed to the study of the topics at an advanced level.

Brouncker, William (Castle Lyons, Ireland 1620 – London 1684)
William, Viscount Brouncker of Castle Lyons obtained his doctorate in physics at Oxford in 1647. Lord Brouncker was one of the founders and first president of the Royal Society. In 1662, he was appointed chancellor to Queen Catherine. Brouncker was especially interested in the rectification of the parabola and the cycloid and in the quadrature of the hyperbola and the circle. He was the first English mathematician to have used continuous fractions.

Budan, François-Désiré (Limonade, Haiti 1761 – Paris 1840)
Budan's parents were plantation owners in the western part of the island of Santo Domingo, now Haiti. At the age of eight, Budan was sent to France to study under the Oratorians. After joining the order, he taught in Nantes at the Collège royal. In 1803 he obtained his doctorate

in medicine, while at the same time working on numerical equations. In 1807, he was appointed deputy to Mauduit at the Collège de France and undertook the duties of Inspector general of studies.

Bürgi, Joost (Lichtenstein 1552 – Kassel 1632)
Bürgi had not, it seems, received much of an education: he had no Latin, the scientific language of the day. From 1579 he was court watchmaker to a number of emperors. In 1603, Bürgi became Kepler's assistant in Prague. Kepler was impressed with his calculating ability when working on the astronomical observations of Tycho Brahe. It was at this time that Bürgi issued his tables of logarithms, published in 1620. The lack of scientific activity in Prague prevented them being more widely known.

Carcavi, Pierre de (Lyon *c.* 1600 – Paris 1684)
Carcavi was the son of a banker. In 1632 he became a member of the Toulouse *parlement* and four years later a member of the Grand Council of Paris. In order to repay debts to his father, Carcavi entered the service of the Duc de Liancourt in 1648. In 1663 he was appointed royal librarian, a post he held until 1683 when Colbert died. Carcavi rendered considerable service to the scientific community of his day: Huygens, Fermat, Pascal, Mersenne, Descartes, Galileo and Torricelli all figure among his correspondents. He was probably the first to recognise the worth of fellow Toulouse mathematician Pierre de Fermat.

Cataldi, Pietro Antonio (Bologna 1552 – 1626)
Little is known of the life of Cataldi. He taught in Florence, Perugia and Bologna. His *Trattato del modo brevissimo di trovar la radice quadra delli numeri* of 1613 was an important contribution to the development of infinite algorithms: the square root of a number is found from an infinite sequence of continued fractions. Cataldi also worked on the problem of Euclid's fifth postulate and he edited the first six books of Euclid's *Elements*.

Cauchy, Augustin Louis (Sceaux 1789 – 1857)
A student at the École Polytechnique and then the engineering college, École des Ponts et Chaussées, Cauchy began his career as engineer for a number of public works. In 1815 he taught at the École Polytechnique, then at the Faculté des Sciences and at the Collège de France. In 1816 he became a member of the Académie des Sciences. Following the 1830 revolution, Cauchy went into exile, first to Turin and then to Prague, where he became tutor to the grand-son of Charles X. His considerable output of work belongs to many branches of mathematics and mathematical physics.

Cavalieri, Bonaventura (Milan *c.* 1598 – Bologna 1647)
Cavalieri entered the Jesuit order while still very young. At Pisa, the Benedictine monk Castelli initiated him into geometry through the reading of the Greek classics. Castelli introduced Cavalieri to Galileo. After teaching theology in a number of monasteries, Cavalieri obtained the post of professor of mathematics at the University of Bologna on the recommendation of Galileo. Author of a number of books on astronomy and spherical trigonometry, Cavalieri's name is associated with the Mean Value Theorem and the integral method of indivisibles.

Cayley, Arthur (Richmond 1821 – Cambridge 1895)
Son of a merchant family established in St. Petersburg, Cayley lived in Russia until the age of eight. He entered Trinity College, Cambridge, where he studied mathematics and law and he was called to the bar in 1849. From 1863 Cayley held the chair of mathematics at Cambridge. He excelled in numerous fields although he only published one work, *Treatise on elliptic functions* (1876). The most important of Cayley's contributions were to the theory of invariants and to matrices.

Chasles, Michel (Epernon 1793 – Paris 1880)
Entering the École Polytechnique in 1812, Chasles was enlisted two years afterwards for the defence of Paris. His *Aperçu historique* (1873) established his reputation as a geometer and historian of mathematics. He taught at the École Polytechnique from 1841 to 1851 and at the Sorbonne from 1846. He became a member of the Académie des Sciences in 1851. His *Traité sur les sections coniques* and his *Rapport sur les progrès de la Géométrie* confirmed him as one of the great geometers of the 19th century.

Chebyshev, Pafnouti (Okatovo 1821 – St. Petersburg 1894)
Chebyshev went to Moscow at the age of 11 to study mathematics. He was appointed assistant at the University of St. Petersburg in 1847 and professor in 1860, a post which he held until his retirement in 1882. Chebyshev became a member of the St. Petersburg Academy of Sciences in 1853 and founded the St. Petersburg School of Mathematics. He is best remembered for his work on the theory of numbers, the laws of probability and polynomial approximations of functions.

Cholesky, André-Louis (Montguyon 1875 – 1918)
After studying at the École Polytechnique, Cholesky entered the Artillery. Allocated to the Section de Géodésie of the Service Géographique, he took part in mapping the island of Crete and directed surveying operations in Algeria and Tunisia. Killed during the Great War, he ended his brief career with the rank of Commandant. In his work on geodesics, Cholesky used original ideas to handle the solution of conditional equations by the method of least squares.

Chomsky, Noam (Philadelphia 1928 –)
After studying linguistics, mathematics and philosophy, Chomsky worked with Zellig Harris from 1950 on syntactical structure and presented his Ph. D. thesis on transformational analysis in 1955. All his work strove to identify the mathematical and logical underpinning of syntactical and structural analysis. He revolutionised linguistics by proposing a 'generative' description of all sentences. From the beginning of the Vietnam war, Chomsky denounced American imperialism and asked intellectuals to face up to their responsibilities with respect to what became known as the military industrial complex.

Chuquet, Nicolas (Paris ? – Lyon 1488)
Indications are that Chuquet was born in Paris and came to Lyon in 1480. There he practised the craft of being a 'writer' – he taught writing to children – and he was a master of algorism. In addition he was a bachelor of medicine. His manuscripts contain many mathematical texts including the *Triparty en la science des nombres, des problèmes et applications*, a geometry entitled *Comment la science des nombres se peut appliquer aux mesures de géométrie* and a practical arithmetic *Comment la science des nombres se peut appliquer au fait de marchandises*.

Church, Alonzo (Washington 1903 – 1995)
Church taught mathematics at Princeton and edited the *Journal of Symbolic Logic*. He held chairs in both mathematics and philosophy at UCLA. He was first a mathematician, logician and philosopher, and then a historian of logic. 'Church's theorem' is that no decision procedure exists for arithmetic, and the well-known 'Church's thesis' is the hypothesis that a function is recursive if and only if it is effectively computable. Photograph by kind permission of Paul R. Halmos

Clavius, Christophore (Bamberg 1537 – Rome 1612)
Entering the Jesuit order in 1555, Clavius studied at the university of Coimbra, Portugal. He taught mathematics, first at the Collegio Romano, then at the University of Rome. He was an excellent teacher who produced two highly esteemed books *Arithmetic* and *Algebra*. His ma-

jor work was a richly annotated edition of Euclid's *Elements*. Under the direction of Pope Gregory XIII, Clavius participated in the 1582 reform of the calendar.

Collins, John (Wood Eaton, near Oxford 1625 – London 1683)
Following the death of his father, Collins abandoned his education and was briefly apprenticed to an Oxford bookseller. In 1642 he went to sea. On his return to London, seven years later, he taught mathematics and then occupied a number of minor government posts. His written works, *Merchants accompts, Decimal Arithmetick, Geometricall Dyabling* and *Mariners Plain Scale* indicate his areas of interest. In 1667 Collins became a member of the Royal Society and promoted the publishing of scientific works, in particular the works of Barrow and Wallis. He was in correspondence with numerous authors and Barrow called him the 'English Mersenne'

Coriolis, Gustave de (Paris 1792 – 1843)
Descendant of an old Provencal family of jurists, ennobled in the 17th century, Coriolis was educated at the École Polytechnique. He spent several years in the East of France as a civil engineer. Because of his fragile health he took up a post of tutor at the École Polytechnique. In 1829, Coriolis was appointed to the chair of mechanics at the École Centrale des Arts et Manufactures. In 1836 he succeeded Navier as professor of applied mechanics at the École des Ponts et Chaussées, and took his place in the Académie des Sciences. In 1838, Coriolis became director of studies at the École Polytechnique. He is remembered by the name given to the (apparent) Coriolis Force.

Cotes, Roger (Burbage 1682 – Cambridge 1716)
Originally going to Leicester School, Cotes showed such flair for mathematics that his uncle took him into his own home to supervise his studies. He later went to St. Paul's School, London and then to Trinity College, Cambridge. In 1706 he was appointed professor of astronomy and natural philosophy. In 1711 Cotes became a member of the Royal Society. He died at the age of 33. Aside from his works on astronomy Cotes published just one book on mathematics, *Logometria* (1714), where he deals with, in particular, logarithms to the base *e*, and where he evaluates the surface of revolution of an ellipsoid. The extent of his research on integration is evident from his posthumous writings. Photograph by kind permission of Cambridge University Press.

Cramer, Gabriel (Geneva 1704 – Bagnols sur Cèze 1752)
The son of a doctor, Cramer studied at Geneva. At the age of twenty, he shared the chair of mathematics at the Calvin Academy with Calandrini: both advocated the use of French instead of Latin. From 1727 to 1729 Cramer travelled to Basle, England, Leyden, and to Paris: there he met the major thinkers of his day. He was in correspondence with numerous scientists and participated in the publishing of the works of Jean and of Jacques Bernoulli. Cramer was active in political life. His most significant scientific contribution was his *Introduction à l'analyse des lignes courbes algébriques*.

D'Alembert, Jean le Rond (Paris 1717 – 1783)
Natural born son of the Chevalier Destouches and Mme de Tencin, d'Alembert studied law, medicine and mathematics at the Collège des Quatre-Nations. In 1741, he became a member of the Académie des Sciences. He published a *Traité de dynamique* (1743), a *Traité de l'équilibre et du mouvement des fluides* (1743) and a *Théorie générale des vents* (1745). In the 1750s he collaborated with Diderot in the production of the *Encyclopédie*, taking responsibility for the mathematics section, and he wrote the preliminary *Discours*. After 1760, he published eight volumes of *Opuscules mathématiques*.

Dedekind, Julius Wilhelm Richard (Braunschweig 1831 – 1916)
Dedekind studied at Göttingen where he received his doctorate under the supervision of Gauss. He was appointed Privatdozent at Göttingen and became friends with Dirichlet and Riemann. He taught at Zurich and finally in his native town of Braunschweig, preferring family life to well-deserved honours. From his mathematical education he developed an interest in the foundations of mathematics: he wrote *Stetigkeit und irrationale Zahlen* (Continuity and Irrational Numbers) (1872) and *Was sind und was sollen die Zahlen?* (The Nature and Meaning of Numbers) (1888). Dedekind played an important role in the development of the Cantorian idea of sets and his work on the algebraic theory of numbers is fundamental.

De Forest, Lee (Council Bluffs 1873 – Hollywood 1961)
An American engineer, De Forest obtained his Ph.D. from Yale and then worked for the US Army and the US Navy. He was a prolific inventor, registering over three hundred patents. De Forest's primary interest was in radio-telegraphy. In 1906 he invented the triode valve and participated in the development of electronics, vital for the invention of the modern computer. His contribution to linking sound with pictures was highly important.

De Lagny, Thomas Fantet (Lyon 1660 – Paris 1734)
After studying with the Jesuits in Lyon and then the Faculty of Law at Toulouse, de Lagny was appointed professor at Rochefort, then pensionnaire at the Académie Royale des Sciences. He collaborated with l'Hospital on the approximation of irrationals, which was the subject of his first publication. De Lagny was essentially a calculator, devising mathematical tables and, in 1717, working out the first 127 decimals of π. It is told how, in a comatose state on his death bed, he answered a mathematical question put to him by Maupertuis before finally extinguishing.

Delambre, Jean-Baptiste Joseph (Amiens 1749 – Paris 1822)
After attending the local school at Amiens, Delambre continued his studies in Latin and history at the *collège* at Plessis. He lived a poor life and took a post as tutor in Compiègne. In 1771 he became tutor to the son of Geoffrey d'Assy (*receveur général des finances*) in Paris and was recognised there by Lalande, who took him as an assistant. From 1788, Delambre worked with Méchain on measuring the length of the terrestrial meridian from Dunkirk to Barcelona. He became a member of the Académie des Sciences and of the Bureau des Longitudes. In 1807 he succeeded Lalande to the chair of astronomy at the Collège de France.

De La Vallée-Poussin, Charles (Louvain 1866 – 1962)
Unable to stomach the teaching of philosophy at the Jesuit college at Mons, De la Vallée-Poussin turned to technology and then to mathematics. In 1891 he became assistant to Gilbert at the University of Louvain and succeeded him in 1892. He remained there for the whole of his life, except for the years 1914-1918, when he taught at the Sorbonne and the Collège de France. Besides his research into the distribution of prime numbers, De la Vallée-Poussin published a number of works, among which were his *Cours d'Analyse infinitésimale* and his *Leçons sur l'approximation des fonctions d'une variable réelle*. Photograph by kind permission of Jean Mawhin and the Université Catholique de Louvain.

De Morgan, Augustus (Madura 1806 – London 1871)
Born in India, De Morgan came to England at the age of seven months. He was privately tutored in Latin, Greek and Hebrew and entered Trinity College, Cambridge in 1823. After first considering a career in medicine or law, he accepted the newly created chair of mathematics at University College, London. He resigned in 1831 in principle against the unfair dismissal of the professor of anatomy. He returned in 1836 and remained there until 1866. De Morgan exercised considerable influence, both through his teaching and through his various works dealing with logic, arithmetic, algebra, trigonometry, probability and the history of mathematics. His name is remembered by De Morgan's Law in the algebra of sets.

Descartes, René (La Haye en Touraine 1596 – Stockholm 1650)
Descartes began his education under the Jesuits at La Flèche where he got to know Mersenne, some seven years his senior. Descartes studied mathematics in Paris under Mydorge and Mersenne. After experience of army life, he settled in Holland in 1628. In 1649 he accepted an invitation from Queen Christiana of Sweden to come to Stockholm where he died from pneumonia soon after his arrival. The unity of Descartes' philosophy resides in the supremacy of reason, which underlies his work on optics, physiology and mathematics. In his *Geometry* he lays the foundations of analytical geometry.

Dinostratus (Ath ns *fl.* 350 BC)
According to Proclus "Amyclas of Heracles, one of the friends of Plato, Menaechmus, a pupil of Eudoxus, but in manner a partisan of Plato, and Dinostratus, brother of Menaechmus, brought even greater perfection to the whole of geometry". Pappus links Dinostratus with Nicomedes and other mathematicians who used the quadratix to find the area of a circle. This curve had been invented by one Hippias, probably Hippias of Elis. The construction is described by Pappus.

Diophantus of Alexandria (*fl.* 250)
Practically nothing is known of the life of Diophantus, and his *Arithmetica*, containing thirteen books, comes to us through numerous translations and interpretations. Recently four of the books have been discovered in a 10th century Arab translation, the first of which has the title *Fourth book of the work of Diophantus on squares and cubes translated from the Greek into Arabic by Qusta ibn Luga of Baalbeck*. The *Arithmetica* is not a work on the theory of arithmetic in the Pythagorean sense; it deals instead with calculations and Diophantus solves numerous problems, most often indeterminate ones.

Dirichlet, Gustav Peter Lejeune (Duren 1805 – Göttingen 1859)
From the age of twelve, Dirichlet would use his pocket money to buy mathematics books. He studied at Bonn, and then at the Jesuit College in Cologne. As tutor to the family of General Foy in Paris, he was able to take courses at the Collège de France and at the Faculté des Sciences. He held chairs at Breslau, Berlin and Göttingen, where he succeeded Gauss in 1855. In 1831 he became a member of the Berlin Academy of Sciences. He worked on the algebraic theory of numbers, analysis and mathematical physics.

Eratosthenes of Cyrene (Cyrene *c.* 276 BC – Alexandria *c.* 195 BC)
At the age of thirty, Eratosthenes left his home town of Cyrene to go to Alexandria, where he took over responsibility for the famous library. He was at the same time tutor to the son of Evergetes and in favour at the royal court. Author of works of geography, philosophy, grammar, chronology and mathematics, not to mention his poetry, he was one of the greatest scholars of his time. Cleomedes tells how Eratosthenes had measured the circumference of the Earth and Nicomachus of Gerasa credits him with sieve that bears his name.

Euclid (Alexandria *fl.* 300 BC)
Little is known of the life of Euclid. Some historians even consider that the works which bear his name are the fruit of collective effort. The Euclidean corpus was written between the death of Plato (347 BC) and the birth of Archimedes (287 BC). Throughout two millennia the *Elements* have constituted the bible of mathematics *par excellence*. The *Elements* differ significantly from Babylonian and Egyptian works, both by the primacy given to geometry and in the use of proof based on deductive reasoning.

Euler, Leonhard (Basle 1707 – St. Petersburg 1783)
Euler learnt his mathematics from his father, a Pastor who was passionately keen on science.
Euler, originally destined to study theology, followed courses given by Jean Bernoulli at the
University of Basel and became a doctor of philosophy. From 1726, he taught physics and
mathematics at the St. Petersburg Academy of Sciences. In 1738 he lost the sight of his right
eye. In 1741, Euler moved to Berlin where, for five years, he taught mathematics. Following a
dispute with the King of Prussia, he returned to St. Petersburg. He became blind soon afterwards but still continued to write.

Fatou, Pierre (Lorient 1878 – Pornichet 1929)
Educated at the École Normale Supérieure, Fatou received his doctorate in 1907 and was appointed to the Paris Observatory. He worked on the determining the absolute positions of
stars and planets, on instrumental constants and on measurements of twin stars. In mathematics, Fatou made important contributions to Taylor's series, Lebesgue integrals, iterations
of rational functions and functions of a complex variable.

Fermat, Pierre de (Beaumont de Lomagne 1601 – Castres 1665)
As a merchant in leather, Fermat's father was of sufficient means for his son to be able to attend the University of Toulouse. Fermat then went to Bordeaux, where he became acquainted
with the work of Viète. He purchased a seat in the Toulouse parliament and, from 1648, was a
member of the *Chambre de l'Edit* at Castres. His work on the theory of numbers was little
known until it was discovered by Euler. Fermat was a precursor in a number of fields: probability, infinitesimal calculus, analytical geometry and optics.

Fibonacci or Leonardo of Pisa (Pisa *c.* 1170 – after 1240)
The father of Fibonacci, as Secretary to the Republic of Pisa, was sent to Bougie, Algeria,
where Leonardo received an excellent mathematics education. In particular, he learnt the Indian calculation methods based on the decimal system. Following his education, Fibonacci
extended his knowledge through travels in Egypt, Syria, Byzantium, Sicily and Provence. On
his return to Pisa, he worked on his *Liber abaci*, the *Practica Geometriae* and the *Liber quad-
ratorum*.

Fontenelle, Bernard Le Bovier (Rouen 1657 – Paris 1757)
At the age of 7, Fontenelle was sent to the Jesuit College at Rouen, where his uncles Pierre and
Thomas Corneille had studied before him. Fontenelle had a brilliant mind, frequenting the
salons and receiving many academic distinctions. In 1697, he was appointed permanent secretary to the Académie des Sciences, of which he wrote a history in forty volumes. His *Entre-
tiens sur la pluralité des mondes* was a brilliant work of popular science. He probably wrote
Éléments de géométrie de l'infini after reading the preface to l'Hospital's *Analyse des infini-
ments petits*. Fontenelle died a centenarian.

Forcadel, Pierre dit de Béziers (Béziers ? – Paris 1576)
Little is known of the life of Forcadel. For some time he worked as a pharmacist. He travelled
to Rome and other cities in Italy. Through Ramus, to whom he explained Euclid, he obtained a
chair of mathematics at the Collège de France. Forcadel produced French translations of the
first books of Euclid's *Elements* and parts of the works of Proclus, Archimedes and other
authors. His books on arithmetic were important books to appear in France in the 16th century: *Arithmétique par les gects* (1558) and *Arithmétique entière et abrégée* (1565).

Fourier, Joseph (Auxerre 1768 – Paris 1830)
Fourier was a fatherless orphan. In 1789, he taught in Auxerre. Successor to Lagrange at the
École Polytechnique, he went on the expedition to Egypt with Monge. On his return to France, he
was appointed prefect of the *département* of Isère. In 1817, Fourier was elected a member of

the Académie des Sciences, of which he became the permanent secretary in 1822. In his *Théorie analytique de la chaleur*, Fourier introduced the series and transformations that bear his name. He also made contributions to the solution of algebraic equations and to statistics.

Frege, Gottlob (Wismar 1848 – Bad Kleinen 1925)

Frege was educated at the universities of Wismar, Jena and Göttingen, where he presented, in 1873, his doctorate *On the geometric representation of imaginary figures in the plane*, followed in 1874 by his memoir *Methods of calculation based on an extension of the concept of magnitude*. His first work of logic *Begriffschrift* (1879) marks the beginning of mathematical logic and was instrumental in his obtaining the post of professor of mathematics at Jena. In the *Foundations of Arithmetic* (1884), Frege gives the first logical definition of a cardinal number.

Frénicle de Bessy, Bernard (Paris 1605 – 1675)

Councillor of the Cour des Monnaies in Paris, Frénicle was an amateur mathematician, in correspondence with the greatest minds of his time, including Descartes, Fermat, Huygens and Mersenne. In 1666, he became a member of the Académie des Sciences. His interests were in physics, mathematics and astronomy. His *Traité des triangles rectangles en nombres* was published posthumously in 1676. Other works by his appear in the *Mémoires de l'Académie Royale des Sciences*, notably "Des quarrez ou tables magiques".

Gaṇesá (Nandod 1507 – after 1554)

Gaṇesá was born into a Brahmin family of astronomers and astrologers. He was taught mathematics and astronomy by his father. It would appear that Ganesá never left his birthplace and that he lived at least until 1554, the date of one of his last works on astronomy or astrology. His *Grahalaghava* was very popular and many commentaries were written on it. In 1554, Gaṇesá published an important commentary on the *Lilāvatī* of Bhāskara.

Gauss, Carl Friedrich (Braunschweig 1777 – Göttingen 1855)

Of humble origins, Gauss went to the local school where his exceptional brilliance brought him to the attention of the Duke of Brunswick who then financed the rest of his education. He went to the Collegium Carolinum in Braunschweig (Brunswick) in 1792, and the University of Göttingen from 1796 to 1800. He obtained his doctorate in 1799, and in 1807 was appointed as director of the Observatory in Göttingen. Often referred to as the 'prince of mathematics' Gauss worked on the theory of numbers, algebra, analysis, geometry and probability, but also in many other disciplines, such as astronomy, mechanics, geodesy and magnetism.

Genaille, Henri (*d.* 1903)

Genaille was a railway engineer for the Tours region of France. He invented a number of ingenious calculating instruments, revolutionary in their time, but now forgotten. He also devised numerous rules for multiplication, division, and calculating interest, as well as a curious "peg rule" for determining whether numbers of the form $2^n - 1$ are prime.

Gerbert d'Aurillac (Auvergne *c.* 933 – Rome 1003)

Educated at the Abbey Saint-Géraud d'Aurillac, Gerbert was noticed by Borel, the Count of Barcelona, who took him to Spain where he had the opportunity of learning Arabic science. Becoming cathedral schoolmaster at Rheims, he was considered the greatest mind of his time. Among his pupils, was the future king of France, Robert the Pious, and the Holy Roman Emperor Otto II entrusted Gerbert with the education of his son, the future Emperor Otto III. Gerbert became Archbishop of Rheims, then of Ravenna, and was elected Pope in 999, taking the name Sylvester II. Of an open minded disposition, he engaged in important scientific correspondence.

Gerling, Christian Ludwig (Hamburg 1788 – Marburg 1864)
Doctor in philosophy, Gerling taught mathematics at the Gymnasium in Kassel from 1812 to 1817, and was then appointed professor of mathematics, astronomy and physics at the University of Marburg. His work was mostly in astronomy. With Bessel, Olbers, Schumacher and von Humbold, Gerling belonged to a group of collaborators who were friends or correspondents of Gauss.

Gödel, Kurt (Brno 1906 – Princeton 1978)
Gödel was born in Brno, the former capital of Moravia, where his German speaking family owned a small textile factory. Originally intending to study physics at the University of Vienna in 1924, he took courses by Furtwängler in number theory and became interested in mathematics. With Hahn, one of his professors, he attended the Philosophy Circle in Vienna. In 1929 he became an Austrian citizen and with his incompleteness theorem he was appointed privatdozent, in 1933, at the University of Vienna. Fleeing the Nazi regime, he went to Princeton where he had previously visited the Institute for Advanced Study. Besides logic, Gödel worked on physics and philosophy of mathematics.

Goldbach, Christian (Königsberg 1690 – Moscow 1764)
Goldbach studied mathematics and medicine at Königsberg. He made extensive journeys in Europe, where he met the major mathematicians of the day: Leibniz, the Bernoullis, de Moivre, etc. In 1725, Goldbach was appointed professor of mathematics and historian to the St. Petersburg Imperial Academy, becoming director in 1737. At the same time, Goldbach was politically active, and in 1742 he was appointed Minister of State for foreign affairs. Goldbach's mathematical works dealt with analysis and arithmetic: more than 200 years after his death, the Goldbach hypothesis still defies mathematicians.

Grassmann, Hermann (Szczecin (Stettin) 1809 – 1877)
Grassman was brought up in a profoundly religious and cultivated atmosphere: his father, a one time Protestant minister, was a teacher of mathematics and physics at the Szczecin Gymnasium. The young Grassman studied variously theology, languages, mathematics and physics, which allowed him later to teach these subjects. He played an active part in the 1848 revolution. His *Theory of Extension* (1844) was considered too abstract and confused, and he was unable to obtain a university post. The Leopoldina Academy elected him to membership in 1864 for his achievements in physics rather than in mathematics. Towards the end of his life, he devoted himself to linguistics.

Gregory, James (Drumoak, near Aberdeen 1638 – Edinburgh 1675)
Gregory received his first lessons from his mother. Later he was sent to Aberdeen, then to London and to the University of Padua, where he learnt mathematics, mechanics and astronomy. Gregory returned to London, where he was elected a member of the Royal Society. He became professor of mathematics at St. Andrews, then at Edinburgh. He invented a reflecting telescope, the so-called Gregory telescope, and he originated the photometric method for estimating the distance between stars. In mathematics, he gave the series for the tangent and the arctangent.

Halley, Edmund (London 1656 – Greenwich 1743)
Son of a rich city merchant, Halley carried out a number of voyages of study as a young man. He became interested in astronomy at an early age. A member of the Royal Society, he edited its *Philosophical Transactions*. He was in correspondence with most of the learned men of his time. He gave financial support to the publication of Newton's *Principia*. In addition to his work in mathematics and geophysics, he is most remembered for his prediction of the return of a comet, in December 1758 (fifteen years after he had died), which has since carried his name.

Harriot, Thomas (Oxford *c.* 1560 – London 1621)
After Oxford University, Harriot became a cartographer and expert in maritime navigation for Sir Walter Raleigh, who was intent on establishing a colony in Virginia. In 1605 he was imprisoned for some time with his patron. From 1610 to 1613 Harriot carried out numerous astronomical observations. He was interested in optics and prisms. Harriot was an established mathematician, whose improved notation contributed to easier handling of algebraic equations.

Herbrand, Jacques (Paris 1908 – La Bérarde 1931)
Herbrand entered the École Normale Supérieure at the age of 17 and presented his doctorate four years later. After a year's military service, he obtained a Rockefeller scholarship to enable him to study in Germany: he worked with von Neumann in Berlin, Artin in Hannover and Emmy Noether in Göttingen. On holiday in the Alps, he suffered a fatal accident while climbing. Despite a short period of mathematical activity, Herbrand made important contributions to mathematics, notably in mathematical logic (theory of proof) and in modern algebra (class field theory).

Hermite, Charles (Dieuze 1822 – Paris 1901)
Hermite entered the École Polytechnique in 1842 but was refused admittance the following year on account of a congenital deformation of his right foot. He therefore turned to teaching: École Polytechnique, then Faculté des Sciences de Paris. He was elected to the Académie des Sciences in 1856. He worked on elliptical and hyperbolic functions, quadratic forms and algebraic number theory, approximation of functions and the transcendence of *e*. Hermite exercised a considerable influence over contemporary mathematicians, both through his publications and through correspondence.

Heron of Alexandria (*fl.* AD 62)
We know nothing of the life of Heron. A certain number of works in Greek, Arabic and Latin bearing his name have come down to us. These include *Mechanics*, the *Pneumatica*, *On the Dioptra*, which deal with mathematics and physics, and the *Metrica*, which is a work for engineers. There are also geometrical works as well as works on constructing various machines. In *Metrica* Heron breaks with the Greek tradition of associating numbers with geometrical magnitudes, preferring to calculate with numbers directly.

Hilbert, David (Königsberg 1862 – Göttingen 1943)
Hilbert studied first in his home university of Königsberg, then in Heidelberg, Leipzig and Paris. Appointed privatdozent at Königsberg, he went on to obtain, in 1895, a chair at Göttingen, which he held till his retirement in 1930. Hilbert made significant contributions to many branches of mathematics: algebraic forms, algebraic theory of numbers, foundations of geometry, analysis, theoretical physics and foundations of mathematics. The twenty three problems that he put before the International Congress of mathematicians in Paris in 1900 continue to stimulate research.

Hippias of Elis (*c.* 425 BC)
Certainly a contemporary of Socrates. None of his works have come down to us. Proclus credits him with the discovery of the quadratrix, the curve used for trisecting an angle. Hippias was a sophist, that is someone who was a paid itinerant expert on various subjects including public speaking, grammar, ethics, literature, mathematics and elementary physics.

Horner, William George (Bristol 1786 – Bath 1837)
Horner was educated at Kingswood school where he later became headmaster. His most significant contribution to mathematics was what is now called Horner's method for solving algebraic equations, published in 1819. This technique became widely known in England by its

inclusion in their own works by Young and De Morgan. Horner type methods appear in earlier Chinese writings and the algorithm had already been published in 1804 by Ruffini.

Huygens, Christian (The Hague 1629 – 1695)
Up to the age of 16, Huygens was educated by his father and private tutors. He then studied law and mathematics, under Van Schooten at the University of Leyden, and under Pell at the Orange College in Breda. Thanks to an allowance from his father, he was able to devote himself entirely to science. When the Académie des Sciences de Paris was set up, he accepted a paid post there, but on Colbert's death, Huygens returned to The Hague. He was an astronomer and physicist, and in mathematics he produced the first complete account of probability.

Hypatia (c. 370 – Alexandria 415)
Daughter of Theon of Alexandria, and numbered among the Alexandrian Neo-Platonists, Hypatia is the only known woman of Antiquity to have worked in the exact sciences. Her works, which have been lost, were mathematical and astronomical commentaries, notably on Appolonius and Diophantus. Synesius of Cyrene figured among her disciples. She was said to have been mistress of the whole of pagan science, and by her eloquence and authority to have attained such influence as to be considered a threat to Christianity. In 415 she was put to death by a fanatical mob.

Ibn al-Bannā, Abu l-ʿAbbās Aḥmad ibn ʿUthmān al-Azdī al Murrākushī
(Marrakesh 1256 – 1321)
Ibn al-Bannā studied in Marrakesh. He spent most of his life teaching mathematics, as well as other subjects. He appears to have written over 80 works on subjects as diverse as mathematics, astronomy, linguistics and grammar. Among the thirteen surviving works are: the *Summary of the Operations of Calculation*, the *Book of the Foundations and Preliminaries of Algebra* and *Lifting the Veil on the Operations of Calculation*, a philosophico-mathematical treatise which contains results of combinations.

Ibn al-Haytham, Abū al-Ḥasan ibn al-Ḥasan (Basra 965 – Cairo c. 1040)
Biographical references are not consistent. Better known under the name of Alhazen, he wrote several books on logic, ethics, politics, poetry, music and theology. Ibn al-Haytham was above all recognised for his contributions to optics, astronomy and mathematics. In his *Treatise on Optics* he analyses astronomical refraction and also gives an exact description of the workings of the eye. Several mathematical works have survived, in particular a treatise on geometric curves.

Ibn al-Majdī (1359 – Cairo 1447)
A mathematician and astronomer, al-Majdī wrote two dozen books, mostly on astronomy. The most important mathematical work is the *Book of Substance*, a voluminous commentary on al-Bannā's *Summary of the Operations of Calculation*.

Ibn Qunfudh or Aḥmad ibn Ḥasan ibn ʿAlī ibn al-Khaṭīb al-Qasanṭīnī
(c. 1330 – Constantine 1407)
Ibn Qunfudh came from a family of scholars and jurists; his grandfather taught the Hadith (the corpus of traditions about Mohammed) at Constantine. Qunfudh was himself an imam, a scholar, a judge and a specialist in the Hadith. In 1357 he went to Morocco, where he improved his education among the best scholars of the day. In 1374 he went to Tunis, and finally returned to Constantine, where he practised as mufti (expounder of Muslim law) and judge. Exiled in 1401, he remained in disgrace until his death. In addition to writings on law and history, Qunfudh published works on astronomy and mathematics. These last were commentaries on manuals or mathematical poems of the 12th and 13th centuries.

Ibrāhīm Ibn Sinān Ibn Thābit Ibn Qurra (Baghdad 908 – 946)
Ibrāhīm ibn Sinān was the son of Sinān Ibn Thābit, a physicist, astronomer and mathematician of repute, and the grandson of Thābit ibn Qurra. Despite a short career, he left an important work covering several areas of research including tangents to circles, quadrature of the parabola, apparent movement of the sun, solar time, the astrolabe, as well as other astronomical instruments. His *Al-Tahlīl wa'l Tarkib* (analysis and synthesis) can be considered to be the first Arabic work on the philosophy of mathematics.

Jacobi, Carl Gustav (Potsdam 1804 – Berlin 1851)
Son of a Jewish banker, Jacobi received his early education from his mother and entered the Potsdam Gymnasium at the age of 12, where he excelled in Greek, Latin, history and mathematics. After submitting his thesis to the University of Berlin, he had to convert to Christianity to obtain a university post. In 1827 he was invited to the University of Königsberg where he instituted the practice of research seminars. In 1843 illness obliged him to leave and go to Italy. On his return to Germany, he accepted a chair at Berlin. Jacobi's research concerned elliptic functions, analysis, number theory, geometry and mechanics.

Jacquard, Joseph Marie (Lyon 1752 – Oullins 1834)
A weaver like his father, Jacquard started to design mechanisms for weaving thread automatically in 1790. He constructed a first machine, which he patented in 1801. This was improved and a new model built in 1806. Perforated cards were used to control the design. One Jacquard loom could replace five manual weavers, which explains the violence of the opposition of the workers to its introduction.

Jevons, William Stanley (Liverpool 1835 – Hastings 1882)
Jevons interrupted his study at University College, London in order to earn his living in Australia. His interests were in meteorology, botany and geology. He returned to England to complete his education and obtained a post as part-time professor of logic and political economy at Queen's College, Liverpool, in 1865. The following year he accepted the chair of logic and philosophy at Manchester. In 1872 he became a member of the Royal Society. In 1876 he moved to University College, London, but four years later retired in order to devote himself to writing. Jevons wrote on logic, economics and the philosophy of science.

Jones, William (Anglesey 1675 – London 1749)
Jones was tutor to wealthy London families. In 1702 he published a work on navigation and in 1706 his *Synopsis palmariorum matheseos*, a work which attracted the attention of Halley and Newton. It was in this work that Jones introduced the symbol π for the ratio of the circumference of a circle to its diameter. He was one of the few scientists who were allowed access to Newton's manuscripts.

Julia, Gaston (Sidi Bel-Abbes 1893 – Paris 1978)
After being educated at the Lycée in Oran, Julia went to Paris to study at the École Normale Supérieure (1911). Mobilised in 1914, he received a severe facial injury, from which he suffered for the rest of his life. His thesis, submitted in 1917, was on number theory. In 1918, Julia published an important paper on the iteration of rational fractions. In 1934 he entered the Académie des Sciences and two years later became professor at the École Polytechnique.

Kausler, Christian (Tübingen 1760 – Stuttgart 1825)
Kausler taught French at the Karlsschule in Stuttgart and was then appointed director of the school. He wrote many books, on chemistry as well as arithmetic. He provided a solution to the problem of the decomposition of non-square integers into two, three or four squares. He also indicated improved methods for finding the factors of integers.

Kepler, Johannes (Weil der Stadt 1571 – Regensburg 1630)
Kepler was educated at Tübingen and then taught mathematics at Graz. He was forced to resign because of his Protestant faith. He went to Prague to study under Tycho Brahe in 1597 and in 1601 succeeded him as astronomer to Emperor Rudolph II. He was one of the first supporters of Copernicus' heliocentric theory of the solar system, and noted that planetary orbits were elliptical, a result which appears in *Astronomia Nova* (1609). His *Rudolphine Tables* (1627) was the first almanac to be based on his three laws. Kepler's scientific thought was characterised by a deep sense of order and harmony tied to his religious beliefs. Despite holding official positions, he died exhausted and destitute.

Kleene, Stephen Cole (Hartford 1909 – Madison 1994)
Photograph by kind permission of C. Smorynski.

Kraïtchik, Maurice Borisovitch (Minsk 1882 – Brussels 1957)
Kraïtchik presented his doctorate in physical sciences and mathematics to the University of Liège. In 1915 he worked as an engineer in Brussels. From 1921 to 1941, and again from 1946 to 1957 he directed the mathematical sciences section of the Institut des Hautes Études de Belgique. From 1941 to 1946 he was associate professor at the New School for Social Research in New York. Kraïtchik's main interest was in number theory.

Kummer, Ernst Eduard (Sorau 1810 – Berlin 1893)
Educated at the Gymnasium in Sorau and the University of Halle, Kummer chose mathematics instead of his intended course in Protestant theology. He considered mathematics to be a sort of preparatory science for the study of philosophy. He taught at the Gymnasium at Legnica (Liegnitz) where both Kronecker and Joachimsthal were his students. Recommended by both Dirichlet and Jacobi, Kummer was appointed to the University of Breslau. When Dirichlet left Berlin, Kummer succeeded him and, with Weierstrass, founded the first seminar of pure mathematics. Kummer's work was on the theory of functions, number theory and geometry.

Kutta, Martin Wilhelm (Pitschen 1867 – Munich 1944)
Kutta was educated at Breslau (1885–1890) and Munich (1890–1894). He spent two years at Cambridge, from 1898. After his doctorate he taught at the Technische Hochschule, Munich (1894–1897) and from 1899 he occupied a number of posts in Universities and institutions of higher education in Munich, Jena, Aachen and Stuttgart. Kutta's mainly worked on numerical integration of differential equations and in the fields of hydrodynamics and aerodynamics. He also wrote a history of geometry.

Lagrange, Joseph-Louis (Turin 1736 – Paris 1813)
From a Touraine family, Lagrange was appointed in 1755 as professor of geometry in the Royal Artillery School at Turin. In 1766 he accepted the post of Director of Mathematics at the Berlin Academy of Science and, in 1787, he became pensionnaire of the Académie des Sciences in Paris. Finally, he taught mathematics at the École Normale in year III of the Revolution and at the École Polytechnique from 1794 to 1799. Lagrange derived elegant results in number theory and published an important memoir *Sur la résolution des équations numériques*. His treatment of isoperimetric problems laid a foundation for the calculus of variations, which he applied in particular to mechanics.

Lambert, Johann Heinrich (Mulhouse 1728 – Berlin 1777)
A self-taught mathematician who earned his living as a tutor, Lambert was in correspondence with Euler from 1758, and became a member of the Berlin Academy in 1765. He wrote numerous works on philosophy, astronomy, photometry, cartography and population statistics. He is famous for a treatise on perspective, for proving that e and π are irrational, for his conjecture that e is transcendent, and for his intuitive speculation of the existence of non-Euclidean geometries. A unit of luminance (the lambert) is named after him.

Laplace, Pierre Simon Marquis de (Beaumont sur Auge 1749 – Paris 1827)
From a poor background, Laplace attended the local military school. In 1783 he became examiner to the artillery corps and in 1785 was elected to the Académie des Sciences. At the Revolution, he participated in the organisation of the École Polytechnique and the École Normale. His reputation survived the fall of successive regimes: he was for six weeks Napoleon's Minister of the Interior, later became Chancellor of the Senate, was ennobled both under the Empire and by Louis XVIII, and was elected President of the Académie Française. Laplace exercised considerable authority over the scientific community. The core of his work lay in the application of analysis to mechanics and probability.

Le Verrier, Urbain Jean Joseph (Saint Lô 1811 – Paris 1877)
Of humble origins, Le Verrier's father sold the family home to enable him to prepare for entry to the École Polytechnique. He became an engineer in the production of tobacco. Le Verrier went on to work on chemistry with Gay-Lussac. He was tutor in chemistry and then in astronomy at the École Polytechnique. In 1846 he entered the Académie des Sciences and in 1854, on the death of Arago, he became director of the Paris Observatory. Following the discovery of the planet Neptune, predicted by Le Verrier, he held the chair of celestial mechanics and then of astronomy at the Sorbonne.

Lebesgue, Henri (Beauvais 1875 – Paris 1941)
Lebesgue came from a modest background and entered the École Normale Supérieure on a scholarship. He taught at the Lycée in Nancy and then at the Universities of Rennes, Poitiers and Paris. In 1931 he left the Sorbonne for the Collège de France; he also taught at Écoles Normales Supérieures (Ulm and Sèvres). For him, teaching was "to think aloud in front of one's pupils". Lebesgue was elected to the Académie des Sciences in 1932 and to the London Royal Society. His major contribution was to the theory of integration: the Lebesgue integral generalises the Riemann integral.

Legendre, Adrien-Marie (Paris 1752 – 1833)
From a comfortable background, Legendre had a relatively advanced scientific education from the Collège Mazarin. His modest personal fortune allowed him to devote himself entirely to learning. He taught at the École Militaire de Paris from 1755 to 1780. In 1783 he was elected to the Académie des Sciences. In 1894 he directed the Commission for public education. He was an examiner at the École Polytechnique from 1799 to 1815 and in 1813 he replaced Lagrange at the Bureau des Longitudes. His interests were primarily in analysis, number theory, geometry and mechanics.

Lehmer, Derrick Henry (Berkeley 1905 – 1991)
Lehmer's father was a mathematics professor at Berkeley and had a considerable influence on his son, particularly in encouraging him to work on arithmetic algorithms. After his Ph.D. Lehmer went to Stanford, Pasadena, Princeton and to Cambridge, England. From 1940 he was at Berkeley. He edited *Mathematics of Computation* and *Acta Arithmetica*. In 1928 he married Emma Trotskaia, with whom he wrote a number of articles. Photograph by kind permission of Paul R. Halmos.

Leibniz, Gottfried Wilhelm (Leipzig 1646 – Hannover 1716)
Leibniz initially studied law at Leipzig where his father was professor of morals. He left Leipzig for ever when he was denied a doctorate of law on the grounds that he was only 20; however he was immediately awarded the degree at Nürnberg where he was also offered a chair. Refusing this, he entered the service of the Elector of Mainz, for whom he travelled widely. He discovered the infinitesimal calculus, independently of Newton. His wide intellectual research into the different areas of philosophy, logic, languages and mathematics were stimulated by a desire to define a unifying universal language of thought. He died in solitude.

Li Chunfeng (602 – 670)
Li Chunfeng was a celebrated scholar, appreciated in his time for his multiple talents in esoteric arts, history, the calendar and computational techniques. In 656 he collaborated on the publication of the official Historical Annals and soon after the Emperor commanded him to compile and edit a series of manuals of ancient mathematics. This became the *Suanjing shishu* (Ten chapters on computational prescriptions) and were used in the "Upper Schools for the Sons of the State" throughout the Tang Dynasty (618 – 907).

Lindemann, Ferdinand (Hannover 1852 – Munich 1939)
Lindemann studied at Göttingen, Erlangen and Munich. He was a pupil of Klein. He travelled to Oxford, Cambridge, London and Paris, where he met Chasles, Bertrand, Jordan and Hermite. Lindemann taught at Freiburg, Königsberg and Munich. He supervised Hilbert's doctoral thesis. He was one of the founders of the German education system; he foresaw the development of seminars and lectures. The most original contribution by Lindemann was the proof in 1882 of the transcendence of π, thus putting an end to speculation on the quadrature of the circle.

Lipschitz, Rudolf (Königsberg 1832 – Bonn 1903)
Lipschitz started to study mathematics at the University of Königsberg at the age of 15. He went on to Berlin, where Dirichlet taught. Lipschitz taught at the Gymnasien in Königsberg and Elbing, and the Universities of Berlin, Breslau and Bonn. He mainly worked on number theory, Bessel functions, Fourier series, differential equations, analytical mechanics and potential theory. Lipschitz was also interested in the foundations of mathematics and in university mathematics education. Of particular note is his *Grundlagen der Analysis*.

Liu Hui (*fl.* 263)
Liu Hui was the foremost commentator on the oldest known Chinese mathematical work, the *Jiuzhang suanshu* (Calculation procedures in Nine Chapters). At first sight a merely literary commentary, but it contains demonstrations intended to justify the methods presented without proofs in the original work.

Lucas, François-Edouard-Anatole (Amiens 1842 – Paris 1891)
Educated at the École Normale d'Amiens, Lucas was employed at the Paris Observatory. After serving as an officer in the Franco-Prussian war, he taught in Paris at the Lycée Saint Louis and the Lycée Charlemagne. In number theory, Lucas worked on tests for primality and factorisation problems. In 1876 he showed that the Mersenne number $2^{127} - 1$ is prime. Lucas produced original research in the fields of arithmetic and elliptical functions. His *Récréations mathématiques* is a classic of its genre.

Lukasiewicz, Jan (Lvov 1878 – Dublin 1956)
Lukasiewicz studied mathematics and philosophy at Lvov, now in the Ukraine. He obtained his doctorate in 1902 and *Habilitation* in 1906. He taught at Lvov and then travelled in Europe. From 1915 to 1939 he taught at the University of Warsaw, except for one year while he was with the Polish Ministry of Education. After the Second World War he left Warsaw for Münster, then Brussels and finally settled in Dublin where he was professor at the Royal Irish Academy. With Lesniewski he founded the school of Mathematical Logic at Warsaw which Tarski made famous. Lukasiewicz also renewed interest in classical logic.

Machin, John (1680 – London 1751)
An astronomer who became a member of the Royal Society in 1710 and secretary to the Society from 1718 to 1747. He sat on the commission appointed in 1712 to adjudicate on the priority dispute between Newton and Leibniz. In 1713 he became professor of astronomy at

Gresham College, a post he held until his death in 1751. In 1706 he had computed the first 100 decimals of π. Otherwise, his work mainly concerned lunar theory.

Maclaurin, Colin (Kilmodan 1698 – Edinburgh 1746)

Maclaurin was 12 when he entered the University of Glasgow, where Robert Simson taught. In 1717 Maclaurin was elected to the chair of mathematics at Marischal College, Aberdeen. On Newton's recommendation, he was appointed to a professorship at Edinburgh in 1725. Maclaurin organised the defence of Edinburgh during the 1745 Rebellion, and when the city fell to the Jacobites he was obliged to flee to York. In 1720 Maclaurin published *Geometrica organica* which contains many proofs of results stated by Newton. In 1742, he published *Treatise of Fluxions* in response to Bishop Berkeley's criticisms of Newton. Portrait by kind permission of the Scottish National Portrait Gallery.

Maupertuis, Pierre Louis Moreau de (St. Malo 1698 – Basle 1759)

Maupertuis went to Paris at the age of 16, where he began to study music, and then mathematics under Nicole. He was made a member of the Académie des Sciences in 1723. Following a visit to London he became an ardent defender of Newtonian mechanics. He directed the Lapland expedition, 1735 to 1737, intended to determine the length of a meridian degree. He was elected to the Académie Française in 1743 and, at the request of Frederick the Great, became president of the physical class in the Berlin Academy in 1746. Maupertuis had a deep interest in biology and the origin of languages, as well as mathematics.

Méchain, Pierre-François-André (Laon 1744 – Castellon de la Plana 1804)

As a child prodigy, Méchain attracted the attention of local dignitaries, who sent him to study at the Paris École des Ponts et Chaussées. Becoming a tutor, he came into contact with Lalande who procured him a post as hydrographer at the Dépôt des Archives de la Marine at Versailles. Méchain was also an accomplished astronomer and for this reason was made a member of the Académie des Sciences in 1742. He undertook geodesy measurements: the difference in longitude between Greenwich and Paris and the length of the meridian from Dunkirk to Barcelona, so as to determine the standard metre.

Mellema, Elcie Edouard Léon (Leuwarden c. 1552 – 1622)

The Dutch writer Mellema is only known through his works. In his *Arithmetica* of 1586, he describes himself as a citizen of Anvers and schoolmaster at Haarlem. In the two volumes of this work, Mellema follows his Dutch colleagues Gemma Frisius and Simon Jacob in solving quadratic equations by the method of double false position. He was also the author of a large and detailed *Dictionnaire ou Promptuaire Flamand-Français*. Mellema took the part of the French against Philip II.

Mercator, Nicolaus (Niklaus Kauffman) (Eutin 1619 – Paris 1687)

Niklaus Kauffman was taught by his father Martin Kauffman and then attended the Universities of Rostock and Copenhagen, staying in Copenhagen until 1653. He then went to London, where he was known by his Latin name Mercator, which he adopted from then on, and became one of the first members of the Royal Society. In 1683 Colbert invited him to draw up plans for the water system at Versailles. Mercator studied astronomy, cosmography, trigonometry and analysis. His most important work was *Logarithmotechnia* (1688) in which he gives the expansion of $\ln (1 + x)$.

Mersenne, Marin (Oizé 1588 – Paris 1648)

Educated at La Flèche from 1604 to 1609, he went on to the Sorbonne to study theology and, in 1611, entered the Minimes, a Franciscan order. In 1619 he returned to Paris, to the Convent de l'Annonciade, and apart from a few brief absences, he remained in Paris for the rest of his life. From 1623, Mersenne was in correspondence with the greatest minds of his time. In 1665 he founded the *Academia Parisiennis*. Mersenne was one of the first scientists to have a laboratory, including a 'physics room', and began the study of quantitative physics. His most important works were on acoustics.

Metius, Adriaen Anthonisz (Alkmaar 1543 - 1620)
Some historians connect Metius with the town of Metz. He was a Dutch cartographer and as military engineer he was responsible for the constructions of fortifications. He was both mayor of his native town and publisher of town maps. In addition, Metius was interested in astronomy. His son, of the same name, published his father's approximation to π, namely 355/113.

Moschopoulos, Manuel (14th century)
Two Greek linguists, uncle and nephew, had the same name of Manuel Moschopoulos and it is not easy to determine the precise author of the works attributed to them. This is particularly the case with the *Treaty on Magic Squares*, translated into French for the first time by La Hire in 1691. Tannery considers the author to have been Moschopoulos the Cretan, who was born in Crete *c*. 1392. His nephew was among the many Greeks who sought refuge in Italy following the fall of Constantinople in 1453.

Moulton, Forest Ray (Osceola County 1872 - Wilmette 1952)
Moulton started teaching when just sixteen at the school where he had been taught. One of his pupils was his own brother, Harold, who later became director of the Brookings Institute. Two years later he entered Albion College and obtained his Ph. D. in astronomy from Chicago University in 1894. Moulton held many administrative positions and was a pioneer of distance learning. In 1920 he founded the Society for Visual Education. He worked with Chamberlin on the origins of the Earth; their research was based on observations of a solar eruption during an eclipse in 1900.

Mourraille, Jean-Raymond-Pierre (Séon-St.-Henri 1720 - 1808)
A mathematician and astronomer, Mourraille became a member of the Marseille Academy in 1767 and its permanent secretary for sciences. He was elected to the Convention in 1792 but did not take up his seat. From November 1791 to April 1793 he was mayor of Marseille. Arrested in 1793, he was acquitted by the tribunal and thereafter remained in obscurity. He published the first part of a *Traité sur la résolution des équations en général*, which contained improvements to approximate numerical solutions of algebraic equations, but the work appears to have been overlooked.

Napier, John (Edinburgh 1550 - 1617)
Napier, Laird of Merchiston, near Edinburgh, was a theologian and amateur mathematician, best known for his invention of logarithms which he published in *Rhabdologiae* (1617). He attended St. Andrews University but left without taking his degree. He probably also studied abroad. As well as his work on logarithms, he contributed to the study of spherical trigonometry and the adoption of decimal notation. His theological writings defended the Protestant cause during the time of the greatest strife between Protestantism and Catholicism in Scotland.

Nekrasov, Alexander Ivanovich (Moscow 1883 - 1957)
Graduating from the Faculty of Physics and Mechanics at the University of Moscow in 1906, Nekrasov taught there from 1912 to 1917. He was professor and director of a number of major institutes: the Central Institute of Aerodynamics, the Sergo Orjonikidze Aviation Institute and the Mechanics institute of the Academy of Sciences of the USSR. Nekrasov was an outstanding contributor to mathematics applied to hydrodynamics and mechanical aeronautics. He also contributed to research on non linear integral equations.

Neville, Eric Harold (London 1889 – Reading 1961)
Neville first became interested in mathematics at William Ellis School. He went up to Trinity College, Cambridge in 1907. In early 1914, while teaching in India, he acted on Hardy's request to persuade Ramanujan to come to Cambridge. In 1919 Neville was appointed professor of mathematics at Reading University, a post which he held until 1954. His first works were on geometry and his most important work was on elliptic functions. Neville was also interested in the history and teaching of mathematics.

Newton, Isaac (Woolsthorpe 1642 – London 1727)
From a farming family, Newton was born prematurely, his father having recently died. He was a sickly child. Educated at Trinity College, Cambridge, he returned home during the plague years of 1665–66 where he produced his fundamental work on the calculus. Newton became Lucasian professor of mathematics at Cambridge in 1669, in place of Barrow. He left Cambridge in 1696 to take charge of the Royal Mint in London. He was a fellow of the Royal Society, then its president, a position he held until he died. During the later period of his life, he turned more and more to alchemy, mysticism and theology.

Nicomachus of Gerasa (*c.* 100)
Nicomachus, the best known of the Greek writers on arithmetic, was born in Gerasa in Palestine. The probable dates of his work come from indications given in the work of Nicomachus himself and other commentators. Nicomachus wrote a number of books, of which only two survive: *Enchiridion Harmonices* which gives details of the Pythagorean and Aristotelean schools of music, and *Introductio Arithmeticae* which is marked by a Pythagorean philosophy rather than Euclidean rigour.

Nīlakaṇṭha (Kuṇḍapura *c.* 1444 – after 1501)
Nīlakaṇṭha was a Nampūriti Brahmin. An astronomer of repute, he published a number of works in which he followed the system developed by Paramesvara, although he sometimes used different parameters. In 1501 he was requested by the religious leader of the Nampūriti to write a commentary on the Āryabhatiya, published by Āryabhata in 1499. Nīlakaṇṭha gives a particular method for computing the value of π.

Oldenburg, Henry (Bremen 1618 – London 1677)
Little is known of the life and career of Oldenburg. Educated at the Gymnasium in Bremen and the University of Utrecht, he then travelled in Europe, earning his living as a tutor. He came to England to attend Cromwell, in 1653, as diplomatic representative of the town of Bremen. Deciding to settle in England, Oldenburg became a member of the Royal Society in 1661, and became its secretary. In 1665 he founded the *Philosophical Transactions of the Royal Society*, the first purely scientific publication, which became the vehicle for the exchange of views between Britain and the Continent.

Ozanam, Jacques (Bouligneux 1640 – Paris 1717)
Ozanam came from a Jewish family that had converted to Catholicism. Destined for the priesthood, he preferred chemistry and mechanics to theology. Following his father's death, he went to teach mathematics, first in Lyon and then in Paris. He wrote a number of popular books on mathematics dealing with practical uses and recreations. The *Nouveaux éléments d'algèbre* (1702) was noted by Leibniz. Ozanam also produced a very full algebraic commentary on the *Arithmetica* of Diophantus, the manuscript being recently discovered in the library of the Turin Academy of Sciences.

Pacioli, Luca (Borgo San Sepulcro 1445 – Rome 1517)
At a very young age, Pacioli, also called Luca di Borgo, worked as a private tutor in Venice where, in 1470, he dedicated to his employer a book on algebra, since lost. There he entered the Franciscan order and became professor of mathematics at Perugia in 1475. He moved later to Florence, Pisa and Bologna. He met Leonardo da Vinci, who contributed magnificent illustrations for his friend's *On Divine Proportion*. Pacioli's most important work is the *Summa de Arithmetica*, a sort of encyclopaedia of all the mathematical knowledge known at the time.

Painlevé, Paul (Paris 1863 – 1933)
On leaving the École Normale Supérieure, Painlevé was undecided between a career in politics, engineering, or pure research. He chose to go to Göttingen to study under Schwarz and Klein and then returned to France to teach, first at Lille, then in Paris at the Faculté des Sciences, then the École Polytechnique, the Collège de France and the École Normale Supérieure. He did not neglect politics, being elected Deputy for Paris in 1910 and then occupying a number of ministerial posts. He became an early enthusiast of aviation and, with Wright and Farman, the holder of a flight duration record, albeit as a passenger.

Pascal, Blaise (Clermont-Ferrand 1623 – Paris 1662)
The young Pascal started his study of mathematics through reading Euclid's *Elements*. At the age of sixteen he took part in the activities of Mersenne's Academy. In 1640 he followed his father to Rouen, where the whole family converted to the austere Jansenist religious persuasion. Pascal returned to Paris in 1647 after falling ill. Here began the "worldly" period of his life which was rich in scientific activity. His *Essai sur les coniques* (1640) was written when he was only 17 and in 1645 he provided the first definitive description of a calculating machine.

Peano, Giuseppe (Cuneo 1852 – Turin 1932)
Peano's university career was entirely confined to Turin, where he had lived from the age of 12. Assistant, then suppliant, to professor Genocchi, Peano succeeded him to the chair of infinitesimal calculus, a post he held from 1890 to 1932. He also taught at the Military Academy in Turin (1886–1901). He was a member of the Turin Academy of Sciences. Peano's first works were on the differential calculus. From 1894 to 1908 he worked with a number of his pupils to produce his *Formulario Mathematico*, which used formalist language to describe logic, the foundations of arithmetic, analysis and algebra.

Peirce, Charles Sanders (Cambridge 1839 – Milford 1914)
Peirce said that he was raised in a laboratory: his father was professor of mathematics and natural sciences at Harvard. After an MA at Harvard (1862) he obtained a Sc.B, *summa cum laude*, in chemistry (1863). Peirce then turned to logic and methods. Feeling the need for practical experience, he took a post with the US Geodesic Service, where he remained until 1891. Despite the efforts of James, no permanent teaching post could be found for Peirce, and he taught logic at Harvard until retirement thanks to funds provided by his friends. A logician and philosopher, Peirce was a pioneer in the field of thought processes.

Pell, John (Southwick 1611 – London 1685)
Pell was admitted to Trinity College, Cambridge at the age of 13. He studied later at Oxford. He was competent in many languages: Latin, Hebrew, Arabic, Italian, French, Spanish and German. He became professor of mathematics at Amsterdam, then at Breda. Pell represented Cromwell in Switzerland. He took Holy Orders in 1661 and thereafter his time was divided between mathematics and his religious duties. He was a member of the Royal Society. Due to an error on Euler's part, the solution to the equation $x^2 - ay^2 = 1$ is falsely attributed to Pell.

Pellos, Francesco (Nice 14th century)
The facts surrounding the life of Pellos are few. He describes himself as a *gentleman* from Nice. His *Compendiom del abaco* (1460) was the first mathematics book to have been printed in the Occitan language. The Occitan manuscript arithmetic to be found in the Bibliothèque Nationale, ascribed to Pamiers and written about 1430, would have been a source for Pellos. In both these works we find the first acceptance of a negative solution in an extant work, either manuscript or printed.

Pépin, Jean-François Théophile (Cluses 1826 – Lyon 1904)
Pépin studied at the Petit Séminaire at Mélan, not far from his birthplace. After literary studies, he became a Jesuit in 1846 and dedicated the first years of his religious life to the study of philosophy and religion. He taught canon law at Aix, then at Lyon. Attracted by the exact sciences, he was sent by his superiors to Paris to study under Cauchy and Liouville. On his return he was appointed professor to the Collège St. Michel at St. Etienne. In 1873 he became editor of the *Revue des Études religieuses*. In correspondence with Boncompagni, Kronecker, Brocard and notably Lucas, Pépin was particularly interested in arithmetic.

Picard, Charles Émile (Paris 1856 – 1941)
Together with Poincaré, Picard was one of the foremost French mathematicians of his generation. He entered the École Normale Supérieure in 1874 and by 1877 he had graduated as a Doctor of Science. He taught at the University of Toulouse, the Sorbonne, the École Normale Supérieure and at the École Centrale. Member of the Académie des Sciences in 1889, Picard became its permanent secretary in 1917. He was also President of the Bureau of Longitudes and a member of the Académie Française. Picard's *Traité d'analyse* quickly became a classic text. Photograph by kind permission of CNRS-Éditions, Paris.

Poincaré, Henri (Nancy 1854 – Paris 1912)
Henri Poincaré's family belonged to the upper bourgeois society of Lorraine. At lycée, he won the mathematics prize at the Concours Général. He entered the École Polytechnique in 1873 and studied in the École des Mines. He worked in this area briefly while completing his thesis. In 1887 he was elected to the Académie des Sciences and, in 1908, to the Académie Française. A universal mathematician, he engaged in all branches of mathematics and physics. Poincaré's final writings were concerned with the philosophy of science: *La Science et l'Hypothèse*, *La Valeur de la Science*, *Science et Méthode* and *Dernières Pensées*.

Poisson, Siméon-Denis (Pithiviers 1781 – Paris 1840)
Being inapt and unwilling to become a surgeon, Poisson was sent to the École Centrale de Fontainebleau and then the École Polytechnique, where he later taught. In 1808 he was appointed astronomer to the Bureau des Longitudes and, in 1809, he became professor of mechanics at the Faculté des Sciences in Paris. He was elected to the Académie des Sciences in 1812 and to the Conseil Royal of the University in 1820. Poisson was the author of numerous works on mechanics, the probabilistic calculus and mathematical physics.

Post, Emil Leon (Augustow 1897 – New York 1954)
As a child Post was passionate about astronomy but the loss of his left arm at the age of 12 prevented him taking up a professional career in that field. Instead he turned to mathematics, and then logic. He obtained his PhD from the University of Columbia in 1920, then taught at Cornell, and then at a number of New York institutions. He was a member of the American Mathematical Society from 1918 and of the Association for Symbolic Logic from its foundation in 1936. His doctoral thesis was based on Russell and Whitehead's *Principia* and marks the beginning of the modern theory of proof.

Prestet, Jean (Chalon sur Saône 1648 – Marines 1690)
Son of a bailiff, Prestet went into the service of Père Malebranche, who encouraged his inter-est in mathematics and arranged for him to enter the Jesuit Congrégation de l'Oratoire in 1675. Engaged to teach at Nantes, Prestet had to decline the post which might prejudice the chair of hydrography, recently created by the Jesuits; he then worked in Angers. At the age of 27, Prestet published the first edition of his *Eléments de mathématiques*; the second, com-pletely revised, edition appeared in 1689.

Ptolemy, Claudius (*c.* 100 – *c.* 170)
What we know of the life of Ptolemy comes from what we can deduce from his own writings and those of other, later, writers. He carried out observations at Alexandria from 127 to 141. In his *Almagest*, Ptolemy proposed a geocentric universe, summarising recent theories, in-cluding those of Hipparchus, but also expanding and complementing them in an important way. He also published an important work on astrology, known as the *Tetrabiblos*, as well as the *Analemma*, an important work on mapping, and books on optics and music.

Puiseux, Victor (Argenteuil 1820 – Fontenay 1883)
Puiseux was educated at the École Normale Supérieure and taught at the Collège Royal de Rennes, the University of Besançon, and then at the École Normale Supérieure. He succeeded Cauchy to the chair of mathematical astronomy at the Collège de France. He took part in the work of the Bureau des Calculs of the Paris Observatory and worked for the Bureau des Lon-gitudes. In 1871 Puiseux was elected to the Académie des Sciences. He was an expert rock-climber. His main work was in mechanics, celestial mechanics, and the theory of functions of complex variables.

Qin Jiushao (Sichuan *c.* 1200 – Guangdong *c.* 1261)
The author of *Shushu jiuzhang* (Mathematical treatise in nine chapters), a work which ap-peared in 1247, based on a succession of problems in astronomy, architecture, military logis-tics, building, etc. with a 'realism' often pushed to its limits (one example dealing with taxes contained 175 unknowns). The text has become famous for its algorithm for a general method for solving simultaneous first order congruences to any modulus, not necessarily co-prime (the generalised 'Chinese remainder theorem').

Qusṭā ibn Lūqā, al-Baᶜlabakkī (*d.* Armenia *c.* 912)
A Christian from Baalbeck, Qusṭā ibn Lūqā was a mathematician, astronomer, doctor, phi-losopher and translator of Greek texts into Arabic. He was also interested in music. He wrote an *Introduction to Geometry* in catechism form and a treatise entitled *Proving by using the method of two errors*. Among his many translations, the most celebrated was the *Arithmetica* of Diophantus.

Raphson, Joseph (London 1678 – 1715?)
Raphson became a member of the Royal Society in 1698. He published *Analysis aequationum universalis* in 1690 and *Methodus ad resolvendas aequationes algebricas* in 1697. His *History of Fluxions*, appearing posthumously in 1717, honouring the work of Newton, appeared at a sen-sitive time during the controversy surrounding the Newton-Leibniz priority dispute.

Rheticus, Georg Joachim (Feldkirch 1514 – Kaschau 1574)
His father having been beheaded for sorcery, Georg Joachim took the name of Rheticus, from the name of his home region, and left for Zurich, and then Wittenberg, where he became professor of arithmetic and geometry in 1536. He visited Copernicus in 1539 whom he en-couraged to publish his discoveries. Rheticus became professor at Leipzig in 1542, but was obliged to resign the post. He then went to study medicine at Prague, and settled in Krakow

where he practised medicine and astrology. A table produced by Rheticus contains for the first time the six trigonometric ratios.

Richardson, Lewis Fry (Newcastle 1881 – Kilmun 1953)
From a Quaker family, Richardson was educated at Durham College of Science, Newcastle and King's College, Cambridge. He worked at the National Physical Laboratory for six years and then in the electrical industry. In 1913 he took charge of the Eskdalemuir Geophysical Observatory. During the First World War Richardson worked with the Friends Ambulance Unit in France. Having already derived meteorological predictions using numerical procedures, he returned to the Meteorological Office in 1919, but since this became attached to the Air Ministry, he resigned the following year. Richardson obtained the post of head of the physics department at Westminster Training College. In 1929 he became principal of Paisley Technical College.

Riemann, Bernhard (Breselenz 1826 – Selasca 1866)
Riemann received his early education from his father, a Protestant pastor, then went on to the gymnasium at Hannover. In 1846 he went to Göttingen to study theology and philology. He also attended classes in mathematics which he then decided to study. He received his doctorate at Göttingen in 1851, his dissertation being considered as a genuine contribution to the theory of functions, and he taught at the university. Falling ill with tuberculosis, Riemann travelled to Italy and during his third visit he died there. Riemann showed great originality in his work in a number of fields, particularly in complex variable theory, representation of functions and geometry. He also contributed to physics.

Roberval, Gilles Personne de (Senlis 1602 – Paris 1675)
Educated privately, Roberval came to Paris in 1628 where he frequented the circle of scientists surrounding Mersenne. In 1632 he became professor of philosophy at the Collège de Maître Gervais. In 1634 Roberval won the competition for the Ramus chair at the Collège Royal, a post he held for the rest of his life. In 1655 he succeeded Gassendi to the chair of mathematics. He was one of the founding members of the Académie des Sciences in 1666. He worked on algebra, physics and the astronomy of Aristarchus, but concerned with protecting his chair, Roberval published little. He was the founder of kinematic geometry based on movement.

Rolle, Michel (Ambert 1652 – Paris 1719)
Son of a retailer, Rolle only received an elementary education. At the age of 23 he set himself up in Paris as a teacher of writing and arithmetic. He taught himself algebra and the methods of Diophantus. Colbert recognising his talents, obtained a pension for him. In 1685, Rolle entered the Académie des Sciences, first as a student of astronomy, then as a pensionnaire in geometry. His *Traité d'Algèbre* (1690) shows Rolle's main interest, namely the algebra of equations. It was Giusto Bellavitis who, in 1848, ascribed Rolle's name to the well-known theorem.

Ruffini, Paolo (Valentano 1765 – Modena 1822)
Ruffini studied medicine, philosophy, literature and mathematics at the University of Modena. While yet a student he taught the foundations of analysis. Under Napoleonic occupation of the city, Ruffini lost his post. He continued to practise medicine and to work on mathematics, his *Teoria generale delle equazioni* appearing in 1799. After the fall of Napoleon, Ruffini was to occupy the chairs of medicine and mathematics at Modena. He contributed to the transition between classical algebra and abstract algebra.

Runge, Carl (Bremen 1856 – Göttingen 1927)
Coming from a comfortable family background, Runge studied with Max Planck at the University of Munich. Both became fellow students at Berlin. Runge obtained his doctorate under the supervision of Weierstrass and qualified for a university post under Kronecker. After marrying a daughter of Emile du Bois-Reymond he accepted a post of professor of philosophy at the Hannover Technische Hochschule. His research was directed towards spectometry and applied mathematics. Considered to be a physicist by mathematicians and a mathematician by physicists, Klein nevertheless succeeded in obtaining him a chair of mathematics at Göttingen. Photograph by kind permission of C. Reid.

Russell, Bertrand (Trelleck 1872 – Plas Penrhyn 1970)
Since the time of the Tudors the Russell family had played an important role in the political, intellectual and social life of Britain. Russell lost both parents at an early age and received a private education before entering Trinity College, Cambridge in 1890. There he met the young Whitehead, with whom he later collaborated in writing the *Principia*. From 1910 he taught philosophy and mathematics at Trinity, until he was excluded for expressing his pacifist views in a pamphlet; he was even imprisoned in 1918. Readmitted in 1925, Russell remained at Cambridge for the rest of his academic life. He was made a Fellow of the Royal Society in 1908 and obtained the Nobel Prize for literature in 1950. The final years of his life were dedicated to campaigning for nuclear disarmament.

Saint-Vincent, Grégoire de (Bruges 1584 – Ghent 1667)
Entering the Jesuit College at Bruges in 1595, Grégoire de Saint-Vincent went on to study philosophy and mathematics at Douai. In 1607 he returned to the Jesuits. On the death of his professor, Christopher Clavius, Saint-Vincent completed his theological studies at Louvain and was ordained a priest. He taught Greek, then mathematics. Saint-Vincent developed a theory of conic sections from the editions by Commandin of the works of Archimedes, Apollonius and Pappus. He also developed a method of infinitesimals in his *Opus geometricum*.

Schreier, Otto (Vienna 1901 – Hamburg 1929)
Son of an architect, Schreier was educated in Vienna, at the Gymnasium and at the Faculty of Philosophy. He submitted his thesis on groups in 1923. He moved to Hamburg where he participated in the work of the mathematics seminar and was appointed assistant in 1925. He also taught at Rostock. He died, very young, of septicaemia. With his colleague Emil Artin, Schreier laid the foundations of the theory of real fields, a branch of abstract algebra.

Seidel, Philipp Ludwig von (Zweibrücken 1821 – Munich 1896)
Seidel went to the University of Berlin in 1840, studying under Dirichlet and Encke, for whom he carried out astronomical calculations. Seidel then went to Königsberg, where Jacobi and Bessel taught. He passed his doctorate at Munich, where he spent the remainder of his academic career. The photometric measurements of planets and stars that he made were the first true measures of this type. In mathematics the concept of non-uniform convergence is due to Seidel. His lectures dealt with probability, the method of least squares and dioptrics.

Sharp, Abraham (Little Horton 1651 – Bradford 1742)
After an apprenticeship with merchants in York and then Manchester, Sharp abandoned this work in order to go to Liverpool to learn mathematics, to which he then committed himself completely. He was taken on by Flamsteed c. 1684 at the new observatory at Greenwich. In 1694 he retired to Little Horton where he carried out calculations, made astronomical instruments and conducted scientific correspondence. His book on geometry, published in 1717, is notable for the calculations it contains.

Sheppard, William (Sydney 1863 – Berkhamsted 1936)
After studying at Trinity College, Cambridge from 1887 to 1895, Sheppard was appointed examiner there. From 1914 to 1921 he worked for the Ministry of Education. In 1928 he was president of the British Mathematical Association. The main part of Sheppard's mathematical work was with statistics and he dealt with the method of finite differences and their applications.

Simpson, Thomas (Market Bosworth 1710 – 1761)
The self-taught Simpson worked as a weaver by day and teacher in the evenings. In London, struggling against poverty, he gave lessons and published a number of works. In 1743 he was appointed to the Royal Military Academy at Woolwich. He became a member of the Royal Society in 1746. Simpson was a disciple of Newton: his principal works were *A New Treatise of Fluxions* (1737), *Algebra* (1745), *Geometry* (1747) and *Trigonometry* (1748).

Simson, Robert (West Kilbride 1687 – Glasgow 1768)
From a comfortable background, he went to Glasgow University with the intention of taking Holy Orders. There he studied Latin, Greek, logic and natural philosophy. He specialised in the study of theology and Semitic languages. Simson also acquired a knowledge of botany. With no previous education in mathematics, he went to study in London, where he met Halley. He was appointed professor of mathematics at Glasgow in 1711. Simson's principal works concerned conic sections, plane and three-dimensional loci of Apollonius and an edition of Euclid's *Elements*.

Stevin, Simon (Bruges 1548 – The Hague 1620)
In his younger days, Stevin was in government service for the town of Bruges. He travelled to Prussia, Poland and Norway. He entered the University of Leyden in 1583. Ten years later he became a quatermaster general in the Dutch army. At the same time he was a tutor in the service of the Prince of Orange. Stevin published a number of works in mechanics, astronomy, military science, statics and hydrostatics. Of particular note is an arithmetic, published in Flemish at Leyden (1585), and in French translation in the same year (*La Disme*), which uses the decimal point for the first time. He also published the first translation of the *Arithmetica* of Diophantus into a modern language. Photograph by kind permission of G.L. Alexanderson and Birkhäuser-Verlag.

Stifel, Michael (Esslingen 1487 – Jena 1567)
A monk at the monastery at Esslingen, Stifel became a priest in 1511. He was an early supporter of Luther but he was also interested in cabbalism, going so far as to make a prediction about the end of the world. Resigning his post as pastor, Stifel turned to mathematics. He taught theology and mathematics at Königsberg, then arithmetic and geometry at Jena. In his *Arithmetica Integra* (1544) and his edition (1553-54) of the *Coss* by Rudolff, he sets out the mathematical knowledge of his time, with original contributions.

Stirling, James (Garden, Stirlingshire 1692 – Edinburgh 1770)
Stirling's family espoused the Jacobite cause and, his father being imprisoned for refusing to take an oath of allegiance, James was obliged to quit Balliol College, Oxford without graduating. A first piece of work by Stirling won him the support of Newton for a professorship at Venice. After possibly learning the secrets of glass making there, he returned to London and, in 1717, was elected to the Royal Society. In later life (1735) he became the manager of a mining company in Lanarkshire and made a great success of this venture. His principal work, *Methodus differentialis* (1730), deals with sums of series and interpolation.

Sturm, Charles-François (Geneva 1803 – Paris 1855)

Sturm studied mathematics and physics at the University of Geneva. In 1823 he was tutor to the son of Madame de Staël and the Duc de Broglie introduced him into the Paris scientific world. He carried out research with his student friend Colladon. In 1829 he took charge of the mathematics section of the *Bulletin des sciences et de l'industrie*. He taught at the École Polytechnique and the Paris Faculté des Sciences and became a member of the Académie des Sciences in 1836. Two of his works, *Cours de mécanique* and *Cours d'analyse* enjoyed the status of classics for half a century.

Sunzi (5th century?)

Sunzi was the author of *Sunzi suanjing* (Master Sun's Arithmetical Canon), an arithmetic which is important for being the earliest known work to deal with remainders. The most famous problem is Problem 26, usually known as 'Master Sun's problem' which states: there are an unknown number of things; counting by threes, 2 remain; by fives, 3 remain; by sevens, 2 remain. How many things? Answer: 23.

Sylvester, James Joseph (London 1814 – 1897)

Educated at St. John's College, Cambridge, Sylvester was not allowed to take a degree there because of his Jewish faith, and was also barred from a fellowship. For a degree he went to Dublin. In 1837 he was appointed professor of natural philosophy at University College, London and two years later (1839) was elected a Fellow of the Royal Society. After a short stay in the United States, Sylvester returned to London and took up actuarial work. He became professor of mathematics at the Military Academy in Woolwich (1855) and later returned to work in the United States. He founded the *American Journal of Mathematics*. His chief line of study was algebra, particularly determinants and the theory of invariants.

Takebe, Katahiro (1664 – 1739)

A disciple of Seki Takakazu, the 'father' of traditional Japanese mathematics, Takebe was an advisor to Yoshimune, a Shogun of unusually open views, interested in both western science and science coming from China. In such a favourable setting, Takebe was able to blossom in the pursuit of geography, astronomy and creative mathematics, producing results hitherto unknown in Japan. His work included, in particular, a map of Japan, a history of Chinese mathematics under the Yuan (Mongol) and Ming dynasties, and a series expansion of the square of the arcsine as well as research into continued fractions.

Tannery, Paul (Mantes la Jolie 1843 – Pantin 1904)

After graduating from the École Polytechnique, Paul Tannery worked as engineer and administrator. In Bordeaux, at the same time as overseeing tobacco manufacture, he started to study the history of mathematics. He continued to have a successful career in the tobacco industry at Le Havre, Paris, Tonneins and at Pantin, and he also taught at the Collège de France. In addition to his research which filled five of the seventeen volumes of his *Mémoires scientifiques*, Paul Tannery collaborated in the publication of the works of Diophantus, Fermat and Mersenne as well as Descartes' correspondence.

Tartaglia, Niccolò (Brescia 1500 – Venice 1557)

Tartaglia – that is the stammerer – owes his name to his handicap that is said to have resulted from a sabre cut to the face at the time Brescia was taken by Gaston de Foix (1512). From a poor family, he was self-taught and achieved considerable skill in science. He was a teacher of the abacus, then professor of mathematics at Venice. He died poor and abandoned. He uncovered the solution to cubic equations found by Scipione del Ferro. His *General Trattato* (1556) was the most important Italian arithmetic in the 16th century. Tartaglia was the first to apply mathematics to artillery. He published an edition of Euclid's *Elements*.

Taylor, Brook (Edmonton 1685 - London 1731)
Taylor was educated privately before entering St. John's College, Cambridge in 1701. He became a member of the Royal Society in 1712 and its secretary from 1714 to 1718. He travelled in France and entered into scientific correspondence with Montmort. In his later years he turned to religion and philosophy. His *Methodus incrementorum* (1715) contains a number of important results, notably what we call Taylor's Theorem and its use in calculating finite differences. Taylor also wrote *Linear Perspective* (1717) which served as a basis for Lambert's work.

Thābit Ibn Qurra, al-Sābī al-Harrānī (Harran 836 - Baghdad 901)
A Sabean who knew Greek and Arabic as well as his mother tongue Syrian. Thābit ibn Qurra's writings played an significant role in making important discoveries known, for example, in extending the concept of a real number, integral calculus, spherical trigonometry, coordinate geometry and non-Euclidean geometry. In astronomy he was one of the first reformers of Ptolemy's system. He was also a physician and philosopher of repute. He translated numerous Greek authors, such as Euclid, Archimedes and Ptolemy.

Theon of Alexandria (late 4th century)
From historical sources Theon of Alexandria appears to have lived during the reign of Theodosius I (379–395). He observed two eclipses at Alexandria in 364. His daughter Hypatia (cf. p. 497) was murdered by a fanatic Christian mob. Theon is the last recorded member of the Museum at Alexandria. All his works were written for students. The most important is the *Commentary on the Almagest* in which he explains, in particular, algorithms for calculating with sexagesimal fractions. He also produced an edition of Euclid's *Elements*.

Theon of Smyrna (*c.* 125)
Theon of Smyrna is only known to us for his *Expositio* (Exposition on the mathematics needed for reading Plato). The work is important for the number of sources cited but contains little original mathematics. It is intended as a work for students of philosophy to show how arithmetic, geometry, stereometry, music and astronomy are connected to each other. The arithmetic section deals with different types of numbers according to Pythagorean philosophy. It is similar to the work of Nicomachus of Gerasa, though less systematic.

Tucker, Robert (Walworth 1832 - Worthing 1905)
Tucker began school in Newport and entered St. John's College, Cambridge in 1851. After working as a private tutor, he taught at Brighton College and then University College School, London. He became a member of the London Mathematical Society when it was founded in 1865, and became secretary of the society from 1867 to 1902. He then left London for Worthing, but had to sell the contents of his library to maintain himself. Tucker edited the works of Clifford and published a number of historical notes, on Gauss, Chasles, Sylvester and others. He was interested in the geometry of the triangle, astronomy and the teaching of mathematics.

Turing, Alan (London 1912 - Wilmslow 1954)
Educated at King's College, Cambridge, Turing went to Princeton to work with Church from 1936 to 1938. This was when he published his celebrated article on Hilbert's *Entscheidungsproblem* and *On computable numbers ...*, which describes the so-called Turing machine. Returning to England, he worked for the communications department of the Foreign Office during the war on code breaking and after 1945 he worked on producing an automatic calculating machine, first at the National Physical Laboratory, then at the University of Manchester. His death from cyanide poisoning while conducting experiments was claimed to be accidental, but is now generally accepted as suicide.

Tycho Brahe (Scanie 1546 – Prague 1601)
The famous Danish astronomer is usually referred to by his first name Tycho. Coming from an aristocratic family, he had no need of university titles in order to establish himself. After studying at the Lutheran University of Copenhagen, he went to Leipzig in 1562. Tycho combined travels with astronomical observations. In 1576, Frederick II had two observatories built for him on the island of Hven, where he worked from 1580 to 1597. On the death of his protector, Tycho went to Prague, where Kepler became his assistant. Tycho is considered as the founder of modern astronomy.

van Ceulen, Ludolf (Hildesheim 1540 – Leyden 1610)
Ludolf van Ceulen left Germany at an early age to settle in Holland, where he received his education and spent the greater part of his life. He taught mathematics at Breda, Amsterdam and Leyden. He is famous for his calculation of the first thirty six decimals of π, sometimes called Ludolf's number. The result, published in 1621 by his disciple Snell, is engraved on his tombstone.

Vandermonde, Alexandre Théophile (Paris 1735 – 1796)
Being a sickly child, Vandermonde only had lessons in music, on the advice of his father. When he was in his thirties, Fontaine told him of his own enthusiasm for mathematics. Vandermonde became a member of the Académie des Sciences in 1771 and the whole of his contribution to mathematics, concerning the solution of equations, comes from this period. In 1776, with Bézout and Lavoisier, he experimented with low temperatures, and in 1786, with Berthollet and Monge, he wrote an article on the manufacture of steel. He became director (1782) of the Conservatoire des Arts et Métiers. An ardent revolutionary, and friend of Monge, he belonged to the Paris Commune and the Jacobins.

Viète, François (Fontenay le Comte 1540 – Paris 1603)
The greatest of all French mathematicians of the 16th century, François Viète, Seigneur de la Bigotière, is better known by the semi-Latin name of Vieta. He practised law as a young man in his native town, afterwards taking up a political career and becoming a member of the Bretagne parliament. He was advocate general of the Paris parliament, master of requests, and a member of the king's privy council. During the conflict with Spain, Viète deciphered intercepted encoded letters. He intervened in the reform of the calendar by Pope Gregory XIII, but without success. Vieta's first mathematical writings related to cosmology and astronomy, some remaining in manuscript form. But Vieta's most important work dealt with algebraic symbolism

Vlacq, Adriaan (Gouda 1600 – The Hague 1666)
Of a good family, Vlacq received an excellent education. Being interested in mathematics, he worked with a local professor, Ezechiel de Decker, to produce Dutch translations of several arithmetics written in Latin, notably those of Napier and Briggs on logarithms. Vlacq himself computed the logarithms of numbers from 20 000 to 90 000 and so filled the gap left by Briggs in his publication of logs from 1 to 100 000. Vlacq's tables were published in England, Germany, France and even in China, which assured his celebrity.

Von Neumann, John (Budapest 1903 – Washington 1957)
Von Neumann received his mathematics education from private teachers. He taught at Berlin, then Hamburg. In 1930 he moved to the United States and taught at Princeton. From 1943 he was an advisor to the US Army on the Atomic Bomb project. From the start of his career, von Neumann was interested in logic, set theory, measure theory and spectral theory. Then he turned to axiomatisation of quantum mechanics, hydrodynamics, automata theory and game theory.

Wallis, John (Ashford 1616 – Oxford 1703)
Son of a clergyman, Wallis studied theology at Emmanuel College, Cambridge. Ordained in 1640, he worked in London as a chaplain. He became a fellow of Queen's College, Cambridge in 1644. His interests lay, however in mathematics, and in 1649 he was elected to the Savilian chair of geometry at Oxford, a position he held until his death. He was one of the founders of the Royal Society. His wide interests included mechanics, astronomy, botany, physiology, music and cryptology. Among his many works should be mentioned *Arithmetica infinitorum* (1655) and *De Algebra Tractatus; Historicus & Practicus* (1673) of which an English edition appeared in 1685. He was the first to use the symbol ∞ for infinity.

Wantzel, Pierre Laurent (1814 – 1848)
A student at the École Polytechnique from 1832 to 1834, Wantzel was a tutor and then examiner there. He wrote a number of articles for the *Journal de l'École Polytechnique*, among which were: On incommensurable numbers, On the theory of complex numbers and On the integration of differential equations. Wantzel was also interested in physics.

Waring, Edward (Shrewsbury 1736 – Plealey, near Shrewsbury 1798)
Little is known of Waring's early life but he was senior wrangler at Cambridge in 1757 and Lucasian professor of mathematics there from 1760. He graduated in medicine but seems not have followed that career. He became a member of the Royal Society in 1763. Mathematics was in decline in England at the time but Waring's *Meditaniones algebraicae* attracted the attention of Lagrange. The work contains many important (unproved) assertions, including Goldbach's conjecture.

Weierstrass, Karl (Ostenfelde 1815 – Berlin 1897)
Weierstrass began his education at the Catholic Gymnasium in Paderborn and studied law and finance at the University of Bonn in 1834. He abandoned his studies and did not sit examinations until 1841 at Münster. He taught at a number of Gymnasien in a variety of disciplines: German, botany, geography, history, gymnastics and even handwriting. He obtained an honorary doctorate from the University of Königsberg in 1854 and went to Berlin in 1856 as a teacher at the Gewerbe-Institut. He was a member of the Berlin Academy of Sciences from 1856. He became a professor of mathematics at the University of Berlin in 1864. He is considered to be the father of modern arithmetisation of analysis.

Whitehead, Alfred North (Ramsgate 1861 – Cambridge 1947)
Whitehead spent the years 1880–1910 at Trinity College, Cambridge, first as a student then as a teacher of mathematics. Russell was one of his students, with whom he wrote *Principia Mathematica* (1910–1913). In 1910 Whitehead moved to London, occupying a number of posts at University College and Imperial College. He worked on the experimental foundations and concepts of physics. In 1924 Whitehead went to teach philosophy at Harvard, turning principally to the philosophy of science, education and religion.

Wu Jing (*c.* 1450)
Wu Jing was financial governor of Zhejiang at the time of the Ming dynasty. He is known for his arithmetical encyclopaedia of 1450, the *Jiuzhang suanfa bilei daquan* (Complete Description of the Nine Chapters of Arithmetical Techniques), containing problems solved by using model examples.

Zhu Shijie (*c.* 1300)
A native of Yanshan, near the present day Beijing, Zhu Shijie travelled around China, at the beginning of the period of the Mongol invasion, for some twenty years, earning his living as a mathematics teacher. We know of two of his works: the *Suanxue qimeng* (1299) (Introduction

to Mathematical Studies) and the *Siyuan yujian* (1303) (The Jade Mirror of Four Unknowns). The first is noteworthy for its positional notation for the unknown and the second for the generalisation of the same technique to four unknowns. Forgotten in China for five centuries, Zhu Shijie had an astonishing recognition in Japan, where his *Suanxue qimeng*, edited and republished from 1658, was responsible for the revival of Japanese mathematics.

Zu Chongzhi (Tsu Ch'ung-Chih) (429 – 500)
From a renowned family of high functionaries, Zu Chongzhi was in the service of the emperor Hsiao-wu (r. 454–464) of the Liu Sung dynasty. On the death of the emperor (464) he left the imperial service to devote himself entirely to science. His most important astronomical work was his attempt to reform the calendar, which was never put into effect. He was famous for constructing ingenious devices, such as the 'south pointing charot' and a self-propelled motor-boat. He is particularly remembered for his value of 355/113 as an approximation to π.

A History of Algorithms

General Index

A History of Algorithms

Index of Names

Printed in Italy by Legoprint S.p.A., Lavis (Trento)